S0-DUE-672

Theory and Research in Lubrication

Theory and Research in Lubrication

Foundations for Future Developments

Mayo Dyer Hersey

Visiting Professor of Engineering (Research)
Brown University

JOHN WILEY & SONS, INC. *New York · London · Sydney*

Library of Congress Catalog Card Number: 66-21058
Printed in the United States of America

Dedicated to the
Research Committee on Lubrication
American Society of Mechanical Engineers

Preface

This book is a sequel to the earlier one, *Theory of Lubrication*, published thirty years ago, which was the first American book on that subject.

There was actually a question at the time, "why a whole book on lubrication? Just put in enough oil!" Yet the book had to be reprinted in two years; and now there are over a dozen good books published in this country alone on the various branches of Lubrication.

With such a wealth of information available, the old question takes a new form—"why still another book on Lubrication?" Mainly because the present work offers a unique guide to the many aspects in a single volume, and indicates where future research is needed.

Out of a lifetime study of friction and lubrication I refer to more than a thousand investigations selected from all over the world. Numerous references were regretfully left out to keep the book down to a reasonable size. We have attempted to show under each topic not only the principal results of theory and research, but how our present understanding has been built up; how one discovery led to another.

The book is addressed to research engineers, machinery designers, teachers of applied mechanics and fluid dynamics, advanced students and librarians. It should especially be of aid to the growing number of physicists and engineers qualified in other fields, who are suddenly confronted with lubrication problems.

For the worker who wishes to delve still further into the literature, additional references to journal bearings can be found in two books by Professor Fuller (1956 and 1958). Earlier work on piston ring lubrication is listed in Poppinga's book of 1942 and in the NACA bibliography (Hersey, 1944). Further literature on pressure-viscosity is cited in the monograph by Hersey & Hopkins (1954); on dynamically loaded bearings in papers by Hersey & Snapp (1957) and in the Navy bibliography by Snapp & Dray (1958). Boundary lubrication is documented in the treatise of Bowden & Tabor (1950, 1954, 1964); and more fully in a critical appraisal being compiled by the ASME. In line with our efforts to condense the book, diagrams and examples have been held to a minimum.

The arrangement of subjects can be seen from the table of contents. Thus the first five chapters are devoted to basic facts and principles, while the next eight go more into applications. Topics are listed compactly at the head of each chapter. Most of the chapters and topics begin easily but gradually get more difficult. No one reader will be equally interested in all topics. There are 113 numbered topics not counting summaries. This arrangement enables the reader to look up the literature in his own chosen field. He will usually find that the discussions are also of value, as, for example, in the ASME Transactions and IME Proceedings.

Abbreviations are explained in Appendix A. We have not converted data from metric to English units, or vice-versa, but have left such values unchanged. Today, most European readers can think directly in English Units. If any cannot, they need only remember that a pound is approximately 0.454 of a kilogram while an inch is exactly 2.54 centimeters.

Alphabetical listing of authors' names will be found at the end of each chapter in which they are cited. Some of the names are repeated in the index at the back of the book, which is primarily a subject index. The book concludes with a list of mathematical symbols frequently used, all of which are defined when they first appear in the text.

The pleasure remains of acknowledging some of the information and assistance received. My indebtedness to Dr. W. H. Kenerson, first Chairman of the Division of Engineering at Brown University, still continues. The earlier book was dedicated to him. I am especially grateful to the members of the ASME Research Committee on Lubrication, to whom this book is inscribed. Timely encouragement and technical data came from conversations with Aurel Stodola, Edgar Buckingham, Albert Kingsbury, A. G. M. Michell, P. W. Bridgman, W. H. Herschel, G. B. Karelitz, B. L. Newkirk, and A. E. Norton over a long period; and more recently with F. T. Barwell, Harmen Blok, Earle Buckingham, Ronald Bulkley, J. A. Cole, Markus Reiner, Richard Vieweg, and others.

Some of the first chapters were read by Professors W. N. Findley and Joseph Kestin of Brown University and B. G. Rightmire of M.

I. T.; also by Dr. W. A. Gross of Ampex, Inc. and H. A. Hartung of West Collingswood, N. J. Chapter VI on thrust bearings was kindly read by Nelson Ogden and S. J. Needs of Kingsbury Machine Works, Inc. as well as by John Boyd and A. A. Raimondi of Westinghouse. Chapter IX on gas-film bearings was read by Dr. Gross; Chapter XI on gear lubrication by Professor Earle Buckingham of Springfield, Vt., Dr. R. S. Fein and G. H. Benedict of Texaco and B. W. Kelley of Caterpillar Tractor Company. The notes on vibration and whirl in Chapter XII were looked over by Professor B. L. Newkirk of R. P. I. and Dr. D. F. Wilcock of Mechanical Technology Incorporated. Chapter XIII was read by Professor Rightmire, and discussed in part with Professor W. E. Campbell of R. P. I.

Professor Dudley D. Fuller of Columbia University and the Franklin Institute read a first draft of the entire manuscript. I am much indebted to him for numerous improvements, and to the publishers for continual cooperation. I have tried to apply the recommendations from all sources except where they would lengthen the book.

Thanks are due Miss Margaret E. Drewett, Physical Sciences Librarian, and her talented staff, for aid in compiling the bibliographies. Acknowledgment is made to Mrs. Marion K. Craven of the Division of Engineering for her understanding and skill in typing the manuscript. I am fortunate, after retiring from government service, to enjoy the hospitality and congenial surroundings of Brown University.

When the book was nearing completion, John A. C. Warner of the Society of Automotive Engineers asked if I could not gaze into the crystal ball and take a look ahead. This prompted my wife (John's sister) to suggest that my story of Mike and the Oil Shed, written thirty years ago, would make a good starting point. If we can go as far in the next thirty years, we'll really be moving along. Hence Chapter XIV, *Epilog: Looking Back and Looking Ahead.*

M. D. H.

Providence, Rhode Island
April 1966

Contents

Theory and Research in Lubrication

chapter 1 Friction in Machinery

1. Frictional resistance. 2. Some friction demonstrations. 3. Coefficients of friction. 4. Frictional power loss. 5. Frictional heating. 6. Rolling friction. 7. Lubrication research. 8. Publications. 9. Summary.

The science of lubrication grew out of the need for reducing the friction in machinery. The detrimental effects of friction are chiefly:

1. The resisting force opposing the motion of bodies.
2. Loss of power owing to work done against friction.
3. Temperature rise and consequent surface damage.

These effects are discussed below, together with useful applications of friction.

1. FRICTIONAL RESISTANCE

Friction is the name given to the force resisting the relative motion of two bodies that are initially at rest or moving without acceleration. The frictional resistance is called *starting* or *static* friction while the bodies are at rest. After motion has begun it is called *kinetic* friction. The two bodies may be in direct contact, or in indirect contact through the medium of a lubricant.

Frictional resistance must be distinguished from inertia. If we pull harder than necessary to get a stationary body moving, it starts with a noticeably accelerated motion. The force applied in the direction of motion is equal to the sum of the frictional resistance and the product of mass by acceleration. The torque applied to a rotating body is equal to the sum of the frictional torque and the product of moment of inertia by angular acceleration.

Static friction on a horizontal surface can be measured by the minimum pull required for barely perceptible motion; or better, by the mean between that value and the maximum pull failing to start motion. Static friction may be measured either by a direct pull or by an adjustable inclined plane (Fig. 1). Precise values of static friction are hard to obtain since static friction increases with the duration of the load, as discovered by Coulomb, and even depends on the rate of application of the pull. Incipient sliding can be difficult to recognize because of elastic deformation and creep effects.

The measurement of kinetic friction offers no such difficulty except under low-speed conditions where "stick-slip" may occur. This action is familiar in the lubricated ways or guides of grinding machines, for example. It depends upon elasticity and inertia factors and is met in the speed range where friction diminishes with increasing speed (Chapter XIII). Kinetic friction in machinery has been measured by various methods, including direct observation of torque applied to the rotor; deceleration of the rotor; and measurement of frictional heat carried off in the oil.

Frictional resistance is of special interest when starting heavy machinery from rest and moving it slowly. Good examples are the starting of hydroelectric generators supported by large thrust bearings; cold starting of diesel engines; and assembling of freight cars on level track to make up a train. Our mention of the grinding machine, with its horizontal ways, illustrates a requirement for low friction combined with smooth motion. Another such example would be the steady rotation of a telescope structure in following the motion of the stars, as described by Professor Fuller (Chapter 1 of 1956). He showed how starting friction can be avoided by hydrostatic support, using externally pressurized bearings.

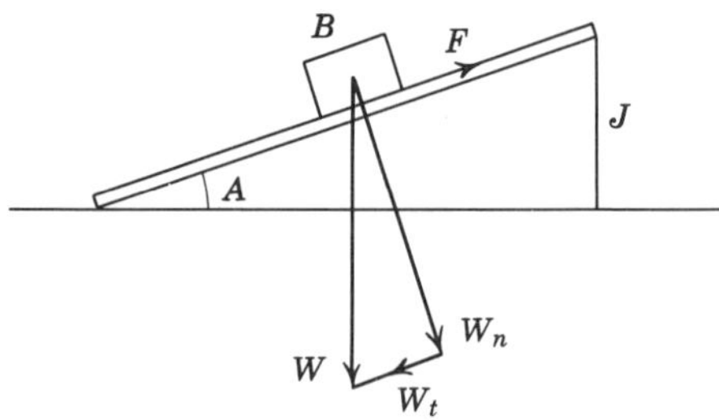

Fig. 1. Friction F on block B sliding down inclined plane (W_n, W_t: normal and tangential components of weight W).

That part of a rotating shaft supported in a bearing is called the *journal*, and the combination is called a *journal bearing*. The journal transmits a radial load to the bearing. Axial movement of the shaft is prevented by a *thrust bearing*. Either the end of the shaft or a collar on the shaft exerts an axial load on the thrust bearing.

Better control of frictional resistance in the future should facilitate plans for harnessing the power of the winds and the tides, or of solar energy, or any other source where friction losses are likely to represent a large part of the power available. Frictional resistance is discussed at length in books by Stanton (1923), Gümbel (1925), and Gemant (1950); also by Kragel'skii & co-authors (1955–56–62); and in a continued article by Palmer (1945). Vibration helps to offset the effects of static friction, as shown by Goodier (1945) in his theory of nuts and bolts. Michell notes that the size of a railway locomotive is fixed by the amount of starting friction, since that determines the size of the propelling units. Thus in the traction industry, capital cost—not just operating cost—depends on bearing friction (page 188 of 1950). The kinetic friction met by projectiles penetrating metal was determined by Krafft (1955). On friction of rigid bodies see Drescher (1959).

Historical Notes. Leonardo da Vinci discovered that the static and low-speed frictional resistances of ordinary bodies were proportional to their weights and independent of the nominal area of contact. A solid block had the same friction whether sliding on a broad flat face or on a narrow side. Smoothing or lubricating the surfaces reduced their friction. Leonardo's experiments are described by Beck (1906), Hart (1925), and Bowden & Tabor (1950). After two hundred years, Amontons (1699) rediscovered the same facts. When Coulomb (1785) learned of Amontons' observations nearly a century later, he extended them systematically. Coulomb found friction independent of speed over the short range investigated, except that kinetic friction was less than static. Hence the Amontons-Coulomb law, or "Coulomb's law" for short, namely that frictional resistance is proportional to the load and unaffected by area or speed. Coulomb set up an exponential equation for static friction as a function of the time elasped under load.

Rennie (1829) noticed that friction was greater for soft than for hard substances but depended more on the lubricant than upon the solid materials. General Morin (1832) confirmed Coulomb's law as an approximation for journal bearings. He observed that static and kinetic friction differed most for compressible materials, and discovered that starting friction could be reduced by vibration. Hirn (1854) discovered the effect of "running-in." He concluded that the friction of lubricated surfaces might be taken as roughly proportional to the square roots of the load and speed.

To anyone familiar with machinery and having a normal degree of scientific curiosity, there can hardly be a more fascinating study than the history of friction and lubrication. The subject has been outlined from several points of view by Benton (1926); the present writer (1933); Vogelpohl (1940); Fuller (1954); and by Courtel & Tichvinsky (1963).

Useful Friction. Without friction solid objects would not stay put on sloping surfaces. Wedges would slide out, corks pop out of bottles, screws and bolts would lose their hold. We should hear no more violin music; even walking would be difficult or impossible. When chromium plate became popular, one of our railroads installed shiny horizontal rods for footrests in passenger cars. There was nothing restful about those slippery rods! Foot pressure had to be precisley at right angles, for want of frictional resistance.

Friction is usefully applied in forced fits and in brakes, in belt drives and friction transmissions, and in providing locomotives and automobiles with the necessary traction to permit acceleration. See, for example, Swanger on shrink fits (1934); McCune on braking high-speed trains (1939); Thomas (1954) and Hewko (X: 1962) on friction drives with rolling contact. Friction sawing is described by Chamberland (1946); friction welding, a more recent development, by Vill' and others (1959).

2. SOME FRICTION DEMONSTRATIONS

The block sliding down an inclined plane is a classical lecture-table demonstration. In Fig. 1 a wooden plank, or metal plate, is raised to an angle A by jack J, provided with graduated drum, or other means for reading elevation. At starting or constant speed, the friction F is given by W_t or $W_n \tan A$, where W_n is the load or the normal component of the weight of the block. A striking example of very low friction is seen in the Crookes radiometer. Here the rotaging arms, tipped with mica vanes, are carried by an inverted glass cup or thrust bearing resting on the point of a needle. Molecular bombardment in a high vacuum exerts greater force on the blackened side of each vane. This slight differential force is enough to overcome friction. A flash of light sets the vanes into rapid rotation. This demonstration goes far toward answering the ancient question, "how many angels can dance on the point of a needle?"

A transparent journal bearing, hand operated, oil lubricated, was devised by John Boyd (1948) to show the sudden drop in friction that accompanies oil-film formation. Pointer and scale indicate static friction coefficients of 0.1, 0.2, or more when the crank is first gently turned. The pointer drops back practically to zero as soon as a thick oil film is dragged into the clearance space. Capillary-size manometer tubes drilled in the bearing member register the film pressure. A transparent slider-block exhibits similar results when pushed along an oiled surface, all because of a taper concealed in the base. Both experiments demonstrate the principle of the convergent film, due to Reynolds (1886), discussed in the next chapter. These demonstrations and others have been described in recent articles (Tichvinsky, 1954, Hersey, 1964).

Frictional effects are suprisingly shown in the "tippe-top" or "toupie magique," which has a globular form with tall stem at upper end. When rapidly spun and then left to itself, the top soon turns a half somersault and continues to spin upside down. Explanations were given by Braams and Hugenholtz (1952) and confirmed by Pliskin (1954).

The Friction Pendulum. Instructive experiments can be made with the friction pendulum in various forms. It is commonly used for determining the average friction of an oscillating bearing. The pendulum is hung from the test shaft instead of from a knife-edge. The bearing is firmly inserted in the top end of the pendulum, which may be simply a flat stick of wood having a weight fixed to the lower end.

1. When the shaft is stationary, the oscillations of the pendulum are rapidly damped by friction in the bearing, whether it be of the plain or rolling type. The drop in height of the center of gravity of the pendulum from its initial amplitude until it comes to rest, multiplied by its weight, gives the potential energy lost. This equals the work done against friction, which is the product of the unknown frictional resistance times the sum of the distances moved over by the bearing member.

2. If the shaft is rotated at a steady speed while the pendulum is restrained from oscillating, it will stand out at a small angle. When two opposed partial bearings are pressed against the journal by a loading spring in the pendulum, the frictional torque on the journal will be measured, and the journal axis may be taken as the point of support. This method was used by Thurston and others in their early measurements on the friction of plain bearings with different lubricants and with different bearing metals at varying speeds, loads, and temperatures.

3. If the pendulum is now allowed to oscillate while supported on a rotating shaft, we have a combination known as the *Froude pendulum.* It was discovered by William Froude that such a pendulum can oscillate indefinitely without damping! Under some conditions the amplitude of the oscillations will increase. A demonstration can be made by forcing a smooth wooden sleeve onto the end of a motor shaft and letting the bearing consist in a smooth hole drilled in the top of the wooden stick. The undamped action is startling and suggests perpetual motion. It was explained qualitatively by Lord Rayleigh in his *Theory of Sound.* He likened it to bowing a violin string, which depends on the fact that the friction of rubbing surfaces decreases with increase of speed when the speed is not too high. Mathematical solutions would be desirable in order to find the limiting and the optimum conditions for a successful demonstration. See Fig. 4 in (Hersey, 1964).

The Froude pendulum was cited independently by two investigators in explaining the failure of a bascule bridge over the Hackensack River. Each leaf in such a bridge is counterweighted so that it can be raised slowly while the whole weight is supported on a trunnion bearing. While the trunnion or journal rotates steadily during the lift-up, the bearing member, nominally stationary, oscillates somewhat owing to the elasticity of the structure and the mass of the counterweight. It seems that the designers were counting on bearing friction to damp the oscillations, if anything; but instead, friction had just the opposite effect, whereupon the structure collapsed, causing one leaf of the bridge to fall into the river. The Froude pendulum, with references to the bridge investigation, was brought to our attention by L. B. Tuckerman (1938).

3. COEFFICIENTS OF FRICTION

Frictional resistance in machine elements is commonly expressed by the *coefficient of friction* f, or ratio of the frictional force F to the load, or force normal to the surface, W_n. Thus in Figs. 1 and 2, $f = F/W_n = \tan A$. In applying this ratio to fluid film bearings, F is defined as the equivalent tangential friction, and W_n may be taken as the resultant load W on the bearing. To determine F experimentally we measure the friction torque, or moment M on the member in question, say the rotor; and then write $F = M/r$, where r is the radius of the journal, or the mean radius in a thrust bearing. In either case the coefficient is given by $f = M/rW$. In fluid film bearings the journal displacement is in the direction of motion U, opposite to that for a dry bearing, as explained in Chapter II.

Frictional forces are of such a nature that the formula $F = fW$ gives only a maximum or limiting value. The frictional resistance may be anywhere from zero to fW, depending on conditions. The story is told of a civil engineering instructor who drew a bridge truss on the blackboard. An arrow pointing to the right was marked "friction." Another pointing to the left was marked "wind." The sketch was to illustrate equilibrium. "Then if the wind stops blowing," asked a student, "why doesn't the friction push the truss off the piers?" Paradoxically, it doesn't because friction never exceeds the opposing force. It builds up so as to just balance that force unless the latter exceeds fW, in which event the motion will be accelerated.

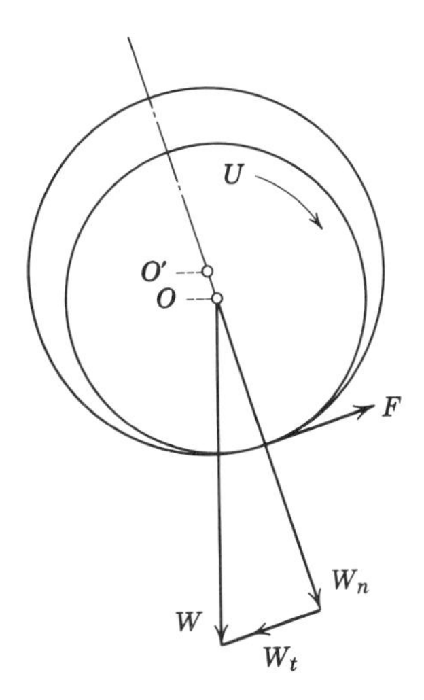

Fig. 2. Friction on the journal in a dry bearing (O, O', journal and bearing centers).

Leonardo da Vinci found $f = \frac{1}{4}$ for polished surfaces, while Amontons reported an average value of 0.3. Four-hundred years after Leonardo, his value was closely confirmed by Douglas Galton (1878), who found coefficients from 0.24 to 0.29 for the adhesion of railway wheels to the track.

During most of the nineteenth century it was customary to treat engineering problems on the assumption of a constant coefficient of friction. Great collections of examples have been worked out mathematically by Jellet and others (1872) on that assumption. Even today it is not uncommon to find calculations for elaborate mechanisms like gear wheels based on the slender assumption of Coulomb's law.

Coulomb had accepted Amontons' explanation attributing solid friction to the interlocking of asperities. It remained for the twentieth century to come up with new concepts like that of Tomlinson on the action of molecular forces (XIII: 1929), or those of Bowden and coworkers on the shearing of welded junctions (Chapter XIII). Although these investigations into the cause of friction tend to confirm the Amontons-Coulomb law, it is known that the friction coefficient ranges from practically zero to infinity under various conditions.

The lowest values are found in heavily loaded hydrostatic bearings; the highest between clean, dry solids or in fluid films at high speeds under light loads, as in vertical shaft guide bearings. These are journal bearings provided only to position the shaft and protect it from accidental radial loads. The coefficient of friction of lubricated surfaces necessarily approaches infinity as the load approaches zero, since the quotient of any finite quantity, such as the shearing resistance of an oil film, divided by zero is infinite. Values found for the coefficient with different solids and lubricants under nonhydrodynamic conditions are set forth by Dudley Fuller (1957) in his compilation for the American Institute of Physics. See also the coefficients found by Boyd & Robertson (1945) under high contact pressures, including data on solid lubricants. The variability of friction values with duration of loading and other test conditions is convincingly shown by Schmidt & Weiter (1957).

The conventional coefficient of friction is most useful when frictional resistance is nearly proportional to the load, as in dry or heavily loaded bearings. It is least useful in *concentric*, or lightly loaded bearings, where the journal axis practically coincides with the bearing axis. Here, friction is nearly independent of load. These limitations were emphasized by Dennison in his study of engine-bearing design (1936). He proposed using a different characteristic. Dennison's coefficient is proportional to the quotient of the frictional torque divided by the product of viscosity and speed. Both coefficients are dimensionless. The conventional coefficient approaches constancy when Coulomb's law is followed. Dennison's approaches constancy when Petroff's equation can be applied, that is, under hydrodynamic conditions with a lightly loaded bearing, as described in Chapter II. Styri preferred torque to coefficient in reporting friction of ball and roller bearings (Chapter X).

Another variant is the "incremental" coefficient of friction introduced by Burwell & Strang (1949). This coefficient is defined by the slope, at any point, of a graph showing frictional resistance against load. It would reduce to the conventional coefficient if the graph happened to be a straight line passing through the origin. The incremental coefficient has been useful in analyzing data on imperfect lubrication (Chapter XIII).

A distinction must be made between the coefficient of friction on the journal and that on the bearing member. There is no difference in a dry bearing, since the surfaces are in direct contact there. But in a fluid film bearing with journal appreciably eccentric, the friction torque on the journal is greater than that on the bearing (Chapter II). For coefficients less than 0.001 the difference can be as great as 50 per cent. When not otherwise specified, it will be understood that f refers to the journal coefficient. Since forces acting on a stationary body do no work, the bearing coefficient is not needed for computing power loss. Experimenters often measure the friction moment on the bearing member only; hence the need for a correction.

4. FRICTIONAL POWER LOSS

Robert H. Thurston, who became the first President of the American Society of Mechanical Engineers, was impressed by the need for greater economy in the use of power. His first book on friction and lubrication (1879) was republished under the title *Friction and Lost Work in Machinery and Millwork* (1885). In this form the book went through seven editions, stimulating widespread interest among engineers in the understanding and reduction of friction losses.

Petroff, in the meantime, was conducting a similar campaign of education in Russia. His long paper on "Friction in Machines and the Effect of the Lubricant" (1883) was awarded the Lomonosoff Prize of the Imperial Russian Academy of Sciences. It showed how the Russian supplies of petroleum could be made into suitable lubricants for reducing the waste of mechanical power, and contained formulas needed for estimating power loss.

An experienced mechanical engineer in Pittsburgh was asked by the management of a steel mill to make a complete analysis of their friction losses (page 196 of Hersey, 1936). His report was an eye-opener: total power loss 90 per cent for all machinery in the mill. From 40 to 50 per cent of the power delivered to the roll stands was consumed in the roll-neck bearings alone—a figure that has been lowered by hydrodynamic bearing design. Dr. Georg Vogelpohl (1951), a European authority, presented a survey of lubrication problems before the Third World Petroleum Congress at the Hague. Vogelpohl estimated that from one-third to one-half of the world's energy production is consumed in friction. See also the opening pages in Professor Fuller's book (1956). More study of power loss might well be included in engineering education (*Mech. Eng.* 1934; Hersey & Hopkins, 1949).

Several investigators, beginning with Thurston, who made a special study of the steam engine, have tried to pinpoint these friction losses. Sparrow & Thorne (1927) did it with some success for piston-type aviation engines; Lichty & Carson (1933) for the automobile engine, operating different parts separately; and Dutcher (1938) for various reciprocating engines. Professor Marks, on a lecture tour, compared the efficiencies of a large power windmill, a new water turbine, high-temperature steam turbines, the mercury-steam turbine, diesel engine, and gas turbine (1942). See also Takahasi & co-author (1951).

It seems that the electric utilities charge higher rates during peak load periods. Accordingly, the Baltimore Street Railways, we are told, effected a saving in power cost by reducing the viscosity grade of the lubricants used. The practice spread to other cities. Following up this lead, the American Electric Railway Engineering Association instituted experiments on friction loss in street railway reduction gears. The results made it possible to level off some of the peak-load power demands. These tests were conducted by S. A. McKee at the National Bureau of Standards, as described in Chapter XI.

The question of power loss and its distribution formed the subject of a Symposium in the Society of Automotive Engineers (1956). It was concluded that some 30 per cent of the engine power goes into friction in the normal operation of a modern automobile. Now according to Newton's first law of motion, if it were not for frictional resistance, no power would be needed to maintain a uniform speed on the road. Fuel would be required only to bring the car up to speed, and the kinetic energy of the car could be recovered and stored in some other form when stopping. A large part of the resistance at medium and high speeds is, of course, due to windage rather than friction in the mechanism.

An appraisal of the foregoing studies, aimed at greater experimental accuracy, was offered by R. E. Gish and co-authors (1958). It was followed up by Vasilica & Nica (VIII: 1963). In the course of determining power loss in engines these authors found that the pumping cycles account for about half the total loss.

Fuel Economy. Since fuel consumption parallels friction loss, it is of interest to note several investigations aimed expressly at fuel economy. Recommendations by W. H. Graves (1933) for reducing crankcase viscosity have been widely adopted with good results; and are justified by the improved construction standards, closer fits, and lower wear rates of later model automobiles.

Dr. William H. Kenerson gave us a first-hand account of a nationwide economy contest in which he and Professor Lockwood of Yale were picked to run the tests with an air-cooled engine, in their respective states. They achieved 30 to 35 miles per gallon, aided by the following procedure:

1. Choose level country and good roads.
2. Pump tires up "hard as rocks."
3. Lower the top and remove windshield.
4. Use very low viscosity oil in crankcase.
5. Accelerate at full throttle to 25 mph, then coast to a walk, and repeat.
6. Minimize air cooling, as engine is more efficient when hot.

Lockwood won the competition with Kenerson second, but the winner may have had fewer hills to climb. Another trick sometimes used, according to N. MacCoull, is to drain the transmission. These procedures reflect some of the factors responsible for power loss.

More detail is given by Greenshields (1950) in reporting a later economy contest. We are not surprised that it was won by a Studebaker, owing to the prior experience gained by leading engineers of the Company in related research at the National Bureau of Standards. Actually a lighter car went far beyond 150 mpg, but was penalized under the rules because of its light weight. New data on power loss in automotive engines will be found in papers by Wilford (1957) and by Clayton (1960). See also the references on piston and ring lubrication in Chapter VIII.

Railroad fuel consumption is discussed by W. M. Keller in a conference arranged by the National Research Council of Canada (1962). He informs us that "the railroads of North America have a fuel bill of over a million dollars a day" (p. 180). Since most of the fuel is needed for overcoming friction, he suggests that if the loss could be reduced only 1 per cent, it might save $10,000 a day.

5. FRICTIONAL HEATING

Without friction, who could strike a match? Or even start fire by twirling a pointed stick? Friction sawing and welding, previously mentioned, are useful effects of the temperature rise caused by friction. Yet, we are more frequently aware of detrimental effects. For example, we often read of people severely burned by sliding down ropes. Describing frictional heat, Edward Turner, author of an early chemistry textbook (1832), states that "The axle-tree of carriages has been burned from this cause, and the sides of ships have taken fire by the rapid descent of the cable."

The man on the street tries to keep well informed on the "re-entry" problem as affected by atmospheric friction, but he may not know how much of the temperature rise is due to sudden compression and how much to shearing the air film. He probably does not know that every 778 ft-lb of work done against friction generates nearly 1 Btu of heat. Where does each Btu go? If absorbed by a pound of water, it makes the temperature rise 1 deg F. Absorbed by an equal weight of oil, the rise will be over 2 deg; of metal, from 5 to 12 deg, the latter figure for steel.

When a lightly loaded full journal bearing 3 in. long by 3 in. in diameter and of customary clearance operates at 1800 rpm with a medium viscosity oil, the frictional power loss will come to about $\frac{2}{3}$

hp, or $\frac{1}{2}$ kw. If the heat generated thereby were to remain in the oil, where it originates, and if the oil were to stay put in the clearance space, the film temperature would rise initially at the rate of 750 deg F/sec! In practice the heat is removed by oil flow, by conduction into the shaft and bearing, and otherwise—a very central problem in the theory of lubrication (Chapter III).

Extremely high temperatures can be reached in dry friction at high speeds. Railway brake shoes were tested up to surface speeds of a mile a minute, or 88 ft/sec by Galton in his historic investigation. The decrease in friction with increasing speed found in those tests may have been due in part to air-film lubrication. Bowden & Ridler (1936) reported early experiments on surface temperatures in which surface melting was observed at speeds of the order of 50 ft/sec. Many investigations to 110 ft/sec were conducted by the National Advisory Committee for Aeronautics, and described by E. E. Bisson and co-authors (1957, 1964). Friction was measured at speeds to 2000 ft/sec by W. W. Shugarts, Jr., and others (1953) at the Franklin Institute incident to interior ballistics. Speeds of 1000 ft/sec were reached by J. M. Kraft (cited previously) in his experiments on ballistic penetration. A new type of friction apparatus was described by Bowden & Freitag (1955), by which the friction of metals was investigated in a vacuum at speeds approaching 3000 ft/sec. The characteristic drop in friction with increasing speed was confirmed without the complication of an air film. The drop was attributed to lack of time for propagation of plastic deformation, together with the high temperatures developed.

Hotboxes. Whenever the frictional heat in a bearing accumulates faster than it can be removed at a maximum safe temperature, an unstable condition is reached known as a "hotbox." Axles have been broken and railroad cars set afire, thus leading to many investigations. The wrecking of the Congressional Express with 79 lives lost is an extreme example (Associated Press, 1943). Whether caused by lack of oil or mechanical faults, hotboxes came to be recognized as a definite problem. See Andersen (1953) and Hawthorne (1953). Automatic detectors have been invented and usefully applied in freight and passenger service, although Downes (1947) advocated roller bearings as the only certain remedy.

Surface Damage. The temperature rise caused by friction aggravates wear and surface failure in machine elements. These conditions are well described by Barwell (1956), Bisson and co-authors (1957, 1964), and Wilcock (1957). Goodzeit (1956), Roach (1956), and others experimented on the friction and surface damage of a large number of different metals in sliding contact with steel. Friction and wear tests have been carried to 1000 C (over 1800 F) in a study by E. P. Kingsbury & Ernest Rabinowicz (1959). Further references to frictional damage are given in Chapters X–XIII.

6. ROLLING FRICTION

Starting friction can often be reduced by substituting rolling in place of sliding contact; hence the use of wheels, rollers, and rolling bearings. According to Coulomb's law of rolling friction, the resistance to rolling will be proportional to the load, independent of the speed, and inversely proportional to the radius of the rolling body. This approximate relation remains in common use even today, with the aid of handbook data, because of its convenient form. In symbols, $F = \text{const}\ (W/r)$, where F is the force needed to overcome rolling resistance when applied in the direction of motion at the axis of the roller, W is the total load, and r the radius.

But consider the meaning of the constant of proportionality. In order that a body shall experience rolling friction, it must either be deformable, or traveling over a deformable surface, or both. The resultant upward force W' that supports a deformable body moving in response to horizontal forces will act off center. It will be displaced *forward* of the load line by some small distance e that may be called the eccentricity (Fig. 3). That distance is the moment arm of a couple formed by the load W and the equal and parallel supporting force

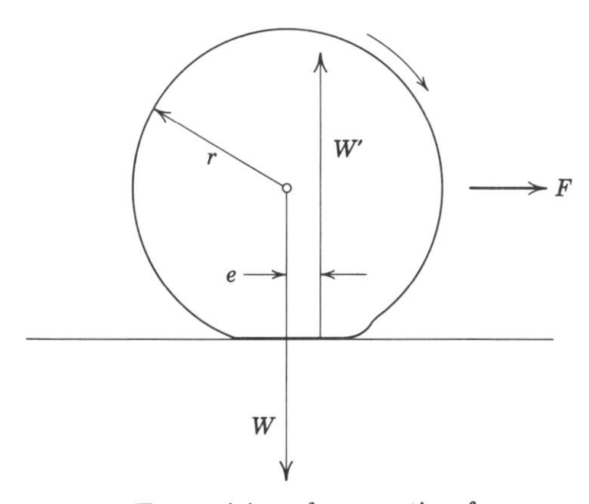

Fig. 3. Eccentricity of supporting force on deformable roller, on rigid surface.

W'. The couple is the source of the rolling resistance. It is balanced by the moment of the pull F acting at a height r above the contact area. Under equilibrium conditions, $F = e(W/r)$.

The constant in Coulomb's law will therefore be recognized as the eccentricity of the supporting force, or moment arm of rolling friction. Some textbooks call this length e the "coefficient of rolling friction." It would seem more logical to define the coefficient of rolling friction by the ratio F/W. To avoid confusion we shall not use the term "coefficient of rolling friction" for the time being; speaking, instead, of the *rolling friction ratio* for F/W, and the *eccentricity* for the moment arm e. Handbooks, in the light of Coulomb's law, usually give values of e on the assumption that they are constant for any given pair of materials, regardless of the roller radius r. The fallacy of that assumption can readily be seen from the fact that the moment arm cannot possibly exceed a fraction of the radius; hence with diminishing values of the radius, the eccentricity must continually decrease. It is easy to see that e must approach zero as r approaches zero.

Dupuit (1842) was apparently the first to offer a rational formula in place of Coulomb's. He confirmed the proportionality of rolling friction to load, but found it inversely proportional to the square root of the radius. This is equivalent to finding e directly proportional to the square root of the radius. Experiments conducted with cast-iron car wheels on steel rails by H. E. Wetzel (1924), M. S. Downes (1925), and the writer tend to support Dupuit's law with respect to the influence of the radius.

The rolling friction ratio F/W diminished slightly with increasing load, and increased appreciably with increasing speed over the limited range of our tests. A value of about 0.004 was found by interpolation for 10-in. diameter wheels with machined treads, running from 3 to 4 mph under a load of 750 lb per wheel (Wetzel, 1924). This comes to nearly $\frac{1}{3}$ of the total friction in a roller-bearing mine car truck, under the conditions stated. Further data are given by the Engineering Foundation (1946), and a more complete report is in preparation. The tests were run at the Pittsburgh Experiment Station of the U. S. Bureau of Mines as part of a cooperative program on mine car friction (Hersey, et al., 1925). See also Chapter X.

An amusing experience in this connection illustrates the significance of rolling friction (Hersey, 1936, p. 194). A ball-bearing car was submitted by the bearing manufacturer in competition with three types of roller-bearing cars. To the surprise of everyone, the ball-bearing car, No. 1, showed the highest friction of the lot. One of the roller-bearing cars, No. 2, gave almost double the friction expected. It was revealed that the ball-bearing wheels had been unskillfully cast in a local foundry and came out exceptionally rough. The wheel surfaces of car No. 2 had been turned down accurately in a lathe, thus eliminating the roughness, but removing the harder skin of the casting. The softer meal exposed created more rolling resistance, with the result that car No. 3, fitted with ordinary wheels, won the test. The true relative merits of the respective bearings were concealed by differences in rolling friction between wheel and rail.

7. LUBRICATION RESEARCH

The efficiency of machinery has been greatly improved since Thurston's time, partly because of the commercial availability of ball and roller bearings, but mainly by research in lubrication (Vogelpohl, 1960). Cumulative experience in scientific "trouble shooting," as described in the final chapter of Wilcock & Booser (1957), has also had a hand in the evolution.

Hirn (1854) may well be considered the father of lubrication research and the discoverer of fluid film lubrication. He was so regarded by Professor Thurston. More commonly, the first lubrication research is credited to Tower in England and Petroff in Russia. Both began publishing in 1883 (Chapter II). Thurston offered empirical formulas for load capacity and friction and discovered the friction minimum point. Yet his chief contribution was not so much in research as it was in his dramatic and persistent way of calling attention to the need for reducing friction in machinery.

Biographies of lubrication pioneers are presented in the articles by F. R. Archibald (1955). The next important experimenter, following Tower, was Albert Kingsbury. His thrust bearing development (Chapter VI) and gas-lubrication experiments (Chapter IX) set a high level of attainment in research. Kingsbury was appointed first Chairman of the ASME Research Committee on Lubrication. This committee, known as the RCL, was organized in 1915. Some of its earlier work in described in a paper by A. E. Flowers and the writer (1931). The RCL aims to investigate the fundamental problems of lubrication and to encourage similar investigations in every way possible.

8. PUBLICATIONS

Present-day interest in friction and lubrication is well shown by the *Proceedings of the First National Symposium,* American Society of Lubrication Engineers (1954). Recent investigations abroad have been reported by the Institution of Mechanical Engineers (1957), British Association for Advancement of Science (1959), and the Soviet Academy of Sciences (Akademiia Nauk, 1960). The Institut Mashinovedeniia (Institute for the Study of Machines) publishes an annual collection of selected papers on *Friction and Wear in Machinery.* They have been translated and republished by the ASME at the United Engineering Center, New York, beginning with the 1956 volume. Returning to home territory, we recall two symposium volumes of special interest, edited by Robert Davies. The first is on *Friction and Wear,* comprising eight contributions with discussion (1959). The second, cited in Chapter X, is entitled *Rolling Contact Phenomena* (X: 1962). Investigations by the National Advisory Committee for Aeronautics and the National Aeronautics and Space Administration have been reviewed by Bisson & Anderson (1964). The regular publications of the technical societies should be scanned for current research on lubrication, including papers presented at the ASME-ASLE International Lubrication Conference, Washington, D. C., in October, 1964. The proceedings of an international symposium on *Lubrication and Wear,* held at the University of Houston, were edited by Muster & Sternlicht (1965).

International conferences are the order of the day. A high standard was set by the conferences on *Lubrication and Wear* and on *Gearing* held by the Institution of Mechanical Engineers at London in 1957 and 1958, and by the first international symposium on *Gas-Lubricated Bearings* organized by the Franklin Institute and held in Washington in 1959. Of these three, the first is cited below, the second in Chapter XI, the third in Chapter IX.

Many good books on lubrication utilizing the results of research have appeared since our first book came out in 1936. Several have been mentioned above, and others will be referred to in later chapters.

A prophetic look into the future has been ventured by Professor Fuller (1965), with notes on work in progress.

9. SUMMARY

Friction can be useful, for example, in brakes, clutches, or friction-welding. It is commonly considered detrimental because it creates difficulty in starting machines under load, and causes "power loss," "hotboxes," and surface damage. Dry friction can be reduced by substituting rollers for sliding contact, or by using a lubricant.

Thurston here, and Petroff in Russia, knew these facts and spoke up like prophets crying out in the wilderness.

Theory and research in lubrication aim to reduce friction and wear in machinery.

REFERENCES[1]

Akademiia Nauk SSRR (1960), *Conference on Friction and Wear in Machinery* (Russian). Moscow, 288 pp. (a) transl. FTD-TT-61-449, OTS, Dept. Commerce, Wash., D. C., 225 pp. (1962).

American Society of Lubrication Engineers (1954), *Fundamentals of Friction and Lubrication in Engineering*. First National Symposium, Chicago, 1952; 196 pp.

Amontons, G. (1699), "De la résistance causée dans les machines." Acad. Roy. des Sci., *Mémoires de mathématique et de physique*, 259–86.

Andersen, G. R. (1953), "Hotboxes and Train Operation." *ASME* 53-A-124. (a) *ME* **76,** 536.

Archibald, F. R. (1955), "Men of Lubrication." *LE* **11,** 9–11, 55; (a) 84–5, 128–9; (b) 162–3, 197–8; (c) 228–9, 283; (d) 304–5, 346; (e) 375, 420–2; (f) **12,** 15, 72 (1956); (g) 91, 157–8; (h) **13,** 12–13, 63 (1957); (i) combined reprint, iv + 21 pp., ASLE, Chicago (1957).

Associated Press (1943), "Flaming Hot Box Seen Too Late to Avert Congressional Wreck." *Boston Herald,* Sept. 10, p. 24.

Barwell, F. T. (1956), *Lubrication of Bearings*. Butterworths, London, 292 pp.

Beck, Th. (1906), "Leonardo da Vinci, 1452 bis 1519. Vierte Abhandlung: Codice Atlantico." *ZVDI* **50,** 524–31, 562–9, 645–51, 777–84 (see 524–5).

Benton, W. A. (1926), "Some Aspects of Friction." *The Engineer* **141,** 403–5, 430–1, 458–9.

Bisson, E. E. & Anderson, W. J. (1964), *Advanced Bearing Technology*. NASA, Wash., D. C., 511 pp.

Bisson, E. E., Johnson, R. L., & Anderson, W. J. (1957), "On Friction and Lubrication at Temperatures to 1000 F." *ASME* 57-Lub-1. (a) Paper 23, *CLW,* 348–54; 319, 363, 758, 805, 822; Pl. 3.

Bowden, F. P. & Freitag, E. H. (1955), ". . . Friction of Solids at High Speeds." *Nature* **176,** 944–6. (a) Paper 44, *CLW,* 652–4; 241, 812; Pl. 21–3 (1957).

Bowden, F. P. & Ridler, K. E. W. (1936), "The Surface Temperature of Sliding Metals—The Temperature of Lubricated Surfaces." *PRS* **154,** 640–56.

Bowden, F. P. & Tabor, D. (1950), *The Friction and Lubrication of Solids*. Oxford University Press, London, 337 pp. (a) Revised reprint, 1954, 372 pp. (b) German transl., Springer, Berlin, 1959, 430 pp. (c) Part II, 1964, 554 pp. See F. F. Ling, *ME* **87,** 74 (July 1965).

[1] Abbreviations are listed in the Appendix.

Boyd, John (1948), "Lubrication Phenomena. Use of Transparent Models for Purposes of Demonstration and Study." *ME* **70,** 589–92.

Boyd, John & Robertson, B. P. (1945), "The Friction Properties of Various Lubricants at High Pressures." *T* **67,** 51–6.

Braams, C. M. (1952) "On the Influence of Friction on the Motion of a Top." *Physica* **18,** 503–14. (a) Hugenholz, N. M., 515–27.

British Association (1959), Symposium on "Modern Problems in Friction and Lubrication." Papers by Archard & Hirst; Barwell, Bowden, Tabor & Williams at Glasgow, 1958. *Adv. of Sci.* **15,** 270–1. (a) *Sci. Lubn.* **10,** No. 11, 11–23 (Nov. 1958).

Burwell, J. T. & Strang, C. D. (1949), "The Incremental Friction Coefficient." *JAP* **20,** 78–89.

Chamberland, H. J. (1946), "Friction Sawing." *Western Mach. & Steel World* (Aug.). (a) *ME* **68,** 1076.

Clayton, P. M. (1960), "Short Cut to Fuel Economy Data." *SAE* 199 B. (a) *JSAE* **68,** 30–2 (Nov.).

Coulomb (C. A. de) (1785), "Théorie des machines simples, en ayant égard au frottement de leur parties et à la roideur des cordages." Acad. Roy. des Sci., *Mémoires de math, et de phys.* **10,** 161–332.

Courtel, R. & Tichvinsky, L. M. (1963), "A Brief History of Friction," I. *ME* **85,** 55–61 (Sept.); II. 33–7 (Oct.). (a) *Nav. Engs. J.* 451–60 (1964).

Davies, Robert (editor) (1959), *Friction and Wear.* General Motors Symposium, 1957; eight contributors. Elsevier, Amsterdam; 191 pp.

Dennison, E. S. (1936), "Film Lubrication Theory and Engine Bearing Design." *T* **58,** 25–36.

Downes, M. S. (1947), "Roller Bearing Applications for Modern Freight Cars." *ASME* 47-A-135; 3 pp. (a) *ME* **70,** 255 (1948).

Dwnes, M. S. & Shore, Henry (1925), *Mine Car Friction with Special Reference to Rolling Friction and Bearing Friction.* M. S. Thesis, Carnegie Inst. Tech., Pittsburgh; 62 pp.

Drescher, H. (1959), "Zur Mechanik des Reibung zwischen festen Körpern." *ZVDI* **101,** 697–707.

Dupuit (1842), "Mémoire sur le tirage des voitures et sur le frottement du roulement." *Annales des ponts et chausées* **3,** 261–335.

Dutcher, F. H. (1938), "The Friction of Reciprocating Engines." *T* **60,** 225–34.

Engineering Foundation (1946), "Rolling Friction." Project 89, *Annual Report* 1945–6, New York, 28–30. (a) *ME* **69,** 90 (Jan. 1947). (b) *Annual Report* 1946–7, New York, 24–5.

Flowers, A. E. & Hersey, M. D. (1931), "Contributions apportées par l'American Society of Mechanical Engineers à l'étude du graissage." *Compte-rendu du congrès du graissage,* Strasbourg; 419–23. (a) *ME* **54,** 269–70.

Fuller, D. D. (1954), "Historical Development of Hydrodynamic Lubrication." Pages 7–30 in *Fundamentals of Friction and Lubrication in Engineering,*" First National Symposium (1952), ASLE, Chicago.

Fuller, D. D. (1956), *Theory and Practice of Lubrication for Engineers.* Wiley, New York, 432 pp. (a) German transl., W. Maier & H. Paul, Berliner Union, Stuttgart, 1960; 429 pp. (b) Polish transl., F. Matczynski & M. Arkuszewski, Wydawnictwa Techniczne, Warsaw, 1960; 352 pp. (c) Spanish transl., A. C. Fernandez & J. L. P. Puga, Ediciones Interciencia, Madrid, 1961; 544 pp.

Fuller, D. D. (1957), "Coefficients of Friction," pp. 39–44 of Sect. 2-d, *Am. Inst. Physics Handbook,* McGraw-Hill, New York. (a) 2nd. ed. (1961), *op. cit.* 42–7.

Fuller, D. D. (1965), "Lubrication." *Internat. Sci. and Technol.* No. 37, 18–27.

Galton, Douglas (1878), ". . . Experiments Upon the Coefficient of Friction between Surfaces Moving at High Velocities." *Rep., Brit. Assoc. Adv. Sci.,* Dublin, 438–41. (a) *Eng.* **26,** pt. 2, 153–54.

Gemant, Andrew (1950), *Frictional Phenomena.* Chem. Pub. Co., Brooklyn; 497 pp.

Gish, R. E., McCullough, J. D., Retzloff, J. B., & Mueller, H. T. (1958), "Determination of True Engine Friction." *TSAE* **66,** 649–67.

Goodier, J. N. & Sweeney, R. J. (1945), "Loosening by Vibration of Threaded Fasteners." *ME* **67,** 798–802.

Goodzeit, C. L. Roach, A. E., & Hunnicutt, R. P. (1956), "Frictional Characteristics and Surface Damage of Thirty-Nine Different Elemental Metals in Sliding Contact with Iron." *T* **78,** 1669–76.

Graves, W. H. (1933), "Automobile Passenger-Car Motor-Oil Recommendations." *Proc. API* (III), Fourteenth Annual Mtg., 4 pp.

Greenshields, R. J. (1950), "150 Miles Per Gallon is Possible." *SAE* 416; 10 pp. (a) *JSAE* **58,** 34–8 (Mar.).

Gümbel, L., ed. by Everling, E. (1925), *Reibung und Schmierung im Maschinenbau.* Krayn, Berlin, 240 pp.

Hart, I. B. (1925), "Friction." Pages 140–1 in *The Mechanical Investigations of Leonardo da Vinci.* Chapman & Hall, London.

Hawthorne, J. W. (1953), "Hot Boxes—Some Fundamental Problems." *ASME* 53-A-104; 9 pp. (a) *ME* **76,** 535.

Hersey, M. D. (1933), "Notes on the History of Lubrication." *JASNE* **45,** 411–29; **46;** 369–85 (1934).

Hersey, M. D. (1936), "The Oil-Shed Fallacy. Attacking the Problems of Lubrication by Rational Methods." *Technology Review* **38,** 181–2, 192, 194, 196, 198. (a) Editorial, "Fighting Friction," 107 (Dec. 1935).

Hersey, M. D. (1936), *Theory of Lubrication.* Wiley, New York, 152 pp. (a) Second printing, 1938, 175 pp. (b) Spanish transl. by R. M. de Vedia, additions by C. Pasqualini; El Ateneo, Buenos Aires, 1947; 202 pp.

Hersey, M. D. (1964), "Research on Friction and Lubrication." *Brown University Engineer,* No. 5, 14–19.

Hersey, M. D., Golden, P. L., Shore, Henry, & Downes, M. S. (1925), *Mine Car Friction with Six Types of Trucks.* Bull. 25, Mining and Metallurgical Investigs., Carnegie Inst. Tech., 33 pp.

Hersey, M. D. & Hopkins, R. F. (1949), "Observations on Educating the Engineer." Pages 204–9 of IKIA, *Internationaler Kongress für Ingenieur Ausbildung* (1947). Ed. Roether Verlag, Darmstadt.

Hirn, G. A. (1854), "Sur les principaux phénomènes qui présentent les frottements médiats . . . ." *Bull. Soc. Indust. de Mulhouse* **26**, 188–277.

Institution of Mechanical Engineers, (1957), *Proc. Conf. on Lubrication and Wear* (=*CLW*), London, 911 pp.

Jellet, J. H. (1872), *A Treatise on the Theory of Friction.* Macmillan, London, 220 pp. (a) German transl. by Lüroth, Leipzig (1910).

Kingsbury, E. P. & Rabinowicz, (1959), "Friction and Wear to 1000 C." *JBE, T* **81,** 118–22.

Kraft, J. M. (1955), "Surface Friction in Ballistic Penetration." *JAP* **26,** 1248–53.

Kragel'skii, I. V. (1962), *Friction and Wear* (Russian). Mashgiz, Moscow, 383 pp. (a) Transl. by L. Ronson with J. K. Lancaster, Butterworth, Inc., Washington, D. C., 1965; 346 pp.

Kragel'skii, I. V. & Schedrov, V. S. (1956), *Dry Friction in a Scientific Survey of Friction* (Russian). AN, SSSR, Moscow; 235 pp.

Kragel'skii, I. V. & Vinogradova, I. E. (1955), *Coefficients of Friction.* Mashgiz, Moscow, 188 spp. (a) 2nd ed., 220 pp. (1962). (b) *Wear* **6,** No. 4, 326 (1963).

Lichty, L. C. & Carson, G. B. (1933), "Friction Losses in Engines." SAE Automotive Eng. Cong., Chicago; *Automotive Ind.* **69,** No. 9, 236. (a) *ME* **55,** 696–7.

Marks, L. S. (1942), "Conserving Power." *Science* **95,** *Supplement,* 8, 10 (Aug. 17).

McCune, J. C. (1939), "Braking High Speed Trains as an Engineering Problem." *ME* **61,** 583–8, 657–60.

Mechanical Eng. (1934), "Lubrication in Engineering Schools," **56,** 289 (May).

Michell, A. G. M. (1950), *Lubrication: Its Principles and Practice.* Blackie, London, 317 pp.

Morin, Arthur (1832), *Nouvelles expériences sur le frottement . . . ,* Bachelier, Paris, 128 pp.; 1834, 104 pp.; 1835, 143 pp. (a) Carilian-Goeury, Paris, 1838, 100 pp.

Muster, D. & Sternlicht, B. (1965), *Proc. Internat. Sympsium on Lubrication and Wear,* University of Houston, Texas (1964). McCutchan Pub. Corp., Berkeley; 974 pp.

National Research Council, Canada (1962), *Some Aspects of Friction and Wear in Mechanical Engineering* . . . Proc. of Conf. at Queen's University, Kingston, Ottawa; 180 pp. See Keller, W. M., 177–80.

Palmer, Frederick (1949), "What About Friction?" *Am. J. Phys.* **17,** 181–7, 327–42.

Petroff, N. P. (1883), Friction in Machines and the Effect of the Lubricant (Russian). *Eng. J.,* St. Petersburg, 71–140, 228–79, 377–436, 535–64. (a)

German transl. by L. Wurzel; Verlag L. Voss, Hamburg, 1887; 187 pp. (b) *Ostwald's Klassiker der Exakten Wissenschaften,* No. 218; L. Hopf, Leipzig, 1927; 1–38, 220.

Pliskin, W. A. (1954), "The Tippe Top (Topsy-Turvy Top)." *Am. J. Phys.* **22,** 28–32 (Jan.). (a) *Phys. Today* **7,** 7 (Apr.).

Radzimovsky, E. I. (1959), *Lubrication of Bearings.* Ronald Press, New York, 338 pp.

Rennie, G. (1829), "Experiments on the Friction and Abrasion of the Surfaces of Solids." *Phil. Trans. Roy. Soc. London,* **119,** pt. 1; 143–70.

Roach, A. E. (1956), Goodzeit, C. L., & Hunnicutt, R. P., "Scoring Characteristics of Thirty-Eight Different Elemental Metals in High Speed Sliding Contact with Steel." *T* **78,** 1659–67.

Schmidt, A. O. & Weiter, E. J. (1957), "Coefficients of Flat Surface Friction." *ME* **79,** 1130–6.

Shugarts, W. W., Jr. (1953), "Measuring Friction at High Speeds." *JFI* **256,** 187–9, (a) Robinson, George, **257,** 243–6 (1954). (b) *ASLE* 55-AM-4C4, with H. G. Clarke, Jr.; 15 pp.

Society of Automotive Engineers (1956), *Where Does All the Power Go?* Symposium, 36 pp. I. Burke, C. E. & Nagler, L. H., "The Engine—the Power Source." II. Campbell, E. C., "The Accessories—the First Bite." III. Zierer, W. E. & Welch, H. L., "Effective Power Transmission." IV. Lundstrom, L. C., "Wind and Rolling Resistance." V. Kosier, T. D. & McConnell, W. A., "What the Customer Gets." Papers 779–83. (a) Summary by W. S. James, *TSAE* **65,** 737–8 (1957).

Sparrow, S. W. & Thorne, M. A. (1927), "Friction of Aviation Engines." *NACA* Rep. 262; *13th Annual Report,* 177–203.

Stanton, T. E. (1923), *Friction.* Longmans, Green, London, 183 pp.

Swanger, W. H. (1934), "Frictional Resistance of Steel and Brass in Shrink Fits." Notes from U. S. Bur. Stds., *JFI* **218,** 99–100.

Takahasi, Y. & Tominaga, H. (1951), "An Analysis on the Running Resistance of Automobiles." *TJSME* **17,** No. 63, 137–9.

Thomas, W. (ed. by Niemann, G.) (1954), *Reibscheiben-Regelgetriebe.* No. 4 of Schriftenreihe Antriebstechnik, Vieweg, Braunschweig; 79 pp.

Thurston, R. H. (1879), *Friction and Lubrication.* The Railroad Gazette, New York; 212 pp.

Thurston, R. H. (1885), *Friction and Lost Work in Machinery and Millwork.* Wiley, New York, 365 pp. (a) 7th ed., 1903; 385 pp.

Tichvinsky, L. M. (1954), "Models for the Study of Lubrication Phenomena." Pages 339–49 in *Mémoires sur la mécanique des fluides* (Riabouchinsky Jubilé) Pub. Sci. et Techn. du Ministère de l'Air, Paris.

Tuckerman, L. B. (1938), "Pseudofrictionless Undamped Vibration." *J. WAS* **28,** No. 8, 374–5.

Turner, Edward (1832), "Sources of Caloric." Page 68 in *Elements of Chemistry,* Grigg & Elliot, Philadelphia, 4th American ed.

Vill', V. I. (1959), *Friction Welding of Metals* (Russian). Mashgiz, Moscow, 86 pp. (a) Am. Welding Soc., New York, 1962; 14 pp.

Vogelpohl, Georg (1940), "Geschichtliche Entwicklung unseres Wissens über Reibung und Schmierung." *Oel u. Kohle* **36,** Nos. 9, 13; 89–93, 129–34.

Vogelpohl, Georg (1951), "Lubrication Problems." *Proc. Third World Petroleum Cong.,* E. J. Brill, The Hague; Sect. 7, pp. 298–303. (a) *Sci. Lubn.* **3,** 9 (Sept.).

Vogelpohl, Georg (1960), "Reibung und Verschleiss in wirklichen Maschinen als Massstab zur Beurteilung von Forschungsergebnissen." *Forschung* **26,** No. 4, 108–16.

Wetzel, H. E. (1924), *Mine Car Friction as Influenced by Wheel Diameter and Other Variables.* M.S. Thesis, Carnegie Inst. Tech.; 71 pp. (a) with M. D. Hersey, Bull. 13, *Coal Mining Investigs.,* C.I.T., Pittsburgh; 37 pp.

Wilcock, D. F. & Booser, E. R. (1957), *Bearing Design and Application.* McGraw-Hill, New York, 464 pp.

Wilford, A. T. (1957), "Lubrication of Road Vehicle Engines and Worm-Driven Axles with Particular Reference to Vehicle Fuel Consumption." Paper 29, *CLW,* 524–8; 322, 429, 493, 789, 794, 797, 850.

chapter II Fundamentals of Lubrication

Lubrication, in the technical sense, refers to the action of the lubricant in separating the rubbing surfaces of machine elements. A distinction is made between this concept and that of conveying the lubricant to the right spot, preferably called *lubricating.*

1. TYPES OF LUBRICATION

It is customary to distinguish thick-film from thin-film lubrication. When the mean thickness of the lubricating film is large compared to the height of the rough spots, thick-film lubrication is said to occur. For practical purposes we can then forget that the surfaces are not perfectly smooth, and a simple geometrical description will serve. Modern high-speed machinery usually operates under thick-film lubrication.

On the other hand, when machines have been standing still for a while, the lubricant, if fluid, will be squeezed out to a great extent. Upon starting a machine from rest while under load, only thin-film lubrication can be expected. Most of the load is supported by high spots in contact, and during the few seconds before a good film is established, some slight wear can easily occur. Thin-film lubrication also prevails more or less between piston ring and cylinder liner, between gear teeth, and in metal cutting, as discussed in Chapters VIII, XI, and XIII.

For the present our attention will be concentrated on thick-film lubrication. This can be subdivided into at least four basic types, hydrodynamic, hydrostatic, thermal, and hydromagnetic lubrication. In hydrodynamic lubrication, the dominant type in well-built machinery, a fluid pressure is generated by the motion of the bearing surfaces. This pressure supports the load, maintaining a film thick enough to prevent rubbing contact. Bearings that depend only on hydrodynamic lubrication are called "self-acting." Hydrostatic lubrication is accomplished by introducing the fluid from outside, under a sufficient pressure to separate the surfaces and balance the load, even when there is no relative motion. Bearings that depend on hydrostatic lubrication are called "externally pressurized." The hydrostatic principle is especially useful in starting under load, and is often relied upon for that purpose in bearings that are otherwise self-acting (Fuller, 1947). Thermal or thermodynamic lubrication depends on fluid pressures being generated by temperature inequalities in bearings that could not otherwise support any appreciable load (Chapter III). Hydromagnetic or "magnetohydrodynamic" bearings are lubricated by electrically conducting fluids such as hot liquid metals or an ionized plasma. Load capacity is enhanced by the action of an external magnetic field (Snyder, 1962). According to Seegal (1961), a light bearing load can be supported by the pressure of sound waves. Is this a scientific novelty or a fifth basic type?

The principles of hydrodynamic and hydrostatic lubrication apply to gas-film as well as liquid-film bearings (Chapter IX), although the results differ in details. Plastic lubricants like greases follow the same basic principles, but again have somewhat different characteristics. These conditions have been referred to as "rheodynamic" and "rheostatic" lubrication (Osterle, 1956). The action of solid lubricants like talc, graphite, or molybdenum disulfide is mentioned in Chapters IV and XIII. Various types of bearings and of lubrication are described in the later chapters; but the principles of hydrodynamic lubrication will be outlined here (see Christopherson, 1951, Hersey, 1954; Slezkin, 1955; and Theyse, 1964).

2. ENOUGH OIL

The first principle of lubrication is that enough oil must be supplied (the term "oil" is often applied for brevity to any lubricant). Tower had the right idea when he advocated "profuse lubrication" as a result of his experiments (1883). He noted that a practical recognition of its value was seen in the trial trips of large marine engines when there was a premium for extra horsepower.

How much oil is needed to keep two surfaces apart? That would be an interesting question for calculation and experiment, taking surface roughness into account. Even without exact knowledge we can avoid the mistake made in a Connecticut factory, where the line-shafting was installed without oil holes. The story was told by the sales engineer from an oil company whose product was blamed for the hotboxes.

In his experiments of 1909 at M.I.T. the writer passed an electric current through a journal bearing while measuring its friction (Chapter VII). At an average oil feed as low as 2 drops per minute, the insulating effect of the oil could be found by recording the voltage drop across the film. It was surprising to see how the voltage would jump when a few extra drops were admitted, and how it would fall when a few drops were missed (Fig. 2 of Hersey, 1954). The film thickness at point of nearest approach increased, at first rapidly and then more slowly, as the oil feed was stepped up from 2 to 60 drops per minute. Over the same range the coefficient of friction decreased, at first rapidly, but only slightly after reaching 10 drops per minute (Hersey, 1914). The question has been further investigated by Fuller and Sternlicht (1956) and by Wilcock (1957). See also pages 264–8 in Professor Fuller's book (I: 1956).

As a corollary to the need for enough oil it is important to admit nothing *but* the lubricant into the clearance space. Abrasives and other contamination, even aeration, should be avoided. The size of grit particles limits the safe film thickness and consequent load capacity—see McKee (1927), as well as papers by Howarth (1935) and by Ambler (1955) cited in Chapter VI, and by Rylander (1952) in Chapter VII.

3. NEWTON'S LAW OF VISCOUS FLOW

Sir Isaac Newton discovered that the shearing stress in a fluid is proportional to the rate of shear at any fixed temperature (1687). This proposition underlies the hydrodynamic theory of lubrication except in the case of "non-Newtonian" lubricants (Chapter IV). It is analogous to Hooke's law for the shearing deformation of an elastic body. In lubricating films the rate of shear may be taken as the velocity gradient at right angles to the direction of motion, although a more general statement is needed in other problems (Stokes, 1849).

Let u, v, and w denote fluid velocities at a fixed point in the x-,

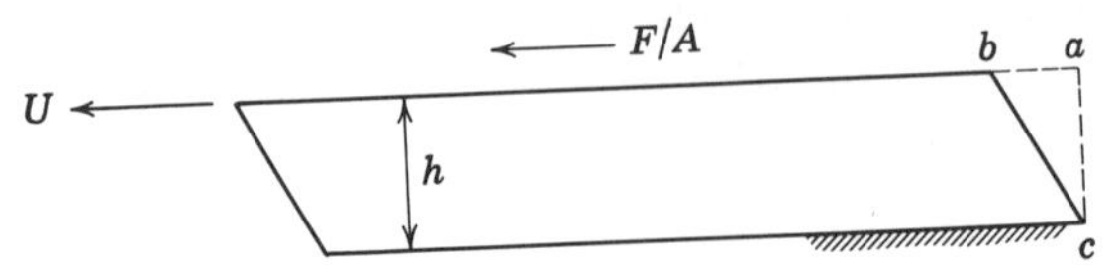

Fig. 1. Shearing under uniform pressure (rate of shear U/h due to shear stress F/A).

y-, and z-directions. When the flow is in the x-direction, the velocity gradient is expressed by du/dy. Let the tangential force per unit area, or shearing stress, be denoted by τ. Newton's law may then be written

$$\tau = \text{const}\,(du/dy)\,. \tag{1}$$

Suppose two flat, parallel surfaces are separated by some small distance h, as in Fig. 1, and that the space between them is filled with a viscous fluid at uniform pressure and temperature. Let F be the tangential force required over an area A to maintain the constant velocity U of one plate relative to the other. The shearing stress is F/A, the velocity gradient U/h; so by Newton's law,

$$F/A = \text{const}\,(U/h)\,. \tag{2}$$

The constant in Newton's law is different for different fluids, or for the same fluid at different temperatures, but is independent of the shear stress and the rate of shear. It is therefore a property of the fluid. This property is called *viscosity* and defined as the ratio of the shearing stress to the rate of shear. It is commonly represented by Z for "Zähigkeit" (Hagenbach, 1860). For mathematical purposes the Greek μ is often used, although experimentalists prefer Z.

Figure 1 has been simplified by omitting other forces that come into play to balance F and maintain equilibrium. They are not needed in the statement of Newton's law. A particle at a moves to b while the fluid is being uniformly sheared. Point c remains fixed. The little triangle *abc* is, in fact, the "abc" of lubrication. As an example, imagine two areas A, each of 1 sq in., separated by a distance h of 1 mil, or one-thousandth of an inch. Suppose a force F of 1 lb maintains a velocity U of 300 in./sec. Then Z will be F/A divided by U/h, or 3.3 millionths of a "pound-second per square inch." This is just about the viscosity of a medium lubricating oil at the operating temperature of a bearing.

Ordinary liquids and gases seem to follow Newton's law faithfully, except as noted by Truesdell (1952), Reiner (1960), and others interested in second-order effects. Plastic lubricants and oils containing high molecular weight ingredients do not conform to Newton's law. The viscosity of such materials may still be defined as the ratio of shearing stress to rate of shear, but the ratio decreases with increasing shear stress, so they are called "non-Newtonian." It is interesting that Newton's law, Eq. (1), a mere statement of proportionality, does not require any mention of viscosity; and that viscosity may be defined without reference to Newton's law.

4. PETROFF'S LAW

The laws of Petroff, Poiseuille, and Stefan provide simple applications of Newton's law of viscous flow, yet they illustrate the physical principles needed for the most advanced study of lubrication.

Petroff's law for the concentric journal bearing is an adaptation of Eq. (2). If C, D and L denote the diametral clearance, diameter of the journal, and axial length of the bearing (Fig. 2), we have only to replace A by πDL and h by $C/2$. If N is the speed of the journal in revolutions per unit time, U is equal to πDN. Let the constant in Newton's law be represented by Z, the viscosity of the fluid. Let M_0 denote the moment of friction on the journal, which is equal to the effective tangential force F multiplied by the radius $D/2$. Equation (2) then becomes

$$M_0 = \pi^2 D^3 LZN/C \,. \tag{3}$$

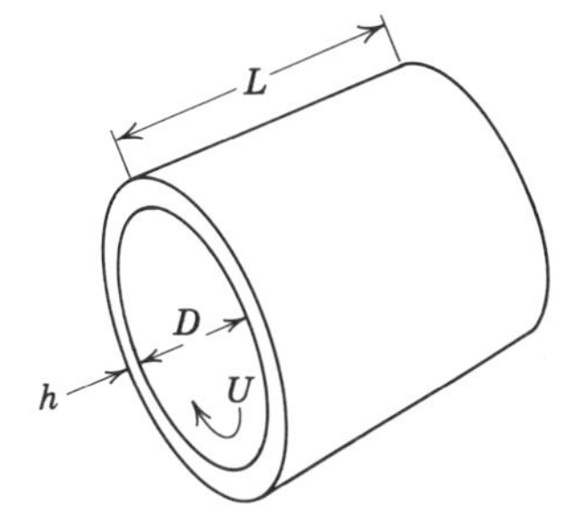

Fig. 2. Journal surrounded by oil film of uniform thickness h (Petroff bearing).

A more rigorous derivation might start with an element of torque, $(D/2)\ dF$, where dF is the tangential force on a surface of area L times $(D/2)\ d\theta$, subtending a small angle $d\theta$. Equation (3) follows by integration. An expression for the coefficient of friction, f, may be derived from Eq. (3) by substituting $FD/2$ for M_0 and remembering that f is equal to F divided by PDL, where P is the load per unit of projected area. With these changes, Petroff's law takes the form

$$f = f_0 = 2\pi^2 \left(\frac{D}{C}\right) \frac{ZN}{P} \,.$$

The application of the law is limited to high values of ZN/P, since a practically concentric journal has been assumed, as implied by the subscript.

To indicate the orders of magnitude involved, consider, as an example, a 2-in. diameter bearing 4 in. long with $D/C = 1000$, $N = 50$ rps (3000 rpm), $P = 100$ psi, and $Z = 3$ millionths of a lb-sec/sq in. This might be the viscosity of an SAE 10W oil at 130 F. It is seen that ZN/P will be 1.5×10^{-6}, so from Eq. (4) the coefficient of friction is 0.0296, or say 0.030. No use was made of the length or diameter! The power loss $H = fWU$, where $W = PDL$ and $U = \pi DN$; so $H = 7440$ in.-lb/sec, or 1.13 hp. This may be checked by starting from Eq. (3), since power loss is the product of friction moment and angular velocity. Note that the power loss in a Petroff bearing is independent of load.

Historically, Petroff (I: 1883) arrived at Eq. (3) as a special case of Margules' formula (1881) for the concentric cylinder viscometer. He allowed for slippage between the cylindrical walls and the fluid by including a term for "external friction," later found negligible. Perfect adhesion of liquid to solid has been confirmed by many investigations Campbell et al., 1657).

Variants from Petroff's Law. An interesting variant is the solution for small eccentricities. Instead of a uniform thickness we now have

$$h = c(1 + \epsilon \cos \theta) , \tag{5}$$

where ϵ is the ratio of eccentricity e to radial clearance c, and θ an angular coordinate. With the velocity gradient equal to U/h as before, the relative friction moment is found to be

$$M/M_0 = (1 - \epsilon^2)^{-1/2}. \tag{6}$$

This result was first published by Couette (1890), who included the case of nonparallel axes. It was rediscovered by Ocvirk (1952) through treatment of the "short bearing." As ϵ approaches zero, (6) approaches $1 + \epsilon^2/2$. Inglis (1939) found $1 + 2\epsilon^2$ for the infinitely long bearing, so the truth must be close to $1 + \epsilon^2$ for bearings of finite length.

Another variant or modification was found experimentally by S. A. and T. R. McKee (1929). Their result may be expressed as a correction term Δf to be applied to Petroff's law, the observed coefficient being given by $f_0 + \Delta f$. This correction depends on L/D, the length-to-diameter ratio. It is practically independent of C/D and ZN/P, being equal to 0.0018 for $L/D = 1$, and 0.0059 for $L/D = \frac{1}{2}$. The correction rises rapidly for L/D values below $\frac{1}{2}$, and falls slightly above $L/D = 1$, provided the journal is free from bending or edge-loading. A similar correction averaging 0.002 in round numbers had been found by the writer in his experiments on a bearing with L/D equal to 3 (Eq. 29 of 1914).

5. POISEUILLE'S LAW

The law governing viscous flow in horizontal capillary tubes was discovered experimentally by Poiseuille in the course of a long continued research (1840); and in part by Hagen a year earlier. The rate of flow was found proportional to the pressure gradient and to the fourth power of the bore diameter, with a proportionality constant that was different for different liquids, and for the same liquid at different temperatures.

Stokes (1849) was the first to attempt a mathematical derivation. He set up the differential equation by means of Newton's law, taking into account the effect of gravity acting upon the fluid in a vertical or inclined tube. Hagenbach (1860), following Wiedemann, completed the integration. Poiseuille's law may be written

$$Q = \frac{\pi}{8}\left(\frac{P}{L}\right)\frac{R^4}{Z}, \tag{7}$$

where Q is the flow rate in volume units per unit time, P the pressure drop in length L, R the radius of the capillary, and Z the viscosity of the fluid. It is understood that the fluid is incompressible, or not much compressed; that the capillary is stationary and horizontal, no body forces acting in the direction of flow; and that the tube is long enough to permit neglecting kinetic energy and end-effects. A body force like weight having an axial component g per unit of mass requires the addition of a term ρg to the factor P/L, where ρ is the density of the fluid.

The simplest derivation starts with a coaxial cylindrical portion of the fluid of radius y having the full length of the capillary. This portion is in equilibrium under the pressure drop P acting upon the end area πy^2; and the shearing stress τ acting uniformly over the surface $2\pi yL$. From Newton's law, τ equals $-Z\, du/dy$; the minus sign because u diminishes as y increases. Here u is the velocity at any radius y, such that $u = 0$ at $y = R$. As in Petroff's law, perfect adhesion is assumed.

For statical equilibrium, P times πy^2 is equal to τ times $2\pi yL$.

Substituting $-Z\ du/dy$ for τ leads to a differential equation from which u is found to be a parabolic function of y. Consider the flow dQ through an elementary section of width dy and circumference $2\pi y$. The flow is equal to the product of u by the area of the element. Integration from $y = 0$ to $y = R$ leads directly to Eq. (7).

Poiseuille's law may also be derived from the "principle of minimum dissipation," which will be discussed later. This derivation offers a concrete introduction to the calculus of variations in a form applicable to lubrication theory.

Variants. The best known variant of Poiseuille's law is the transition from streamline to turbulent motion, as determined by Reynolds (1883). Another is the widely used formula for streamline flow through a slot of breadth b and height h, where b is large compared to h (Mathieu, 1863). The slot formula may be written

$$Q = \frac{1}{12}\left(\frac{P}{L}\right)\frac{bh^3}{Z}, \tag{8}$$

where the remaining symbols are the same as in (7). Equation (8) has been used in deriving Reynolds' equation (Hersey, 1935) and in hydrostatic lubrication (Fuller, 1947). In such applications the flow under a variable pressure gradient may be calculated with $(-dp/dx)$ in place of P/L. Note that bh^3 has the dimensions of R^4.

Formulas for flow through straight and curved pipe lines with varying roughness are found in treatises on hydraulics, including transport of mud, sludge, and liquids containing granular matter in suspension. We have even read of oil being lubricated by water, the pipes being rifled to throw the water outward by centrifugal force.

Other variants are the solutions for short capillaries (Kreith, 1957); for streamline and turbulent flow through annular clearance spaces (Buckingham, 1923; Lamb, 1932; Yamada, 1962); for flow through eccentric clearances (Becker, 1907; Tao, 1955; Bachman, 1959); through elliptical and polygonal cross sections (Mathieu, 1863; Boussinesq, 1868; Paschoud, 1924; Cornish, 1928; Allen, 1937; Purday, 1949; Nuttall, 1954); and through slightly conical tubes (Henny, 1921). A "wetted perimeter" approximation for flow through a duct of any cross section is given by Wilcock & Booser (I: 1957, page 216). For compressible flow see Mathieu (1863); for plastic flow through capillaries, Buckingham (1921); for high-pressure flow see Hersey & Snyder (1932) and Exline (1946); for non-Newtonian flow, Tiabin (1956). Heat effects in capillary flow are reviewed in Chapter III.

6. STEFAN'S LAW

Josef Stefan's curiosity finally got the better of him (1874). He was determined to find out whether clean, flat, circular metal plates that seemed to adhere when pressed together really did so, or were merely held by a slowly yielding air film. The latter conjecture proved correct. His experiments were conducted both with air and liquid films. The upper plate was held fast and the lower one loaded so as to pull the two plates apart. He then derived a formula to explain the observations. As applied to compression rather than tension in the film, Stefan's law for the rate of approach may be written

$$-\frac{dh}{dt} = \frac{2}{3}\frac{PR}{Z}\left(\frac{h}{R}\right)^3. \qquad (9)$$

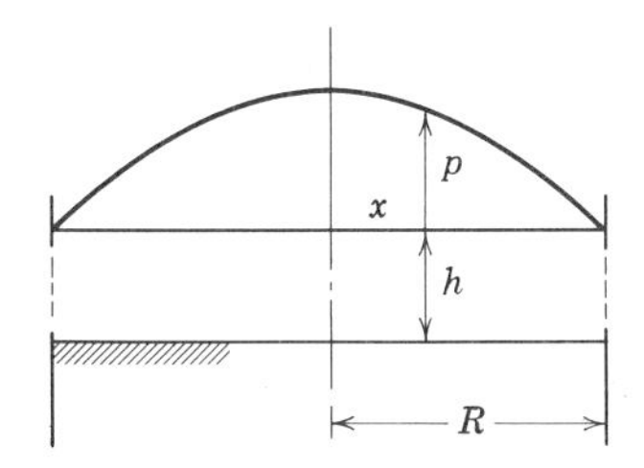

Fig. 3. Derivation of Stefan's law, with pressure p at x.

Here P is the load per unit area on a disk of radius R, and h is the film thickness at a given instant t (Fig. 3). Integrating gives an equation for the time required to squeeze the film down from h_0 to h. When h_0 is large compared to h, the equation reduces to

$$\frac{h}{R} = \frac{1}{2}\sqrt{3}\left(\frac{Z}{Pt}\right)^{1/2}. \qquad (10)$$

Equation (10) gives the relative film thickness as a function of Z/Pt.

The derivation of Eq. (9) illustrates the principle of continuity as well as Newton's law. At any radius x, the oil is being squeezed out radially at the rate

$$Q_x = -\frac{1}{12}\left(\frac{dp}{dx}\right)\frac{2\pi x h^3}{Z}. \qquad (11)$$

This is an adaptation of the slot formula, Eq. (8), with the pressure gradient expressed by $-dp/dx$ in place of P/L, and the breadth of the slot by $2\pi x$ in place of b. The principle of continuity is applied by equating Q_x to the decrease in volume per unit time within the cylinder of radius x, namely $-2\pi x dh/dt$. The resultant equation may be integrated once to express the fluid pressure p as a function of x, and once more to evaluate the total $\pi R^2 P$. This evaluation takes the form of Eq. (9) when transposed to provide an expression for $-dh/dt$.

Stefan's law is a solution of Reynolds' equation (reviewed in the next section) when there is no sliding velocity. It was discussed by Fuller in his articles of 1947 and more recently by Archibald (1956). For experimental confirmation see Needs' investigations (1940) and Bikerman's report (1957). Variants of Stefan's law comprise its extension to plastic and other non-Newtonian materials (Peek, 1932; Osterle, 1956); to collisions through liquid films (Eirich, 1947), and, especially, to a variety of geometrical forms. These include the elliptical disk, the half-bearing, the narrow rectangle and square plate approximations, and the approach of the cylinder and sphere to a plane surface (Reynolds, 1886; Gümbel I: 1925; Michell, I: 1950; Casacci, VI: 1951; Fuller I: 1956; and Hays, 1963). Archibald's paper of 1956 contains ten additional solutions. Butler (1960) studied the effects of high pressure. Kuzma (1964) showed how a magnetic field can improve the squeeze-film action.

The term "squeeze-film" was introduced by Arthur Underwood (1944) to explain the operation of engine bearings. A good review was published by D. F. Moore (1965).

7. REYNOLDS' EQUATION

Although Petroff discovered how to calculate the friction of a lightly loaded journal bearing, his formula does not apply to heavily loaded journals, nor to thrust bearings, nor could he predict load capacity. All these problems can be solved by the more complete hydrodynamic theory originated by Professor Osborne Reynolds (1884, 1886, 1902).

Consider, in Fig. 4, a rigid upper surface of unlimited extent moving with velocity U_1 above a surface of velocity U_0. Fix coordinate axes OX, OY, and OZ relative to the lower surface, the OZ axis being at

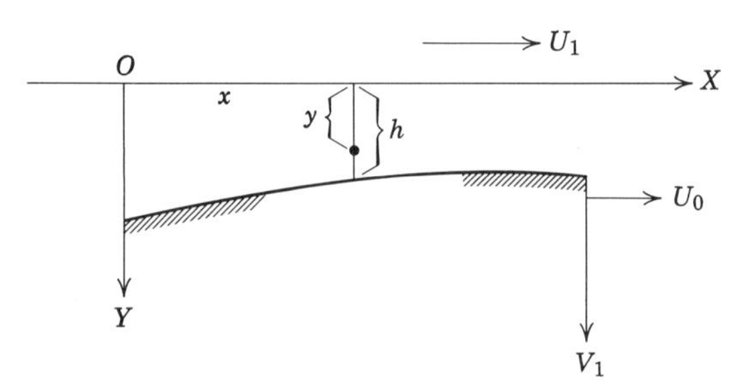

Fig. 4. Stationary and moving surfaces; velocities U_0, U_1, V_1.

right angles to the other two. In general the surfaces need not be flat or parallel, although the angle of inclination may be imperceptible. Let p denote the pressure at any point (x,y,z) where the surfaces are separated by a small distance h. Let V_1 denote the velocity of separation. The space between the solid surfaces is filled with an incompressible fluid of uniform viscosity μ and negligible inertia. Assume that the pressure p is independent of y. The basic problem of hydrodynamic theory is to determine p as a function of x and z.

Reynolds chose for his starting point the Navier-Stokes equations (1823, 1849). He showed that when stripped of needless complications, these equations could be reduced to the form

$$\frac{\partial p}{\partial x} = \frac{\partial^2 u}{\partial y^2}, \qquad \frac{\partial p}{\partial z} = \frac{\partial^2 w}{\partial z^2}. \tag{12}$$

The equations were to be applied in conjunction with the equation of continuity

$$\frac{\partial u}{\partial x} + \frac{\partial v}{\partial y} + \frac{\partial w}{\partial z} = 0, \tag{13}$$

where u, v, and w are the respective velocity components. The boundary conditions are $u = U_0$, $v = V_1 + U_0\ \partial h/\partial x$, and $w = 0$ at the lower surface, where $y = h$; and $u = U_1$, $v = 0$, and $w = 0$ at the upper surface, where $y = 0$. The two Eqs. (12) are now integrated for u and w as functions of y, since the pressure gradients $\partial p/\partial x$ and $\partial p/\partial z$ are independent of y. Taking into account the boundary conditions, Reynolds found that

$$u = \frac{1}{2\mu}\frac{\partial p}{\partial x}(y - h)y + U_1\left(\frac{h - y}{h}\right) + U_0\left(\frac{y}{h}\right)$$

and

$$w = \frac{1}{2\mu}\frac{\partial p}{\partial z}(y - h)y. \tag{14}$$

These two expressions are now differentiated by x and z, respectively, for substitution into the equation of continuity (13). This done, the resultant equation may be solved for $\partial v/\partial y$ and integrated from $y = 0$ to h, keeping in mind again that $\partial p/\partial x$ and $\partial p/\partial z$ are constants with regard to y. The result is the merging of the two equations (14) into a single equation connecting the pressure gradients and their two derivatives with the film thickness h, its slope dh/dx, and the four constants U_0, U_1, V_1, and μ. Collecting terms, we have

$$\frac{\partial}{\partial x}\left(h^3\frac{\partial p}{\partial x}\right) + \frac{\partial}{\partial z}\left(h^3\frac{\partial p}{\partial z}\right) = 6\mu\left[(U_0 + U_1)\frac{\partial h}{\partial x} + 2V_1\right] \cdot \qquad (15)$$

This is Reynolds' equation in its original form—a differential equation for the pressure distribution, taking into account the velocities of the solid surfaces and the profile or contour of the film, $h(x)$. In the usual case of a stationary lower surface, $U_0 = 0$ and the subscripts on U_1 and V_1 can be dropped, thus making a neater looking result.

The Navier-Stokes equations appeal to theoreticians and offer a foundation for diverse applications.They have been discussed at length by Lamb (1879), Tipei (1957), Pinkus & Sternlicht (1961), Gross (IX: 1962), Langlois (1964), and others. There is, however, an easier way to derive Reynolds' equation, without climbing up the Navier-Stokes mountain and down again. It is based on Kingsbury's thought that Reynolds' equation is another way of stating the requirement of continuity (1931).

Thus in Fig. 4 construct a mathematical box at x, with a height h, and a base of relatively small area $\Delta x\, \Delta z$. Let the box remain fixed in position, while the fluid adhering to the two surfaces is dragged through it. Call the rates of flow into the box Q_x and Q_z from the two directions and the rates out of it $Q_x + (\partial Q_x/\partial x)\,\Delta x$ and $Q_z + (\partial Q_z/\partial z)\,\Delta z$, respectively. The rate of loss of fluid volume from the box will be $(\partial Q_x/\partial x)\,\Delta x + (\partial Q_z/\partial z)\,\Delta_z$. For continuity this must equal the rate of decrease in the volume of the box, $-V_1\,\Delta x\,\Delta z$. If there is no pressure gradient in the direction of motion, fluid will be dragged in at the rate $h\,\Delta z$ multiplied by the mean velocity $(U_0 + U_1)/2$. If there is a gradient, we know from the slot formula that there will be an additional flow rate $(h^3\,\Delta z)/12\mu$ times $(-\partial p/\partial x)$. Thus

$$Q_x = \left(\frac{U_0 + U_1}{2}\right) h\,\Delta z - \frac{h^3}{12\mu}\left(\frac{dp}{dx}\right)\Delta z\,, \qquad (16)$$

from which

$$\left(\frac{\partial Q_x}{\partial x}\right)\Delta x = \left(\frac{U_0 + U_1}{2}\right)\frac{\partial h}{\partial x}\,\Delta x\,\Delta z - \frac{\Delta x\,\Delta z}{12}\frac{\partial}{\partial x}\left(\frac{h^3}{\mu}\frac{\partial p}{\partial x}\right) \cdot \qquad (17)$$

Similarly,

$$\left(\frac{\partial Q_z}{\partial z}\right)\Delta z = -\frac{\Delta z\,\Delta x}{12}\frac{\partial}{\partial z}\left(\frac{h^3}{\mu}\frac{\partial p}{\partial z}\right), \qquad (18)$$

since there is no surface velocity in the z-direction. Adding together the right-hand sides of Eqs. (17) and (18) and equating the sum to $-V_1 \Delta x \Delta z$ leads immediately to Reynolds' equation in the form

$$\frac{\partial}{\partial x}\left(\frac{h^3}{\mu}\frac{\partial p}{\partial x}\right) + \frac{\partial}{\partial z}\left(\frac{h^3}{\mu}\frac{\partial p}{\partial z}\right) = 6(U_0 + U_1)\frac{\partial h}{\partial x} + 12V_1 . \tag{19}$$

Use of the slot formula implies, as before, that the viscosity is the same up-and-down at any one x, z position; but the result is no longer limited to a uniform viscosity in the x and z directions.

Under steady-state conditions with lower surface stationary, the right-hand side reduces to $6U\, dh/dx$. The lower surface usually represents the tapered land or pivoted shoe of a thrust bearing or a portion of the stationary curved surface of a journal bearing. The upper surface, moving tangentially with a relative speed U, then represents the shaft collar or "runner" of a thrust bearing, or the cylindrical surface of the journal. When U_0 and U_1 are zero but V_1 is finite, Eq. (19) leads to Stefan's law.

Several investigators have generalized Reynolds' equation so as to apply it, for example, to dynamically loaded or gas-lubricated bearings, as described in later chapters (Trumpler, XII: 1966).

8. INTEGRATIONS

Two integrations give the film pressure as a function of x and z. A third is needed to find the load capacity, or relation between film thickness and total load in the bearing.

Although it is difficult to integrate Reynolds' equation taking all terms into account, relatively simple solutions have been obtained (1) by setting the $\partial/\partial z$ term equal to zero, and (2) by setting the $\partial/\partial x$ term equal to zero. The first method, used by Reynolds himself, neglects end leakage. It treats the bearing as one of infinite length at right angles to the motion. The second method, perfected by Ocvirk (1952), neglects the flow due to pressure gradients in the direction of motion. It is called the "short-bearing approximation" because, when applied to journal bearings, the approximation is closer the shorter the bearing.

As examples of the first method, we may cite the solution by Reynolds for plane surfaces (1886), and that of Sommerfeld for the full journal or "sleeve" bearing (1904). Both are commonly called two-dimensional solutions, since the results can be represented on a diagram with two coordinates.

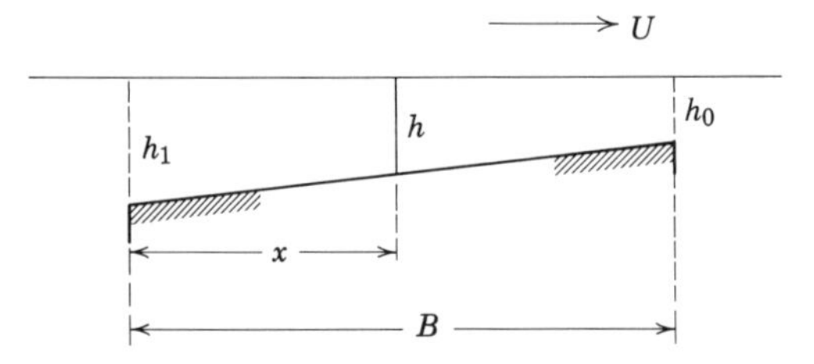

Fig. 5. Convergent film between two plane surfaces; $h = h_1$ at leading edge (from Fig. 4 with $U_1 = U$, $U_0 = 0$, $V_1 = 0$).

Plane Surfaces. Any two-dimensional solution requires an expression for h as a function of x. The film thickness between plane surfaces can be written

$$h = h_0 + (x/B)(h_1 - h_0) \, , \tag{20}$$

where h_0 is the minimum thickness, h_1 the film thickness at the entering edge, and B the breadth of the bearing (Fig. 5). If we put U_0 and V_1 equal to zero in Eq. (19), write U for U_1, and drop the $\partial/\partial z$ term, Reynolds' equation reduces to

$$\frac{d}{dx}\left(\frac{h^3}{\mu}\frac{dp}{dx}\right) = 6U\frac{dh}{dx} \cdot \tag{21}$$

Integration gives

$$p = (\mu UB/h_0^2)q \tag{22}$$

where

$$q \equiv \frac{6(a-1)(1-b)b}{(a+1)(a-ab+b)^2} \cdot \tag{23}$$

Here a denotes the film ratio h_1/h_0, and b stands for the fractional distance x/B. The pressure p should be interpreted as the gage pressure, or pressure above ambient at $x = 0$ and $x = B$. The dimensionless factor q was named the "pressure function" by Norton (1942). It may be shown that the peak pressure and the center of pressure are somewhat aft of the middle of the bearing shoe, looking from the leading toward the trailing edge.

Integrating (22) along x from $b = 0$ to 1.0 gives for the load capacity, or load per unit area of the bearing

$$P = (\mu UB/h_0^2)C_P \, , \tag{24}$$

where

$$C_P = \frac{6}{(a-1)^2}\left[\ln a - 2\left(\frac{a-1}{a+1}\right)\right] \cdot \tag{25}$$

Here "ln a" denotes the natural, or base e logarithm. Note that C_p, called the "load factor" by Professor Norton, is a dimensionless function of a.

Reynolds showed, and it may easily be verified, that C_P is a maximum when the film ratio a equals 2.18. That is why designers are constantly trying to arrange matters so that a will be very near 2 in round numbers; although a slightly different optimum may be found when side leakage is considered. Putting $a = 2$ in (23) and (25) and differentiating (22) with respect to b shows that the peak pressure is 1.62 times the mean pressure, or load per unit area, P. Similarly, the center of pressure, or theoretical pivot location for a tilting shoe, is found to be $0.58B$, measuring aft from the leading edge. These figures neglect end leakage (often called side leakage) and assume a uniform viscosity.

For parallel surfaces $h =$ const, so Eq. (21) gives $p = 0$. Thus no load-carrying pressure can be generated in parallel surface bearings under the conditions described. Equations (21), (22), and (24) lead to Reynolds' principle that the converging or wedge-shaped film is essential to hydrodynamic lubrication. Lord Rayleigh emphasized the same principle in a less restricted form when he said at Montreal (1884) that "the layer should be thicker on the ingoing than the outgoing side." Many confirmatory experiments have been reported; see, for example, Charron (1946), Huetz (1948), and Strong (1951). A simple demonstration with transparent block was devised by John Boyd (I: 1948).

Long Journal Bearing. Sommerfeld's integration for the full journal bearing of infinite length starts from Eq. (5). This equation gives the film thickness in terms of an angular coordinate θ measured clockwise from the line of centers, with $\theta = 0$ at the location of maximum thickness (Sommerfeld, 1904). Substitute from (5) into (21) with x equal to $r\theta$, where r is the radius. Integrating twice, after dropping the $\partial/\partial z$ term, leads to a pressure distribution

$$p = \frac{6\mu U r}{c^2}\left(\frac{\epsilon}{2 + \epsilon^2}\right)\frac{(2 + \epsilon \cos \theta) \sin \theta}{(1 + \epsilon \cos \theta)^2}. \tag{26}$$

Here ϵ is the eccentricity ratio and c the radial clearance, as in Eq. (5). Theoretically, p is the pressure over and above any arbitrary film pressure assumed when the journal is concentric with the bearing. Practically, when applying Sommerfeld's equation to bearings that are long but not infinite, there will be an ambient pressure feeding oil to the bearing at the two ends, or at an oil inlet near the location of maximum film thickness ($\theta = 0$). In Eq. (26), p may be taken as the pressure excess above such ambient value.

A striking feature of Sommerfeld's solution, which will be apparent on plotting p against θ, is the symmetry of the positive and negative pressure loops. This has been confirmed for submerged bearings, and for any case where the maximum pressure is less than one atmosphere, or whatever the ambient is. But in general it is not realistic, as pointed out by Sommerfeld himself, since liquids cannot ordinarily support a tensile stress. Various approximations have been proposed, to which we can return later.

Integrating pressure components parallel to the load line gives for the load capacity per unit of projected area

$$P = \frac{6\mu Ur}{c^2}\left(\frac{\epsilon}{2 + \epsilon^2}\right)\frac{\pi}{(1 - \epsilon^2)^{1/2}}. \tag{27}$$

This value is larger than can be realized in the ordinary case where end leakage occurs and negative pressures are ineffectual. It is a useful upper limit.

As an illustration of Eq. (27) consider a journal bearing of 1-in. radius, having a radial clearance $r/1000$, running at 3000 rpm with an oil of $\mu = 3 \times 10^{-6}$ lb-sec/sq in. What load per square inch can be supported on a minimum film thickness of $\frac{1}{2}$ mil? The eccentricity (27) comes to 0.813. Now $U = \pi DN = 100\ \pi$ in./sec; so finally $p = 4560$ psi. Imagining half this capacity lost by end leakage, and half again by the absence of negative film pressure, we could still count on 1140 psi. The same bearing, speed, and oil were taken for the example under Petroff's law, where the loading was only 100 psi—light enough for nearly concentric operation.

Equation (27) can be converted into a relation between the relative film thickness h_0/c and the "Sommerfeld number" S, and plotted as shown by Fig. 6. To do this we replace ϵ by $1 - (h_0/c)$ and U by $2\pi rN$, letting S denote the product of r^2/c^2 and $\mu N/P$. Thus h_0/c approaches unity, the concentric position, as S approaches infinity. In the above example, where $h_0/c = \frac{1}{2}$ the data give $S = 0.033$, in agreement with Fig. 6.

Under Sommerfeld conditions the eccentricity or journal displacement takes place at right angles to the load. The minimum film thickness h_0 or point of nearest approach is swept past the load line in the direction of motion. Ordinarily it is not displaced as much as 90

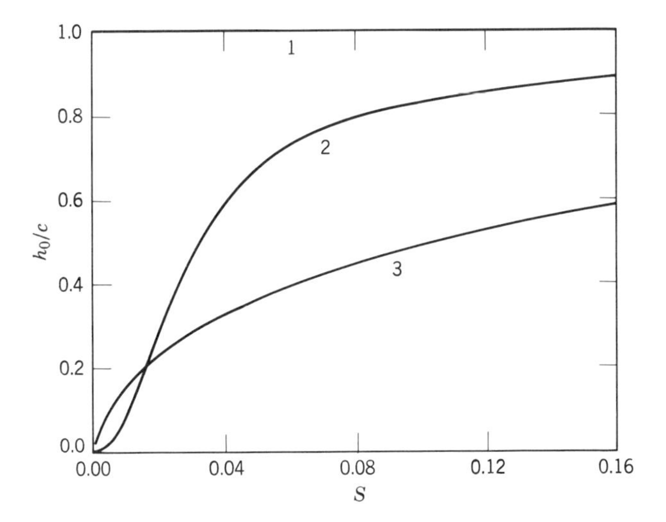

Fig. 6. Relative film thickness in full journal bearing; 1, Petroff; 2, Sommerfeld; 3, Ocvirk, $L/D = 1$ (h_0, minimum thickness; c, radial clearance; S, Sommerfeld number).

deg, but more nearly as shown in Fig. 7. Here Q is the oil feed and W the total load. The eccentricity in a hydrodynamic bearing is on the opposite side from the eccentricity in a dry bearing (Fig. 2 of Chapter I).

It will be seen from Fig. 7 that the convergent film is formed automatically as a result of eccentricity. Further details of the Sommerfeld theory can be found in Norton (1942), Radzimovsky (I: 1959), and Booker (1965).

The Narrow or Short Bearing. The second method of integrating, where Reynolds' equation is simplified by dropping the $\partial/\partial x$ term, was proposed by Michell (1905, 1929). He applied it to plane surfaces. The method was extended to full journal bearings by Cardullo (1930), who derived the pressure distribution

$$p = 3\frac{\mu U}{r}\left(\frac{L}{c}\right)^2\left(\frac{1}{4} - \frac{z^2}{L^2}\right)\frac{\epsilon \sin \theta}{(1 + \epsilon \cos \theta)^3}. \tag{28}$$

Here L is the total length of the bearing; and z is the axial coordinate from midpoint to either end, where $z = \pm L/2$. The remaining symbols are the same as in Eq. (26).

Cardullo made some practical applications of Eq. (28). It was integrated the following year by G. F. Rouse (1931), in an unpublished report to the ASME Research Committee on Lubrication. His study, mentioned in discussing the McKee experiments on pressure distribution (1932), leads to the equation for load capacity,

$$P = \pi^2 \mu N \left(\frac{L}{D}\right)^2 \left(\frac{D}{C}\right)^2 \frac{\epsilon}{(1 - \epsilon^2)^2} (1 + 0.62\epsilon^2)^{1/2}, \tag{29}$$

independently derived by Ocvirk (1952). The value 0.62 has been written for $16/\pi^2$ less 1. Thus at a given eccentricity ratio, the load capacities of short bearings that are otherwise geometrically similar are proportional to the squares of the length/diameter ratio. Eqs. (28) and (29) carry a built-in side leakage correction. They have been found useful for L/D ratios less than unity. The approximation is closer the smaller L/D.

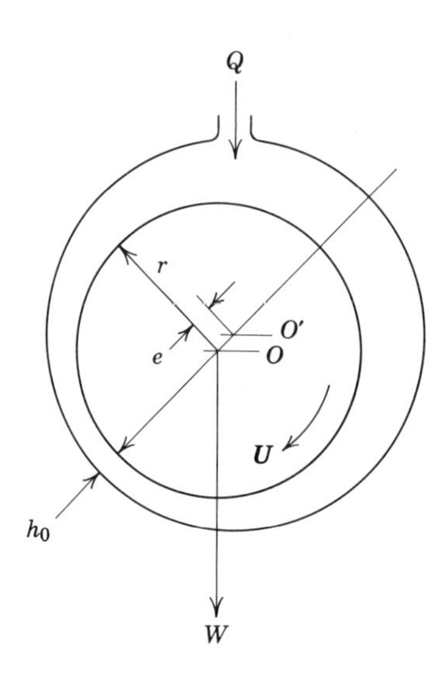

Fig. 7. Hydrodynamic journal bearing with eccentricity e (journal center at O; bearing center at O').

Three-Dimensional Solutions. Michell (1905) worked out the first general solution of Reynolds' equation for the case of flat surfaces. He considered two planes inclined at some small angle of c radians, and intersecting along the z-axis, where x and y are zero. Thus he could take $h = cx$ instead of using Eq. (20). By substituting cx for h in (19) and performing the differentiations, he reduced Reynolds' equation to a more convenient form. It could then be integrated by expressing p as a series of terms, each being the product of X and $\sin mz$ divided by mx, where m is an integer and X is a function of x alone. The width of the block, L, at right angles to the motion was taken equal to π for mathematical reasons. His solutions were restricted to the film ratio $a = 2$.

Michell worked out four cases: the square block, for which $B/L = 1$; a narrower block with $B/L = 3$; the infinitely narrow block, $B/L =$ infinity; and the case of the infinitely wide block, $B/L =$ zero, which Reynolds had solved. The results may be put into tabular forms analogous to Eqs. (23) and (25). When that is done the "pressure function" q will be expressed in terms of three ratios b, z/L, and B/L for any constant value of a. Similarly, the "load

Table 1. Pressure Distribution on a Square Block
(Values of the pressure function $q \times (10)^3$ for a film ratio $a = 2$, at fractional distances b from leading edge)

b	Relative distance z/L from side of block				
	0.1	0.2	0.3	0.4	0.5
0.1	12.4	20.7	25.4	28.0	29.0
0.3	33.4	56.6	72.2	80.8	83.9
0.5	52.4	89.2	114.0	127.9	132.8
0.7	65.3	106.5	137.0	152.0	157.3
0.9	49.8	76.2	92.8	100.6	103.7

Note: $q = 0$ at the leading and trailing edges, where $b = 0.0$ and 1.0; and at the two sides, where $z/L = 0.0$.

factor" C_P will be expressed as a function of B/L for a given a. It will be recalled that B is the breadth in direction of motion and L is the length of the block at right angles thereto.

Table 1, plotted in Fig. 8, shows the q values computed from Michell's theory for points equally spaced over the surface of a square block or "slider." Table 2 gives the maximum q, and its location b, together with the load factor C_P for four values of B/L.

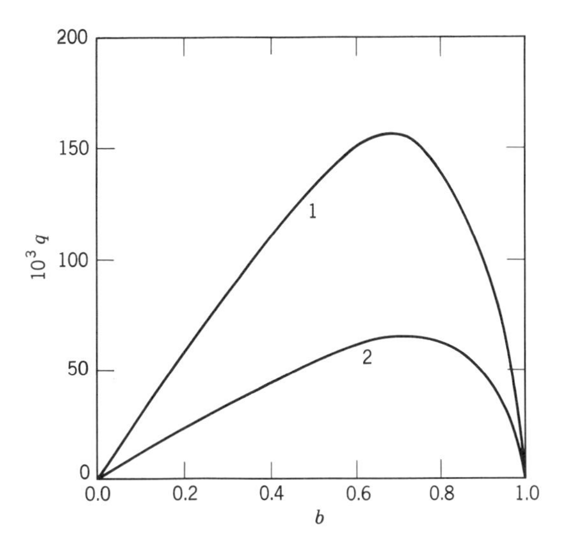

Fig. 8. Film pressures on a square slider (1, $z/L = 1/2$; 2, $z/L = 1/10$; b, fractional distance from leading edge).

Table 2. Peak Pressures and Load Factors for Rectangular Blocks
(Maximum values of 10^3q and corresponding location b with values of the load factor C_P)

B/L	10^3q	b	C_P
0	250	0.66	0.159
1	157	0.69	0.067
3	13.2	0.83	0.013
∞	30.8	1.00	0.000

Note: As before, $a = 2$, and b is measured back from the leading edge as a fraction of the breadth B.

It will be seen from Table 1 for the square block that the maximum or peak pressure in the film occurs on the axis of the block ($z/L = 0.5$), at a fractional distance of approximately $b = 0.7$ from the leading edge. More precisely, from Table 2, $b = 0.69$. It is interesting to note from this table how the load factor C_P drops from Reynolds' value of 0.159 for $B/L = 0$ to less than half as much for the square bearing, $B/L = 1$. A lucid mathematical review of Michell's integration, with additional data, can be found in Norton's book (1942).

As an example of a square slider bearing operating under the favorable condition $a = 2$, we note first, from Table 2, that $C_P = 0.067$. What load can be supported with a minimum film thickness $h_0 = 10^{-3}$ in. (1 mil), if the dimensions are $B = L = 3.0$ in.; $U = 1000$ in./sec (5000 fpm) and the lubricant is an oil of 3×10^{-6} lb-sec/sq in. viscosity? Equation (24) gives 603 psi. Film thickness will be $h_1 = ah_0 = 2 \times 10^{-3}$ in. (2 mils) at the leading edge. The inclination or slope $c = h_1 - h_0$ divided by B, or one part in 3000.

An elastic-deformation model was used by A. G. M. Michell (1905) to demonstrate load capacity in thrust bearings. It would seem that such methods could be further developed, in view of the fact that the shearing strains in an elastic solid are mathematically analogous to the rates of shear in a lubricant.

Albert Kingsbury (1931) obtained three-dimensional solutions by means of an electrical analogy. Kingsbury saw that a perfect analogy exists between the pressure distribution in a lubricating film and the potential distribution in a suitably designed electrical conductor: an electrolytic bath geometrically equivalent to the oil film. The electrical system lends itself readily to the precise measurements required. It is fully described in Norton's book (1942, pages 169–75). Kingsbury and Needs applied this method to certain thrust and journal bearing problems. Morgan and co-authors (1940) extended it to the full or 360-deg journal bearing.

Other methods of integration are described in later chapters; see, for example, Kotlyar (1959) and Fernlund (1962). One method, minimum dissipation, should be mentioned here because of its general interest and wide applicability.

9. MINIMUM DISSIPATION

It was shown by Helmholtz (1868) and Kortweg (1883) that the pressure and velocity distribution in an incompressible viscous fluid must be such as to render the dissipation of energy a minimum subject to the given conditions. The Helmholtz-Kortweg theorem was somewhat clarified by Rayleigh (1913) and Lamb (1932 edition, page 618). Poiseuille's law has been mentioned as a simple application. Stokes' law is another (DuPont, 1963).

The principle of minimum dissipation is analogous to that of minimum potential energy, used in predicting the equilibrium configuration of elastic bodies under load. It may be a more general principle than is commonly appreciated, since it applies also to the distribution of electrical currents in a network, and is consistent with the principle of minimum entropy production (Prigogine, 1947).

Weibull (1925) was the first to apply this principle to a lubrication problem. Other such applications are described by Schiebel (1933) and by Vogelpohl (1935, 1937, 1943). The finite-width thrust and journal bearing solutions by D. F. Hays (1958, 1959) will be reviewed in Chapters VI and VII. These integrations employ the principle of minimum dissipation, with the calculus of variations. A series, or other formal solution, is arbitrarily set up. The coefficients are then evaluated so as to satisfy all boundary conditions as well as the principle of minimum dissipation (Ritz, 1909; Habata, 1961). The Reynolds equation, not assumed at the start, turns up as the "Euler equation" of the variation method.

Discussions on minimum dissipation were published by Herivel (1954), Korovchinskii (1954), Philipzik (1956), and Serrin (1959). Glansdorff and co-authors (1962) offer a generalization applicable to liquids at a nonuniform temperature. Tao modified the Helmholtz principle to include compressible fluids (1964).

10. FLUID FRICTION

Reynolds showed that when two surfaces are in relative motion with velocity U, the shear stress opposing the motion, or friction per unit area on the moving surface, will be

$$\tau = \frac{\mu U}{h} + \frac{h}{2} \cdot \frac{\partial p}{\partial x}\,; \tag{30}$$

provided the x-axis is so chosen that the terms on the right are additive. The equation is easily derived from Newton's law of viscous flow and the conditions for static equilibrium. A knowledge of the pressure distribution is required in the general case, in order to evaluate the last term. The corresponding shear stress on the stationary surface is found by subtracting, instead of adding, the pressure gradient term. Thus the friction is always greater on the moving surface. Tangential forces on the stationary member do no work and are of less interest. Petroff's law will be recognized as a special case of Eq. (30) when there is no pressure gradient.

The principle of minimum dissipation makes use of the fact that the rate of dissipation of energy into heat per unit volume of the fluid in simple shear is equal to the product of viscosity by the square of the rate of shear; see, for example, Lamb (1932). The expression for power loss in a Petroff bearing,

$$H = 2\pi^3 \left(\frac{D}{C}\,\frac{L}{D}\right) \mu N^2 D^3\,, \tag{31}$$

follows directly since the rate of shear is U/c, or $2\pi DN/C$; and the volume of oil in the clearance space is $\pi DCL/2$.

Continuing the review of fundamentals, it may be of interest to cite the Sommerfeld and Ocvirk friction equations. Sommerfeld showed that the coefficient of friction for the 360-deg bearing with clearance filled with oil is given by

$$f = \frac{C}{D}\left(\frac{1 + 2\epsilon^2}{3\epsilon}\right). \tag{32}$$

When the journal contacts the bearing at zero speed so that $\epsilon = 1$, $f = C/D$. When the journal is concentric, $\epsilon = 0$ and f is infinite. Equations (27) and (32) may be combined graphically to eliminate the eccentricity. Doing so leads to the curve for f against S, shown

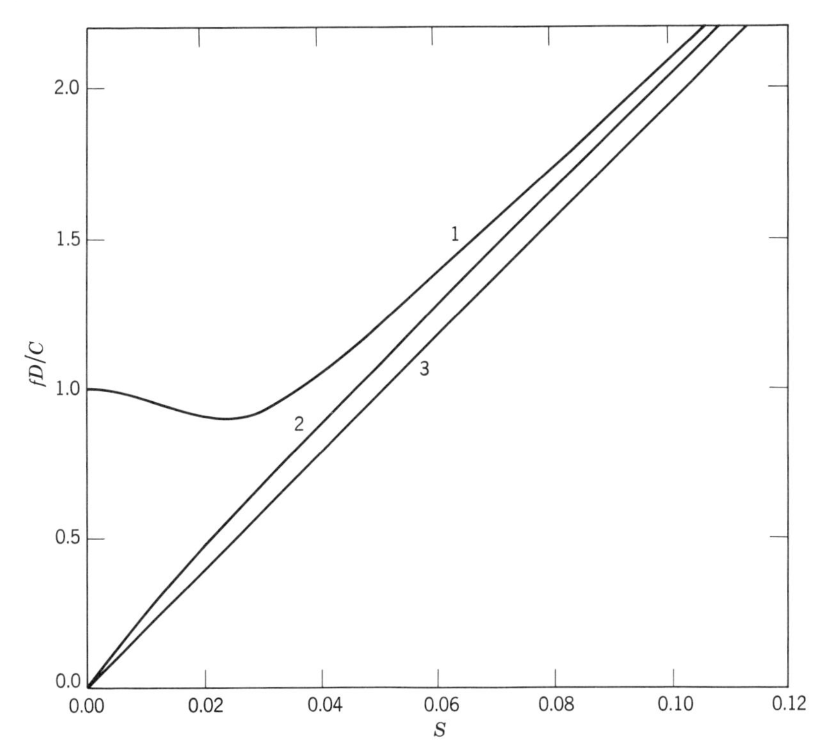

Fig. 9. Friction coefficients in full journal bearing: 1, Sommerfeld; 2, Ocvirk, $L/D = 1$; 3, Petroff (D, diameter; C, clearance).

in Fig. 9. The "Sommerfeld minimum" is clearly seen. Sommerfeld's curve approaches the Petroff line as an asymptote.

Ocvirk's load equation (29) may be expressed in terms of the Sommerfeld number

$$S\left(\frac{L}{D}\right)^2 = \frac{(1 - \epsilon^2)^2}{\pi^2\epsilon(1 + 0.62\epsilon^2)^{1/2}}; \tag{33}$$

and his friction equation, from (6) above, may be written

$$f = 2\pi^2\left(\frac{C}{D}\right)\frac{S}{(1 - \epsilon^2)^{1/2}}. \tag{34}$$

This reduces to Petroff's equation when $\epsilon = 0$. It assumes that the clearance space is completely filled with oil. Eliminating S between the last two equations leads to an alternative form of Ocvirk's equation

$$\frac{D}{C}\left(\frac{L}{D}\right)^2 \cdot f = \frac{2}{\epsilon} \frac{(1 - \epsilon^2)^{3/2}}{(1 + 0.62\epsilon^2)^{1/2}} \tag{35}$$

for comparison with Sommerfeld's (32) above.

Ocvirk denotes the left-hand side of Eq. (33) by C_n, which he calls the "capacity number"; and the left-hand side of (35) by f_v, the "friction variable." These new variables expressed in terms of L/D enable Ocvirk's theory to provide an approximate solution for the influence of the length/diameter ratio (see Chapter VII). Ocvirk's theory was first published by Burwell (VIII: 1951), who expressly credited it to Ocvirk.

Eliminating the eccentricity between Eqs. (33) and (35) permits plotting a curve for f_v against C_n. When $L/D = 1$, a common design, $C_n = S$. Comparison can readily be made in this case between the Ocvirk, Sommerfeld, and Petroff graphs (Fig. 9). Ocvirk's friction curve, unlike Sommerfeld's, shows no minimum point. It starts from the origin, and then lies everywhere above the Petroff line, but below the Sommerfeld curve. In like manner, it can be shown that Ocvirk's h_0/c curve, for the case $L/D = 1$, crosses Sommerfeld's at $S = 0.015$. At the right of this intersection, Ocvirk's curve of relative film thickness falls well below Sommerfeld's (Fig. 6).

The difference between the friction coefficient on the bearing member, f_b, and that on the journal, f_j, mentioned in Chapter I, must still be explained. This important difference was pointed out by Harrison (1913) and more fully analyzed by Boswall (1928). Thus in treating the general case of a partial bearing—one not completely surrounding the journal—Boswall isolates the oil film, as if it were a solid body, and considers the equilibrium of the forces acting upon it. See Fig. 37, page 203 of the Boswall reference. Now since the fluid pressures on the bearing surface act radially, their resultant passes through the bearing center. The reaction to this resultant is an equal and opposite force acting on the oil film. Similarly, the resultant of fluid pressures on the journal surface must pass through the center of the journal. Thus the film is acted upon by a couple equal to the product of the load and the component of eccentricity at right angles thereto. Again, the bearing is acted on by the moment M_b of frictional forces exerted by the oil film. The reaction is a moment M_b exerted on the film. The journal is acted on by the moment M_j of frictional forces exerted by the oil. This creates a reaction M_j on the film. The difference in these two moments, $M_j - M_b$, is balanced by the load couple $e_h W$, where W is the load and e_h the eccentricity component at right angles thereto. This couple may also be expressed as the vector product of eccentricity and load, $e \times W$.

Writing down the implied equation and dividing through by Wr, where r is the journal radius, gives

$$f_j = f_b + \frac{e_h}{r}. \tag{36}$$

This formula shows how to determine the coefficient on the journal from measurements of friction on the bearing member, together with a knowledge of the direction and magnitude of the journal eccentricity.

In his review of friction and lubrication, Vogelpohl (1954) emphasized the fact that hydrodynamic theory presupposes a knowledge of the viscosity of the film, which is highly sensitive to temperature. Accordingly, we turn next to a study of frictional temperature rise.

11. SUMMARY

Although hydrodynamic lubrication is the prevailing type in modern machinery, other varieties are mentioned here.

Petroff's equation, Poiseuille's law, Stefan's law, and Reynolds' equation are derived from Newton's law of viscous flow. Reynolds' integration for the load capacity of flat inclined surfaces of infinite width is given to indicate the principle of the "convergent film," along with solutions for finite width. Sommerfeld's theory of the infinitely long journal bearing, and Ocvirk's for the short bearing, are described and compared.

The principle of minimum dissipation is reviewed, as well as formulas for the coefficients of friction. The mathematical treatment is kept simple in order to bring out the underlying physical principles.

Tower discovered, and Reynolds explained, hydrodynamic load capacity. Sommerfeld, Michell, and Kingsbury showed how to apply hydrodynamic principles in bearing calculations.

REFERENCES

Allen, J. & Grunberg, M. D. (1937), "Resistance to the Flow of Water Along Smooth Rectangular Passages, and the Effect of a Slight Convergence of the Boundaries." *Phil. Mag.* (7) **23,** 490–503.

Archibald, F. R. (1956), "Load Capacity and Time Relations for Squeeze Films." *T* **78**, 29–35.

Bachman, C. H. & Hottenrott, H. C. (1959), "Flow Through Capillary Tubing with Eccentric Annular Sections." *Rev. Sci. Instrs.* **30**, 126–31.

Becker, Ernst (1907), *Strömungsvorgänge in ringförmigen Spalten and ihre Beziehungen zum Poiseuilleschen Gesetz.* Forschungsarbeiten No. 48; 42 pp. (a) *ZVDI* **51**, pt. 2, 1133–41.

Bikerman, J. J. (1957), "The Tackiness of Liquid Adhesives." *Trans. Soc. Rheol.* **1**, 3–14.

Booker, J. F. (1965), "A Table of the Journal Bearing Integral." *JBE, T* **87,** 533–4. (a) Trumpler (XII: 1966).

Boswall, R. O. (1928), *The Theory of Film Lubrication.* Longmans, London, 280 pp.

Boussinesq, J. (1868), "Mémoire sur l'influence des frottements dan les mouvements réguliers des fluides." *J. Math. Pure et Appliquées,* Liouville (2) **13,** 377–424. (a) *CR* **158,** 1743–49, 1846–50 (1914).

Buckingham, Edgar (1921), "On Plastic Flow through Capillary Tubes." *Proc. ASTM* **21,** pt. 1, 1154–6.

Buckingham, Edgar (1923), "Leakage through Thin Clearance Spaces." *Eng.* **115,** 225–6. Cf. (a) Egli, A., *JAM4, T* **59,** A63–A67 (1937).

Butler, L. H. (1960), "Hydrodynamic Effect between Approaching Surfaces with Interposed Fluid Films, and Its Influence on Surface Deformations." *JIP* **46,** 63–73 (Mar.).

Campbell, W. E., Pilarczyk, K., and Papp, C. A. (1957), "Effect of Wettability of a Lubricant on Journal Bearing Performance." Paper 30, *CLW,* 46–52; 12, 743–4, 832; Pl. 8.

Cardullo, F. E. (1930), "Some Practical Deductions from the Theory of the Lubrication of Cylindrical Bearings." *T* **52,** MSP-52-12, 143–53.

Charron, Fernand (1946), "Justification des formules du 'coin d'huile'." *Pub. sci. et tech. du Ministère de l'Air,* No. 198; 28 pp.

Christopherson, D. G. (1951), "Hydrodynamic Lubrication: General Survey." *Brit. JAP,* Sup. 1, 1–7.

Cornish, R. J. (1928), "Flow in a Pipe of Rectangular Section." *PRS* **120,** 691–700.

Couette, M. (1890), "Études sur le frottement des liquides." *Ann. chim. et phys.* (6) **21,** 433–510.

DuPont, K. C. & Rosen, Philip (1963), "Solution of the Stokes Flow of a Sphere by Minimization of the Dissipation of Energy." *Am. J. Phys.* **31,** 294–6.

Eirich, F. R. & Tabor, D. (1947), "Collisions through Liquid Films." *Proc. Camb. Phil. Soc.* **44,** pt. 4, 566–80 (+ addendum). (a) Tabor, D., *Eng.* **167,** 145–8 (1949). (b) Klorig, W. N. & Gatcombe, E. K., *Wear* **9,** 93–102 (1966).

Exline, P. G. (1946), "Formulas for Leakage in Capillary Seals." *Product Eng.* **17,** 290–6.

Fernlund, Ingemar (1962), *On Solving Reynolds' Equation for Bearings of Finite Width with Fourier Sine Transforms.* No. 261; 28 pp.

Fuller, D. D. (1947), "Hydrostatic Lubrication." *Mach. Design* **19**: "I. Oil-Pad Bearings," June, 110–6; "II. Oil Lifts," July, 117–22; "III. Step Bearings," Aug., 115–20; "IV. Oil Cushions," Sept., 127–31, 188, 190.

Fuller D. D. & Sternlicht, B. (1956), "Preliminary Investigation of Minimum Oil-Feed Rates for Fluid Film Conditions in Journal Bearings." *T* **78**, 1193–1200.

Glansdorff, P., Prigogine, I., & Hays, D. F. (1962), "Variational Properties of a Viscous Liquid at a Non-Uniform Temperature." *Phys. Fluids* **5**, 144–9.

Habata, Kichye (1961), "Theoretical Pressure Distribution in Journal Bearings." *JAM28, T* **83**, 497–506.

Hagen, G. (1839), "Über die Bewegung des Wassers in engen zylindrischen Röhren." *Ann. Physik und Chemie,* **46**, 423–442.

Hagenbach, Ed. (1860). "Über die Bestimmung der Zähigkeit einer Flüssigkeit durch den Ausfluss aus Röhren." *Ann. Physik* **109**, 385–426. Cf. (a) G. Wiedemann, **99**, 177–233 (1856).

Harrison, W. J. (1913), "The Hydrodynamical Theory of Lubrication with Special Reference to Air as a Lubricant." *T. Camb. Phil. Soc.* **22**, 39–54.

Hays, D. F. (1958), "Plane Sliders of Finite Width." *TASLE* **1**, 233–40.

Hays, D. F. (1959), "A Variational Approach to Lubrication Problems, and the Solution of the Finite Journal Bearing." *JBE, T* **81**, 13–23.

Hays, D. F. (1963), "Squeeze Films for Rectangular Plates." *JBE, T* **85**, 243–6.

Helmholtz, H. von (1868), "Zur Theorie der stationaren Ströme in reibenden Flüssigkeiten." *Verh. d. naturhist.-med. Vereins zu Heidelberg* **5**, 1–7 (Oct. 30). (a) *Wiss. Abhandlungen* **1**, 223–30; Barth, Leipzig.

Henny, A. S. (1921), "Flow of Viscous Fluids through Slightly Conical Capillary Tubes." *Proc. Phys. Soc. (London)* **34**, pt. 1, 22–6 (Dec. 15).

Herivel, J. W. (1954), "A General Variational Principle for Dissipative Systems." *Proc. Roy. Irish Acad.* **56A**, 37–44; 67–75.

Hersey, M. D. (1914), "The Laws of Lubrication of Horizontal Journal Bearings." *JWAS* **4**, 542–52 (Nov. 19).

Hersey, M. D. (1935), "A Short Account of the Theory of Lubrication." *J. Franklin Inst.* **219**, 677–702; **220**, 93–119, 187–214, 305–31.

Hersey, M. D. (1949), "Basic Principles of Lubrication. I. Fundamentals. II. Engineering Applications." Pages 19–42 of *Diesel Lubricating Oils and Basic Principles of Lubrication.* ASME, New York.

Hersey, M. D. (1954), "Fundamentals of Hydrodynamic Lubrication." Pages 54–81 in *Fundamentals of Friction and Lubrication in Engineering.* First National Symposium, ASLE, Chicago. (a) *J. Am. Soc. Naval Eng.* **67**, 51–68 (1955).

Hersey, M. D. & Snyder, G. H. S. (1932), "High Pressure Capillary Flow." *J. Rheol.* **3**, 298–317.

Huetz, Jacques (1948), "Contribution à l'étude du coin d'huile." *CR* **227**, 956–8.

Inglis, D. R. (1939), "Viscous Forces between Almost-Coaxial Cylinders." *Phys. Rev.* **56**, 1041.

Kingsbury, Albert (1931), "On Problems in the Theory of Fluid-Film Lubrication, with an Experimental Method of Solution." *T* **53**, APM-53-5, 59–75.

Korovchinskii, M. V. (1954), On the Variational Method in the Hydrodynamic Theory of Lubrication (Russian). *TIM* **9**, 114–42.

Kortweg, D. J. (1883), "On a General Theorem of the Stability of the Motion of a Viscous Fluid." *Phil. Mag.* (5) **16**, 112–8.

Kotlyar, Y. M. (1959), The Possibility of Obtaining Exact Integrals of Reynolds' Equation in Closed Form (Russian). *Dokl. AN SSSR* **127**, 59–62 (July 1).

Kreith, F. & Eisenstadt, R. (1957), "Pressure Drop and Flow Characteristics of Short Capillary Tubes at Low Reynolds Numbers." *T* **79**, 1070–8.

Kuzma, D. C. (1964), "Magnetohydrodynamic Squeeze Films." *JBE, T* **86**, 441–4.

Lamb, Horace (1879), *Hydrodynamics*. Cambridge University Press; 258 pp. (a) 6th ed., 1932; 738 pp. (b) Reprinted, Dover, New York, 1945.

Langlois, W. E. (1964), *Slow Viscous Flow*. Macmillan, New York; 229 pp.

Margules, Max (1881), "Über die Bestimmung des Reibungs-und Gleitungscoefficienten aus ebenen Bewegung einer Flüssigkeit." *Sitzber. Kaiserlich Akad. Wiss., Math.-Naturw. Kl.*, 2A, 83, pt. 2, 588–602.

Mathieu, Émile (1863), "Sur le mouvement des liquides dan les tubes de très petit diamètre." *CR* **57**, 320–4.

McKee, S. A. (1927), "Performance Characteristics of Journal Bearings When an Abrasive is in the Lubricant." *T SAE* **22**, pt. 1, 73–77.

McKee, S. A. & McKee, T. R. (1929), "Friction of Journal Bearings as Influenced by Clearance and Length." *T* **51**, APM-51-15, 161–171.

McKee, S. A. & McKee, T. R. (1932), "Pressure Distribution in the Oil Films of Journal Bearings." *T* **54**, RP-54-8, 149–165.

Michell, A. G. M. (1905), "The Lubrication of Plane Surfaces." *Z. für Math und. Phys.* **52**, 123–37. (a) *Ostwald's Klassiker der exakten Wissenschaften,* No. 218. Akademische Verlagsgesellschaft m.b.H., Hopf, Leipzig, 1927; pp. 202–219, 224.

Michell, A. G. M. (1923), "Viscosity and Lubrication." Chapter V, pp. 97–151 in *The Mechanical Properties of Fluids*. Blackie, London. (a) French transl. by A. Troller, Gauthier-Villars, Paris, 1927; 68 pp. (b) 2nd ed., London, 1937.

Michell, A. G. M. (1929), "Progress in Fluid-Film Lubrication." *T* **51**, MSP-51-21, 153–63.

Milne, A. A. (1957), "An Alternative Approach to the Theory of Hydrodynamic Lubrication." *Ninth ICAM,* Brussels, 1956; *Actes 4,* 284–92.

Moore, D. F. (1965), "A Review of Squeeze Films." *Wear* **8**, 245–63.

Morgan, F., Muskat, M., & Reed, D. W. (1940), "Studies in Lubrication, VI. Electrolytic Models of Full Journal Bearings." *JAP* **11**, 141–52.

Navier, C. L. M. H. (1823), "Mémoire sur les lois du mouvement des fluides." *Mémoires de l'Acad. Roy. des Sci* (2) **6**, 389–440.

Needs, S. J. (1940), "Boundary Film Investigations." *T* **62,** 331–45.

Newton, Isaac (1687), *Philosophiae naturalis principia mathematica.* London; see book 2, sect. 9, prop. 51. (a) English transl. by Andrew Mott, London, 1729, revised by Florian Cajori, University of California Press, Berkeley, 1934; see pp. 385–7.

Norton, A. E., ed. by Muenger, J. R. (1942), *Lubrication.* McGraw-Hill, New York, 244 pp.

Nuttall, Henry (1954), "Flow of a Viscous Incompressible Fluid. Expressions for a Uniform Triangular Duct." *Eng.* **178,** 298–300.

Ocvirk, F. W. (1952), "Short-Bearing Approximation for Full Journal Bearings." *NACA TN* 2808; 61 pp.

Osterle, F., Charnes, A. & Saibel, Edward (1956), "The Rheodynamic Squeeze Film." *LE* **12,** 33–6.

Paschoud, Maurice (1924), "Sur le problème du régime uniform dans un tube cylindrique fin à section en triangle rectangle isoscèle." *CR* **179,** 379–81. (a) ". . . à sections polygonales regulières," 451–4.

Peek, R. L., Jr. (1932), "Parallel Plate Plastometry." *J. Rheol.* **3,** 345–72.

Philipzik, W. (1956), "Zur hydrodynamische Theorie der Schmiermittelreibung." *ZAMM* **36,** 51–60.

Pinkus, Oscar & Sternlicht, Beno (1961), *Theory of Hydrodynamic Lubrication.* McGraw-Hill; 465 pp.

Poiseuille, J. M. L. (1840), "Recherches expérimentales sur le mouvement des liquides dans les tubes de très petits diamètres." *CR* **11,** 961–7, 1041–8; **12,** 112–5 (1841). (a) *Mémoires presentées par divers savants à l'Académie Royale des Sciences de l'Institut de France, Sciences Math. et Phys.* **9,** 433–545 (1846). (b) Transl. by W. H. Herschel, *Rheological Memoirs,* E. C. Bingham, Easton, Pa., **1,** No. 1, 1940; 101 pp.

Prigogine, I. (1947), Étude thermodynamique des phénomènes irreversible. Desoyer, Liège. (a) Second English ed., Wiley, New York, 1961; 119 pp.

Purday, H. F. P. (1949), *Streamline Flow. An Introduction to the Mechanics of Viscous Flow, Film Lubrication, the Flow of Heat by Conduction, and Heat Transfer by Convection.* Constable, London; 185 pp. (a) Reprint by Dover, New York.

Radzimovsky, E. I. (1959), *Lubrication of Bearings.* Ronald, New York, 338 pp.

Rayleigh (1884), "President's Address." Pages 13–14 in *Report, British Association for the Advancement of Science,* Montreal, 1884.

Rayleigh, Lord (1913), "On the Motion of a Viscous Fluid." *Phil. Mag.* (6) **26,** 776–86. (a) *Sci. Papers,* **VI,** 187–96; Cambridge University Press (1920).

Reiner, Markus (1960), "Cross-Stresses in the Laminar Flow of Liquids." *Phys. Fluids* **3,** 427–32.

Reynolds, Osborne (1883), "An Experimental Investigation of the Circumstances Which Determine Whether the Motion of Water Shall Be Direct or Sinuous." *Phil. Trans. Roy. Soc. London,* **174,** 935–82. (a) *Papers* *on Mechanical and Physical Subjects,* Macmillan, London, **2,** 51–105 (1901).

Reynolds, Osborne (1884), "On the Action of Lubricants," p. 622; "On the Friction of Journals," p. 895 in *Report, Brit. Assoc. Adv. of Sci.*, Montreal (titles only).

Reynolds, Osborne (1886), "On the Theory of Lubrication and its Application to Mr. Beauchamp Tower's Experiments, Including an Experimental Determination of the Viscosity of Olive Oil." *Phil. Trans. Roy. Soc. London,* **177,** pt. 1, 157–234. (a) *Papers on Mechanical and Physical Subjects,* Cambridge University Press, 1901, 2, 228–310. (b) Ostwald's *Klassiker der exacten Wissenschaften,* No. 218, Akademische Verlagsgesellschaft m.b.H., L. Hopf, Leipzig, 1927; pp. 39–107, 221–3.

Reynolds, O. (1902), "Lubrication." *Encyclopedia Britannica,* Suppl. 30, 10th ed., 372–4; reprinted, 11th to 13th eds.

Ritz, W. (1909), "Über eine neu Methode zur Lösung gewisser Variations Probleme der mathematischen Physik." *Crelle's J. Rein. u. Angew. Math.* **135,** 1–61.

Rouse, G. F. (1931), *Discussion of the Cardullo Theory of Film Pressure in Journal Bearings.* A Study for the ASME-RCL; 5 figs. (a) *T* **54,** RP-54-8, 162 (1932).

Schiebel, A. (ed. by Körner, K.) (1933), *Die Gleitlager.* Springer, Berlin; 70 pp. + plates.

Seegal, M. I. (1961), "Accoustic Radiation Pressure Bearing." *J. Acoust. Soc. Am.* **33,** 566–74.

Serrin, James (1959), "The Helmholtz-Rayleigh Dissipation Theorem." Pages 258–60 in *Handbuch der Physik,* ed. by S. Flügge & C. Truesdel; Springer, Berlin, vol. 8/1.

Slezkin, N. A. (1955), Hydrodynamic Theory of Lubrication (Russian). Pages 190–224 in *Dynamics of Viscous Incompressible Fluids,* Gostekhteoretizdat, Moscow.

Snyder, W. T. (1962), "The Magnetohydrodynamic Slider Bearing." *JBE, T* **84,** 197–204.

Sommerfeld, Arnold (1904), "Zur hydrodynamische Theorie der Schmiermittelreibung." *Z. Math. u. Physik* **50,** 97–155. (a) *Archiv Elektrotech.* **3,** 1–5 (1914). (b) *Z. techn. Physik* **2,** 58–62, 89–93 (1921). (c) *Ostwald's Klassiker der exakten Wissenschaften,* No. 218 (cited under Reynolds, 1886) pp. 202–19, 224. (d) *GDLL* **1,** 297–301. (e) Pages 253–62 in *Lectures on Theoretical Physics* **2,** transl. by G. Kuerti from 2nd German ed., Academic Press, New York, 1950.

Stefan, J. (1874), "Versuche über die scheinbare Adhäsion." *Sitzber. Kaiserlichen Akad. Wiss., Math.-Naturw.*, Wien, **69,** pt. 2, 713–35.

Stokes, G. G. (1849), "On the Theories of the Internal Friction of Fluids in Motion, and of the Equilibrium and Motion of Elastic Solids." *Trans. Camb. Phil. Soc.* **8,** 287–319 (a) *Math. and Phys. Papers,* Cambridge University Press **1,** 75–129 (1880).

Strong, John (1951), "New Johns Hopkins Ruling Engine." *J. Opt. Soc. Am.* **41,** 3–15. (a) Ingalls, A. G., *Sci. Am.* **186,** 45–54 (June 1952).

Tao, L. N. (1964), "On the Variational Principle and Lagrange Equations in Studies of Gasdynamic Lubrication." *JAM* **31,** *T* **86,** 43–6.

Tao, L. N. & Donovan, W. F. (1955), "Through-Flow in Concentric and Eccentric Annuli of Fine Clearance with and without Relative Motion of the Boundaries." *T* **77,** 1291–1301.

Theyse, F. H. (1964), "Fundamentals of Hydrodynamic Lubrication and Their Consequences in Design Engineering." *Wear* **7,** 419–34, 477–97.

Tiabin, N. V. (1956), "Anomalous Viscous Flow between Two Plates and between Two Coaxial Cylinders." *Soviet Phys.* (*Tech. Phys.*) **1,** 1929–37.

Tipei, Nicolae (1957), *Hydro-Aerodynamic Lubrication* (Rumanian). Editura Academiei Repub. Rom., Bucarest; 695 pp. (a) *Theory of Lubrication,* transl. ed. by W. A. Gross, Stanford University Press, Stanford, 1962; 566 pp.

Tower, Beauchamp (1883), "First Report on Friction Experiments." *Proc. IME* 632–659; 1884, pp. 29–35. (a) "Second Report . . . ," 1885, pp. 58–70; (b) "Third Report . . . ," 1888, pp. 173–205.

Truesdell, C. A. (1952), "A Program of Physical Research in Classical Mechanics." *ZAMP* **3,** 79–85.

Underwood, A. F. (1944), "Rotating Load Bearings—A New Concept of Lubrication and a Frictionless Support." *ASME* 44-A-29; 4 pp. (a) *Automotive and Aviation Industries* **92,** 26–28, 60, 64 (1945).

Vogelpohl, G. (1935), "Zur hydrodynamischen Theorie der Lagerreibung." *ZAMM* **15,** 378.

Vogelpohl, G. (1937), "Beiträge zur Kenntnis der Gleitlagerreibung." *VDI Forschungsheft* No. 386, 1937; 28 pp.

Vogelpohl, G. (1943), "Zur Integration der Reynoldschen Gleichung für das Zapfenlager endliche Breite." *Ing. Archiv* **14,** 192–212.

Vogelpohl, G. (1954), "Die Grundlagen der Reibung und Schmierung in die laufenden Maschine nach dem gegenwärtigen Stand unseres Wissens." *Schmiertechnik* No. 4; 113–20. (a) *RUM* (9) 10 (June), 386–96.

Weibull, Waloddi (1925), "Bidrag Till Glid-Lagrets Teori." (Contribution to Plain-Bearing Theory) *TTM* **55,** 56–7.

Wilcock, D. F. (1957), "Predicting Performance of Starved Bearings." *LE* **13,** 348–52.

Yamada, Yutaka (1962), "Resistance of a Flow Through an Annulus with an Inner Rotating Cylinder." *Bull. JSME* **5,** No. 18, 302–10.

chapter III Frictional Temperature Rise

A new dimension, temperature, is studied in this chapter. Only the simplest applications are considered, in order to bring out the physical principles. More complicated conditions met in practice are reviewed in later chapters, together with full-scale test results. Temperature will be denoted here by θ, temperature rise by T, and time by t.

1. EXAMPLES OF POWER LOSS

The frictional power loss H, or work per unit time dissipated into heat, is given by FU, the product of frictional resistance by speed.

Plane Surfaces. In the case of parallel surfaces of area A separated by an oil film of thickness h and a uniform viscosity μ, without pressure gradients, the power loss per unit area, H/A, will be

$$H_1 = \mu U^2/h\,. \tag{1}$$

Referring to Fig. 1 of Chapter II, take $A = 25$ sq in., $\mu = 3.3 \times 10^{-6}$ lb-sec/sq in., $U = 300$ in./sec, and $h = 10^{-3}$ in. as before. From these data $H = 7430$ in.-lb/sec; or, dividing by 6600, $H = 1.125$ hp $= 840$ watts.

Now take an example represented by Fig. 5 of Chapter II, where the surfaces are inclined at a slight angle so as to provide the convergent film. Suppose the surfaces are each of 25 sq. in. area, and separated by a mean film thickness of 1 mil, so that $(h_1 + h_0)/2 = 10^{-3}$ in. Assume the optimum film ratio $a = h_1/h_0 = 2$. Let the length L at right angles to the motion and the breadth B in direction of motion each be 5 in., but neglect side leakage, μ and U remaining as before. The load capacity is given by Eqs. (24) and (25) of Chapter II, from which $C_P = 0.159$ and $P = 354$ psi. By an intermediate step, it can be seen that $h_0 = \frac{2}{3}$ of a mil. The parallel surface bearing of the original example had no load capacity.

We still need formulas for the frictional resistance of the wedge-film bearing. Reynolds' Eq. (30) of Chapter II will meet this need as soon as we evaluate the pressure gradient and integrate over the moving surface. Norton put the result in a convenient form, from which we can write for the resisting force per unit area

$$F_1 = C_F \mu U / h_0 \, . \tag{2}$$

The dimensionless "friction factor"

$$C_F = \frac{1}{a-1} \left[4 \ln a - 6 \left(\frac{a-1}{a+1} \right) \right] \cdot \tag{3}$$

When $a = 2$, $C_F = 0.772$. Substituting in Eq. (2), with $h_0 = 6.67 \times 10^{-4}$ in. and μU as before, gives $F_1 = 1.15$ psi; only 15 per cent greater than for parallel surfaces. The power loss H will be 1.15×7430, or 8540 in.-lb/sec.

In a square bearing, the correction for side leakage, Table 2 of Chapter II, makes $C_P = 0.067$, from which the load capacity will be reduced in the ratio of 0.067 to 0.159, or 42 per cent. How much is the friction reduced? Norton shows that the friction factor C_F, based on Michell's solution, now equals 0.725 as against 0.772, a reduction of 6.1 per cent. Both friction and load capacity are reduced by side leakage, but the reduction in friction is less. The new power loss in the wedge-film example is 8030 in.-lb/sec.

Journal Bearings. A Petroff bearing was considered in Chapter II. In the example given, $D = 2$ in., $L = 4$ in., $D/C = 1000$, $N = 50$ rps, and the viscosity was 3.0×10^{-6} lb-sec/sq in. The power loss is given by

$$H = 2\pi^3 \left(\frac{D}{C} \right) \left(\frac{L}{D} \right) \mu N^2 D^3 \, . \tag{4}$$

It is seen from the data that $H = 7440$ in.-lb/sec. Suppose, however, that the bearing has an eccentricity ratio of $\frac{1}{2}$; then by Couette's formula, Eq. (6) of Chapter II, the torque and power loss will be increased 15 per cent. It was shown that by Sommerfeld's theory, a heavy load would be required for $\epsilon = \frac{1}{2}$. Next try the McKee formula with $\Delta f = 0.0018$ or a bit less. Choose a loading of 1000 psi for this example. Then by Eq. (4) of Chapter II the Petroff coefficient is $f_0 = 0.0030$. Adding Δf brings f up to 0.0048, an increase of 60 per cent. The power loss will be increased in the same ratio, making $H = 1.6 \times 7440$, or 11,900 in.-lb/sec. At a lighter load the effect of Δf would be less marked. Even so, the McKee formula is more realistic than either Petroff's or Sommerfeld's and is widely used in practice for small bearings.

Here then are six numerical examples of power loss from which to calculate temperature rise.

2. ADIABATIC HEATING

If all the heat generated remained in the oil and the oil stayed in the clearance space, the temperature would rise at the rate H/Vq, where V is the volume of the film and q the heat capacity of the oil per unit volume. Here H/V and q are expressed in the same system of units, preferably mechanical units rather than Btu (British thermal units) or calories. The value of q for an ordinary petroleum oil at 130 F may be taken equal to 140 in.-lb/cu in, deg F, or more simply 140 psi/dF, in round numbers (Chapter IV). The abbreviation dF is used for an interval or difference in temperatures as distinguished from an actual temperature. Heat capacity per unit volume is the product of density and specific heat. It increases with rising temperature, but not so rapidly as specific heat.

In our plane surface examples, $V = 25 \times 10^{-3}$ cu in. In the first example the rate of temperature rise, $d\theta/dt$, will be 297,000 divided by 140, or 2120 dF/sec. In the journal bearing examples $V = 25.1 \times 10^{-3}$ cu in. Thus in the Petroff application the rate would be 296,000 divided by 140, or 2110 dF/sec, and greater in the other two journal bearings. These amazing rates are fictitious, mainly because adiabatic surfaces cannot be realized; but they provide an upper limit, and show the necessity for cooling by natural or artificial means.

3. EXTERNAL HEAT TRANSFER

Heat escapes from the bearing housing mostly by atmospheric convection. Radiation is appreciably only at the higher temperatures. Newton's law of cooling states that the rate of heat transfer to the atmosphere, H'_a, is proportional to the exposed area A and to its temperature elevation T_a above the ambient air at some distance. Thus

$$H'_a = C_a A T_a \,, \tag{5}$$

where Newton's constant C_a, the heat transfer coefficient, has the dimensions of power per unit area and unit temperature difference. The symbol H has been used to denote frictional power loss, or heat generated by friction in unit time; hence the addition of a prime to distinguish the rate of heat transfer from the rate of heat generation. Here $C_a A = C_0$ of earlier publications.

A convenient constant for Newton's law in still air is found to be $C_a = 2$ Btu/hr, sq ft, dF, or 0.036 in.-lb/sec sq in., dF (Heilman, 1929). This value is in agreement with experiments by Karelitz (1930, 1942) on self-contained bearings. In some of his work, Professor Karelitz used a dummy journal with electric heater to simulate friction, as proposed by the writer (1915). It appears from his study that the temperature difference T_a, measured from the outer surface of the housing, is from one-half to four-fifths of the temperature elevation T of the oil film itself above ambient. Equation (5) may therefore by approximated by the relation

$$H'_a = C_1 A T \,. \tag{6}$$

The Karelitz ratio T_a/T or C_1/C_a is roughly $\frac{2}{3}$, making $C_1 = 0.024$ in.-lb/sec, sq in., dF. A factor of 2 is sometimes applied for moving air. See Elenbaas on free convection (1948).

For temperature elevations greater than about 50 deg F, radiation plays a noticeable part, as found by Lasche (1902). His frequently cited experiments on bearings for high-speed electric railways may be described by the formula

$$H'_a = C_n A_0 T^n \,. \tag{7}$$

Here A_0 is the bearing surface πDL, not the housing area. For temperature elevations from 50 to 200 deg F, the exponent n may be taken equal to 1.3. Above 200 F it increases noticeably, and below 50 we can go back to Newton's law with $n = 1$. Note that C_n has the dimensions of power per unit area and temperature difference to the nth.

The heat-transfer coefficient C_n in Eq. (7) was found by Lasche to be about 17 for bearings with large housings in still air when H'_a is expressed in in.-lb/sec, sq in., (deg F)$^{1.3}$. Lasche states that for very thin shells, the values may be as little as one-sixth of the foregoing; while for ordinary bearings cooled by the motion of adjacent machinery, they may be three or four times as great. A systematic investigation by Professor Hanocq at Liège led to a two-term formula for the heat transfer from self-contained bearings (Hanocq, 1947; Leloup, 1954). The first term is proportional to T and the second to T^4. The constants in the formula are carefully chosen to reflect the influence of shaft diameter, L/D ratio, geometry of the housing, and velocity of air currents. Both Hanocq, 1947) and Burr (1960) took into account the heat transfer from exposed parts of the shaft.

4. OIL FLOW AND WATER CIRCULATION

A major part of the heat generated by friction in the oil film, as it is dragged along from leading edge to trailing edge of a thrust shoe, is retained in the oil and thereby removed from the bearing proper. A similar process occurs in journal bearings, where a certain amount of oil would be discharged from the ends of a bearing owing to its natural pumping action. The flow rate can be accelerated by forced lubrication.

In either type of bearing the heat removed by the oil per unit time is given by the product of Qq into the temperature rise $T = \theta_o - \theta_i$. Here Q is the rate of oil flow in volume units per unit time; q the heat capacity of the oil per unit volume; θ_o the outlet, and θ_i the inlet temperature of the oil. Care should be taken in numerical applications to employ the same system of units on both sides of the equation—for example, mechanical units, by which the heat removed per unit time is expressed in inch-pounds per second; Q in cubic inches per second; and q as above. See also Hirano & Shōdai (1958). The difficult factor to evaluate is Q.

Oil Film between Flat Surfaces. The rate of flow Q in volume units per unit time, in the case of parallel surfaces with no pressure gradient or side leakage, is equal to the mean velocity $U/2$ multiplied by the cross section of the film, Lh. In the example where $U = 300$ in./sec, $L = 5$ in., and $h = 10^{-3}$ in., it follows that $Q = 0.75$ cu in./sec. If all heat generated is carried off in the oil, $H = QqT$ and the temperature rise will be H/Qq. It was shown that $H = 7430$ in.-lb/sec, while q may be taken as 140 psi/dF. The calculated temperature rise would therefore be about 71 deg F.

In the more practical case of the wedge-film the effective film thickness h_e occurs at the location of maximum pressure, since at this point there is no pressure gradient and the velocity distribution is again a straight line. From Eqs. (22) and (23) of Chapter II this location is given by $b = a/(a + 1)$. For the near-optimum value of $a = 2$, $b = \frac{2}{3}$. Again taking $h_1 = \frac{4}{3}$ and $h_0 = \frac{2}{3}$ of a mil, we see that $h_e = \frac{8}{9}$ mil, so Q is now $\frac{8}{9}$ of the previous value (still disregarding side leakage). The temperature rise, then, would be $\frac{9}{8}$ of 71; or say 80 deg F. The effects of side leakage in reducing this estimate are considered in Chapter VI.

Oil Flow in Journal Bearings. The temperature rise due to oil flow in full journal bearings depends upon the amount of end leakage. Both the natural pumping action of such bearings and the effect of forced lubrication were demonstrated in Barnard's experiments (1925). A 1-in. diameter bearing 2 in. long was used, having C/D ratios of 0.6 and 1.1 per thousand. Oil was supplied through a hole about $\frac{3}{16}$ in. in diameter on the unloaded side, and fed to the journal through an axial groove $1\frac{5}{8}$ in. long. Viscosity ranged from 13 to 44 cp, load from 40 to 270 psi, and speed from 200 to 2000 rpm. Pumping efficiency E was defined as the ratio of oil flow per revolution to the clearance volume, or Q/N divided by $\pi DCL/2$. A linear relation was found between E and the oil-inlet gage pressure p for constant values of ZN/P, with p ranging from zero to nearly 70 psi. For convenience, ZN/P was expressed in the conventional mixed units, ranging from 50 to 500 cp rpm/psi, here abbreviated "cmu."

By means of a slight extrapolation, data were obtained for the pumping efficiency E_0 at zero inlet pressure and plotted against ZN/P. The effect of practically doubling the clearance was imperceptible. Presumably L/D variations would have a marked effect. Since the variables chosen are dimensionless, the same curve may be applied to geometrically similar bearings of a larger size. It shows a drop in E_0 from 0.43 at 50 cmu to 0.32 at 100, 0.21 at 200, and 0.16 at 500 cmu.

Greater oil flows are obtained by forced lubrication, especially in bearings designed with axial grooving or relief at the sides, or with extra clearance under the cap. Generalization of Barnard's data, plotted against inlet pressure, is difficult because the loads were not given. If we assume a constant load P of 125 psi in round numbers, E/E_0 will exceed unity by $2p/P$. Pumping efficiencies at any ZN/P may then be estimated with the aid of his curve. The rate of oil flow, Q, is half the product of E by $\pi DNCL$, where C is the diametral clearance.

Barnard's data were analyzed by Orloff (1935), who treated the oil flow as consisting of two components, one due to hydrodynamic action, the other to the inlet pressure. Oil flow calculations have been published by Muskat and Morgan (1939) for full journal bearings with L/D greater than $\pi/2$, operating at eccentricity ratios less than $\frac{1}{2}$. Their findings were confirmed experimentally by Boyd & Robertson (1948) in an investigation of oil-flow and temperature relations, using oil inlet pressures from 10 to 70 psi. The test bearing was 2 in. in diameter by $1\frac{7}{16}$ in. long, with a clearance ratio $C/D = 2.6$ per thousand, and oil hole $\frac{1}{4}$ in. in diameter.

The rate of oil flow through a full journal bearing supplied from a circumferential groove in the middle is given by

$$Q = \frac{pC^3}{\mu}\left(\frac{\pi}{24}\right)\frac{1 + 1.5\epsilon^2}{\lambda}, \tag{8}$$

where λ denotes L/D. Equation (8) comes from Becker's formula for eccentric annular flow (II: 1907), and was confirmed in Boyd's investigation.

McKee (1952) found experimentally that the oil flow Q in plain bearings could be separated into Q', the part depending on forced feed, and Q'', the part depending on hydrodynamic action. To this extent, McKee's tests confirmed Orloff's theory. Like Barnard, McKee found that the total flow is linear with inlet pressure p for any constant value of ZN/P. The intercept of such a graph with QZ plotted against p is equal to $Q''Z$. After separating out Q' and Q'', two dimensionless plots are constructed, one for $Q'Z/pC^3$ against P/ZN, the other for $Q''Z/PC^3$ against ZN/P. Note that Q'/p appears in the first diagram and Q''/P in the second. It may be assumed that Q' is governed primarily by the inlet pressure p, and Q'' by the load P. These diagrams serve as a framework for comparing three types of bearings, one with a single oil hole on the unloaded side; a second with an axial groove half the length of the bearing on the unloaded side; and a third type with a complete circumferential groove in the middle. The graphs show unmistakably that the forced flow Q' is greatest with the circumferential groove, and least with the oil-hole bearing; the hydrodynamic flow is greatest for the axial groove bearing, and vanishingly small for the circumferential groove. The bearings tested were 2 in. in diameter with L/D about 1.3, C/D from 1.6 to 2.5 per thousand.

It would be interesting to compare McKee's results with those found by David Clayton in 1946 (Table 10, Chapter VII). Other data are usefully reviewed by Barwell (I: 1956) in his chapter on "Oil Flow in Journal Bearings." Wilcock and Rosenblatt (1952) tested a 4-in. diameter bearing with L/D equal to 1.0, and six bearings 8 in. in diameter having L/D ratios from 0.5 to 1.5, C/D's from 1.1 to 2.5 per thousand. These bearings were of a split design having an axial oil-supply groove along each side, closed at the ends. In effect, each half is a partial bearing of 150 degrees arc. The calculations offered were substantially confirmed experimentally. Conclusions regarding oil flow are consistent with McKee's data.

Oil flow experiments by Cole and Hughes (1956) serve to clarify questions raised in earlier investigations. The photographs shown are especially interesting in revealing the effects of oil grooving, of oil starvation, and the separation of the film into filaments on the unloaded side. A study of oil transfer and cooling in ring-oiled bearings was reported by Özdas and Ford in 1955. Oil-ring performance is reviewed in Chapter V.

As an oil flow example, consider the journal bearing with eccentricity ratio $\frac{1}{2}$, for which $D = 2$ in., $L = 4$ in., $N = 50$ rps and $\mu = 3 \times 10^{-6}$ lb-sec/sq in. It was found that $H = 8560$ in.-lb/sec. From Eq. (8) above, assuming 2-atm gage pressure at the inlet, $Q = 1.76$ cu in./sec. With q again 140 psi/dF, the temperature rise T comes to about 35 deg F. In the case of zero inlet gage pressure, but other factors unchanged, a solution by Hays (Fig. 1 of 1961) leads to a flow rate of 3 cu in./sec, from which T is about 20 deg F.

Water Circulation. Heat may also be removed by water-jacketing the housing or submerging copper cooling coils in the oil bath (Kingsbury Machine Works, 1922 and later catalogs; R. A. Baudry, 1945; Needs, VI: 1954, and Fuller, I: 1956, Fig. 68). The safety of the installation is then dependent on reliable water pressure. The rate of heat transfer by water cooling is fixed by the size and shape of the passages or tubing, the mean velocity of the water, and the temperature difference between cooling water and adjacent metal. Calculations may be found in the literature of heat transfer by convection; see, for example, McAdams (1933 and later). In general, water cooling is needed only for the larger bearings operating at high surface speeds.

5. THERMAL EQUILIBRIUM

Let it be required to calculate the steady running temperature of a bearing, a temperature such that the rate of heat transfer, H', is just equal to the rate of generating heat by friction, H. Since H' is a function of the temperature elevation T, we can write symbolically

$$H' = H'(T) , \tag{9}$$

other factors being treated as constant. Similarly,

$$H = H(Z) , \tag{10}$$

the power loss being expressed as a function of the viscosity alone. When necessary, the right-side of (10) may be multiplied by a known function of T as a correction factor for thermal expansion which tends to reduce the clearance between journal and bearing. Then under equilibrium conditions,

$$H'(T) = H(Z) . \tag{11}$$

Here $H'(T)$ and $H(Z)$ are known functions, but Z and T are unknown quantities. Another equation between Z and T is needed, for which we can use the viscosity-temperature relation for the lubricant,

$$Z = Z(T) . \tag{12}$$

The simultaneous solution of (11) and (12), either by graphical or analytical means, provides the required values of Z and T. This general approach was outlined by the writer in 1915 and further discussed in 1939 and 1942. Previously the temperature problem was attacked only by special assumptions or cut-and-try methods. Petroff understood the problem, but apparently did not complete his solution.

The temperature elevation T in the foregoing equations denotes $\theta - \theta_1$, where θ and θ_1 are defined according to the application. If there is no appreciable oil cooling, so that Eq. (9) is an expression for heat transfer from the housing, θ should be defined as the mean temperature of the oil film and θ_1 as the temperature of the ambient air. In the case of forced lubrication with negligible heat loss from the housing, θ may be defined as the oil outlet temperature and θ_1 as the inlet temperature. The assumption is then made that the outlet temperature is equal to the mean film temperature. This assumption had been used by the writer (II: 1949, p. 21) and confirmed by Wilcock & Rosenblatt (1952). It is believed more reliable than averaging the inlet and outlet temperature. Wilcock showed that a large bearing operating with forced lubrication acts like a mixing apparatus, yielding an outflow mixture that is intermediate in temperature between that of the inlet oil and the hottest areas in the bearing.

Mathematical Theory. Analytical solutions offer a convenient means for examining the separate influence of the operating and design factors. The same formal solution can be applied to both cases mentioned above—in one of which the heat is removed wholly by transfer from the housing, in the other wholly by forced feed. Consider a lightly loaded, high-speed bearing operating under Petroff's law with heat transfer directly proportional to the temperature elevation. By disregarding thermal expansion Eq. (11) becomes

$$KT = 2\pi^3(LD^3/C)ZN^2 , \tag{13}$$

where K is a constant. In the air-cooled bearing, K is C_1A; in the forced-feed bearing, K is Qq. For simplicity, represent the viscosity-temperature curve by an equilateral hyperbola

$$Z = \frac{\text{const}}{\theta - \theta_0} = \frac{T_0 Z_1}{T + T_0} \cdot \tag{14}$$

Here θ is the effective film temperature and θ_0 the "apparent solidifying temperature"; T_0 denotes the difference $\theta_1 - \theta_0$; Z_1 is the ambient or inlet viscosity; and T is $\theta - \theta_1$, the elevation of film temperature above the ambient or inlet temperature θ_1. Equations (13) and (14) are to be solved simultaneously to find the equilibrium temperature rise T. Substituting from (14) into (13) to eliminate Z leads to a quadratic, from which

$$T/T_0 = [(1 + k_1 N^2)^{1/2} - 1]/2 , \tag{15}$$

where

$$k_1 = 8\pi^3 \left(\frac{LD^3}{C}\right)\left(\frac{Z_1}{KT_0}\right) \cdot \tag{16}$$

Thus T approaches near-proportionality with N at high speeds, and drops to zero when $N = 0$.

As an example, take the air-cooled Petroff bearing of 2-in. diameter by 4-in. length with $D/C = 1000$, lubricated with an oil of 3.0×10^{-6} lb-sec/sq in. viscosity at 100 F, running at an ambient temperature of 77 F. If the oil has an apparent solidifying temperature of $\theta_0 = 32$ F, $T_0 = 77 - 32 = 45$ dF. To find Z_1, put $Z = 3 \times 10^{-6}$ in. Eq. (14) with $T = 100 - 77 = 23$ dF, and T_0 as above. Thus $Z_1 = 4.53 \times 10^{-6}$ lb-sec/sq in. Equation (14) reduces to $Z = 204 \times 10^{-6}$ divided by $T + 45$, from which $Z = 0.734 \times 10^{-6}$ lb-sec/sq in. at $\theta = 210$ F or $T = 210 - 77 = 133$ dF. These viscosities at $\theta = 77$, 100, and 210 F lie close to a straight line on the ASTM Chart (Chapter IV).

Substituting the known values of C, D, L, T_0, and Z_1 in Eq. (16) makes $k_1 = 0.40/K$. With cooling in still air, take $C_1 = 0.024$ in.-lb/sec, sq in., dF, as above. For the housing area, an estimate of 4 times the journal area may be taken as a probable minimum, making A equal to $4\pi DL$, or roughly 100 sq in. Therefore, $K = C_1A = 2.4$ in.-lb/sec dF, and $k_1 = 0.4/2.4 = \frac{1}{6}$. With slightly moving air and a larger housing area, K could easily be doubled. Figure 1 shows a plot of T/T_0 against N for two values of K, 2.4 and 4.8. Thus for an oil with $T_0 = 45$ dF the equilibrium temperature rise at 10 rps (600 rpm) would lie between 45 and 70 dF.

Suppose it is desired to increase K tenfold by forced lubrication,

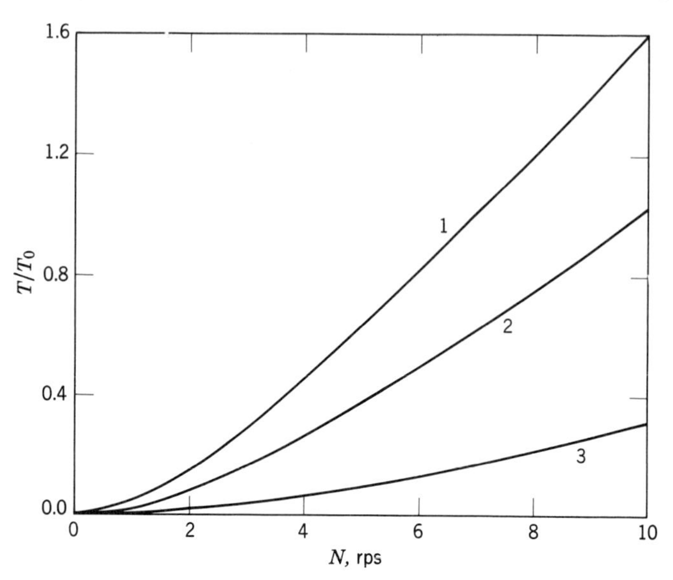

Fig. 1. Equilibrium temperature rise in a Petroff bearing (1, 2, air cooling; 3, forced feed; T_0 = 45 deg F).

say from 2.4 to 24. The required flow rate is now given by $Qq = 24 - 2.4 = 21.6$ in.-lb/sec dF. With $q = 140$ psi/sec dF as before, $Q = 0.154$ cu in./sec. Since $K = 24$ in.-lb/sec dF, $k_1 = 0.40/24 = 0.0167$ sec^2. Equation (15) now indicates a temperature rise of only 14.2, as against 70 dF by air cooling above.

The same reasoning can be extended to find the coefficient of friction f at the equilibrium temperature $\theta = \theta_1 + T$. Let f_1 denote the coefficient at inlet viscosity Z_1; then by Eq. (4) of Chapter II, $f/f_1 = Z/Z_1$. Substituting for Z from Eq. (14) and then for T/T_0 from Eq. (15) shows that

$$\frac{f}{f_1} = \frac{2}{1 + (1 + k_1 N^2)^{1/2}} \cdot \tag{17}$$

The curve for f/f_1 against N starts from unity at $N = 0$ and falls toward zero as N approaches infinity, passing through a point of inflexion. Since f_1 is given by Petroff's equation with $Z = Z_1$, Eq. (17) shows that under equilibrium conditions

$$f = \frac{k_2}{P} \cdot \frac{2N}{1 + (1 + k_1 N^2)^{1/2}}, \tag{18}$$

where

$$k_2 \equiv 2\pi^2(D/C)Z_1\,. \tag{19}$$

From (18) it is seen that the f curve starts out proportional to speed, but bends over and approaches an asymptotic or maximum value,

$$f_\infty = \frac{2}{P}\cdot\frac{k_2}{\sqrt{k_1}}\,. \tag{20}$$

For the conditions chosen in the air-cooled example, $k_1 = \frac{1}{6}$ sec^2 and $k_2 = 0.179$ lb-sec/sq in. Thus for a load of 100 psi, the limiting coefficient would be $f_\infty = 0.0083$. Higher values would be reached with cooling by forced lubrication.

The solution for a 3-constant viscosity-temperature formula in place of our 2-constant equation (14) will be found in the discussion of Wilcock's investigation (1952). The examples given show the need for more reliable data on heat-transfer coefficients and on effective housing areas. The formulas show the influence of design factors like D, L, and C; operating conditions like N, P, θ_1, and Q; and the properties of the lubricant, especially Z_1, T_0, and q.

It is recognized by Eq. (18) that the film viscosity depends on temperature and therefore on speed. Only the ambient or inlet viscosity Z_1, not Z itself, is shown as an independent variable. Such equations have been called "working" equations to distinguish them from the basic or "characteristic" equations in which the viscosity appears explicitly. They are more realistic than any equation which treats the viscosity as a known quantity.

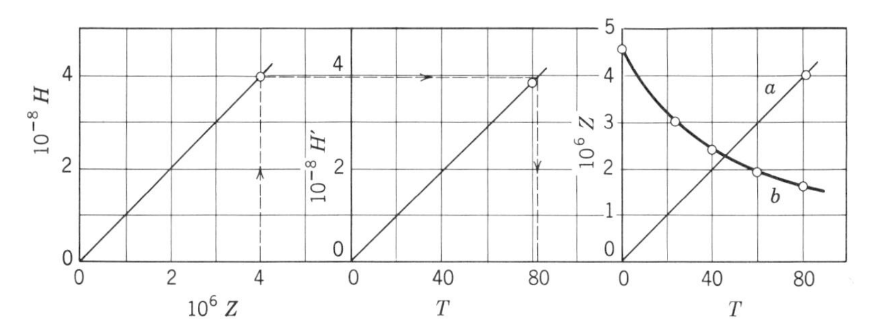

Fig. 2. Graphical solution for curve 2 of Fig. 1 (Z in reyns; H, H', in. lb/sec; T, deg F; graph a constructed from $H = H'$; b from viscosity-temperature curve of the oil).

Table 1. Data for Graphical Method (Z, lb-sec/sq in.; H, H', in.-lb/sec; T, deg F)

At $10^6\,Z = 4.0$, $10^{-8}\,H = 3.97$		At $10^{-8}\,H' = 3.97$, $T = 82.7$	
T	Z	T	Z
-45	∞	40	2.40
0	4.53	60	1.94
23	3.00	80	1.63

Graphical Methods. A graphical solution is shown in Fig. 2, based on Table 1. Starting at any point on the H, Z curve, draw a horizontal line until it hits the H', T curve. Record the Z, T coordinates of the intersection. Repeat until enough values are obtained to plot a Z, T curve. This will be a curve in which Z increases with T. It represents the viscosity at equilibrium temperatures. Superpose the original Z, T curve of the oil, in which Z decreases with increasing T. The intersection of the two Z, T curves gives the equilibrium temperatures rise T. The data plotted in Fig. 2 are from the example of an air-cooled Petroff bearing at 10 rps, but the method is not limited to such a bearing.

Similar diagrams have been devised by Boyd (1948), Wilcock (1952, 1957) and co-authors; and by Vogelpohl, Fig. 39 of the Houston lecture (VII: 1965). Wilcock named it the "operating line" method. Graphical solutions are more accurate and useful than analytical methods, except when trying to probe into the influence of the separate factors. A graphical method has been described by Leloup (1954) in which the equilibrium temperature is found by interpolation.

General Discussion. Thermal equilibrium problems have been further discussed by Kiesskalt (1928), Falz (1933), Hersey (II: 1935), Orlov (1935), McKee (1937, 1940), Thoma (1938), Vogelpohl (1940 and 1958), Cameron (1951), Heidebroek (1951), and Stephan (1953); as well as by Blok, Leloup, and Steller (all in 1954). Özdas (1955) reported on oil-ring bearings and Keller (1958) on railroad bearings.

Under nonhydrodynamic conditions where Coulomb friction predominates, the frictional resistance being independent of speed, H is proportional to U and H' to T, making T proportional to U after equilibrium is reached.

It was noted by Wilson & Barnard (V: 1922) that an unstable condition exists at the left of the minimum point when friction is plotted against viscosity. Accidental local heating, say at "high spots," would decrease the film viscosity and raise the friction, thereby releasing more heat, and so on. Thus a vicious circle could be initiated that might lead to a catastrophic failure. Stable conditions prevail at the right of the minimum. Questions of instability of equilibrium at low ZN/P have been further discussed by Jean Baudry (1948) and H. Blok (1954).

6. TRANSIENT CONDITIONS

It would be useful to have means for estimating the time required to reach approximate thermal equilibrium. A temperature-time curve might be idealized by assuming that the rate of temperature rise at any time t is proportional to the difference $T - \theta$ remaining between the instantaneous temperature elevation θ and the equilibrium temperature rise T. Integration would then give

$$\theta = T(1 - e^{-\beta t}) , \tag{21}$$

where β is the constant of proportionality relating $d\theta/dt$ to $T - \theta$. The slope at the origin is βT. The curve approaches $\theta = T$ asymptotically (Fig. 3). The time at which θ reaches 99 per cent of thermal equilibrium, that is, the value of t for which $T - \theta = T/100$, is $4.60/\beta$. Thus if θ rises initially 10 deg F/hr, and has to climb say to $T = 100$ dF, it may be expected to reach 99 F in 4.6 hr.

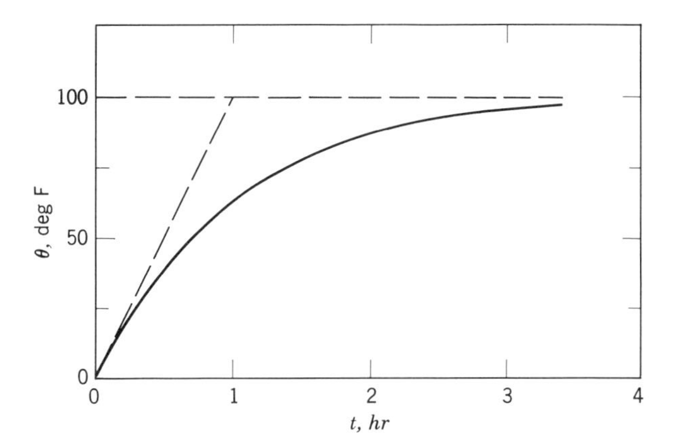

Fig. 3. The approach to thermal equilibrium (temperature rise θ at time t).

Muskat & Morgan's Curve. A relation of the foregoing type was derived by the above authors (1943) by integrating their heat balance equation

$$M(d\theta/dt) = H - C_3\theta^n \,, \tag{22}$$

where M is the effective heat capacity of the system; C_3 a constant for Lasche's formula, equal to the factor $C_n A_0$ of Eq. (7); and n is Lasche's exponent. The power loss H is assumed constant, and calculated from Petroff's law. It follows that

$$\beta = \frac{C_3 T^n}{M}\left(\frac{n}{T} - a\right), \tag{23}$$

where a is the temperature coefficient of viscosity, taken at the equilibrium temperature. Since M, H, and a have been assumed constant, the results obtained must be limited to a modest range of temperature.

Kelvin's Formula. An alternative approach might be to use Kelvin's formula for the interface temperature rise due to a heat source, as applied by Kingsbury (1933) to the tapered-plug viscometer,

$$\theta(t) = C_1 \int_0^t \frac{H(\tau)\, d\tau}{\sqrt{t - \tau}} \,. \tag{24}$$

For the case of a plane interface between semi-infinite solids having the same conductivity k, and the same heat capacity q per unit volume, C_1 is given by 0.282 kq. Here H is the power dissipated, expressed as a function of the time τ from zero to t; while the film temperature θ is a function of the present time t. Since H depends on the viscosity Z, which is a function of θ, an integral equation is involved. The first step would be to express H in terms of θ with the aid of Eq. (14) or its equivalent. A trial expression $\theta(\tau)$ might then be inserted in the integral and refined by successive approximations. Reference to Carslaw & Jaeger (1947) may be useful in applications of Eq. (24) and its variants.

Effects of Thermal Expansion. Falz (1933) and others have noted that the temperature-time curve of a journal bearing can overshoot its equilibrium value, approaching the final asymptote from above if the rate of heating is too rapid. This may be accounted for by thermal expansion. An interesting example of a large shaft bearing that failed by seizure in this manner was described by Albert Kingsbury (1906). Local effects of thermal expansion were investigated by Burton (1965) as well as by F. F. Ling and V. C. Mow (1965). Overall effects on bearing clearance were discussed by Nica (1965). The increasing use of aluminum and other non-ferrous materials for bearing construction draws attention to the factor of thermal expansion.

7. TEMPERATURE DISTRIBUTION

Part of the heat generated by friction is conducted through the oil film into the metal. Call this fraction m, the remainder being carried off in the oil. Fourier's law states that the rate of heat flow through unit area in either material will equal the product of temperature gradient and thermal conductivity, assuming a steady state.

Consider, first, substantially parallel surfaces of area BL, separated by an oil film of mean thickness h. By Eq. (1) the rate of generating heat is approximately $\mu U^2/h$ per unit area, where μ is the mean viscosity and U the relative velocity.

Metallic Conduction. From Fourier's law, if the heat flows only at right angles to the two surfaces and is equally divided between the moving and stationary members (Charron, 1941),

$$mH = -2BLk_m(d\theta/dy)\,, \tag{25}$$

where k_m is the thermal conductivity of the metal, and θ the temperature in the metal at any distance y from the bearing surface. Thermal quantities are again expressed in mechanical units, to avoid unnecessary use of the mechanical equivalent of heat. Substituting for H with the aid of Eq. (1),

$$d\theta/dy = -m(\mu/2k_m)U^2/h\,, \tag{26}$$

where μ, k_m, and h refer to the mean values.

Let θ_0 be the temperature at the oil-metal interface where $y = 0$ and let θ_2 be the outside temperature of the metal reserving θ_1 for the maximum temperature in the film; call the metal thickness h_m. Integrating from θ_2 at $y = h_m$ to θ_0 at $y = 0$ gives for the temperature drop across the bearing member

$$\theta_0 - \theta_2 = m(\mu/2k_m)(h_m/h)U^2\,. \tag{27}$$

In a journal bearing of radius R, if the entire heat flow is radially outward and the wall thickness h_m is a large fraction of the diameter, Eq (25) should be expressed with $2\pi(R - y)$ in place of B before integrating.

To establish an order of magnitude for the temperature drop, take $m = \frac{1}{2}$ and let $h_m = 1$ in. The thermal conductivity of steel, k_m, is about 5.6 lb/sec dF. Imagine a "square" thrust shoe of radial dimension $L = 4$ in., loaded to an effective film thickness of 1 mil while running at $U = 500$ in./sec. This corresponds to a thrust bearing of 15-in. outside diameter with a shaft speed of 800 rpm. Take $\mu = 3 \times 10^{-6}$ lb sec/sq in.; then by Eq. (27) the temperature drop should be about 34 deg F. See also Thoma (1938), Pensig (1939), and Pargin (VI: 1958).

Conduction in the Oil. Under the assumed condition of equal partition the temperature will have its maximum θ_1 near the middle of the film thickness, where $y = h/2$. The heat generated in unit volume per unit time in the layer from y to $h/2$ is μ times the square of U/h. A fraction m flows away from that layer into the metal at the rate $-k\, d\theta/dy$ per unit area, where k is the mean conductivity of the oil. The temperature gradient in the oil is

$$d\theta/dy = m(\mu/2k)(U/h)^2(h - 2y)\,. \tag{28}$$

Integrating from θ_0 at $y = 0$ gives a parabolic temperature distribution in the film. The drop from midpoint of film to the metal surface is

$$\theta_1 - \theta_0 = m\mu U^2/8k\,. \tag{29}$$

Paradoxically, the temperature drop across the film is independent of its thickness. These formulas and variants thereof were discussed by Pohlhausen (1921), Crocco (1931), Barnard (1932), Dickinson (1932), Hersey (II: 1935), and Vernotte (1937). Similar relations, but including temperature-dependent viscosity, were derived by Nahme (1940), Vogelpohl (1949, 1951), Dizioglu (1952, 1955), Dowson (1963), and Hays (1963).

To learn the magnitude of the temperature drop in the film, consider the case $m = \frac{1}{2}$, where half the frictional heat is conducted into the metal. Assume $k = 0.016$ lb/sec dF, a mean value for light petroleum oils; take μ and U as before. Substituting in (29) gives a drop of 2.9 dF from midpoint to either bearing surface. Since the distribution is parabolic, this represents a mean temperature say 2 deg F above the metal temperature. For higher speeds, the temperature rise will vary as the square of the speed.

For steady-state concentric journal bearings, with no appreciable heat flow into the shaft, the temperature is a maximum at the journal surface, instead of in the middle layer of the film. The drop across the film will then be 4 times the amount indicated by Eq. (29); or for $m = 1$,

$$\theta_j - \theta_0 = \mu U^2/2k\,, \tag{30}$$

where θ_j is the temperature at the journal surface. The formulas for concentric bearings have been generalized by Kingsbury (1933), Bratt (1934), and Hagg (1944) to allow for nonuniform viscosity. Kingsbury's solution is a graphical one taking into account the variation of k as well as μ with temperature. The solutions by Bratt and by Hagg are analytical. Hagg's formula was confirmed experimentally up to shear rates above 2 million reciprocal seconds. See also Eckert (1942); Golubev (1958); Bjorkland (1959; and Becker (1962).

The journal bearing formulas were extended by Hummel (1926) to the case of an eccentric, or loaded, half-bearing without side leakage. The temperature distribution in the film now follows a fourth-degree instead of a parabolic equation. The drop across the film from journal to bearing surface for $m = 1$ is given by

$$\theta_j - \theta_0 = \frac{\mu U^2}{2k}\left(1 - \frac{G}{3} - \frac{G^2}{12}\right), \tag{31}$$

where G is a dimensionless pressure gradient equal to the product of dp/dx by $h^2/\mu U$. Thus the film temperature varies along the bearing arc in a manner depending on the film profile and the local pressure gradients. Equation (31) reduces to (30) when $G = 0$. A numerical error led to the impression that the temperature difference is much smaller than actually given by Eq. (31). This fact is said to have retarded the advance of film temperature research. Further studies were published by Pinkus (1957), Purvis (1957), Hughes (1958), and their associates. For experimental evidence see reports by Nücker (1932), Rumpf (1938), Clayton (1948), Özdas (1955), Cole (1957), and co-authors. Melt lubrication is described by Sternlicht & Apkarian (1960).

Temperatures in Capillary Flow. Thermal effects in capillary flow lead to some interesting variants of Poiseuille's law, and must be corrected for before non-Newtonian properties can be established by flow observations. The mean temperature rise at exit under adiabatic conditions can be simply expressed by P/q, where P is the pressure drop and q the heat capacity of the liquid per unit volume. Thus in a typical petroleum oil, every increment of 140 psi is accompanied by a temperature rise of 1 deg F. Professor Bridgman observed that kerosene turned into "smoke" when it leaked out of a crack in a high-pressure cylinder. The foregoing rule is insufficient, however, to predict the flow rate and temperature distribution.

A preliminary study of the problem was described by J. C. Zimmer and the writer (1936, 1937). Philippoff (1942), Fritz & Hennenhöfer

(1947), Hausenblas (1950), and Brinkman (1951) offered more complete solutions. Toor (1957) introduced compressibility, Gee & Lyon (1957) non-Newtonian properties. Kearsley (1962) and Hays (1963) treated the ideal case of infinite length with isothermal wall. More realistic investigations followed by Gerrard and co-authors (1965), leading to computer solutions confirmed experimentally.

The Energy Equation. Theories of temperature distribution in the more difficult lubrication problems have been facilitated by the "energy equation"—a handy form of the first law of thermodynamics. See: Goldstein (1948), Vogelpohl (1949 and 1951), Cope (1949), Charnes (1952), Osterle (1953), Dizioglu (1955), Sternlicht (VI: 1957), and others collaborating. Applications are reviewed later.

Cope's energy equation, often cited, is limited to adiabatic conditions. It may be written

$$\left(1 - \frac{h^2}{6\mu U} \cdot \frac{\partial p}{\partial x}\right)\frac{\partial \theta}{\partial x} - \frac{h^2}{6\mu U}\left(\frac{\partial p}{\partial z}\right)\frac{\partial \theta}{\partial z} = \frac{2\mu U}{qh^2}\left\{1 + \frac{h^4}{12\mu^2 U^2}\left[\left(\frac{\partial p}{\partial x}\right)^2 + \left(\frac{\partial p}{\partial z}\right)^2\right]\right\} \tag{32}$$

where q, as before, is the product of density by specific heat. Charnes and co-authors pointed out that the "flow-work" represented by the pressure gradients on the right had been overlooked in some of the earlier literature. Equation (32) is to be applied in conjunction with the appropriate form of Reynolds' equation. The $\partial/\partial z$ terms may be dropped when there is no appreciable side leakage.

8. THERMAL LUBRICATION

The surprising discovery was made by Fogg (VI: 1946) that a parallel-surface thrust bearing can carry a substantial load. The bearing must be nicely constructed, with radial grooves to supply cool oil.

Theory of the Fogg Effect. At first sight this discovery seems to contradict hydrodynamic theory, but then it must be remembered that the classical assumption of uniform temperature is not quite true to life. Fogg called this new effect the "thermal wedge," but we like to call it the Fogg effect. He saw at once that the oil heats up as you go from the leading edge to the trailing edge of the sector. So he explained the effect by thermal expansion. The oil tries to expand and thereby develops enough fluid pressure to support an appreciable load, at least one-tenth of the load carried by a conventional thrust bearing.

Bower offered calculations in his discussion. Milton Shaw (VI: 1947; V: 1949) developed a simple formula by treating the viscosity as uniform and disregarding side leakage. Since the oil density ρ is variable, the equation of continuity must be modified by introducing ρu, ρv, and ρw in place of u, v, and w. And since the surfaces are parallel, h will be a constant. These changes reduce Reynold's equation to the three-dimensional form

$$\frac{\partial}{\partial x}\left[\rho\left(1 - \frac{h^2}{6\mu U}\cdot\frac{\partial p}{\partial x}\right)\right] = \frac{h^2}{6U}\frac{\partial}{\partial z}\left(\frac{\rho}{\mu}\frac{\partial p}{\partial z}\right). \tag{33}$$

The right-hand side may be set equal to zero when there is no side leakage. Equation (33) then becomes

$$\frac{d}{dx}\left(\frac{\rho}{\mu}\frac{dp}{dx}\right) = \frac{6U}{h^2}\left(\frac{d\rho}{dx}\right). \tag{34}$$

Suppose the density has a linear variation from ρ_1 at the leading edge to ρ_0 at the trailing edge, while the viscosity remains uniform. Integration now gives a load coefficient C_P equal to $0.54\Delta\rho/\rho_1$ for use in Eq. (24) of Chapter II; where $\Delta\rho$ denotes the density decrement $\rho_1 - \rho_0$. If the oil expands 1 per cent every 25 deg F (a common rule); and the temperature rise from leading to trailing edge happens to be 75 deg F; then $\Delta\rho/\rho_1$ is 0.030, and $C_P = 0.0162$. It may be recalled that the optimum value by classical theory is 0.159, about tenfold greater.

Boussages & Casacci (VI: 1948) offered a theory applicable to bearings of finite dimensions, and in which the viscosity is allowed to vary in the direction of motion. A parallel-surface thrust bearing of 320-cm diameter, tested at 12 rpm under 0.60 kg/sq cm load, gave a temperature rise of the order of 40 deg C from leading to trailing edge. The measured film thickness averaged just noticeably more than their calculated value of 0.16 mm. A three-dimensional theory is outlined by Cope with the aid of his adiabatic energy equation, including variable viscosity. After working out special cases for comparison, Cope concludes that viscosity variation has a profound effect; but that when such variation is small and the surfaces close enough together, and the lubricant well filtered, the thermal wedge can be expected to outclass the geometric wedge.

Assuming, with Cope, that the flow will be adiabatic, Osterle and co-authors (1953) take viscosity as a function of both pressure and temperature. Neglecting side leakage, they find that a load of 50 psi, in a typical case, should give 74 dF temperature rise compared to

16 dF with a uniform viscosity. They conclude that the Fogg effect is equivalent to an inclined surface film having the ratio $a = 1.04$, as compared to the ratios of 2.0 or more, commonly used. Calculations have been carried out in detail by Ulukan (1956) taking into account both variable viscosity and finite dimensions. Experiments are reviewed in Chapter VI.

The Cameron Effect. It was noticed by H. Blok that the hydrodynamic action between two rotating disks does not always cancel out, when their surface velocities in the contact zone are equal and opposite (Chapter X). How can that be explained? Cameron (X: 1952) continued the experiments and came up with an idea—namely, that a hydrodynamic pressure will be created when the viscosity of the lubricant varies across the film. He credits the idea initially to Professor C. Weber. Looking into the mathematics of the problem, Cameron found a term in the Navier-Stokes equations that is customarily neglected.

Consider the static equilibrium of a layer of lubricant of thickness dy, acted on by shear stresses τ and τ' at the upper and lower surfaces. Say the layer is of length L at right angles to the motion, and select an element of breadth dx in the direction of motion. The net force on the element due to shear is $(\tau' - \tau)L\ dx$. This is balanced by the force $(p' - p)L\ dy$, where p is the pressure at any point x, and p' that at $x + dx$. Therefore, $d\tau\ dx$ must equal $dp\ dy$, where $d\tau$ has been written for $\tau' - \tau$ and dp for $p' - p$. By Newton's law, τ may be set equal to $\mu\ du/dy$, where μ is the variable viscosity. If there is no side leakage or pressure gradient in the z-direction, dp/dx must equal d/dy of $\mu\ du/dy$. This much is familiar; but if μ is to be a function of y,

$$\frac{dp}{dx} = \mu \frac{d^2u}{dy^2} + \frac{du}{dy}\left(\frac{d\mu}{dy}\right). \tag{35}$$

The first term on the right is the classical one. It appears in Eq. (12) of Chapter II. The remaining term, proportional to the viscosity gradient, represents the Cameron effect; or as he called it, the "viscosity wedge." Integrating Eq. (35) gives the load capacity per unit of length L.

In a second paper Cameron (VI: 1958) applied his calculation to a parallel-surface bearing with stationary shoes of breadth B in the direction of motion. Let θ_1 denote the temperature in deg F of the lubricant and of the two surfaces at the leading edge; frictional heat gradually develops a temperature difference $\Delta\theta$ between stationary and moving surfaces at the trailing edge. The relative film thickness h/B

is shown to approach at least $\frac{3}{4}$ that of a pivoted-shoe bearing when $\Delta\theta/\theta_1$ is as great as 1, although vanishing for $\Delta\theta = 0$.

In the meantime a numerical solution was published by Zienkiewicz (1957) applicable to the Fogg and Cameron effects combined. When both surfaces are maintained at the temperature of the incoming oil, Zienkiewicz finds a negative pressure near the entering edge, but with a net positive load capacity. When one surface is held at a higher temperature than the other, he finds a greater load capacity than before, nearly four-fifths of which is contributed by the Cameron effect. The study was continued by Hunter (1960).

Although we have discussed thermal lubrication only as applied to parallel-surface bearings, the beneficial Fogg and Cameron effects must be present to some degree in the more conventional bearings, as noted in the last reference above; and perhaps in other machine elements, depending on the temperature distribution. Possibly these effects can be artificially controlled in the future, instead of being left entirely to frictional heat.

9. RUBBING CONTACT

The Fogg effect depends on the availability of fresh oil at an entering edge; the Cameron effect on maintaining a temperature difference between the two surfaces.

Suppose that two flat annular disks without radial grooves, or dissymmetry, are rotating with a relative velocity U and are separated by a thin liquid film of uniform thickness h. Then by Newton's law the frictional resistance will be proportional to U/h. If the film thickness diminishes under load, the friction must approach infinity as h approaches zero, unless the viscosity decreases faster than h; which is unlikely for a liquid, and impossible for a gas film. Under these conditions the lubricant will soon get hot enough to vaporize or decompose. Dry or "boundary" friction follows as described in Chapter XIII. We are concerned here only with the temperature rise corresponding to a given amount of friction.

The Steady State of Temperatures. The simplest condition to think about is that of two symmetrical solids in perfect contact at a plane interface, where power is being dissipated at a constant rate H_1 per unit area and heat flows in both directions. It will be assumed that any heat flow parallel to the interface may be disregarded. Fourier's law tells us that there will be a temperature gradient at the interface equal to $-H_1/2k$. To evaluate the temperature θ, here, we need to know the temperature θ_a at one of the exterior surfaces at a distance d. Then $\theta - \theta_a = H_1 d/2k$. Other geometries lead to similar solutions.

Varying States. Suppose H_1 is a function of time and that the temperature is initially uniform with $\theta = \theta_a$. Kelvin's formula (24) can then be applied until the surface temperature θ_a begins to be appreciably affected by the advancing heat wave. In the particular case of a constant heat source, integration gives

$$\theta - \theta_a = H_1 \sqrt{t}/\sqrt{\pi k q}. \tag{36}$$

When two blocks of metal with a relative velocity U are in contact under a load per unit area P with friction coefficient f, we can set $H_1 = fPU$. For example, if the coefficient is 0.10, load 100 psi, and velocity 100 in./sec, the temperature rise will be 564 $\sqrt{t}$ divided by the square root of kq. For steel at say 100 F, $k = 5.6$ lb/sec dF, $q = 280$ psi/dF (twice that of an oil), and $\sqrt{kq} = 40$, all in round numbers. The expected temperature rise approximates 14 $\sqrt{t}$ or 14 dF in the first second, 108 dF in the first minute. It is assumed that the rubbing surfaces are not exposed, and that all frictional heat remains in the two bodies. Actually the temperature rise will be less than indicated by the calculation, until finally a state of equilibrium is reached. It is interesting that conductivity, k, is the only bulk property of the solids governing thermalequil ibrium, whereas one additional property is needed for determining nonsteady temperatures. This may either be the heat capacity per unit volume q or the diffusivity k/q.

Moving Source of Heat. We have been considering a stationary source of frictional heat uniformly distributed over the interface. Suppose, instead, that the heat source is localized and moving with respect to one or both of the bodies, as in gear teeth or metal cutting.

Figure 4 shows two bodies in rubbing contact, the upper one moving to the right with velocity v, the lower one stationary. The upper body carries a flat rectangular protuberance, like an inverted plateau, of breadth B in the direction of motion. Heat is generated by friction at the rate H_1 per unit area. Let θ_b denote the bulk temperature of the stationary body near the surface, and θ the momentary temperature at any point x on the surface. Suppose at time t the leading edge of the plateau has reached the point $x = B$, having passed the origin at $t = 0$. Then $t = B/v$. The maximum temperature rise $\theta_m - \theta_b$ experienced by the stationary surface will be at $x = 0$. It is given by Eq. (36) with θ_b in place of θ_a, and t set equal to B/v.

Our temperature θ is Professor Blok's "flash temperature"

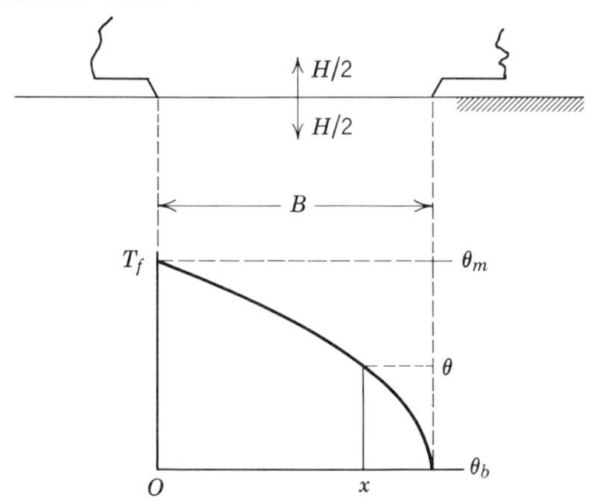

Fig. 4. Temperature flash T_f due to rubbing contact of breadth B.

(1937[1], [2]). The temperature rise $\theta_m - \theta_b$ is the "temperature flash," denoted here by T_f. Now $H_1 = fv(W/B)$, where W is the load per unit width. Substituting into Eq. (36) gives for the temperature flash

$$T_f = fW\sqrt{v}/C_2\,, \tag{37}$$

where C_2 is the square root of πkqB. Blok treated the more general case of two bodies of different materials under combined sliding and rolling on the assumption of a parabolic heat flow distribution in the contact (1937[1], [2], [3]). His solution may be written

$$T_f = 1.17\,\frac{fW(v_1 - v_2)}{(\sqrt{k_1q_1v_1} + \sqrt{k_2q_2v_2})\sqrt{B}}, \tag{38}$$

where v_1 and v_2, taken without regard to sign, denote the velocities of the contact band, of breadth B, along the respective surfaces.

As an example, consider two steel rollers, of 2-in. and 4-in. diameter, pressed into contact by a load $W = 500$ lb/in., each rotating 10 rps. Then $v_1 = 11.2$ and $v_2 = 7.9$ in./sec. As before, we take $\sqrt{kq} = 40$ units in the in., lb, sec, deg F system. Under the stated load, we may expect a contact band of breadth $B = 10.1$ mils (Chapter X). If it be assumed that the friction coefficient f is $\frac{1}{10}$, Eq. (38) makes $T_f = 48$ dF. See also discussion by Bowden & Tabor (I: 1950), and Holm (1952); by H. Blok (1954, 1963); and applications to gear teeth, reviewed in Chapter XI.

Similar calculations have been applied to a variety of problems by Jaeger (1942), Barwell (1957), Chao (1958), and Archard (1959); as well as by F. F. Ling and co-authors (1957, 1963, 1964, 1965) and Korovchinskii (1965).

Experimental Techniques. Thermocouple measurements of the surnique (Bowden & Ridler, I: 1936; Blok, XI: 1937; Ling, 1963; Furey, (1964), Niemann (1965), and others. Difficulties due to the finite size of the hot junction limit the possibilities of that method. Accordingly, frequent applications are made of the "dynamic thermocouple" technique (Bowden & Ridler, I: 1936; Blok, XI: 1937; Ling, 1963; Furey, 1964; Shu, 1964). This method, proposed by L. J. Briggs of the National Bureau of Standards in 1920, was first used by Henry Shore (1925). The two rubbing solids must be of dissimilar materials, so that the interface temperature will generate an electromotive force. Professor Emmons' theory (1940) of the "extended surface" thermocouple applies to both methods.

10. SUMMARY

Formulas are given for heat transfer from bearings both by external cooling and by oil flow, and applied to the calculation of thermal equilibrium temperature. This temperature can be predicted by combining the equations for power loss, heat transfer, and viscosity-temperature variation.

Temperature distribution in the metal and across the oil film is considered. Thermal lubrication, or load capacity resulting from temperature inequalities, is attributed to two conditions: the Fogg effect, caused by thermal expansion of the entering oil, and the Cameron effect, due to a temperature gradient across the film.

Calculations for temperature rise under conditions of solid rubbing contact are reviewed, together with thermoelectric techniques for measuring surface temperatures.

In general, a better knowledge of heat transfer is needed.

REFERENCES

Archard, J. F. (1959), "The Temperature of Rubbing Surfaces." *Wear* **2,** 438–55.

Barnard, D. P., 4th (1925), "Oil Flow in Complete Journal Bearings." *TSAE* **20,** pt. 2, 66–81. (a) *IEC* **18,** 1460–2 (1926).

Barnard D. P., 4th (1932), "A Possible Criterion for Bearing Temperature Stresses." *JSAE* **30,** 192–7.

Barwell, F. T. & Grunberg, L. (1957), "Maximum Temperature Rise in Boundary Friction." *Research* **10**, 165–6.

Baudry, Jean (1948), "Conditions de stabilité de l'équilibre thermique d'un palier lisse fonctionnant en régime du graissage oncteux." *CR* **226**, 1954–5.

Baudry, R. A. (1945), "Some Thermal Effects in Oil-Ring Journal Bearings." *T* **67**, 117–22.

Becker, K. M. & Kaye, Joseph (1962), "Measurements of Diabatic Flow in an Annulus with an Inner Rotating Cylinder." *J. Heat Transfer, T* **84**, 97–105.

Bjorklund, I. S. & Kays, W. M. (1959), "Heat Transfer Between Concentric Rotating Cylinders." *J. Heat Trans., T* **81**, 175–86.

Blok, H. (1937) [1], "Les températures de surface dans des conditions de graissage sous pression extrême." *Second World Petroleum Cong.*, Paris, Sect. 4; 151–82.

Blok, H. (1937) [2], "Measurement of Temperature Flashes on Gear Teeth under Extreme Pressure Conditions." *GDLL* **2**, 14–20.

Blok, H. (1937) [3], "Theoretical Study of Temperature Rise at Surfaces of Actual Contact Under Oiliness Lubricating Conditions." *GDLL* **2**, 222–35.

Blok, H. (1954), "The Dissipation of Frictional Heat." *Appl. Sci. Research* **A5**, 151–81.

Blok, H. (1963), "The Flash Temperature Concept." *Wear* **6**, 483–94.

Boyd, J. & Robertson, B. P. (1948), "Oil Flow and Temperature Relations for Lightly Loaded Journal Bearings." *T* **70**, 257–62. In Eq. (5) for 10^{-3} read 10^{-4}.

Bratt, Donald (1934); cited by Duncan in Kingsbury, 1933 (a).

Brinkman, H. C. (1951), "Heat Effects in Capillary Flow, I." *Appl. Sci. Research A* **2**, 120–4.

Burr, A. H. (1960), "The Effect of Shaft Rotation on Bearing Temperatures." *TASLE* **2**, 235–41.

Burton, R. A. (1965), "Thermal Aspects of Bearing Seizure." *Wear* **8**, 157–72. (a) *ASME* 65-Lub-3 (with Y. C. Hsu).

Cameron, A. (1951), "Heat Transfer in Journal Bearings. A Preliminary Investigation." *Proc. Gen. Dis. on Heat Transfer*, IME, London, 194–7.

Carslaw, H. S. & Jaeger, J. C. (1947), *Conduction of Heat in Solids*. Clarendon Press, Oxford; 386 pp. (a) Kelvin, *Math. and Phys. Papers* **2**, 41–60. (b) *Phil. Mag.* (7) **26**, 473–95 (1938).

Chao, B. T. & Trigger, K. J. (1958), "Temperature Distribution at Tool-Chip and Tool-Work Interface in Metal Cutting." *T* **80**, 311–20.

Charnes, A., Osterle, F., & Saibel, E. (1952), "On the Energy Equation for Fluid Film Lubrication." *PRS* **214**, 133–6.

Charron, F. (1941), "Répartition de la chaleur entre deux corps frottant séparés par un film lubrifiant." *CR* **212**, 695–7.

Clayton, D. & Wilkie, M. J. (1948), "Temperature Distribution in Bush of Journal Bearing." *Eng.* **166**, 49–52.

Cole, J. A. (1957), "An Experimental Investigation of Temperature Effects in Journal Bearings." Paper 63, *CLW*, 111–7; 10, 743, 757, 838, 846.

Cole, J. A. & Hughes, C. I. (1956), "Oil Flow and Film Extent in Complete Journal Bearings." *Proc. IME* **170,** 499–510; 520–34.

Cope, W. F. (1949), "The Hydrodynamical Theory of Film Lubrication." *PRS* **197,** 201–17.

Crocco, L. (1931), On a Maximum Value for the Coefficient of Heat Transmission from a Plane Lamina to a Fluid in Motion (Italian). *Atti Academia Lincei* **14,** 490–6.

Dickinson, H. C. & Bridgeman, O. C. (1932), "Fundamentals of Automotive Lubrication." *JSAE* **31,** 278–82, 304.

Dizioglu, Bekir (1952), "Die mittleren Temperaturen in Schmierschichten zwischen parallelen wärmeundurchlässigen Wänden." *Rev. de la Faculté des Sciences de l'Université d'Instanbul A* **17,** No. 1, 61–5. (a) op. cit. Nos. 2, 3; 159–77, 259–81.

Dizioglu, Bekir (1955), "Temperatur, Zähigkeits-und Reibungsverhältnisse in raschlaufenden Gleitlagern." Pages 236–56 in *50 Jahre Grenzchichtforschung,* ed. by H. Görtler & Tollmien; F. Vieweg & Sohn, Braunschweig.

Dowson, D. & Hudson, J. D. (1963), "Thermo-Hydrodynamic Analysis of the Infinite Slider-Bearing. Part I, The Plane-Inclined Slider Bearing; Part II, The Parallel-Surface Bearing." Papers 4 and 5, *IME Lubn. and Wear Conf.*, 31–41, 42–8.

Eckert, Ernst (1942), *Berechnung des Wärmeübergangs in der laminaren Grenschicht unströmter Körper,* VDI Forschungsheft No. 416; 24 pp.

Elenbaas, W. (1948), "The Dissipation of Heat by Free Convection from Vertical and Horizontal Cylinders." *JAP* **19,** 1148–54.

Emmons, Howard (1940), "The Theory and Application of Extended Surface Thermocouples." *JFI* **229,** 29–52.

Falz, E. (1933), "Der Wärmeausgleich im Gleitlager." *Petrol. Z.* **29,** Sept. 13, Motorenbetrieb 2–5. (a) "Das Lagerspiel bei höheren Temperaturen," **30,** No. 2 (1934).

Fritz, W. & Hennenhöfer, J. (1947), "Strömung zäher Öle in Kapillaren bei hohen Schergeschwindigkeiten und hohen Reibungsleistungen." *Z. Angew. Chem. B* **19,** 135–42.

Furey, M. J. (1964), "Surface Temperatures in Sliding Contact." *TASLE* **7,** 133–46.

Gee, R. E. & Lyon, J. B. (1957), "Nonisothermal Flow of Viscous Non-Newtonian Fluids." *IEC* **49,** 956–60.

Gerrard, J. E. & Philippoff, W. (1965), "Viscous Heating and Capillary Flow." *Proc., Fourth Int. Cong. Rheol.*, Providence, 1963; Wiley, New York, Part 2, 77–94.

Gerrard, J. E., Steidler, F. E., & Appeldoorn, J. K. (1965), "Viscous Heating in Capillaries, I. The Adiabatic Case." *IEC Fundamentals* **4,** 332–9. (a) *Nature* **194,** 1067–8 (1962).

Goldstein, S. (1938), "The Energy Equation." Pages 603–7 of *Modern Developments in Fluid Dynamics.* Oxford University Press, London, vol. 2.

Golubev, A. I. (1958), Influence of Heat on the Fluid Friction in a Non-

Loaded, Annular Oil Film (Russian). *TIM* **12,** 181–204. (a) Transl., *FWM* **12,** 171–95, ASME, New York (1960).

Hagg, A. C. (1944), "Heat Effects in Lubricating Films." *JAM* **11,** *T* **66,** A72–A76; **12** and **67,** A126 (1945).

Hanocq, Ch. (1947), "Recherches sur la loi de dissipation de la chaleur dans les paliers refroidis par convection naturelle." *RUM* **90,** No. 7, 245–58.

Hausenblas, H. (1950), "Die nichtisothermische laminare Strömung einer zähen Flüssigkeit durch enge Spalte und Kapillarröhren." *Ing.-Archiv* **18,** 151–66.

Hays, D. F. (1961), "Oil Flow in a Full Journal Bearing." *JBE, T* **83,** 312–4.

Hays, D. F. (1963), "A Variational Formulation Applied to Couette and Poiseuille Flow." *Bull. de l'Acad. royale de Belgique* (Classes des Sciences), 5th Ser. **49,** No. 6, 576–602.

Heidebroek, E. & Hagedorn (1951), "Zur Berechnung der Reibungswärme und der Kühlölmenge bei schnellaufenden Gleitlagern." *Die Technik* **6,** 21–3.

Heilman, R. H. (1929), "Surface Heat Transmission." *T* **51,** FSP-51–41, 355.

Hersey, M. D. (1915), "On the Laws of Lubrication of Journal Bearings." *T* **37,** 167–202.

Hersey, M. D. (1939), "Thermal Equilibrium in Journal Bearings." Fifth *ICAM,* Cambridge, Mass., 1938; *Proc.*, Wiley, New York, 638–41.

Hersey, M. D. (1942), "Heat Conditions in Bearings. An Outline of Problems for Research." *T* **64,** 445–55.

Hersey, M. D. & Zimmer, J. C. (1937), "Heat Effects in Capillary Flow at High Rates of Shear." *JAP* **8,** 359–63. (a) *Physics* **7,** 403–7 (1936).

Hirano, F. & Shodai, N. (1958), "Oil-Flow Coefficient of Pressure-Fed Journal Bearing." *Bull. JSME* **1,** 184–8.

Holm, Ragnar (1952), "Temperature Development in a Heated Contact with Application to Sliding Contacts." *JAM* **19,** *T* **74,** 369–74.

Hughes, W. F. & Osterle, F. (1958), "Temperature Effects in Journal Bearing Lubrication." *TASLE* **1,** 210–6.

Hummel, C. (1926), *Kritische Drehzahlen als Folge der Nachgiebigkeit der Schmiermittels im Lager.* VDI Forschungsarbeiten No. 287; 48 pp.

Hunter, W. B. & Zienkiewicz, O. C. (1960), "Effect of Temperature Variations across the Lubricant Films in the Theory of Hydrodynamic Lubrication." *J. ME Sci.* **2,** No. 1, 52–8, 372–4.

Jaeger, J. C. (1942), "Moving Sources of Heat and the Temperature at Sliding Contacts." *Proc. Roy. Soc. New South Wales* **76,** 203–24.

Karelitz, G. B. (1930), "Performance of Oil Ring Bearings." *T* **52,** APM-52-5, 57–70.

Karelitz, G. B. (1942), "Heat Dissipation in Self-Contained Bearings." *T* **64,** 463–4.

Kearsley, E. A. (1962), "The Viscous Heating Correction for Viscometer Flows." *Trans. Soc. Rheol.* **6,** 253–61.

Keller, W. M. & Pigman, G. L. (1958), "Analysis of Equilibrium Operating Temperatures of Railroad Journal Bearings." *LE* **14,** 108–15.

Kiesskalt, S. (1928), "Berechnung der Lagertemperatur." *Petrol. Z.* **24,** 1234–5.

Kingsbury, Albert (1906), "Tests of Large Shaft Bearings." *T* **27,** 425–32.

Kingsbury, Albert (1933), "Heat Effects in Lubricating Films." *ME* **55,** 685–8. (a) Duncan, J. A., citing D. Bratt, **56,** 120–1 (1934).

Kingsbury Machine Works (1922), "Lubrication and Cooling." Pages 7–10 in Catalog C, Philadelphia; and later issues. (a) Catalog KG, p. 32 (1948).

Korovchinskii, M. V. (1965), "Plane Contact Problem of Thermoelasticity during Quasi-Stationary Heat Generation on the Contact Surface." *JBE, T* **87,** 811–7.

Lasche, O. (1902), "Die Reibungsverhältnisse in Lagern mit höher Umfangsgeschwindigkeit." *ZVDI* **46,** 1881–90. (a) Forschungsarbeiten No. 9; 59 pp. (1903). (b) Transl., *Traction and Transmission* **6,** 33–64 (1903).

Leloup, Lucien (1954), *Étude de la lubrification et calcul des paliers.* Dunod, Paris, 1954; 294 pp.

Ling, F. F. & Mow, V. C. (1965), "Surface Displacement of a Convective Elastic Half-Space under an Arbitrarily Distributed Fast-Moving Heat Source." *JBE, T* **87,** 729–34.

Ling, F. F. & Pu, S. L. (1964), "Probable Interface Temperatures of Solids in Sliding Contact." *Wear* **7,** 23–34, 36.

Ling, F. F. & Saibel, E. (1957), "Thermal Aspects of Galling of Dry Metallic Surfaces in Sliding Contact." *Wear* **1,** 80–91.

Ling, F. F. & Simkins, T. E. (1963), "Measurement of Pointwise Juncture Condition of Temperature at the Interface of Two Bodies in Sliding Contact." *JBE, T* **85,** 481–7.

McAdams, W. H. (1933), *Heat Transmission.* McGraw-Hill, New York, 383 pp. (a) 3rd ed., 1954; 532 pp.

McKee, S. A. (1937), "Journal Bearing Design as Related to Maximum Loads, Speeds, and Operating Temperatures." *GDLL* **1,** 179–86.

McKee, S. A. (1940), "Friction and Temperature as Criteria for Safe Operation of Journal Bearings." RP 1925, *NBS J R* **24,** 490–508.

McKee, S. A. (1952), "Oil Flow in Plain Journal Bearings." T **74,** 841–8.

Muskat, M. & Morgan, F. (1939), "Studies in Lubrication, III. The Theory of Thick Film Lubrication of a Complete Journal Bearing of Finite Length with Arbitrary Positions of the Lubricant Source." *JAP* **10,** 46–61.

Muskat, M. & Morgan, F. (1943), "Studies in Lubrication, XI. Temperature Behavior of Journal Bearing Systems." *JAP* **14,** 234–44.

Nahme, R. (1940), "Beiträge zur hydrodynamischen Theorie de Lagerreibung." *Ing. Archiv.* **11,** 191–209.

Nica, Al. (1965), "Contributions to the Determination of the Real Clearance in Sliding Bearings." *JBE, T* **87,** 781–4.

Niemann, G. & Lechner, G. (1965), "The Measurement of Surface Temperatures on Gear Teeth." *JBE, T* **87,** 641–54.

Nücker, W. (1932), *Über die Schmierungsvorgang im Gleitlager.* VDI Forschungsheft No. 352; 24 pp.

Orlov, P. I. (1935), Coefficient of Friction, Oil Flow, and Heat Balance in a Complete Cylindrical Bearing (Russian). *Tekhnika Vosdushnogo Flota*, Moscow; **9**, 25–56. (a) Transl. by S. Reiss, *NACA TM* 1165 (1947); 56 pp.

Osterle, F., Charnes, A., & Saibel, E. (1953), "On the Solution of Reynolds' Equation for Slider-Bearing Lubrication, IV. Effect of Temperature . . . on the Viscosity." *T* **75**, 1117–23; ". . . VI. The Parallel Surface Slider-Bearing without Side Leakage." *T* **75**, 1113–36.

Özdas, N. & Ford, H. (1955), "Oil Transfer and Cooling in Ring-Oiled Bearings." *Eng.* **180**, 268–72; 570–3.

Pensig, F. (1939), "Sichtbarmachen von Temperaturfelden durch temperaturabhängige Farbanstriche." *ZVDI* **83**, 69–74 + colored plates.

Philippoff, W. (1942), "Über das Fleissen in Kapillaren bei extrem hohen Schubspannungen." *Physik. Z.* **43**, 373–89.

Pinkus, O. & Sternlicht, B. (1957), "The Maximum Film Temperature Profile in Journal Bearings." *T* **79**, 337–41.

Pohlhausen, E. (1921), "Der Wärmeaustausch zwischen festen Körpern und Flüssigkeiten mit kleiner Reibung und kleiner Wärmeleitung." *ZAMM* **1**, 115–21.

Purvis, M. B., Meyer, W. E., & Benton, T. C. (1957), "Temperature Distribution in the Journal Bearing Lubricant Film." *T* **79**, 343–50.

Rumpf, A. (1938), "Reibung und Temperatur Verlauf im Gleitlager." *Forschung* **9**, 149–58.

Shore, Henry (1925), "Thermoelectric Measurement of Cutting Tool Temperatures." *JWA* **15**, 85–8.

Shu, H. H. H., Gaylord, E. W., & Hughes, W. F., (1964), "The Relation Between the Rubbing Interface Temperature Distribution and Dynamic Thermocouple Temperature." *JBE, T* **86**, 417–22.

Steller, A., (1954), "Die Berechnung von Gleitlagern mit Flüssigkeitsreibung." *ZVDI* **96**, 89–97.

Stephan, H. (1953), *Temperatur und Verlagerung von zylindrischen Gleitlagern bei höher Drehzahl.* VDI Forschungsheft 439; 36 pp.

Sternlicht, B. & Apkarian, H. (1960), "Investigation of Melt Lubrication." *TASLE* **2**, 248–56.

Thoma, H. (1938), "Die Heisslauf der Gleitlager." *Forschung* **9**, 149–58.

Toor, H. L. (1957), "Heat Generation and Conduction in the Flow of a Viscous Compressible Liquid." *Trans. Soc. Rheol.* **1**, 177–90.

Ulukan, von Lütfullah (1956), "Thermischen Schmierkeilbildung." *Bull. Techn. Univ. Istanbul* **9**, 77–101. (a) *Ninth ICAM*, Brussels, 1956; *Actes* **4**, 303.

Vernotte, Pierre (1937), "Les équations de Navier et la fonction de dissipation, en régime hydraulique. Les phénomènes thermiques provoqués dans le fluide par un mouvement rapide." *CR* **205**, 21–3.

Vogelpohl, G. (1940), "Die rechnerische Behandlung des Schmierproblems beim Lager." *Öl u. Kohle* **36**, 9–13, 34–8.

Vogelpohl, G. (1949), *Der Übergang der Reibungswärme von Lagern aus der Schmierschicht in die Gleitflächen.* VDI Forschungsheft 425; 26 pp.

Vogelpohl, G. (1951), "Die Temperaturverteilung in Schmierschichten zwischen parallelen wärmedurchlässigen Wänden." *ZAMM* **31,** 349–56.

Vogelpohl, G. (1958), *Betriebsiche Gleitlager. Berechnungsverfahren für Konstruktion und Betrieb*. Springer, Berlin; 315 pp.

Wilcock, D. F. (1957), "Predicting Sleeve Bearing Performance." Paper 48, *CLW,* 82–92; 10, 11, 754–5, 832.

Wilcock, D. F. & Rosenblatt, M. (1952), "Oil Flow, Key Factor in Sleeve Bearing Performance." *T* **74,** 849–66.

Zienkiewicz, O. C. (1957), "Temperature Distribution within Lubricating Films between Parallel Bearing Surfaces and Its Effect on the Pressure Developed." Paper 81, *CLW,* 135–41; 14, 743, 839. (a) *Ninth ICAM,* Brussels, 1956; *Actes* **4,** 251–8.

chapter IV Physical Properties of Lubricants

The lubricant serves both to separate the rubbing surfaces and to remove heat. What is a lubricant? Any softer material that prevents the two surfaces in relative motion from coming into contact may be called a lubricant, even if it does not reduce friction. The need for a knowledge of the properties of lubricants is evident from the discussion of frictional temperature rise in the preceding chapter. Properties of liquid and solid lubricants are reviewed here, those of gaseous lubricants in Chapter IX.

1. LUBRICATING AND COOLING PROPERTIES

Viscosity, the most important property of a lubricant, is a function of temperature and pressure. To describe the viscosity of a lubricant completely, its value must be given at two temperatures and at two or more pressures; or at least the temperature and pressure coefficients must be stated in addition to its value at a chosen temperature and atmospheric pressure.

Under normal operating conditions, both friction and film thickness increase with increasing viscosity. When a high viscosity is required in order to insure against surface damage, care must be taken to avoid too great frictional resistance and temperature rise. Hence the need for an accurate knowledge of viscosity as affected by operating conditions, temperature especially (Vogelpohl, 1954; Bridgeman, 1960).

Density has little effect except when the lubricant is being sharply accelerated or in turbulent motion. A knowledge of density is needed for converting viscosity readings or calibrating viscometers (Barr, 1931). The effect of a lubricant may be influenced by compressibility and thermal expansion. Surface tension has only an indirect effect. The principal cooling properties of a lubricant are its heat capacity per unit volume and thermal conductivity, the first being the product of specific heat and density. Since specific heat increases with rising temperature, whereas density decreases, the heat capacity per unit volume is more nearly constant than the specific heat. In the event of cooling by vaporization of the lubricant, water especially, the latent heat can be significant. See also Fuller (1949; I: 1956).

Excluded from consideration are properties like color, flash point, and refractive index, useful in identifying materials and controlling refinery processes, but which have no direct effect upon lubrication. Also excluded are characteristics like foaming, oxidation, and thermal instability governing the misbehavior of lubricants; or changes over a period of time; and the effects of nuclear radiation.

Commercial viscosities are described in Appendix C.

2. DEFINITIONS AND UNITS

In general, physical properties are defined by the constants of empirical equations—for example, the initial slope of a curve. We do something to a body, measure what happens, continue over a sufficient range, plot the curve, and represent it by an equation. We repeat with the same form of specimen but a different material. Any constants that are numerically different in the two experiments are said to be properties of the material.

Viscosity. When the observed rate of shear is plotted against the applied shear stress, curves of different shapes are found for different classes of materials. If the curve is a straight line through the origin, the material is called Newtonian. Ordinary petroleum and fatty oils are Newtonian except at the lowest temperatures and highest pressures. Newtonian materials are characterized by a single property, viscosity, the ratio of shear stress to rate of shear—a constant at any fixed temperature and pressure. The term "fluidity" is used for the reciprocal of viscosity.

It follows that the unit of viscosity must be the unit of shear stress divided by the unit of shear rate. In the pound-force, inch, and second system the unit of shear stress is the pound per square inch. The unit of shear rate is the reciprocal second (velocity divided by length). Hence the unit of viscosity will be the pound per square inch multiplied by the second; or more neatly expressed, the pound-second per square inch. In the cgs system, where the dyne is the unit of force, the unit of shear stress is the dyne per square centimeter, and that of viscosity the dyne-second per square centimeter. This unit is called the *poise* in honor of Poiseuille. The name was proposed by R. M. Deeley and P. H. Parr in 1913. It happens that a viscosity of 1 poise is rather high, hence the frequent use of the centipoise, one-hundredth as much.

The pound-second per square inch is an even larger unit, equal to 68,950 poises. It was called the *reyn* by H. A. S. Howarth in honor of Reynolds. The term gradually gained acceptance. The micro-reyn, or millionth part of a reyn, is more convenient for everyday use. The *norton* may be suggested as a name for the new unit. It would honor the late Arthur E. Norton, whose teaching and writing clarified hydrodynamic theory. The proposed unit is equal to 6.9 centipoises within $\frac{1}{10}$ of 1 per cent, and can be abbreviated by the letter n.

Conversion of viscosity data from one system to another may be facilitated by Table 1. Three metric and two English systems are represented here, all based on force, length, and time. Metric units are the centipoise, kilogram-second per square meter, and the Newton-second per square meter. English units are the pound-second per square inch or reyn, and the norton or micro-reyn. The Newton is the unit of force in the mks system favored by electrical engineers, and used by Floberg in his lubrication reports (Chapters VII and X).

TABLE 1. Viscosity Conversion

Unit	cp	kg-sec/sq m	N-sec/sq m	lb-sec/sq in.	n
cp	1	1.02×10^{-4}	10^{-3}	1.45×10^{-7}	1.45×10^{-1}
kg-sec/sq m	9.807×10^{3}	1	9.807	1.422×10^{-3}	1.422×10^{3}
N-sec/sq m	10^{3}	1.02×10^{-1}	1	1.45×10^{-4}	1.45×10^{2}
lb-sec/sq in.	6.9×10^{6}	7.034×10^{2}	6.9×10^{3}	1	10^{6}
n	6.9	7.034×10^{-4}	6.9×10^{-3}	10^{-6}	1

Values good to four significant figures, recorded or not.

Viscosity Standards. The viscometers in everyday use are calibrated in terms of the viscosity of water. Determinations at the National Bureau of Standards (Swindells, 1952) resulted in the value 1.002 centipoises at 20 C or 68 F. This is 0.3 of 1 per cent less than the previously accepted value for the viscosity of water. The change required a revision of published tables (ASTM 1957). Standard viscosity samples are available from the NBS comprising convenient viscosity grades of petroleum oil. More permanent standards that can be sharply defined and reproduced, having higher viscosities than that of water, may be adopted in the future.

Kinematic Viscosity. The ratio of viscosity to density is commonly called the "kinematic viscosity." The ratio is literally kinematic, all trace of force or mass canceling out. The unit of kinematic viscosity may be written square inches per second; or, in the cgs system, square centimeters per second. The name *stoke,* in honor of Sir George Gabriel Stokes, was proposed for the cgs unit by Max Jakob in 1928. The centistoke, one-hundredth part, is an everyday unit of more convenient size, corresponding to the centipoise. The kinematic viscosity of a lubricant in centistokes is equal to its viscosity in centipoises divided by its density in grams per cubic centimeter. Capillary flow measurements under a known head of the test liquid come out in terms of kinematic viscosity, which must then be multiplied by density to obtain the true viscosity. Since kinematic viscosity is not viscosity, it can be referred to as the kinematic reading, or "kinematic value," to avoid confusion. Barr proposed the term "lentor" (1934).

Density and Specific Gravity. Density is mass per unit volume, so the unit of density in the cgs system is the gram per cubic centimeter. In chemical literature, densities are given in grams per milliliter; but for engineering accuracy, the milliliter may be taken equal to a cubic centimeter, since it is only 3 parts in 100,000 greater.

The unit of mass in the pound-force, inch and second system is the mass to which a force of 1 lb would impart an acceleration of 1 in./sec^2. The unit is therefore 1 lb-sec^2/in. Such a mass weighs 12×32.17, or 386 lb. No name has been assigned. Among ourselves, it might be called the *great-slug,* since it is twelves times the mass of a slug. In hydrodynamics, for example, when computing Reynolds' number to go with pipe diameters in inches, velocities in inches per second, and viscosities in reyns, the density of the fluid may be expressed in great-slugs per cubic inch; or in pounds-force seconds-squared per inch-fourth, but not in pounds per cubic inch. Density of water (77 F) = 1.933 slugs/cu ft.

Specific gravity is the ratio of the weight of a given volume of the fluid at a stated temperature to that of an equal volume of water at the standard temperature, usually 60 F. Since water has its maximum density at 4.0 C (39.2 F), where it weighs 1 gram per milliliter, its density will be slightly less at 60 F; in fact, about 0.999 of the maximum. Thus the density of a lubricant in grams per cubic centimeter may be taken equal to 0.999 of its specific gravity.

Heat Capacity. The heat capacity of a lubricant per unit volume is defined as the heat required to raise the temperature of a unit volume of the lubricant one degree, usually at constant pressure (Chapter III). It can be expressed in inch-pounds per cubic inch per degree F, and abbreviated psi/dF by cancellation of units; or in British thermal units per pound mass per degree F; or in ergs per cubic centimeter per degree C; and in still other units. It may be computed by multiplying the specific heat at constant pressure by the density of the lubricant, specific heat being defined as the heat required to raise the temperature of unit mass one degree.

Thermal Conductivity. The time rate of heat conduction across unit area is proportional to the temperature gradient under Fourier's law, the constant of proportionality being called the thermal conductivity of the material. Hence conductivity may be defined as the power transferred per unit area per unit temperature gradient. It can be measured in inch-pounds per second, square inch, and degrees F per inch, and abbreviated to lb/sec dF, as in Chapter III. Conductivity can also be expressed in dynes per second dC; or in other units to fit the associated parts of the problem.

A closely related property is the diffusivity of a material, or ratio of thermal conductivity to the heat capacity per unit volume.

Thermal Expansion and Compressibility. The change in density with temperature and pressure can be expressed by considering its reciprocal, the specific volume v, which is the volume per unit mass. The fractional increase in volume of a material per degree rise of temperature at constant pressure, α, is called its thermal expansivity or volumetric coefficient of expansion. A related property β, defined as the fractional increase of pressure per degree rise of temperature at constant volume, is called the coefficient of pressure increase. Both α and β are expressed in the same unit, the reciprocal deg F or C.

Compressibility, the reciprocal of elasticity, is defined as the fractional decrease in volume per unit increase of pressure. A distinction is made between the adiabatic compressibility K_a and the isothermal compressibility K_t. The adiabatic value is required when the change in volume is too rapid for appreciable heat conduction. Both are expressed in the same unit, the reciprocal of the pressure unit (Cameron, 1945; Chu & Cameron, 1963).

The four properties α, β, K_a, and K_t are not independent, but are linked by thermodynamic relations. Thus

$$\beta = \alpha/pK_t, \tag{1}$$

where p is the pressure; and

$$K_a = K_t - \frac{\alpha^2 T}{q} = \gamma K_t, \tag{2}$$

where T is the absolute temperature, q the heat capacity of the lubricant per unit volume, and γ the ratio of specific heat at constant pressure to that at constant volume.

When a lubricant is subjected to a sudden increase of pressure, a rise in temperature occurs, called the "heat of compression." It is troublesome in high-pressure research because of the time required before the temperature comes back to its initial value. Under isentropic conditions (adiabatic compression without internal dissipation) thermodynamic calculations give

$$dT/dp = \alpha T/q. \tag{3}$$

For mineral oils the temperature rise comes to about 1.3 deg C per 100 atm, or 1.6 deg F per 1000 psi. Equation (3) applies equally to gaseous lubricants, where the effect is many times greater, and more conveniently expressed by the equation for adiabatic compression of an ideal gas. See Bridgeman & Aldrich (1965).

The temperature change accompanying expansion depends on the efficiency with which work is extracted from the lubricant. If there is maximum efficiency and no heat transferred, the process is the reverse of isentropic compression. A cooling effect then occurs in accordance with Eq. (3). If, at the other extreme, no work is extracted from the expanding lubricant and no heat added, the temperature change is known as the Joule-Thomson effect. In liquid lubricants this effect will be a rise in temperature.

The foregoing paragraph is based on correspondence with Dr. F. W. Smith, National Research Council, Canada. The Joule-Thomson effect in liquids has also been discussed by Grubin (X: 1949), Vereshchagin (1956), Sternlicht (X: 1961), and co-authors.

Surface Tension. The surface tension of a lubricant at the air-liquid interface is related to its wetting and spreading tendencies in a bearing, and its rise in a capillary tube. High values are associated with its retention at a given point, as required in watch oils; low values with its spreading and penetrating ability, although other factors such as vapor pressure are then involved (Chapter XIII). It may be defined as the tension across any line in a free surface of the liquid, and is usually measured in dynes per centimeter.

Latent Heat. The energy absorbed in vaporizing unit mass of a liquid, say water, at constant temperature and pressure is called its latent heat of vaporization. This may be expressed in units of energy per unit mass, for example ergs per gram.

3. LUBRICANTS AND VALUES IN COMMON USE

Petroleum is still the principal source of industrial lubricants. The fatty oils are no longer so extensively used, although porpoise-jaw oil is still in favor for watches, lard oil for machining steel, and palm oil for rolling. Water and air are coming into use at the higher speeds. Synthetic oils are offered in great variety for especially high- and low-temperature applications.

Petroleum or mineral oils have been further adapted to modern requirements by the addition of small percentages of chemicals appropriately called "additives." Antirust additives are used in turbine oils; antioxidants and detergents in diesel lubricants; pour-point depressants and viscosity-index improvers in crankcase oils; extreme-pressure agents like tricresyl-phosphate, sulfur, and chlorine in gear lubricants. Mixtures of melted sodium and potassium have been used in reactor pumps; molten glass in extruding metals; solid lubricants in high-temperature and high-vacuum applications.

Mineral Oils. The viscosities of petroleum oils, straight or compounded, at 100 F range from 1.5 cp (about 0.2 n) for kerosene and from 14 cp (2 n) for instrument and household oils at 100 F up to 500 cp (71 n) for heavy steam cylinder and gear oils at 210 F. Automobile crankcase oils range from say 35 cp for SAE 10W to 105 cp for SAE 30, both at 100 F. Viscosities at 210 F run from $\frac{1}{5}$ to $\frac{1}{20}$ of their values at 100 F.

Relative viscosities of liquids such as water, crankcase oils, and castor oil can be demonstrated with glass tubes of about 1-cm bore and $\frac{1}{4}$-in. steel balls. Ask the glassblower to close one end of each tube. The other end will be corked after filling. Roll-time T, by stop-watch, is nearly proportional to the viscosity Z when the tube is held at an angle not greater than 10 or 15 deg from horizontal. For practical use the Z, T relation can be determined by calibration, somewhat as described by Hersey & Shore (1928). Barr's treatise (1931) provides an introduction to the general literature on viscosity measurement.

Specific gravities range from below 0.85 to above 0.95 (60/60 F). Thermal expansivities at 60 to 100 F run from $\frac{1}{20}$ to $\frac{1}{30}$ of 1 per cent/deg F over the specific gravity range from low to high (NBS 1946; Zisman, 1949; Chu & Cameron, 1963). Isothermal compressibilities are nearly the same for all petroleum oils, around 0.40 per cent per 1000 psi at 100 F, and increasing to 0.60 at 210 F. These are not mean or secant values, but initial compressibilities at atmospheric pressure. The mean compressibility decreases with increasing pressure (Jessup, 1930; Dow, 1934, 1940; Grunberg, 1954; Hartung, 1956). Adiabatic compressibility is less than isothermal, say 90 per cent thereof (Talbott, 1935; see refs. in H & H; Klaus, 1964). The coefficient of pressure-rise in a confined volume is of the order of 6.5 to 7.0 psi/dF. The temperature rise due to adiabatic compression has been shown above to be of the order of 1.6 dF per 1000 psi. Surface tensions of mineral oils run from 30 to 33 dynes/cm at room temperature, dropping about $\frac{1}{8}$ of 1 per cent per deg F rise in temperature between 70 and 350 F (Ross, 1950).

Heat capacity of petroleum oils per unit volume at 100 F may be taken equal to 136.7 psi/dF for an oil of nominal (60/60 F) specific gravity 0.90. To correct for temperature, add 0.7 per cent for every 10 degrees F above 100 F. The round number 140 applies at 130 F. It corresponds to a specific heat 0.49 that of water, combined with a nominal specific gravity of 0.90; or a specific heat half that of water, and gravity 0.88. To correct for specific gravity, add 0.6 per cent for each 0.01 above 0.90. The foregoing applies to oils of medium viscosity-temperature slope. For 100 VI (low slope) add 2 per cent; for zero VI (steep slope) subtract that amount. These rules are derived from Cragoe's correlation (1929). Additional data have been published by Kraussold (1932) and by Riedel (1950). Here "VI" refers to the viscosity index as defined in Appendix C.

From Cragoe's study, thermal conductivity may be taken equal to 0.0159 lb/sec dF at 100 F for an oil of 0.90 nominal specific gravity. To correct for temperature, subtract 0.3 per cent for each 10 deg F above 100 F; and for specific gravity, subtract 1.1 per cent for each 0.01 above 0.90. Additional data are given by Riedel (1950) and by Suge (1937).

The latent heat of vaporization of petroleum at a Fahrenheit temperature t, expressed in Btu/per lb mass, was found by Cragoe to be approximately 111 minus $t/11$, all divided by the nominal specific gravity. Thus for an oil of 0.90 gravity, the fraction vaporizing at 500 F would absorb 66 Btu/lb, only about $\frac{1}{15}$ that of water boiling off.

Fatty Oils. In contrast with mineral oils, which can be obtained in practically any viscosity, the properties of the fatty oils are fixed by nature, except for slight variations. The lubricating and cooling properties of some of the fatty oils are collected in Table 2. Values have been averaged from Hyde (1920), Woog (1926), and Suge (1933, H & H, Ref. 66), with some aid from Mitchell (1927) and Keulegan (1930). Vegetable and animal oils in Table 2 are listed in descending order of viscosity within each group. Compressibilities are isothermal, at atmospheric pressure. Suge reports that the compressibilities of fatty oils, shown by his unpublished tests, run about 10 per cent less than for mineral oils. He found the product of compressibility and surface tension nearly constant, and noted that the surface tension drops about $\frac{1}{8}$ of 1 per cent per deg F rise in temperature. The viscosity of porpoise-jaw oil, having a nominal specific gravity 0.925, was reported about 19 cp at 100 F.

It must be assumed that the heat capacities per unit volume shown in Table 2 increase with rising temperature at nearly the same rate for fatty as for mineral oils, 0.07 per cent per deg F. The only data found on thermal conductivity of the oils in Table 2 are from Suge's tests on castor and rape, 22.5 and 21.5 $\times$ 10^{-3} lb/sec dF, respectively. The test temperature may have been 18 C (about 64 F). Suge remarks

TABLE 2. Properties of Some Fatty Oils

	60 F		100 F			
Oil	Specific Gravity	Expansivity, %/dF	Compr., 10^3 × %/psi	Surface Tension, dynes/cm	Viscosity, cp	Heat Capacity, psi/sec, dF
Castor	0.961	0.039	0.33	34.7	245	176
Rape	0.912	0.038	0.37	32.4	42	169
Neatsfoot	0.919	0.038	0.38	36.5[a]	42	167[a]
Lard	0.916	0.039	0.47	30.7	35	167[a]
Sperm	0.882	0.040	0.47	36.5	22	170
Oleic acid	0.896	0.042	0.44[b]	30.5	19	–

[a] Estimated from Hyde's trotter oil.
[b] Calc'd from surface tension.

that the conductivities of vegetable oils are at least 15 per cent greater than those of mineral oils.

Other Lubricants Greases and solid lubricants will be described under "Non-Newtonian Lubricants." Synthetic oils, including the diester, phosphate, silicone, and silicate types; also polyglycols, chlorinated aromatics, polybutenes, and fluorlubes were described by Barnes & Fainman (1957) and Moreton (1964). See also Gunderson & Hart (1962), as well as Chapter III in Michell's book (V: 1950). Temperature and pressure coefficients of viscosity for glycerine, water, and two silicones appear in Table 3. Similar data for phosphate-ester and water-glycol fluids were published by Cordiano and co-authors (1956). Other tests are cited in the following pages, including glass and molten metals as lubricants. Lubricating properties of gases are given in Chapter IX.

4. VISCOSITY TEMPERATURE RELATIONS

The natural way of expressing the effect of temperature upon viscosity is by means of the *temperature coefficient,* or fractional change in viscosity per degree increase of temperature. In symbols, the temperature coefficient of viscosity is written $(1/Z)dZ/dt$ and denoted by a. Note our return to the usual symbol t for temperature since time is not involved here.

The effect of temperature upon viscosity is notably greater than its effect upon any other common physical property. The change in volume of a petroleum lubricating oil per deg F rise in temperature is only about 0.04 of 1 per cent, but the viscosity of a medium petroleum oil can drop 3 or 4 per cent in 1 degree. The temperature coefficient is less for fatty oils of comparable viscosity. Representative values for three petroleum, three fatty, and four other liquids are compared in Table 3. The data for all liquids except the two silicones are taken from Hersey & Hopkins (1954), where the original sources are cited. The data on these two are from the ASME Report (1953).

The three petroleum oils were tested by R. B. Dow (1937, 1941). A comparison of the first two shows the effect of viscosity grade. The California and Pennsylvania oils were chosen to compare the naphthenic and paraffinic crudes. They were of equal viscosity, 40 cp, at about 130 F. The data on fatty oils were averaged from the sources given. The two silicones are identified in Table 3 by ASME sample numbers 53 and 55.

Differences in the viscosity-temperature characteristics of two oils having the same viscosity at room temperature may be demonstrated with the ball and slanted tube. Pack the tubes vertically in a large

TABLE 3. Temperature Coefficients of Viscosity

Lubricant	Specific Gravity, 60/60 F	100 F		210 F		Z Ratio 210 F/100 F
		Z, cp	*a*, %/dF	Z, cp	*a*, %/dF	
Petroleum Oils from Three Sources						
Rumania	0.962	1640[a]	−6.0[a]	32	−2 3	0.02
California	0.933	116	−3.3	8.0	−1.3	0.07
Pennsylvania	0.884	86	−3.0	9.0	−1.3	0.10
Three Fatty Oils						
Castor	0.961	245	−3.6	31	−0.8	0.13
Lard	0.916	35	−2.1	6.5	−1.1	0.21
Sperm	0.882	22	−1.9	3.7[a]	−1.1[a]	0.17
Four Other Liquids						
Silicone (53)	1.078	116	−1.8	32	−1.5	0.18
Silicone (55)	0.978	71	−1.0	27	−0.8	0.37
Glycerine	1.256	205	−3.9	10.3	−1.5	0.05
Water	0.999	0.68	−1.2	0.28	−0.5	0.42

[a] Extrapolated

bottle of snow or ice an hour or more in advance. Roll-times should be observed immediately on withdrawing tubes from the ice bath, to obtain the maximum contrast; and again near the close of the lecture hour, when it may be seen that the viscosities have practically equalized.

Temperature coefficients are useful both to characterize oils and to correct experimental data for small departures from any intended test temperature. For other purposes the mean temperature coefficient from 100 to 210 F may be preferred. The ratio of viscosity loss, 100/210 F, to the viscosity at 100 F was proposed by Wilcock (1944). It was named the "VTC" (viscosity-temperature coefficient). It varies from 0.98 for the Rumanian oil down to 0.58 for water. The ratio of viscosity at 210 to that at 100 F is given in the last column. Note the contrast, reading down from 2 to 42 per cent for the ten liquids in Table 3.

Viscosity-Temperature Formulas. Poiseuille found flow resistance inversely proportional to a quadratic function of temperature (II: 1840). Petroff used this relation as a viscosity-temperature formula,

$$Z = A/(1 + c_1 t + c_2 t^2)\ , \tag{3}$$

in his incomplete discussion of thermal equilibrium (I: 1883). Professor A. W. Duff, writing in 1897, showed that all the viscosity-temperature equations published since the time of Poiseuille were integrals of the equation

$$(dZ/dt)/Z = 1/(c_1 + c_2 t + c_3 t^2)\,, \tag{4}$$

where Z is the viscosity at any temperature t and the c's are empirical constants. Anong the formulas to which Duff's equation applies are those of Reynolds (II: 1886), Slotte (1890), and Vogel (1922). These three formulas are still in common use because of their mathematical simplicity.

Reynolds' formula is a two-constant equation,

$$Z = Ae^{-mt}\,. \tag{5}$$

Here A is the viscosity at $t = 0$ and m is 2.30 times the downward slope of the line obtained by plotting log Z against t. Equations (3) and (5) represent curves of type 1 in Fig. 1, falling toward zero as t approaches infinity.

Slotte's formula contains three constants. It is valid over a wider range, and may be written

$$Z = A/(t - c)^m\,; \tag{6}$$

where c is the cold point or temperature of apparent solidification, since Z becomes infinite when $t = c$. It can be reduced to a two-constant form by dropping c, which is not far from zero F, provided we use only the Fahrenheit scale. This was pointed out by Herschel (1922). Log Z may then be plotted against log t, which gives a straight line with an intercept log A, and having a downward slope m. Note that $Z = A$ when $t = c + 1$. Slotte's equation represents the curve of type 2 in Fig. 1. The usefulness of the equation in Fahrenheit form ($c = 0$) has been demonstrated by Appeldoorn (1963).

The equilateral hyperbola is a special case of Slotte's formula when $m = 1$. An example is seen in Eq. (14) of Chapter III. It was used by the writer in 1914, and by D. F. Hays in 1963, although limited to a small temperature range (Chapters II and III).

Vogel's formula is a three-constant expression,

$$Z = Ae^{m/(t-c)}\,. \tag{7}$$

Again c represents the cold point and A is the viscosity for $t = \infty$. Log A is the intercept, and m the numerical slope of a straight line obtained by plotting log Z against $1/(t - c)$, where c is found by trial. When plotting Z against t, the curve approaches a vertical

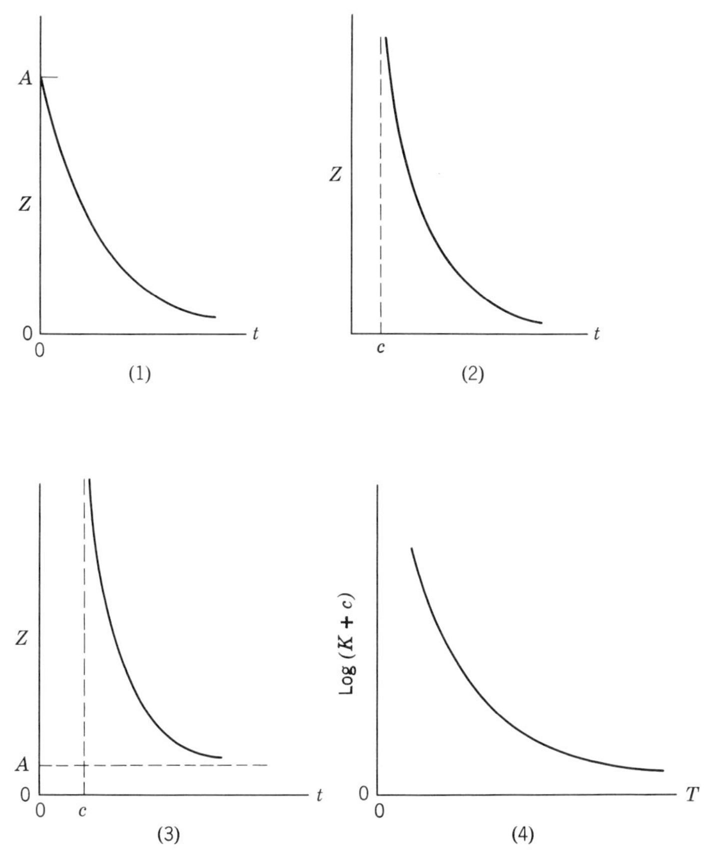

Fig. 1. Four types of viscosity-temperature curves: (1) Reynolds', (2) Slotte's, (3) Vogel's, (4) Walther's.

asymptote at $t = c$ and a horizontal one at $Z = A$. It is usually a closer fit than the other two. Vogel's equation was used to advantage by Cameron (1945). Its molecular interpretation and application are discussed by Rost (1955).

The most widely used formula is probably that of Walther (1931), an expression for the kinematic viscosity K in centistokes at an absolute temperature T:

$$\log (K + c) = A/T^m . \tag{8}$$

Walther's is essentially a three-constant formula, with the "adjustable" constant c frozen at the optimum value for petroleum oils over a chosen temperature range. The value 0.8 cs was originally assigned to this constant. According to Erk & Eck (1936), Vogel's equation may be slightly more accurate. Walther's equation can be represented by curve 4, Fig. 1. Plotting the logarithm of both sides yields a straight line having a downward slope m. The value dK/dt is $-2.3m(K+c)$. Division by K gives the temperature coefficient of kinematic viscosity, as discussed by Kiesskalt (1944). Temperature coefficients may be found from any of the K or Z equations by differentiating. See references in H & H (1954).

Although the five formulas described above are the best known, at least six others should be mentioned. Suge's equation (1933) for the Z, p, t relation may be written as an isobar in the form

$$\log \frac{Z}{Z_0} = \frac{m}{t-r} - \frac{m}{t_0 - r}, \tag{9}$$

where Z_0 is the viscosity at t_0; m and r being empirical constants. The viscosity is infinite at $t = r$, but drops to a finite value when $t = \infty$, like curve 3 in Fig. 1.

A two-constant isobar for "liquidity," L, was offered by Cragoe (1934), in which L is a function of viscosity having a linear relation to temperature. If Z is restricted to centipoises, L may be defined as 1300 divided by log $20Z$. It is then found empirically that L/L_0 equals 1 plus $c(t - t_0)$. Here L_0 is the value of L at $t = t_0$ where $Z = Z_0$, and c is a constant. The viscosity is infinite at a temperature t_1 equal to t_0 minus $1/c$. It drops only toward a finite value $Z_\infty = 0.05$ cp when t approaches infinity, as in curve 3 except for location of asymptotes. The two-constant formula for the kinematic value in centistokes offered by G. Barr (1937),

$$[\log (K + 0.8)]^{0.3} = A + (m/T), \tag{10}$$

was said to be in close accord with experimental data; and possibly better than others at high temperatures. It is apparently a four-constant formula with two constants frozen at 0.8 and 0.3. The curve is like type 4, Fig. 1, with K infinite at $T = 0$, but finite at $T = \infty$. Call this type 5.

Bradbury's equation (1951) for the Z, p, t relation leads to an isobar

$$\log \frac{Z}{Z_0} = c(e^{k/T} - e^{k/T_0}). \tag{11}$$

As before, Z_0 is the viscosity at an absolute temperature T_0, and c and k are empirical constants. The viscosity is infinite at $T = 0$, but finite at $T = \infty$. The curve is also one of type 5. A simpler relation of this type is that of Cornelissen and co-author (1955),

$$\log Z/A = c/T^m . \tag{12}$$

Here Z is infinite when $T = 0$, and drops to the limiting value A as T approaches ∞. See also Ellis (1957).

More recently Roelands and co-authors (1964) proposed a formula that can be written

$$1.200 + \log Z = \frac{G}{\left(1 + \dfrac{t}{135}\right)^S}, \tag{13}$$

provided Z is in centipoises and t in degrees C. If t is in Fahrenheit, read 211 in place of 135. The new formula is not limited to lubricating oils. It leads to an absolute viscosity-temperature chart of wider range than the ASTM, yet said to be of equal accuracy. The slope S of the straight lines on this chart may be taken as a "slope index" having greater simplicity than the conventional VI and free from the faults described below. The lines are parallel for "naturally homologous" liquids.

Viscosity-Temperature Charts. Logarithmic charts for absolute viscosity against temperature were published by Herschel in 1922. Based on Slotte's Fahrenheit relation, they give practically straight lines for petroleum oils. Many such graphs were plotted.

The convenient charts C and D for kinematic viscosity of petroleum oils, standardized by the ASTM, are equivalent to Walther's equation except for the value of c. This constant is taken equal to 0.6 cs for kinematic viscosities above 2.0 cs, increasing to 0.75 at 0.4 cs—the bottom of the low-range chart D. These charts are the result of a long study summarized by Geniesse (1932, 1937) and by Zisman (1949). A similar chart designed by Neil MacCoull had been published without describing the formula used (Leslie, 1927). Since petroleum oils plot as straight lines, the kinematic viscosity can be determined over a wide range from observations at two temperatures. The graphs for non-petroleum oils often show perceptible curvature, requiring three points for the most satisfactory determination. Straight-line charts for kinematic viscosity require a knowledge of density, to be exact, before conversion into true viscosity. However, fair results can be had by plotting centipoises directly on the kinematic scale, ignoring thermal expansion. Charts reading directly in viscosity units, instead of kinematic viscosity, are preferred for lubrication calculations. A chart of this kind, based upon Vogel's equation, was constructed by Herschel in 1931 for an ASTM Committee. Absolute viscosity charts were published by Boyd & Raimondi (VII: 1951). A temperature scale from 30 to 300 F is provided, or say from zero to above 140 C. Viscosity scales range from 0.2 to 10^4 n, and from below 1.5 to above 6×10^4 cp. An accompanying chart shows curves for various fluids, including gases, up to 1000 F. The new chart by Roelands and co-authors (1964) gives straight lines for petroleum oils, synthetics, and thirteen miscellaneous liquids to viscosities of the order of 10^9 cp.

The Viscosity-Index Problem. Petroleum technologists are constantly striving to express the viscosity-temperature characteristics of an oil by a single number. They want a number that is not only scientifically correct but meaningful to experts familiar with the present and future commercial lubricants. The Dean and Davis "VI" or viscosity-index (1940) when defined in kinematic units (ASTM, 1957) did very well until the advent of new lubricants with VI's up to 125 or 150 (western oils run to about 0 VI, Pennsylvanian to 100 without special treatment). Then someone thought of plotting contour lines, or graphs of constant VI, on a diagram having the viscosity at 210 F as its horizontal axis, and that at 100 F as the vertical axis. All graphs start upward toward the right from the origin. The graph labeled 0 VI is practically a straight line. The graph labeled 100 bends over considerably. Those for VI 125 and above pass through a maximum and start down. Now the fat is in the fire—these high VI graphs reveal two oils having the same viscosity at 100 F and the same VI, but with greatly different viscosities at 210 F! The difficulties of the problem have been reviewed by Geniesse (1956), Ellis (1957), and others as noted in Appendix C; but see Klaus (1958), Roelands (1964), and co-authors.

Research at Extreme Temperatures. Fortsch and Wilson (1924) measured the viscosities of petroleum fractions up to nearly 500 F. Needs (1942) investigated the viscosities of seven fatty and mineral oils to temperatures over 400 F, heating some of the oils above 550 F. A precision-type, tapered-plug viscometer was employed, permitting the study of small changes due to thermal instability. To go much beyond these temperatures requires the use of synthetic oils, liquid metals, or solid and gaseous lubricants (see references on synthetic oils and molten glass). Properties of liquid metals for nuclear reactors have been determined by Soda (1948), Ewing (1951), Webber (1955), Shrock (1957), and co-authors. Solid lubricants are briefly described below.

Early investigations of low-temperature properties include a study by H. K. Griffin (1925) on a petroleum oil for airplanes. The oil was observed to have a maximum density near —40 F or C. Keulegan and Houseman (1930) followed with low-temperature damping liquids for aircraft instruments. The flow-resistance of oils at low temperatures was investigated at the Physikalisch-Technische Reichsanstalt (Erk, 1931). Recent experiments on aircraft engine oils have been published by Schrock and co-authors (1957) comprising data on thermal conductivity, specific heat, and latent heat. Extensive research on the viscosity and related properties of oils at low temperatures has been reported by Velikovski and others (1941) at conferences in Russia.

5. VISCOSITY-PRESSURE RELATIONS

Whereas viscosity falls off with rising temperature, it increases with pressure—slowly at first, and then more rapidly, as shown by Fig. 2. The diagram is plotted from test results on Navy Symbol 9250, SAE 30 diesel engine lubricating oil used in investigations of bearing

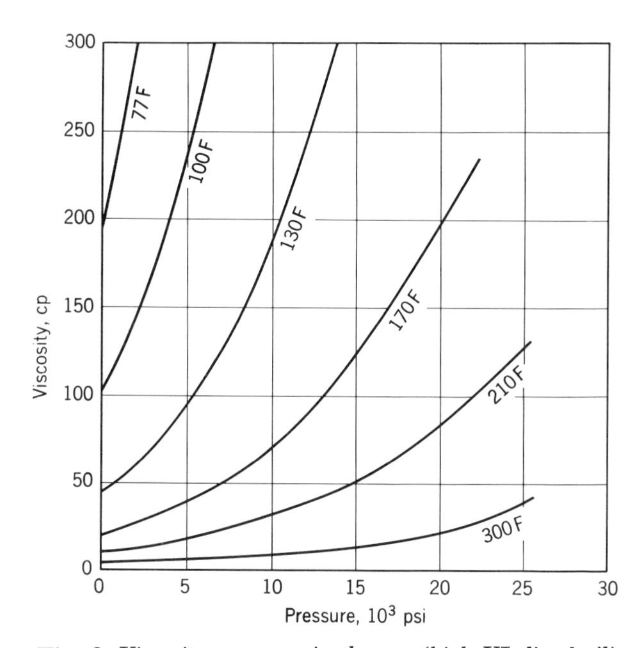

Fig. 2. Viscosity-pressure isotherms (high VI diesel oil).

life. It was a "heavy duty" petroleum oil of 103 VI, being ASME Sample No. 30 (1953).

Following along the isotherm for 170 F, we find a tenfold increase in viscosity upon raising the pressure from atmospheric to 20,000 psi. The same increase in viscosity is produced by dropping the temperature from 170 to 77 F at atmospheric pressure (Fig. 3). Thus in comparing the overall effects of pressure and temperature within a stated range, it is seen that a pressure rise of 20,000 psi is equivalent to

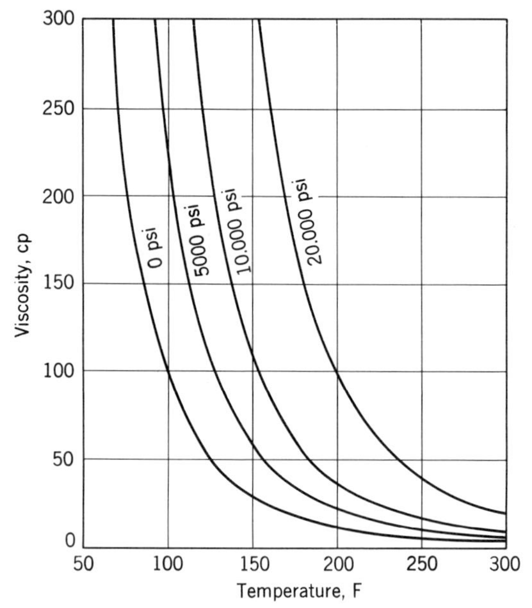

Fig. 3. Viscosity-temperature isobars (oil of Fig. 2).

a temperature drop of 93 dF. The viscosity of this oil, or similar ones, could be maintained at a constant value of 200 cp by stepping up the pressure roughly 200 psi per deg F. Logarithmic viscosity-pressure graphs are shown in Fig. 4.

Pressure Coefficients. The slope of an isotherm, dZ/dp, is less at atmospheric pressure than at elevated pressures; but the fractional change, $(dZ/dp)/Z$, is greater. This fractional or relative change is called the *pressure coefficient* of viscosity. It may be denoted by b at any gage pressure r, and by b_1 at a pressure of 1 atm ($p = 0$). The use of the "initial pressure coefficient" b_1 is one method of expressing the viscosity-pressure characteristic of an oil. The term "pressure coefficient" is often used without qualification, for brevity, to denote the initial value. The pressure coefficient decreases slightly with increasing temperature. Thus from Fig. 3, reading off slopes and dividing by viscosities, it is found that b_1 decreases from 2.0 per cent per 100 psi at 77 F to about 1.7 at 100, 1.2 at 210, and 0.9 at 300 F. See also Fig. 5(a) in H & H (Hersey & Hopkins, 1954).

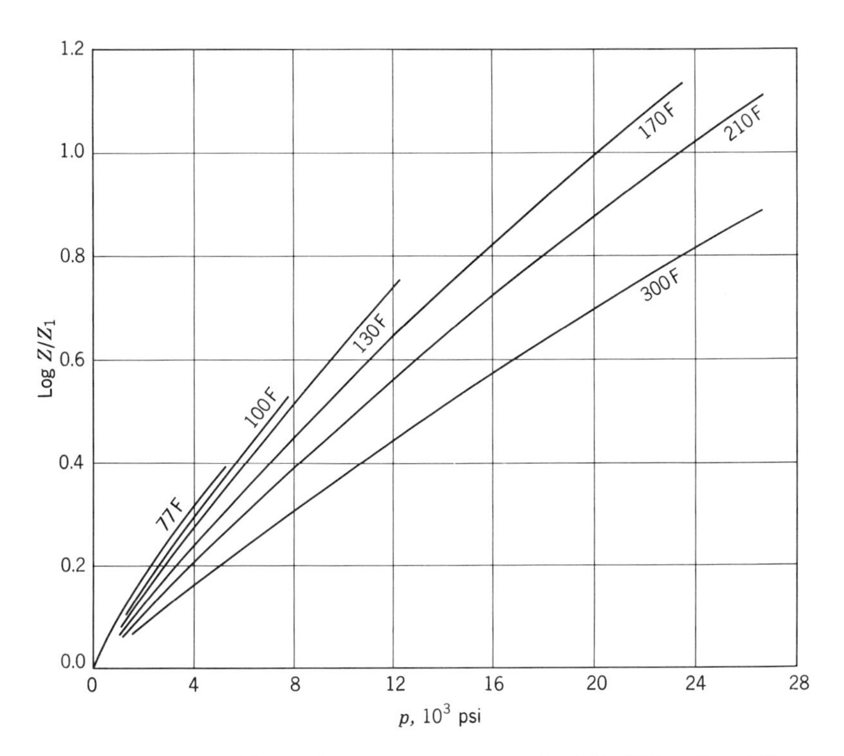

Fig. 4. Logarithmic viscosity-pressure graphs (oil of Figs. 2 and 3).

In a series of oils of any one type the b_1 value increases with increasing viscosity grade. Comparing different types of lubricants in the same viscosity grade, Hyde (1920) discovered that mineral oils as a class have higher pressure coefficients than fatty oils. Table 4 shows the pressure coefficients at atmospheric pressure for the fluids listed in Table 3. It should be useful as a guide in estimating pressure coefficients for similar lubricants that have not been tested. Data for all the liquids in Table 4 except the two silicones are taken or interpolated from H & H. where the original sources are given. Data for

TABLE 4. Pressure Coefficients of Viscosity

Lubricant	100 F		210 F		100 to 210 F	
	b_1	R_1	b_1	R_1	$10^3 c_1$	D
Petroleum Oils from Three Sources						
Rumania	2.6[a]	43[a]	1.8	10.0	−7.3	3
California	2.2	19.5	1.5	7.1	−6.4	9
Pennsylvania	1.4	7.8	1.2	4.2	−1.8	9
Three Fatty Oils						
Castor	1.07	4.1	0.84	2.7	−2.1	12
Lard	1.05	3.9	0.77	2.7	−2.5	7
Sperm	1.10	4.1	0.81	2.8	−2.6	16
Four Other Liquids						
Silicone (53)	1.41	8.6	1.34	4.9	−0.6	14
Silicone (55)	1.39	4.9	1.70	4.5	+2.8	28
Glycerine	0.43	1.8	0.34	1.6[a]	−0.8	6
Water	0.05	1.07	0.03	1.04[a]	−0.2[a]	0

Units of b_1 are %/100 psi; of c_1, %/100 psi dF. R_1 is a ratio, D a percentage, both at 1000 kg/sq cm.

[a] Extrapolated.

the silicones are from the ASME *Pressure-Viscosity Report* (1953), where Samples 53 and 55 are described.

In Table 4, b_1 denotes the pressure coefficient at 1 atm and R_1 is the isothermal relative viscosity at 1000 atm of pressure. Thus the value $R_1 = 4.1$ for castor oil at 100 F shows that its viscosity is 4.1 times as great at 1000 atm as at 1 atm, and the same temperature. To evaluate b_1, a curve is plotted for log R_1 against pressure and a straight line drawn tangent to the curve at its origin. The slope of the line, multiplied by 2.303, is the value of the pressure coefficient. The unit of b_1 is the reciprocal of the pressure unit chosen for the diagram. Multiply by 100 to obtain its percentage value, and by 10^4 for the percentage coefficient per 100 units of pressure. To convert the coefficient from per cent per metric atmosphere to its value in per cent per 100 psi, multiply by 100/14.22 or 7.03. The metric atm, 1 kg/sq cm or 14.22 psi, has been commonly used in high-pressure investigations.

Mean values from 100 to 210 F are shown in the last two columns. Thus c_1 is the mean pressure-temperature coefficient, or rate of change of b_1 with temperature. Note that c_1 would also equal the rate of change of the temperature coefficient with pressure, since the order of partial differentiation is immaterial. Finally D is the percentage drop in the curve for log R_1 against pressure, below a line drawn tangent at the origin. Normally D is evaluated at 1000 metric atm. It is a measure of the curvature of the logarithmic graph, too often assumed to be straight. The D values in Table 4 are averaged from 100 and 210 F. Water is the only liquid in the table that conforms to a straight line as far as 1000.

The testing of grease under high pressure is described by Hersey & Snyder (II: 1932) and on pages 46–47 of H & H. Just as with oils below their pour points (Norton, 1941), the rate of flow passes through a maximum and then, paradoxically, decreases with any further increase of inlet pressure.

Viscosity-Pressure Formulas. The familiar isothermal equation

$$\log R_1 = kp \tag{14}$$

was first proposed by Professor Barus when reporting his observations on marine glue (H & H, Refs. 1 and 2), although a linear equation proved to fit better. Here R_1 denotes the relative viscosity Z/Z_1 as before; p is the gage pressure and Z_1 the viscosity at $p = 0$, k being a constant. A more general equation was favored by H. Blok in his discussion of Hersey and Lowdenslager (V: 1950). It may be written

$$\log R_1 = n \log (1 + mp) . \tag{15}$$

In this equation both n and m are empirical constants. When n is set equal to 2, Eq. (15) reduces to Karlson's parabola of 1926 (H & H, Ref. 32). When p approaches zero, (15) takes the form of (14). Differentiating (15) shows that $b_1 = mn$, and from (14), $b_1 = 2.30k$. Applications of Eq. (15) led to n values from 2 to 12 (castor oil giving $n = 8$), depending on the pressure range over which the best fit is desired. Once b_1 and n are determined, the remaining constant is given by $m = b_1/n$. Conversely, for any arbitrarily chosen n, an optimum m can be found graphically or by trial. The empirical equation of Boelhouwer & Toneman (1957) can be simplified to read

$$\log R_1 = k[(1 + mp)^{1/2} - 1] , \tag{16}$$

where k and m are constants.

Another formula recognizing the upward convexity of the curves for log Z against pressure was introduced by Roelands and co-authors

(1963), Eq. (21) below. The corresponding isothermal equation states that log R_1 varies with some power y of the pressure p slightly less than unity. Here y is an empirical constant little affected by temperature.

Still other isothermal formulas are reviewed in H & H, together with "isometrics"—the pressure-temperature equations for constant viscosity. There, too, are seen many formulas proposed for the complete viscosity-pressure-temperature relation.

The Complete Z, p, t Relation. The complete equations can be written in compact form by letting R denote the ratio of Z at any gage pressure p and temperature t, to its value Z_0 at atmospheric pressure and some standard temperature t_0.

It was shown that a simple form of Z, p, t relation can be written by combining Reynolds' isobar with Barus' isotherm; thus in our present notation

$$R = e^{kp - m\theta}, \tag{17}$$

where k and m are empirical constants and θ denotes $t - t_0$. Such equations have been widely used, but Sternlicht (1958) obtained a closer fit by writing

$$Z = Ae^{\frac{\alpha}{t} + \beta p + \frac{\gamma p}{t}}. \tag{18}$$

Here γ is a third empirical constant, and A is equal to Z_0 times $e^{-\alpha/t_0}$; so that (18) can be expressed in terms of R as well as Z.

Bradbury's equation (1951) is one of the few that are not restricted to a direct proportionality between log Z and the pressure. In the usual case where there is no point of inflexion[1] his formula may be written

$$\log R = ce^{k/T} f(p) - ce^{k/T_0}, \tag{19}$$

where

$$f(p) = 1 + \frac{p}{P} - G\left[\frac{p}{P} - (1 - e^{-p/P})\right]. \tag{19a}$$

Here c, k, G, and P are all empirical constants, and T is the absolute temperature. Equation (19) reduces to the atmospheric isobar (11) upon substituting $p = 0$, which makes $f(p) = 1$, and to the T_0 isotherm upon substituting T_0 for T, allowing p to vary. Bradbury's paper should be consulted for the treatment of reversed curvature.

[1] The 1951 reference should have μ_p instead of μ_0 in its own Eq. (19). Accordingly Eq. (21) thereof should include a negative term with T_0 in place of T. The BP values on page 676 should contain a factor $(10)^7$, not $(10)^6$.

Appeldoorn's equation (1963) may be written, in our notation,

$$\log R = kp + (m + cp) \log \frac{t}{t_0} \cdot \tag{20}$$

This reduces to Slotte's Fahrenheit isobar when $p = 0$, and to the Barus isotherm when $t = t_0$. Correlations were found among the empirical constants k, m, and c by statistical studies of test data such that the number of constants needed in Eq. (20) could be progressively decreased, depending upon the degree of accuracy required. The author emphasized that the potential usefulness of Slotte's equation in Fahrenheit form had not been fully exploited.

Roelands and co-authors (1963) offered a complete equation of the form

$$\log Z = AT^{-x} + Cp^y T^{-x} + Dp^y + B \, . \tag{21}$$

When T is constant, this reduces to the isotherm cited. When $p = 0$, it reduces to the Cornelissen-Waterman isobar (12). It was shown that C and D can be expressed in terms of A, B, and y; together with the pressure and viscosity coordinates of the pole at which the isotherms intersect.

Solidification. Most fatty oils, and petroleum oils with sufficient wax content, can be solidified or rendered plastic by the application of a moderate pressure at constant temperature. Lard oil, for example, solidifies when the pressure in metric atmospheres reaches about 83 times the temperature C. This was first pointed out by Henry Shore (H & H, Ref. 45). The p, t graphs of apparent solidification may be considered isometrics for infinite viscosity. When projected down to atmospheric pressure they fall near the ASTM pour point. The consistency of lubricants while in this condition under high pressure was investigated by Professor Norton (1941).

Historical Note. Professor Barus of Brown University and the U. S. Geological Survey, whose viscosity-pressure formula of 1893 has been cited, was the grandfather of high-pressure research in this country. Professor Bridgman of Harvard was the first to note the influence of high pressure on the behavior of lubricants. As early as 1909 he reported the sluggish action of oil in transmitting hydrostatic pressure. The writer confirmed this effect a few years later in Professor Bridgman's laboratory by means of a rolling ball in a glass tube filled with oil. The tube was a strong piece of Scotch boiler gage-glass, connected by flexible copper tubing to a high-pressure hand pump. It is interesting that measurable effects could be observed without breaking the glass. The experiments were continued with the aid of a thin-walled steel tube; and the data published in 1916 were obtained with this simple equipment. The use of a rolling ball in slanted tube originated with A. E. Flowers (see H & H, Refs. 10, 12). His observations were at atmospheric pressure. A thick-walled viscometer was built in 1920 (Hersey & Shore, 1928).

The history of pressure-viscosity through 1952 has been recorded in H & H, where all original data were reduced to the same units for comparison, and followed by 198 references. This publication was based on 6 progress reports that led to our recommendation for a new investigation, conducted at Harvard University and published as a two-volume ASME report (1953). See also the summary by H. A. Hartung (1958).

Data from at least twelve additional investigations are now available, inviting coordination somewhat after the pattern set by H & H—

1. Data on three oils by Seeder & Boelhouwer, included by H. Blok in a general survey (1951).
2. The ASME Harvard project, forty-six oils (1953).
3. Data on fifty oils by E. Kuss (1953).
4. Testing of seven oils by Lundberg (1954).
5. The study of five polybutenes by Sargent (1955, 1958).
6. Three hydraulic fluids and a mineral oil compared by Cordiano & co-authors (1956).
7. Ten organic liquids investigated by Boelhouwer & Toneman (1957).
8. Fourteen pure compounds under API Project 42 (Griest et al., 1958; Lowitz et al., 1959).
9. One hydrocarbon, two mineral oils, and a polymer blend tested by Appeldoorn et al. (1962).
10. Twenty mineral oil fractions related to chemical composition (Roelands & co-authors, 1963).
11. Effect of pressure on eleven liquids of interest in elastohydrodynamic lubrication (Galvin et al., 1965).
12. Data on two mineral oils and three synthetics by Tabor & Winer (XIII: 1965).

Pressure-viscosity data are also available from several investigations of more limited scope: (*a*) Dane & Birch (1938); (*b*) Charron (1947—H & H p. 82); and (*c*) Niemann & Ohlendorf (XI: 1958). The data on molten glass by Dane & Birch were omitted from H & H because it was not then known that glass would be used as a lubricant. They determined the viscosity of one composition to a pressure of 2000 kg/sq cm using the long-capillary method of Hersey & Snyder (II: 1932). Pressure coefficients were found to be 2.1 per cent per 100 psi at 678 F, and 0.65 at 962 F. Charron compared glycerine with a mineral oil in a study of transient effects. Niemann compared two mineral oils with a polyether synthetic at 50 C, the latter showing a smaller pressure coefficient. See also Klaus and co-authors (1965).

6. CORRELATION OF PROPERTIES

The difficulty of high-pressure experimentation offers an incentive to predict the effect of pressure from observations under atmospheric conditions. Kiesskalt's correlation (H & H, Ref. 49) was notably successful. He plotted pressure coefficients against the viscosity-temperature slopes at atmospheric pressure. All points for the fatty oils tested fell reasonably near a single graph. The points for mineral oils came higher up on the chart, but again, although scattering, were clustered along a single graph. These two graphs took the form of straight lines when a logarithmic scale was chosen for the viscosity-temperature slope. By reference to these graphs, Kiesskalt was able to forecast the pressure coefficients for oils that had been tested only at atmospheric pressure. Note that the correlation found is not between the pressure coefficient and the temperature coefficient, but between the pressure coefficient and the viscosity-temperature slope. The latter is equal to the product of the temperature coefficient by the viscosity. Thus the pressure coefficient depends on two parameters, which merge into one when multiplied together. Kiesskalt's correlation may be written

$$b_1 = k + m \log dZ_1/dt \; ; \tag{22}$$

where for mineral oils, $k = 0.181$ with $m = 0.0625$; and for fatty oils, $k = 0.123$ with $m = 0.027$. These constants give b_1 in per cent per metric atmosphere provided dZ_1/dt is expressed in millipoises per degree F (mp/dF); 1 millipoise being $\frac{1}{10}$ of a centipoise. Only the numerical value of dZ_1/dt, without regard to its negative sign, is to be used in Eq. (22). See H & H, pages 59–63, for further discussion. The above constants are based on Figs. 19 and 21 therein. They can be multiplied by 7.03 to express b_1 in per cent per 100 psi.

Matteson and Vogt correlated the pressure coefficient for petroleum oils against two parameters, the viscosity at atmospheric pressure and the viscosity index. See H & H pages 59 and 70, including Ref. 121. We infer that their coefficient is a mean value up to pressures of about 12,000 psi, since the data are credited to Kiesskalt and Suge. A portion

TABLE 5. Matteson & Vogt's Correlation
(Revised b_1 Values in %/100 psi)

Z_1 cp	Viscosity Index 0	50	100
5	1.0	0.9	0.9
10	1.2	1.2	1.1
50	1.6	1.6	1.4
100	1.9	1.7	1.6
500	2.5	2.1	1.9
1000	2.7	2.3	2.0

of the Matteson & Vogt chart is represented by Table 5. Values were raised 10 per cent to convert from mean to initial coefficients.

The correlation methods of Blok, Bradbury, Cameron, and Cragoe; also Dow, Grunberg, Kadmer, Sanderson, Weber, and a modified "volume correlation," are reviewed on pages 57–58, 69–70 of H & H. New correlations have since been offered by Clark (1956), Grunberg (1954), and Dow (1955); by Wooster discussing A. E. Bingham's paper (1957); by Hartung (1958), Chu & Cameron (1962), Appeldoorn (1963), and by Roelands and co-authors (1963). Hartung's method is based expressly upon the ASME data (1953). Appeldoorn utilized a computer analysis of data from 122 oils. Roelands correlated pressure-viscosity with chemical constitution.

P. W. Bridgman, Bondi, Eyring, and others have studied the relation of viscosity to molecular structure; see H & H pages 72–73, as well as Refs. 30, 94, and 184–185 therein.

7. NON-NEWTONIAN LUBRICANTS

Greases are the best-known non-Newtonian lubricants. Other examples are polymer-thickened oils; oils below their pour points; oils solidified under high pressure; and the solid lubricants, like graphite or talc. Common greases are mixtures of soap and oil, the soap acting as a carrier for the oil, which may separate out, when warm, to act as a liquid lubricant. Grease can serve also as a plastic lubricant. The performance of grease-lubricated bearings is reviewed in Chapters VI, VII, and X.

Several investigators, notably Blott & Samuel (1940), Singleterry & Stone (1951), Mahncke (1955), Brunstrum (1956), and Summers-Smith (1957), have found that greases act very much like an ideal "Bingham plastic" (Bingham, 1922) except at the lowest rates of flow. They are characterized by the yield shear stress τ_0 and mobility m. These properties are the constants of an equation connecting the shear rate R with the shear stress τ, namely

$$R = m(\tau - \tau_0) , \tag{23}$$

or conversely

$$\tau = \tau_0 + Z_0 R . \tag{24}$$

Here Z_0 is the mobility reciprocal $1/m$, also called the "plastic viscosity." It reduces to the ordinary viscosity Z when $\tau_0 = 0$ or $R = \infty$. The theory of capillary flow was developed by Edgar Buckingham (II: 1921) for the Bingham plastic and by B. Rabinowitsch (1929) for non-Newtonian materials in general. The outstanding characteristic of a grease, or other plastic solid, is that little or no flow can be observed until a shear stress greater than the yield value τ_0 is applied. We recall, for example, a pressure-gun grease having a yield stress of about $\frac{1}{20}$ psi. The "worked penetration" value was 360 by ASTM Method D-217-27. Its flow properties under high pressure are shown by Fig. 2 and Table 28 of H & H. Plastic materials are *thixotropic*, losing some of their initial stiffness when sheared or worked, but regaining it after a period of rest (McKee & White, 1948; Beerbower, 1960).

Rheology. The Bingham plastic is the simplest of the many non-Newtonian materials whose investigation is the goal of Rheology, "science of flow." The Ree-Eyring fluid (1955) is favored as a mathematical model by J. C. Bell (1962). Its viscosity, like that of a Bingham material, decreases with shear rate, but it has no initial yield stress. Another of great interest is the Maxwell body as described by Reiner (1932), Mason (1948), Hutton (1954), Bland (1960), and others. Such material reacts elastically to a stress quickly applied, but flows like a viscous liquid after a certain *relaxation time.* Viscoelasticity may be demonstrated with a sample of "silicone putty." Rolled up into a ball and dropped on the lecture table, it bounces readily. Left to itself the remainder of the hour, it spreads flat as a pancake. Any complete survey of the properties of non-Newtonian lubricants would involve a considerable excursion into rheology—see, for example, Hersey (1932), Dow (1947), and Klaus (1955); also Eirich (1956), Reiner (1949–1960), Selby (1958), Bondi (1960), and Wilkinson (1960);

Proceedings of the Fourth International Congress on Rheology (Lee & Copley, 1965); and current Transactions of the Society of Rheology.

A lively problem of lubricant research is to determine the safe limits of shear stress, or rate of shear, beyond which ordinary oils become non-Newtonian, especially under high pressure. A limiting figure proposed by H. Bondi (1945) for straight petroleum oils, and upheld by Professor Blok (1946), is one-half million dynes per sq cm of shear stress, something over 7 psi. The experimental difficulty is to separate true non-Newtonian properties from heat effects (Chapter III). Needs (1951), Barber (1955), Porter (1965) and co-authors, using rotation viscometers with close clearances and especially accurate workmanship, found no deviation from Newton's law in straight petroleum oils. This interesting problem has been further analyzed by Cole and co-authors (1955), indicating optimum conditions for future experiments. All investigators found non-Newtonian characteristics in polymer-thickened oils. Klaus and Fenske (1955) described the breakdown of non-Newtonian lubricants under shear treatment, resulting in a permanent loss of viscosity. Investigators differ in advocating capillary flow, rotational viscometry, or the vibrating cylindrical crystal (Mason, 1947; Barlow & Lamb, 1959; Appeldoorn and co-authors, 1962) as the most promising technique; but all agree on the importance of the problem.

8. SOLID LUBRICANTS

Sometimes incorporated in the bearing material, solid lubricants are usually applied in powdered form with the aid of a "binder," or suspended in a volatile carrier (Boyd I: 1945). They have shown good results as a remedy for "fretting corrosion," and in extending the operation of bearings to high contact pressures and high temperatures where ordinary lubricants would break down, and to high vacuum conditions. Papers by W. E. Campbell (1953), Fortrat (1953), and by Bisson and co-authors (1957) typify the growing literature in this field. Reports by E. P. Kingsbury (1958) and Stupf (1958) describe solid films at high temperature, including molybdenum disulfide and related lubricants. See also Fuller (I: 1956), Braithwaite (1964), Hopkins & Gaddis (1965), and Devine & co-authors (1965).

Few data are available on conventional bulk properties, the research literature being taken up chiefly with the composition and structure of solid lubricants, effective methods of application, and practical operating performance. Summarizing five types of solid film lubricants, Rabinowicz (1965) indicates that:

1. Both graphite and molybdenum disulfide, applied with a binder, make good general-purpose lubricants of low friction and reasonably long life; coefficient f, 0.06 to 0.12.

2. Teflon with binder offers [low friction and] better resistance to detrimental liquids, but shorter life.

3. Graphite and molybdenum disulfide applied directly by rubbing offer lower friction, but still shorter life; $f = 0.04$ to 0.10.

4. Soft metals (lead, indium, cadmium) as lubricants give higher friction and shorter life than resin-bonded types; $f = 0.15$ to 0.30.

5. Phosphate and other chemical coatings help prevent galling, but with high f (around 0.20).

Some investigators consider the action of solid lubricants best explained by boundary friction (Chapter XIII), especially when the solid form results from chemical reaction between liquid lubricant and bearing surfaces. Others picture the solid layer as a relatively thick one, of low internal shear resistance because of its lamellar or laminar structure.

9. IMPROVEMENT OF LIQUID LUBRICANTS

The advent of petroleum oils offered an improvement over the fatty oils in their greater stability at elevated temperatures, and because any desired viscosity grade could readily be obtained; but they are inferior to fatty oils in respect to "oiliness" (Chapter XIII) and viscosity-temperature slope.

Early improvements were effected by blending or "compounding." It was soon discovered in Britain that improved oiliness (reduction of friction) could be obtained by adding as little as 1 or 2 per cent of commercial oleic acid to the mineral oil. It was gradually found that many different qualities could be improved by small additions of meritorious chemicals, which came to be called "additives." Best known are the oxidation inhibitors, "pour-point" depressants, viscosity-index improvers, and EP (extreme pressure) agents (Barwell I: 1956, Chapter 18). At the same time improvements were being made in refining processes, notably by solvent refining.

Temperature Coefficients. The history of VI improvers dates from Albert Einstein's derivation of the formula for the viscosity of a suspension—see the account by Lewis & Squires (1934). Differentiating the formula showed that the temperature coefficient of viscosity of the suspension would be the same as that of the base liquid. It was seen that the temperature coefficient could be kept low by choosing a low viscosity base, while the viscosity level of the suspension could be raised by increasing the aggregate volume of the suspended particles. The next move was to find a source of suitable particles, not necessarily solid, but acting like solids in suspension. The source was found in petroleum itself. High molecular weight polymers were obtained which behaved, qualitatively, like the spherical particles of Einstein's equation, although far from spherical in shape. The evolution of the idea is shown by patent literature originating in Germany, but soon extending to the United States. This was a surprising development since it had previously been believed that we were near the end of the road—that it was inherently impossible to greatly decrease the viscosity-temperature effect without sacrificing viscosity. Is it too much to hope for, that a breakthrough can be found leading to lubricants in which a low viscosity-temperature effect is combined with a high viscosity-pressure effect?

Extreme Pressure Agents. The development of EP additives for protection from scuffing or seizure would, if completely told, make an equally interesting story. Some of the more powerful types include sulfur, chlorine, or phosphorus. Iron compounds are formed by chemical reaction after certain operating temperatures are reached. These agents protect steel surfaces from welding together, much as the blacksmith is frustrated by sulfur in the coal until he makes a clean bed of coke. Gear teeth have been protected, their load capacity increased, and their useful life extended by this discovery. Simple "bench tests" were soon devised in efforts to rate the load capacity of lubricants in advance of full scale operation. The Almen, four-ball, SAE, and Timken tests are among those frequently mentioned. Many are referenced and some described in Chapters X and XI. Little correlation with service performance has been reported, yet bench tests are not without merit for preliminary screening and for scientific study.

The proper use of bench tests has been described by J. O. Almen,[1] after examining the records of 10,000 gear failures:

> First, establish the relative order of merit of a series of reference lubricants by actual service operation. Then run these lubricants through all available bench tests, each under various operating conditions. If any combination is found that rates the reference lubricants in the same order they were rated in service, that is the combination to adopt in the future rating of new lubricants for the particular type of service.

Synthetic Lubricants. The development of improved lubricants is advancing in two directions: (1) new additives, and (2) synthetic

[1] Page 5 in the *Survey on Gear Lubrication,* USN Engineering Experiment Station, Annapolis, Md., July 17, 1953.

lubricants. A collection of chapters by fifteen contributors, edited by Gunderson & Hart (1962), describes the development and use of synthetic lubricants. Types covered are mainly polyglycols, phosphate esters, dibasic acid esters, chlorofluorocarbon polymers, silicones, silicate esters, neopentyl polyol esters, and polyphenyl ethers. The physical properties treated include viscosity, pour point, compressibility, volatility, heat transfer characteristics, and the like. See also the review by Bickerton (1962).

Changing the Oil. Many people seem to think automobiles are the only contraptions that require lubrication. Rare is the lecturer on lubrication research who can dodge the question "How often must I drain the crankcase?" Every thousand miles was the rule in the old days of dusty roads. The first company advertising a 2000-mile oil was frowned upon by the rest of the petroleum industry.

One of our mechanic friends, with a fine record of experience in service stations, rarely changes his oil. "When conditions get that bad," he explained, "it's time to trade in the old bus for a new model." A more sophisticated view was taken by Paul D. Foote (1937) in his chapter entitled "Let the physicist change your oil." The question has been discussed at length by the Society of Automotive Engineers (Kalinowski et al., 1962).

10. SUMMARY

The physical properties of lubricants have been described in this chapter with emphasis on bulk properties, like viscosity as affected by pressure and temperature. Formulas, charts, and correlation methods have been reviewed.

Synthetic and solid lubricants are mentioned. The value of additives has been indicated. Little reference is made to the chemistry of lubricants, or to their misbehavior in service. Such topics occupy a large part of the effort of petroleum technologists and ASTM committees (Klaus, 1965). Readers not limiting their outlook to theory and research will be interested in A. F. Brewer's book (1955) on basic lubrication practice, and in Schilling's (1962) on automotive lubricants. The latter compares American, British, and European experience.

A field of research lies ahead in the further development of improved lubricants combining, say, high pressure-coefficients of viscosity with low temperature-coefficients. And more knowledge is required about non-Newtonian properties, with especial reference to transient or time effects.

REFERENCES

American Society for Testing Materials (1957), *ASTM Viscosity Tables.* STP No. 43-B; 60 pp.

Appeldoorn, J. K. (1963), "A Simplified Viscosity-Pressure-Temperature Equation." *SAE Paper* 709A; 7 pp. (a) *JSAE* **71,** 108.

Appeldoorn, J. K., Okrent, E. H., & Philippoff, W. (1962), "Viscosity and Elasticity at High Pressures and High Shear Rates." *Proc. API* **42** (III), 163–74.

ASME Research Committee on Lubrication (1953), "Pressure-Viscosity Report." *Viscosity and Density of Over 40 Lubricating Fluids of Known Composition at Pressures to 150,000 Psi and Temperatures to 425 F.* ASME, New York; 2 volumes.

Barber, E. M., Muenger, J. R., & Villforth, F. J. (1955), "A High Rate of Shear Viscometer." *Analytical Chemistry* **27,** 425–9.

Barlow, A. J. & Lamb, J. (1959), "The Viscoelastic Behaviour of Lubricating Oils under Cyclic Shearing Stress." *PRS A* **253,** 52–69.

Barnes, R. S. & Fainman, M. Z. (1957), "Synthetic Ester Lubricants." *LE* **13,** 454–8.

Barr, Guy (1931), *A Monograph of Viscometry.* Oxford University Press, London; 318 pp.

Barr, Guy (1934), "Viscosity and Its Expression." *World Petroleum Congress,* London, 1933; *Proc.,* 508–11. (a) Cragoe, 529–33.

Barr, Guy (1937), "The Determination of the Viscosity of Oils at High Temperatures." *GDLL* **2,** 217–21.

Beerbower, Alan (1960), "Effect of Rate of Shear on Viscosity." Pages 17–22 in *RV.* (a) Pavlov & Vinogradov, *LE* **21,** 479–84 (1965).

Bell, J. C. (1962), "Lubrication of Rolling Surfaces by a Ree-Eyring Fluid." *TASLE* **5,** 160–71.

Bickerton, R. G. (1962), "Some Aspects of Synthetic Lubricants." *Sci. Lubn.* **15,** 82–90.

Bingham, A. E. (1957), "Some Problems of Fluids for Hydraulic Power Transmission." *Proc. IME* **165,** 254–61; 266–77. See Table 4, Fig. 21, and discussion by R. C. Wooster.

Bingham, E. C. (1922), *Fluidity and Plasticity.* McGraw-Hill, New York, 440 pp.

Bisson, E. E., Johnson, R. L., & Swikert, M. A. (1957), "Friction, Wear and Surface Damage of Metals as Affected by Solid Surface Films: A Review of NACA Research." Paper 31, *CLW,* 384–91; 363–4, 591–2, 594, 805, 807, 812.

Bland, D. R. (1960), *The Theory of Linear Viscoelasticity.* Oxford University; Pergamon Press, London; 125 pp.

Blok, H. (1946), "Viscosity of Lubricating Oils at High Rates of Shear." *Sixth ICAM,* Paris, Section 3. (a) *De Ingenieur,* No. 21 (1948).

Blok, H. (1951), "Viscosity-Pressure-Temperature Relationships, Their Correlation and Significance for Lubrication." *Proc. Third World Petrol. Cong.*, E. J. Brill, Leyden, Section VII, 304–19.

Blott, J. F. I. & Samuel, D. L. (1940), "Flow Characteristics of Lime-Base Greases." *IEC* **32,** 68–72.

Boelhouer, J. W. M. & Toneman, L. H. (1957), "The Viscosity-Pressure Dependence of Some Organic Liquids." Paper 38, *CLW,* 214–8; 201, 322, 749–50; Pl. 9.

Bondi, A. (1945), "Flow Orientation in Isotropic Fluids." *JAP* **16,** 539–44.

Bondi, A. (1960), "Rheology of Lubrication and Lubricants." Pages 443–78 in Eirich, *Rheology—Theory and Applications,* **3.**

Bradbury, Donald; Mark, Melvin, & Kleinschmidt, R. V. (1951), "Viscosities and Densities of Lubricating Oils from 0 to 150,000 Psig and from 32 to 425 F." *T* **72,** 667–76.

Braithwaite, E. R. (1964), *Solid Lubricants and Surfaces.* Pergamon, New York; 286 pp.

Brewer, A. F. (1955), *Basic Lubrication Practice.* Reinhold, New York; 286 pp.

Bridgeman, O. C. (1960), editor, *The Role of Viscosity in Lubrication.* Proc., Symposium, 1958; ASME, New York; 108 pp.

Bridgeman, O. C. & Aldrich, E. W. (1965), "Estimation of Temperature Rise on Adiabatic Compression of Petroleum Lubricating Oils." *ASME* 65-LUBS-16; 7 pp. (a) *ME* **87,** 60 (Aug.).

Brunstrum, L. C. & Leet, R. H. (1956), "Capillary Viscometry of Lubricating Grease." *LE* **12,** 316–22.

Cameron, A. (1945), "The Isothermal and Adiabatic Compressibilities of Oil." *JIP* **31,** 421–7.

Campbell, W. E. (1953), "Solid Lubricants." *LE* **9,** 195–200.

Chu, P. S. Y. & Cameron, A. (1962), "Pressure Viscosity Characteristics of Lubricating Oils." *JIP* **48,** 147–55.

Chu, P. S. Y. & Cameron, A. (1963), "Compressibility and Thermal Expansion of Oils." *JIP* **49,** 140–5.

Clark, O. H. (1956), "Prediction of Lubricating Oil Viscosities at High Pressures." *T* **78,** 905–8.

Cole, J. A., Petersen, R. E., & Emmons, H. W. (1955), "Experiments with a Rotating-Cylinder Viscometer at High Shear Rates." *NACA TN* 3382; 31 pp.

Cordiano, H. V., Cochran, E. P., Jr., & Wolfe, R. J. (1956), "A Study of Combustion Resistant Hydraulic Fluids as Ball Bearing Lubricants." *LE* **12,** 261–6.

Cornelissen, J. & Waterman, H. L. (1955), "The Viscosity-Temperature Relationship of Liquids." *Chem. Eng.* **4,** 238–46.

Cragoe, C. S. (1929), *Thermal Properties of Petroleum Products.* NBS M **97;** 48 pp. (a) For Cragoe on "liquidity" see Barr, 1934 (a).

Dane, E. B., Jr., & Birch, Francis (1938), "The Effect of Pressure on Boric Anhydride Glass." *JAP* **9,** 669–74.

Dean, E. W., Bauer, A. D., & Berglund, J. H. (1940), "Viscosity Index of Lubricating Oils." *IEC* **32.1**, 102–7. (a) *Chem. Met. Eng.* **36,** pp. 618–9 (1929), by E. W. Dean & G. H. Davis.

Devine, M. J., Lamson, E. R., Cerini, J. P., & McCartney, R. J. (1965), "Solids and Solid Lubrication." *LE* **21,** 16–26.

Dow, R. B. (1947), "The Rheology of Lubricants." *J. Colloid Sci.* **2,** 81–91. (a) *ME* **77,** 46, 1006–7 (1955). (b) Other references in H & H.

Duff, A. W. (1897), "Empirical Formulae for Viscosity as a Function of Temperature." *Phys. Rev.* **4,** 404–10.

Eirich, F. R., editor (1956), *Rheology—Theory and Applications.* Academic Press, New York; vol. 1, 761 pp; vol. 2 (1958), 591 pp; vol. 3 (1960), 680 pp.

Ellis, E. G. (1957), "Viscosity-Temperature Relations." *Petroleum (London),* **20,** 294–8.

Erk, S. (1931), "Des huiles lubrifiants à basse température." *Compte-Rendu, Congrès du Graissage,* Strasbourg, Éditions Techniques, Paris, 258–9. (a) *Physik. Z.* **38** (1937), 449–53.

Erk, S. & Eck, H. (1936), "Über die Temperaturabhängigkeit der Zähigkeit von Schmierölen." *Physik. Z.* **37,** 113–9.

Ewing, C. T. & Grand, J. A. (1951), "Measurements of the Thermal Conductivity of Sodium and Potassium." *Report* 3835, Naval Res. Lab., Washington, D. C. 18 pp.

Foote, P. D. (1937), "Let the Physicist Change Your Oil." Chap. III, 47–92, in *Physics in Industry.* Am. Inst. Phys., New York; 290 pp.

Fortrat, M. R. (1953), "Sur la lubrification solide." *Helv. Phys. Acta* **26,** 191–8.

Fortsch, A. R. & Wilson, R. E. (1924), "The Viscosity of Oils at High Temperatures." *IEC* **16,** 789–92.

Fuller, D. D. (1949), Chmn. Educational Com., *Physical Properties of Lubricants. ASLE,* Chicago; 72 pp. (a) 2nd ed., 1951; 69 pp.

Galvin, G. D., Naylor, H., & Wilson, A. R. (1965), "The Effect of Pressure and Temperature on Some Properties of Fluids of Importance in Elastohydrodynamic Lubrication." Paper 14, *LWG,* 10 pp.

Geniesse, J. C. (1937), "Viscometry Applied to Petroleum Products." *GDLL* **2,** 308–16.

Geniesse, J. C. (1956), "A Comparison of Viscosity Index Proposals." *Bull. ASTM* No. 215, TP 137, 81–4. (a) Wright, W. A., TP 140, 84–6; (b) Klaus, E. E. & Fenske, M. R., *TP* 143, 87–94.

Geniesse, J. C. & Delbridge, T. G. (1932), "Variation of Viscosity with Temperature." *Proc. API* **13** *M* (III), 56–8.

Griest, E. M., Webb, W., & Schiessler, R. W. (1958), "Effect of Pressure on Viscosity of Higher Hydrocarbons and their Mixtures." *J. Chem. Phys.* **29,** 711–20.

Griffin, H. K. (1925), "Density of a Lubricating Oil at Temperatures from −40° to +20° C." *IEC* **17,** 1157–8.

Grunberg, L. (1954), "Viscosity and Density of Lubricating Oils at High

Pressures." *Proc. Second Int. Cong. on Rheology,* Oxford, 1953. Butterworths, London; 437–44.

Gunderson, R. C. & Hart, A. W., eds. (1962), *Synthetic Lubricants.* Reinhold, N. Y., 1962; 497 pp. Collected contributions.

Hartung. H. A. (1956), "Density-Temperature-Pressure Relations for Liquid Lubricants." *T* **78,** 941–7.

Hartung, H. A. (1958), "The Pressure-Viscosity Effect: Background." *T* **80,** 1097–8.

Herschel, W. H. (1922), "The Change in Viscosity of Oils with the Temperature." *IEC* **14,** 715–23.

Hersey, M. D. (1916), "The Theory of the Torsion and the Rolling Ball Viscosimeters and Their Use in Measuring the Effect of Pressure on Viscosity." *JWAS* **6,** 525–30.

Hersey, M. D. (1932), "Future Problems of Theoretical Rheology." *J. Rheol.* **3,** 196–204. (a) Reiner, M., 245–56.

Hersey, M. D. & Hopkins, R. F. (1954), *Viscosity of Lubricants Under Pressure. Coordinated Data from Twelve Investigations.* ASME, New York; 87 pp., with bibliography.

Hersey, M. D. & Shore, Henry (1928), "Viscosity of Lubricants Under Pressure." *ME* **50,** 221–32.

Hopkins, V. & Gaddis, D. (1965), "Development of Solid Film Lubricants for Use in Space Environments." *LE* **21,** No. 2, 52–8.

Hutton, J. F. & Matthews, J. B. (1954), "Viscoelastic Behaviour of Lubricating Greases." *Second Int. Cong. on Rheology,* Oxford, 1953; *Proc.,* 408–13.

Hyde, J. H. (1920), "On the Viscosities and Compressibilities of Liquids at High Pressures." *PRS* **97,** 240–59. (a) H & H 78, Ref. 17.

Kalinowski, M. L. et al. (1962), "When Should Oil Be Changed?" *JSAE* **70,** 73–6; Paper No. 439 C; 4 pp.

Keulegan, G. H. (1930). "Investigation of Damping Liquids for Aircraft Instruments—I." *NACA TR* No. 299, in 14th Annual Report; 24 pp. (a) Housman, M. R. & Keulegan, G. H.—II. No. 398, in *TR* (1931).

Kingsbury, E. P. (1958), "Solid Film Lubrication. . . ." *TASLE* **1,** 121–3.

Klaus, E. E. (1965), "Properties of Lubricants." *JBE, T* **87,** 797–800. (a) "Pressure Viscometer." ASLE **65,** LC-25.

Klaus, E. E. & Fenske, M. R. (1955), "Viscosity-Shear Characteristics. . . ." *LE* **11,** 101–8. (a) Bestul, A. B. & Belcher, H. V., **9,** 1011–4 (1953).

Klaus, E. E., Hersh, R. E. & Pohorilla, M. J. (1958), "Slope Index. . . ." *LE* **14,** 439–48.

Klaus, E. E. & O'Brien, J. A. (1964), "Precise Measurement and Prediction of Bulk Modulus Value of Fluids and Lubricants." *JBE, T* **86,** 469–74.

Kraussold, H. (1932), "Die spezifische Wärme von Mineralölen." *Petrol. Z.* **28,** No. 3, 1–7.

Kuss, E. (1953), "Einige physikalische Eigenschaften von Erdölen bei hohen Drücken." *Erdöl und Kohle* **6,** 266–70. (a) *Fourth World Petrol. Congress,*

Section V, Paper No. 5 with R. G. Schultze, *Proc.*, Carlo Colombo, Rome, 101–17 (1955).

Lee, E. H. & Copley, A. L., editors (1965), *Proceedings of the Fourth International Congress on Rheology* (Providence, 1963). Wiley, New York: Part 1, 384 pp.; part 2, 728 pp.; part 3, 650 pp.; part 4, 648 pp.

Leslie, E. H. & Geniesse, J. C. (1927), "Petroleum, Petroleum Products and Commercial Oils of Mineral Origin." *International Critical Tables*, McGraw-Hill, New York, **2**, 136–62. (a) Kinematic Viscosity Chart, MacCoull, 147.

Lewis, W. K. & Squires, L. (1934), "The Structure of Liquids and the Mechanism of Viscosity." *Proc. API* **15** *M* (III), 29–37.

Lowitz, D. A., Spencer, J. W., Webb, W., & Schiessler, R. W. (1959), "Temperature-Pressure-Structure Effects on the Viscosity of Several Higher Hydrocarbons." *J. Chem. Phys.*, **30**, 73–83.

Lundberg, Sven (1954), "A Method for Approximate Determination of Viscosity-Pressure-Temperature Relationships for Oils. Results from an Investigation of Compressibility and Viscosity." *JIP* **40**, 104–15.

Mahncke, H. E. & Tabor, W. (1955), "A Demonstration of Bingham Type Flow in Greases." *LE* **11**, 22–8.

Mason, W. P. (1947), "Measurement of the Viscosity and Shear Elasticity of Liquids by Means of a Torsionally Vibrating Crystal." **69**, 359–70. (a) *J. Colloid Sci.*, **3**, 147–62 (1948). (b) *Phys. Rev.* **75**, 936–46 (1949).

McKee, S. A. & White, H. S. (1948), "A Worker Consistometer. . . ." *ASTM Bull.* No. 153 (Aug.) 90–8.

Mitchell, C. A. (1927), "Animal and Vegetable Oils, Fats, and Waxes." *Internat. Critical Tables*, **2**, McGraw-Hill, New York; 138–61.

Moreton, D. H. (1964), "Liquid Lubricants." Chap. 7. pp. 175–201 in *ABT*.

National Bureau of Standards (1936), *National Standard Petroleum Oil Tables.* Circular C, Washington, D. C.; 175 pp. (a) *ASTM Standards on Petroleum Products and Lubricants*, Philadelphia; 166, 336–344 (1937) and later issues.

Needs, S. J. (1942), Discussion on "Heat Conditions in Bearings." ***T* 64**, 453.

Needs, S. J. (1951), "The Kingsbury Tapered-Plug Viscometer for Determining Viscosity Variations with Temperature and Rate of Shear." *ASTM STP* **111**, 24–44.

Norton, A. E., Knott, M. J., & Muenger, J. R. (1941), "Flow Properties of Lubricants Under High Pressure." *T* **63**, 631–43.

Porter, R. S. & Johnson, J. F. (1965), "Viscosity Measurements near a Million Seconds^{-1}." *T. Soc. Rheology* **9**, Pt. 2, 49–55.

Rabinowicz, Ernest (1965), *Friction and Wear of Materials.* Wiley, New York, 244 pp. (a) Reviewed by K. L. Johnson, *Eng.* **200**, 509 (1965).

Rabinowitsch, B. (1929), "Über die Viscosität von Solen." *Z. physik. Chem. A* **145**, 1–26.

Ree, T. & Eyring, H. (1955), "Theory of Non-Newtonian Flow. I, II." *JAP* **26** (7), 793–809.

Reiner, Markus (1932), "Outline of a System of Rheological Theories." *J. Rheol.* **3**, 245–56. (a) *AMR* **4**, 202–4 (1951).

Reiner, Markus (1949), *Deformation and Flow; an Elementary Introduction to Theoretical Rheology*. H. K. Lewis, London; 346 pp. (a) *Twelve Lectures on Theoretical Rheology*, North Holland Pub. Co., Amsterdam, 162 pp. (b) 3rd ed., Interscience, New York (1960); 158 pp.

Reiner, Markus (1956), "Phenomenological Macrorheology." Pages 9–62 in Eirich, *Rheology*, vol. 1, Academic Press, New York.

Riedel, L. (1950), "Bestimmung der Wärmeleitfähigkeit und spezifischen Wärme verschiedener Mineralöle." *Chemie-Ingenieur Technik* **22**, 107–8.

Roelands, C. J. A., Blok, H., & Vlugter, J. C. (1964). "A New Viscosity-Temperature Criterion for Lubrication Oils." *ASME* 64-LUB-3; 16 pp. *ME* **86** (Dec.), 74.

Roelands, C. J. A., Vlugter, J. C., & Waterman, H. I. (1963), "The Viscosity-Temperature-Pressure Relationship of Lubricating Oils and Its Correlation with Chemical Constitution." *JBE, T* **85**, 601–10.

Ross, Sydney (1950), "Variation with Temperature of Surface Tension of Lubricating Oils." *NACA TN* 2030; 14 pp.

Rost, U. (1955), "Das Viskositäts-Temperatur-Verhalten von Flüssigkeiten." *Kolloid-Z.* **142**, 132–50.

Sargent, L. B., Jr. (1958), "The Effect of Pressure and Molecular Weight Upon the Viscosity of Polybutenes." *LE* **14**, 298–301, 309. (a) op. cit. **11**, 249–54 (1955).

Schilling, Alphonse (1962), *Les huiles pour moteurs et le graissage des moteurs*. L'Institut Francaise du Petrôle; Editions Technique, Paris; 2 volumes, 351 and 411 pp.

Schrock, V. E. & Starkman, E. S. (1957), "Thermal Conductivity of Aircraft Engine Lubricants at Low Temperature." *LE* **13**, 358, 393–8.

Selby, T. W. (1958), "The Non-Newtonian Characteristics of Lubricating Oils." TASLE **1**, 68–81. (a) LE **14**, 222.

Singleterry, C. R. & Stone, E. E. (1951), "Rheological Properties of a Lubricating Grease." *J. Colloid Sci.*, **6** (2), 171–89.

Slotte, K. F. (1890), "On the Internal Friction of Liquids." *Öfvers af Finska Vetensk. Soc. Förhandl* **32**, 116–49. (a) *Beibl. Ann. Phys. Chem.* **16**, 182–5 (1892).

Soda, N. & Miyakawa, Y. (1948), Molten Metals as Lubricants (Japanese). *Report, Inst. of Science and Technology*, Tokyo, **2**, 80–90.

Sternlicht, B. (1958), "Influence of Oil Pressure and Temperature on Oil Viscosity in Thrust Bearings." *T* **80**, 1108–12.

Stupf, B. C. (1958), "Molybdenum Disulfide and Related Solid Lubricants." *LE* **14**, 159–63.

Suge, Y. (1937), "Physical Properties of Lubricants." *GDLL* **2**, 412–7.

Summers-Smith, D. (1957), "Flow Properties of Lubricating Grease." Paper 25, *CLW*, 519–23; 494, 797.

Swindells, J. F., Coe, J. R., Jr., & Godfrey, T. B. (1952), "Absolute Viscosity of Water at 20° C." RP 2279, *NBS J. Res.*, **48**, 1–31.

Velikovski, D. S. et al. (1941), Viscosity of Oils at Low Temperatures (Russian). *Conference on Viscosity of Liquids and Colloidal Solutions.*

Akademiia Nauk, SSSR, Institut Mekhaniki-Inzhenernyi, Moscow **1,** 161–89; **2** (1944) 128–87, 222–45 (eleven papers). (a) *Petroleum* (London), **10** (Mar., Apr. 1947) 62–3, 75, 87, 89.

Vereshchagin, L. F., Semerchan, A. A., Firsov, A. I., Galakmionov, V. A., & Filler, F. M. (1956), Some Investigations on a Hydrodynamic Jet of Liquid Flowing from a Nozzle at Pressures Up to 1500 atm (Russian). *Zh. tekh. Fiz.* **26,** No. 11, 2570–7 (1956). (a) Transl. No. T 4832, Ministry of Aviation, Grt. Brit.

Vogel, H. (1922), "Die Bedeutung der Temperatur Abhängigkeit der Viskosität f.d. Beurteilung von Oelen." *Z. angew. Chem.* **35,** 561–3.

Vogelpohl, G. (1954), "Die Bedeutung des Viskositätsverhaltens der Schmiermittle für Betrieb und Bau der Maschinen." *Brennstoff-Chemie* **35,** No. 23/24, 363–8.

Walther, C. (1931), "Über die Auswertung von Viskositätsangaben." *Erdöl u. Teer* **7** (Aug. 25) 382–4. (a) *Maschinenbau* **10,** 67–5. (b) *World Petroleum Congress,* London, 1933, *Proc.* **2** (1934) 419–20.

Webber, H. A., Goldstein, David, & Fellinger, R. C. (1955), "Determination of the Thermal Conductivity of Molten Lithium." *T* **77,** 97–102.

Wilcock, D. F. (1944), "Viscosity-Temperature Coefficient." *ME* **66,** 739; **67,** 201–2 (1945).

Wilkinson, W. L. (1960), *Non-Newtonian Fluids.* Pergamon Press, London, 138 pp.

Woog, P. (1926), *Contribution à l'étude du graissage.* Delagrave, Paris, 277 pp.

Zisman, W. F. (1949), "Viscosities and Densities of Lubricating Fluids from −40 to 700 F." *T* **71,** 561–74.

chapter V Dimensions and ZN/P

1. Physical equations. 2. Dimensional homogeneity. 3. The "Pi-theorem." 4. Derivation of ZN/P relations. 5. The Sommerfeld number. 6. Bearing lubrication under various conditions. 7. Application to other machine elements. 8. Thermal applications. 9. General considerations. 10. Summary.

What is so magical about ZN/P that makes it come up every day in lubrication literature? One reason is that it describes the condition of a lubricating film concisely. Another is because it has no dimensions. The magic numbers in the engineering sciences today are the dimensionless numbers.

An important result of hydrodynamic theory was pointed out by Osborne Reynolds when he said (II: 1886, page 178):

> Thus if with speed U, a load W, and friction F a certain thickness of oil is maintained, the same will be maintained with a speed MU, a load MW and the friction will be MF.

Concluding his paper of 1904, Sommerfeld wrote that

> The important thing is not the speed alone or the journal pressure alone, but the ratio of the two; or more precisely put, the ratio $\mu U r^2/Wc^2$, which is an abstract number . . . . If the lubricant is replaced by a more fluid one, the influence on the friction is the same as if the speed is diminished or the journal pressure increased.

In the above expression μ is the viscosity, U the speed, W the load per unit of axial length, and r/c the ratio of radius to radial clearance.

Reynolds' statement applies equally to the case of a limited or unlimited oil supply. Sommerfeld's is restricted to the clearance-type journal bearing of infinite length and without oil grooves, and results from the complicated solution of differential equations. Is it possible to discover any similar relations that are less restricted, and more easily derived? Yes, an affirmative answer is found in the ZN/P relations reviewed in this chapter; but it will pay to brush up, first, on some of the physical concepts needed.

1. PHYSICAL EQUATIONS

Equations that indicate a relation between physical quantities are called physical equations. Usually one quantity is placed by itself on the left-hand side and treated as a dependent variable, the quantities on the right serving as independent variables. The equations met in economics, statistics, and pure mathematics are usually not physical equations. Since, however, length is a physical quantity, geometrical relations follow the rules of physical equations.

Examples of physical equations are readily found in the preceding chapters. One such is the equation for rolling friction cited in the text of Chapter I,

$$F = eW/r\,. \tag{1}$$

Here F is the force, in direction of motion, that will just overcome rolling resistance when applied at the height of the axis of the roller; r being the radius of the roller, W the load, and e the eccentricity, or distance between load line and supporting force. Another is Petroff's equation for the friction moment of a concentric journal bearing,

$$M_0 = \pi^2 D^3 LZN/C\,, \tag{2}$$

where D is the diameter of the journal, L the length of the bearing, Z the film viscosity, N the speed in revolutions per unit time, and C the diametral clearance. A third example is Poiseuille's law for the volumetric flow rate under a pressure gradient P/L in a capillary of inner radius R, discharging a liquid of viscosity Z; namely

$$Q = \frac{\pi}{8}\left(\frac{P}{L}\right)\frac{R^4}{Z}\cdot \tag{3}$$

Equations (2) and (3) are taken from Chapter II.

Physical equations are either complete or incomplete; and either theoretical or empirical. The first distinction is the more important one. Equations (1) to (3) are complete; no independent variable has been left out that could influence the quantity on the left, under the conditions understood or implied. It may be asked regarding Eq. (1), doesn't F depend on the material of the roller, and that of the supporting surface? The answer is no, not when e is fixed; although e can be quite different for different materials. Then doesn't M_0 depend on the load? No, because we are discussing only the factors that can influence the friction while the journal is maintained concentric. But surely Q depends on velocity? Yes, when turbulence enters. The condition of streamline flow should have been specified, unless we can argue that it was implied by the term "capillary."

If instead of presenting Eq. (1) in the complete form above, it had been written $F = \text{const}\ (W/r)$, we should call it an incomplete equation. It was, in fact, so expressed by Coulomb in his report on rolling friction, the constant being assigned various numerical values. Equation (3) was offered by Poiseuille in the form $Q = \text{const}\ (PR^4/L)$ with numerical values for the constant; again what we call an incomplete physical equation. A frequent type of incomplete equation is the engineering handbook formula where some familiar factor like gravity, or atmospheric pressure, has been suppressed, or concealed in a numerical coefficient, on the ground that it will remain constant. "Complete" equations are to be understood when the contrary is not stated or apparent.

Equations (1) to (3) are theoretical rather than empirical in the sense that they were obtained by calculation. They were deduced from accepted principles or laws of experimental origin, but without direct observation. There is seldom any incentive to leave theoretical equations incomplete. Empirical formulas are, however, frequently incomplete. Either an incomplete relation will suffice; or if not, it is too difficult to explore the problem immediately; it is therefore set aside and "recommended for future research." Typical examples of empirical formulas are those seen in the literature on windage of disks and flywheels. The results of preliminary experiments lead to incomplete formulas, later made complete but still empirical. Thus an empirical equation is not necessarily incomplete. The method of dimensions often serves to convert an incomplete equation into a complete equation without additional observations.

An essential feature of any physical equation is that it shall enable us to visualize the relation expressed. It should tell a story in terms that we can understand. It should indicate how the characteristic, or performance variable, on the left is influenced by the operating conditions, material properties, and design factors on the right of the equation. Thus Eq. (2) shows that although the moment of friction is directly proportional to the length of the bearing, it varies as the cube of the journal diameter and inversely with the clearance. It says plainly that the composition of the lubricant makes no difference as long as the viscosity is right. It says that the friction moment will be the same whenever D, L, Z, N, and C are the same.

Some investigators confuse the issue by inserting g where it has no business to appear. An example is found in a paper on turbulence in journal bearings. Here Petroff's equation for laminar flow, correctly given by Eq. (2) above, has been written with g in the denominator. The equation apparently states that the same operating conditions will create a greater friction on the moon than on the surface of the earth, greater in the ratio of gravity on the earth to that on the moon. The present writer (1950), in discussing the paper, referred the case to Lord Rayleigh, whose words might carry more weight per unit volume than any of his own. Rayleigh (1915) had remarked upon the practice of many engineers who put g in their descriptions of phenomena that are in no way influenced by gravity. A second example appears in a formula[1] for the heat Q developed in unit time by sliding friction. For a given load W, the formula states that Q is proportional to g. This time, friction will be less on the moon. The aim of Lord Rayleigh's criticism was to set up equations that will not merely serve for slide-rule work, but will convey a true physical picture of what happens.

2. DIMENSIONAL HOMOGENEITY

Two kinds of units are employed in the measurement of physical quantities: fundamental units and derived units. The *fundamental* units are represented by material standards, directly or indirectly, while the *derived* units are defined in terms of the fundamental units. Three fundamental units are sufficient in measuring mechanical quantities. The units of force, length, and time are found convenient for engineering work, although mass, length, and time lead to the same results. Only two units, those of length and time, are needed for problems in kinematics; only two, the force and length units, are required in statics. In general four fundamental units are needed for problems involving temperature, and five for electrical or electromagnetic problems; although in any particular case, a smaller number may do. Thus the unit of velocity is defined as k times the length unit divided by the time unit; the unit of power as k' times the product of the force and velocity units, and so on. When the proportionality constants k, k', etc., are taken equal to unity, the system of fundamental and de-

[1] Quoted by Bowden & Tabor (I: 1950, page 33). A similar equation but without the g, credited to Jaeger, appears in Part II, page 178.

rived units is called a *normal* system, or a system of normal units. Such a system will be understood when not otherwise stated. A system constructed from fundamental and derived units is called an *absolute* system of units, or a system of absolute units.

There is a great advantage in working with *complete* physical equations based on *normal* units, sometimes called "consistent units," in that the solution of a problem will be independent of the units employed; the same, for example, in English or metric units; the same in the foot, pound-force, minute system as in the inch, pound-force, second system. Hence, when complete equations and normal units are employed in a publication, it is unnecessary to specify the units corresponding to a list of symbols. This fact is a cardinal doctrine of physics, taken for granted in scientific work, but often overlooked in engineering.

The foregoing broad statements are seen in a fresh light when expressed in terms of dimensions. In general, the "dimensions" of a physical quantity are the exponents of the fundamental units that appear in the definition of a derived unit. Since velocity is defined as length divided by time, its dimensions are L/T or LT^{-1}, where L and T symbolize the fundamental units of length and time. Briefly, velocity has the dimension 1 in length and —1 in time. Power has the dimensions FL/T or FLT^{-1}, or the dimension 1 in force, 1 in length, and —1 in time. Mass can be defined as force divided by acceleration, whereby its dimensions are FT^2/L or $FL^{-1}T^2$. Angle has the dimension L/L or L^0; we can either say it has zero dimensions, or the dimensions of unity, which are zero. Heat quantities are most conveniently expressed in work units, so the dimensions of heat are then FL. Heat capacity per unit volume has the dimensions $FL/L^3\Theta$ or $FL^{-2}\Theta^{-1}$, where Θ is our symbol for the temperature unit. The dimensions of force, length, time, and temperature are written F, L, T, and Θ.

Tables 1 and 2 give the dimensions of some familiar mechanical and thermal quantities for convenient reference. It is customary in publications to write dimensional equations with square brackets, thus $[Z] = [FTL^{-2}]$ for viscosity, and $[q] = [FL^{-2}\Theta^{-1}]$ for heat capacity per unit volume of a lubricant. This notation implies that Z has the dimensions FTL^{-2} without numerical equality. But in real life it is simpler to use a colon; thus Z: FTL^{-2}, q: $FL^{-2}\Theta^{-1}$, meaning Z has the dimensions FTL^{-2}, and so on. Use of the colon saves a lot of pencil-and-paper work when experimenting with the many combinations of dimensions that may be available.

All the terms in a physical equation must have the same dimensions, if it is a complete equation. Such an equation is expected to

TABLE 1. Dimensions of Some Mechanical Quantities

Symbol	Quantity	Dim.	Symbol	Quantity	Dim.
E	Elastic modulus	F/L^2	a	Film ratio, h_1/h_0	1
H	Power loss	FL/T	b	Pressure coeff. vis.	L^2/F
K	Kinematic viscosity	L^2/T	e	Roller eccentricity	L
K_a	Compressibility, ad.	L^2/F	f	Coefficient of friction	1
K_t	Compressibility, iso.	L^2/F	m	Mass of a body	FT^2/L
M	Moment of force	FL	p	Film pressure	F/L^2
N	Revs. per unit time	$1/T$	v	Specific volume	L^4/FT^2
P	Load per unit area	F/L^2	w	Specific weight	F/L^3
Q	Flow rate	L^3/T	ϵ	Eccentricity ratio	1
R	Rate of shear	$1/T$	ρ	Density	FT^2/L^4
Z	Viscosity	FT/L^2	τ	Shear stress	F/L^2

describe a fact of nature, some objective relation among the variables. It must tell the same story no matter what units are employed. No one seems to have found any real exception to this rule, which is known as the principle of *dimensional homogeneity*. Mathematical and philosophical proofs have been offered, but are so abstract that they are hardly more convincing than plain common sense to anyone with experience in using physical equations. Hunsaker & Rightmire

TABLE 2. Dimensions of Thermal Quantities

Symbol	Quantity	In Terms of F, L, T, Θ	With Heat Unit Φ
C_a, C_1	Constants	$F/LT\Theta$	$\Phi/L^2T\Theta$
C_n	Constant	$F/LT\Theta^n$	$\Phi/L^2T\Theta^n$
H', Qq	Heat trans. rate	FL/T	Φ/T
J	Mech. equivalent	1	FL/Φ
T	Temp. rise	Θ	Θ
a	Temp. coeff. vis.	$1/\Theta$	$1/\Theta$
c_p	Specific heat	$L^2/T^2\Theta$	$\Phi L/FT^2\Theta$
k, λ	Thermal conduct'y	$F/T\Theta$	$\Phi/LT\Theta$
q	Heat capacity/vol.	$F/L^2\Theta$	$\Phi/L^3\Theta$
α	Expansivity	$1/\Theta$	$1/\Theta$

(1947, page 104) cut the knot by saying, in effect, that we may add and subtract only numerical quantities of the same dimensions, so that physical equations must be dimensionally homogeneous.

An apparent exception has been found when two equations of different dimensionality are added together. But when one equation talking about apples is added to another talking about oranges, you get only a tutti-frutti formula that is hardly a physical equation. In any event the components are immiscible and can be filtered out, bringing us back to the two separate equations. Again, it has been noted that the logarithms of a dimensional quantity *can* appear in a correct equation, apparently violating the rule that only dimensionless quantities have logarithms, sines, or cosines. But a careful inspection is sure to bring to light another such logarithm, whose argument will cancel the dimensions of the first one, when the two logarithms are properly combined and the fracture repaired.

Dimensional homogeneity can be directly applied as an aid in checking and memorizing formulas, in deriving simple formulas, and in normalizing differential equations. Suppose we are trying to recall the formula for the "heat of compression," dT/dp. Is it $\alpha T/q$ or $\alpha q/T$? Our head is in a whirl, but the dimensions are T: Θ, p: F/L^2, α: $1/\Theta$, and q: $F/L^2\Theta$. Therefore, dT/dp has the dimensions $L^2\Theta/F$, and the right-hand side must be the same. The first mentioned expression, $\alpha T/q$, is seen by trial to be correct (Chapter IV).

Poiseuille's Law. Again, we can try deducing Poiseuille's law and Stefan's law, Eqs. (7) and (10) of Chapter II. Let the pressure gradient P/L be denoted by G and assume Q is some product of powers,

$$Q = \text{const}\, G^a R^b Z^c \,. \tag{4}$$

The dimensions are Q: L^3/T, G: F/L^3, R: L, and Z: FT/L^2. For homogeneity,

$$L^3/T = (F/L^3)^a (L)^b (FT/L^2)^c \,. \tag{5}$$

Since the dimensions must be the same on both sides, we are led to three equations for the force, length, and time dimensions, respectively:

$$\begin{aligned} 0 &= a + c, \\ 3 &= -3a + b - 2c, \\ -1 &= c. \end{aligned} \tag{6}$$

From Eq. (6), $a = 1$, $b = 4$, and $c = -1$. Substituting in Eq. (4) makes Q proportional to GR^4/Z, as shown by Eq. (4). The numerical constant $\pi/8$ cannot be found by dimensions. It could, however, be determined from a single test observation. To balance this shortcoming, note that the GR^4/Z result applies to capillary flow through any shape of cross section, R being a chosen linear dimension thereof.

Stefan's Law. A partial derivation of Stefan's law can be made starting from

$$h = \text{const}\ P^a R^b Z^c t^d\ ; \tag{7}$$

but now there will be 4 unknowns, a, b, c, and d, to be evaluated from three equations. The result can be written

$$\frac{h}{R} = \text{const}\left(\frac{Pt}{Z}\right)^d. \tag{8}$$

The constant, 0.866 for circular disks, and the exponent, $d = \frac{1}{2}$, cannot be found by dimensions, but could be determined from a pair of test observations. The results apply to geometrically similar plates of any size defined by R, whether circular or not.

The Series Method. In more complicated problems where it may not be safe to assume a single-term result, we can start from an infinite series of terms like the right-hand side of (7). We should then end up with an unknown *function* of the dimensionless combinations or "products," instead of with an unknown exponent like d in Eq. (8). Even that kind of result proves to be exceedingly useful, as will be seen from applications of the "Pi-theorem."

Normalizing. The principle of dimensional homogeneity permits "normalizing" the terms of a differential equation so that all terms are made dimensionless. This is done by dividing through by appropriate factors and assigning new symbols to the combinations obtained. The process simplifies the appearance of the equation, expedites its solution, and makes it easier to judge the limits of applicability. Examples may be seen in the treatise on *Gas Film Lubrication* by W. A. Gross (IX: 1962), and in practically all research today where differential equations are required.

The concept of dimensions and the principle of homogeneity go back to Fourier in his book on the theory of heat (1822).

3. THE "PI-THEOREM"

It was shown by Edgar Buckingham (1914 [1], 1914 [2], 1915) that any physical equation expressing a relation among n quantities,

$$Q_1 = F(Q_2, Q_3, \ldots, Q_n)\ , \tag{9}$$

can be put in the form

$$\Pi_1 = \phi(\Pi_2, \Pi_3, \ldots, \Pi_i); \tag{10}$$

where each of the Π's is a product of powers of some of the Q's. The number of Π's that can be constructed independently of each other is given by

$$i = n - k, \tag{11}$$

where k is the minimum number of fundamental units needed for measuring the n quantities. Buckingham called this proposition the *Pi-theorem* because of his use of the Greek capital Π to represent a continued product or quotient, a customary symbol in mathematics. It is often called "Buckingham's theorem," although he credited it mainly to Riabouchinsky (1911). See also the earlier publications by Vaschy (1892).

It happened that our government departments, before World War I, had been making occasional use of *dynamical similarity* without understanding it too well. They asked Dr. Buckingham, long with the National Bureau of Standards, to attempt a clarification of the subject. The Pi-theorem was the outcome. It facilitated the work then in progress at model basins and wind tunnels. It was widely acclaimed by mechanical engineers. The method of dimensions had been used in a slight-of-hand manner by geniuses like Lord Rayleigh; but now, any of us can use it.

By applying the Pi-theorem to the examples already discussed, we see that in Petroff's equation $\Pi_1 = M_0/D^3ZN$, $\Pi_2 = L/D$, $\Pi_3 = C/D$. The form of the function ϕ is such that $\Pi_1 = \text{const}\ \Pi_2/\Pi_3$. In Poiseuille's law, $\Pi_1 = QZ/GR^4$ and ϕ is a constant. In Stefan's law, $\Pi_1 = h/R$ and $\Pi_2 = Pt/Z$, while ϕ remains unknown for the moment. Integrations or experiments are required to reveal that $\phi(\Pi_2) = (\Pi_2)^{1/2}$.

Constructing the Π's "independently" means, for example, that in Petroff's equation a fourth product $\Pi_4 = L/C$ would not be counted because it can be obtained merely by dividing Π_2 by Π_3. The i products must be independent in that sense, but Π_1 may still be considered a dependent variable related to the $i - 1$ remaining Π's as independent variables in the experimental sense.

Deductions of the Pi-theorem were given by Buckingham in each of the three references cited, but his proof in the Appendix of the ASME paper (1915) is of special interest to engineers. Any such proof starts from the principle of dimensional homogeneity. Buckingham's proof assumes that a physical equation like (9) can be expressed (if need be) with the sum of a series of terms on the right-hand side, each term being a continued product Q_2^a, Q_3^b, . . . , where the exponents a, b, . . . may be different in successive terms. The equation is then divided through by Q_1 to make all terms dimensionless, after which it is shown that only $n - k$ are independent.

Associated Units. Special interest attaches to Eq. (11) when two of the fundamental units are "associated," not acting separately. In a problem of mechanics it may happen that the n original quantities require only two fundamental units, say F' and T, to specify their dimensions, where F' denotes F/L^2. Thus in Stefan's equation if we take h', Z, P, and t for the originals, where h' denotes h/R, the dimensions can be written h': 1, Z: $F'T$, P: F', and t: T. Effectively, the "minimum" number of fundamental units needed is only $k = 2$, namely F' and T; not the usual 3; hence $i = n - k = 4 - 2$, from which $\Pi_1 = h'$ and $\Pi_2 = Z/Pt$.

Hunsaker & Rightmire (1947) prefer to define k by the difference $n - i$, and to find i by trial. That leads to the same result as our use of the associated units. See also Buckingham's discussion (1921) of the number of fundamental units required.

Methods of Application. There are two methods of applying the theorem. The first, outlined by Buckingham (1921), requires selecting any k of the quantities, and treating them as fundamental. A product is built up around each, and the exponents evaluated by simultaneous algebraic equations so as to render each product dimensionless. If the selection is limited to k quantities, there will never be more unknowns than there are equations. The second is by trial and inspection with no set rule. This method is usually found preferable after gaining a little experience. Equation (11) serves as the control in using this procedure. As soon as $n - k$ dimensionless products have been assembled and shown to be independent one from another, we can stop.

It will soon be learned that not every set of $n - k$ products, determined by either of the foregoing methods, is equally suited for the purpose in view. Among the optional combinations available, some are more attractive than others. We may wish to segregate some variable, Q, so that it appears only once. The situation can be rectified by substituting any quantity or combination for another having the same dimensions. The independent products can be multiplied or combined in any manner desired, since a function of x, y, and z is equally a function of x, xy, and z^2, for example.

Physical Similarity. Two systems are said to be physically similar when all the dimensionless products in the unknown function have the same value in both systems. If we restrict our comparison to systems that are geometrically similar, then length ratios like L/D or C/D, that are needed only to specify the shape of the bodies involved, can be omitted from ϕ since they are certainly constant.

Two or more bodies may be said to have the same *generalized shape* if they have not only the same geometrical shape, but the same distribution of physical properties, such as density or elastic constants. They may then be described as *generally similar*. In comparing two systems of bodies that are "generally" similar, the function ϕ can be further simplified by omitting any ratios needed for specifying the distribution of the properties of materials. This refinement is seldom needed; and quite often in practice there will be only one argument, or product, in the unknown function, so that $\Pi_1 = \phi(\Pi_2)$.

When the products Π_2, etc., in ϕ are respectively equal in both systems, it follows that the values of Π_1 are also equal. The two systems are then said to be "physically" similar. If only mechanical quantities are involved, the systems are called "dynamically" similar. The principle of physical similarity has been widely used in fluid mechanics, heat transfer, structural design, and other engineering fields. The main use of the principle is in conducting model experiments.

Suppose now that a quarter-sized "model" is used in experimenting on turbulent flow through a pipe. The pressure gradient G may be expected to depend on the diameter D, the mean velocity v over the cross section, and the viscosity and density of the liquid, Z and ρ. From the Pi-theorem for geometrically similar systems

$$G = \frac{vZ}{D^2}\,\phi(\text{Re})\,, \tag{12}$$

where Re denotes Reynolds' number, $Dv\rho/Z$. The *condition for similarity* is that Re shall be the same for the model and the original. The *law of comparison* then follows, from which GD^2/vZ will be the same in both. Let G_0, D_0, v_0, ρ_0, and Z_0 refer to the original system and plain symbols to the model. When $D/D_0 = \frac{1}{4}$, the condition for similarity requires that $v\rho/Z = 4v_0\rho_0/Z_0$. If the same liquid is used, it will be necessary for the mean velocity to be four times greater in the model experiment than in the original. The law of comparison then predicts that

$$\frac{G_0}{G} = \frac{v_0 Z_0}{vZ}\left(\frac{D}{D_0}\right)^2 = \frac{1}{64}\,. \tag{13}$$

Thus the pressure gradient in the larger pipe will be only $\frac{1}{64}$ as much as observed in the model experiment. These relations have been confirmed, and the form of the function ϕ determined both for smooth pipes and for specified degrees of roughness. In the limiting case of streamline motion, where density has no effect, ϕ is constant; and, since $D = 2R$ and $v = Q/\pi R^2$, Eq. (12) then reduces Poisuille's law, except for the numerical constant.

If we went too fast in setting up Eq. (12) simply by trial and checking of dimensions, we can start again using Buckingham's step-by-step procedure. There are $n = 5$ physical quantities, G, D, v, ρ, and Z, which require $k = 3$ fundamental units, making $i = 2$ dimensionless products. Select D, v, and Z as fundamental quantities in terms of which G and ρ may be expressed, so that G and ρ shall each appear only once in the result. Then write

$$\begin{aligned} G&: \; D^a v^b Z^c \, , \\ \rho&: \; D^{a'} v^{b'} Z^{c'} \, , \end{aligned} \tag{14}$$

and substitute the respective dimensions F, L, and T. Solving for the six exponents leads to the equivalent dimensions

$$\begin{aligned} G&: \; D^{-2} v Z \, , \\ \rho&: \; D^{-1} v^{-1} Z \, . \end{aligned} \tag{15}$$

From (15) it is seen that $\Pi_1 = GD^2/vZ$ and $\Pi_2 = Dv\rho/Z$, hence Eq. (12).

Empirical Equations. It is not always appreciated that empirical as well as theoretical equations, if complete, are dimensionally homogeneous. They can then be expressed in the most general possible form, that given by the Pi-theorem. All that is necessary is to plot the observed data in terms of dimensionless variables. When empirical equations are found not homogeneous, it is because they are incorrect or incomplete.

An empirical equation that is incomplete can sometimes be made complete after writing down a carefully studied list of all the physical quantities that could reasonably be expected to influence the dependent variable. It does no harm if the list is overcomplete. For this purpose the phenomenon described may be somewhat idealized, as, for example, when we treat the flow of liquid through a pipe as if it were strictly incompressible. However, no pertinent quantity should be left out of the list merely because it will remain constant.

It is not essential in applying dimensional theory that the test data actually be fitted with an empirical equation. Purely graphical representations are acceptable if plotted in dimensionless variables. Ordinarily, when the discussion is restricted to geometrically similar systems, the story of an investigation is sufficiently told by plotting

Π_1 against Π_2. Sometimes a third parameter Π_3 enters the picture, requiring a family of curves, each labeled with a constant value of Π_3; or a curved surface in Π_1, Π_2, and Π_3 space. For example, the coefficient of friction Π_1 may be a function of ZN/P in the role of Π_2, at constant values of p_a/P, the ambient pressure ratio, as Π_3. If another parameter E/P, the ratio of Young's modulus to the load per unit area, is brought in, we shall need a family of surfaces, like a Chinese pagoda; and so on. But such complications cannot properly be blamed on "dimensional analysis." The complications were there to begin with; the analysis opens our eyes to see them.[1]

4. DERIVATION OF ZN/P RELATIONS

Consider the isothermal operation of a journal bearing without forced lubrication, in which the coefficient of friction depends only upon the size and shape of the bearing, speed, load, and viscosity of the lubricant. Any effect of the ambient pressure will be disregarded. Just as in the classical theory, we are imagining a rigid bearing and journal, and a film of uniform viscosity. But we need not assume smooth surfaces, or any particular geometrical shape. The reasoning offered holds good for the most fantastic oil-grooving, as well as for a straight cylindrical interior.

Let f denote the coefficient of friction on the journal, D the diameter, P the load per unit area, and N the speed and Z the viscosity of the lubricant. Then for any fixed geometrical shape, f is some function of D, P, N, and Z. "Shape" includes relative clearance C/D, length-diameter ratio L/D, relative roughness, and any other length-ratios characterizing the design. In symbols,

$$f = f(D,P,N,Z) \,, \tag{16}$$

where the unknown function f on the right will be different for different shapes.

It is essential that the right-hand side of (16) be a complete list of the physical quantities whose variation can affect the coefficient of friction in the idealized case agreed upon. If it is overcomplete, no harm is done; the superfluous quantity will drop out. If incomplete, false conclusions may be drawn. Could variations of gravity make any difference, now is the time to put g in the list. The mere fact that

[1] Captain Lybrand P. Smith USN, one of the founders of the Office of Naval Research, said (1944) that no one "should be permitted to direct any important experiment involving the physical sciences unless he has a thorough grasp of dimensional analysis."

gravity will remain constant is no reason for leaving it out. But P is understood here to include the entire load, whether caused by gravity, springs, or otherwise. If, then, (16) is a complete physical equation, both sides must have the same dimensions.

Equation (16) has no dimensions on the left, f being a pure number, the ratio of one force to another. Hence it can have no dimensions on the right. The function must be of such a form that its dimensions cancel out. It may be seen by checking dimensions that the only combination satisfying this requirement is some function of the single variable ZN/P. Since D drops out, it makes no difference how big or little the bearing is, and (16) reduces to

$$f = \phi(ZN/P) , \tag{17}$$

where ϕ is another unknown function. Although both functions are unknown, ϕ is a function of one variable only, whereas f was a function of four in Eq. (16). Whether ϕ is linear, or proportional, say, to the square root we are not told. The form remains to be discovered by calculation or experiment. What we have done is to reduce the number of independent variables, which can simplify the experiment. Whether we take advantage of the simplification, or vary everything separately and use (17) merely as a check, is optional. The surveyor who remembers his geometry can determine all the angles of a four-sided field by measuring only three of them; or he can measure all four, and use the fixed sum of 360 degrees as a check, dividing any difference between them.

Alternative Derivations. We might have started from

$$f = f(D,L,C,W,N,Z) , \tag{18}$$

instead of from Eq. (16). Here L denotes the length of the bearing and C its diametral clearance, and W is the total load (PLD). The Pi-theorem can now be applied in a formal manner, instead of by inspection. Here $n = 7$, $k = 3$, and by Eq. (11), the number of independent dimensionless products will be $i = 7 - 3 = 4$. We can set down f as the first, L/D and C/D as the second and third, writing

$$\Pi_4 = ZN^aW^bD^c . \tag{19}$$

Substituting dimensions and solving for the exponents gives $a = 1$, $b = -1$, and $c = 2$; from which

$$\Pi_4 = ZND^2/W = ZND^2/PLD , \tag{20}$$

which is ZN/P divided by L/D.

Having seen that the diameter D is superfluous, someone is sure to propose that we check the derivation by omitting D from the start. This time $n = 4$ instead of 5, so we should expect $i = 4 - 3 = 1$. With $\Pi_1 = f$ as before, it would seem that no Π_2 can be allowed. Apparently D served as a catalyzer in the first solution, effecting the desired result but dropping out unchanged! A better explanation is found in the rule of "associated" fundamental units. Under this rule the dimensions are f: 1, P: F', N: T^{-1}, and Z: $F'T$, where F' denotes F/L^2. Therefore, $k = 2$ so that $i = 4 - 2 = 2$ and $\Pi_2 = ZN/P$. This confirms Eq. (17).

Film Thickness. The ZN/P equation for the minimum film thickness ratio h_0/c can be derived in the same manner. Here h_0 is the film thickness at point of nearest approach, and c denotes the radial clearance. Thus for geometrically similar bearings,

$$h_0/c = \psi(ZN/P) , \tag{21}$$

where ψ is an unknown function of the single variable ZN/P. The only difference is our use of c in place of D as the basic linear dimension.

If we now define *load capacity* as the load per unit area, P, that will reduce the film thickness ratio to some arbitrary, safe value, the left-hand side of (21) remains constant, and so the right-hand side must also be a constant. Therefore, ZN/P itself is a constant, and

$$P = \text{const } (ZN) , \tag{22}$$

where the constant is the same for all geometrically similar bearings. Thus the load capacity of a bearing, when defined as above, varies with the product of film viscosity and speed.

In practice it is not found necessary to maintain as great a relative film thickness in large bearings as in small ones. A compromise between the same absolute and the same relative h_0 values is found satisfactory.

Owing to the decrease in mean film viscosity with frictional heating, it is clear from Eq. (22) that the load capacity does not increase as fast as the first power of the speed. If it should happen that Z varies inversely with speed, the load capacity would be independent of speed. For heavy oils of low VI it is possible for the viscosity to fall off so sharply, with increasing speeds, that the load capacity passes through a maximum, and diminishes with further increase of speed, tending to approach the old Thurston rule, load $\times$ speed = a constant (McKee, III: 1937, 1940).

The friction, film thickness, and load capacity equations derived by dimensional reasoning have been confirmed experimentally. They are in general agreement with theoretical equations, but are much less restricted in their application to complicated geometrical forms.

Historical Note. A long series of experiments on full journal bearings, under fluid film conditions, was conducted by the writer at M.I.T. (1909). The coefficient of friction was plotted against speed, load, temperature, viscosity, and rate of oil supply for two mineral and two fatty oils. Difficulty was met in condensing these data into a short enough form for publication. The Pi-theorem provided the answer. The writer's report (II: 1914) appeared immediately following Edgar Buckingham's two papers of 1914. By plotting against the product of speed and viscosity divided by load, $\mu n/p$ as it was then called, all the friction observations could be shown on a single chart, Fig. 1. The principal series could be represented by an empirical equation,

$$f = 0.002 + 6800\ \mu n/p\ , \tag{23}$$

The constant 6800 should be replaced by 16×10^{-6} when customary units are employed as in the figure. Scatter reflects difficulty of measuring film temperature. There is a perceptible trend toward convexity, as in partial bearings, possibly the result of cavitation. Equation (23) was the first of many such equations to appear in lubrication literature.

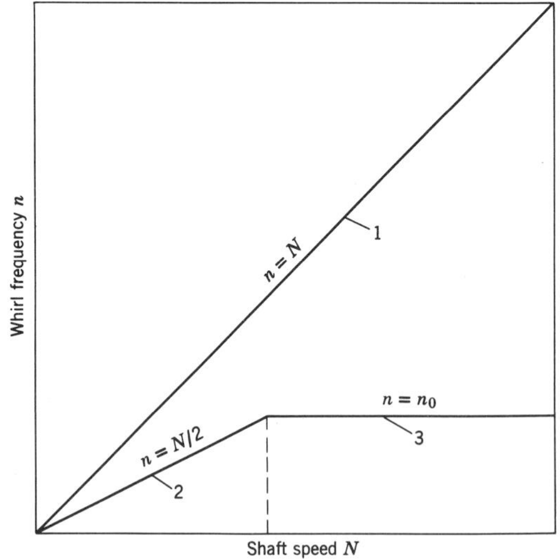

Fig. 1. Coefficient of friction, full journal bearing, from M. I. T. experiments of 1909. (Two points off chart are 1000 f = 27.5 at 1475; 23.5 at 1690.)

Measurements of electrical resistance of the oil film were made, during the same investigation, to gain some idea of how the film thickness varied with speed and load. The principal results obtained can be explained by Eq. (21).

The following year Buckingham (1915) and the writer (III: 1915) presented papers to the ASME at its Buffalo meeting. They were accompanied by lively discussions. The lubrication paper prompted an editorial in the *American Machinist* (1915), which concluded that

> The next steps are plain. Existing experimental data should be again reviewed in the light of this new grouping of the fundamentals of the problem. It will then undoubtedly appear that more experimental work is needed . . . .

To that end the Research Committee of the Society soon organized a Subcommittee on Lubrication, known today as the RCL.

First to carry out the recommendations were Wilson & Barnard (1922). Introducing z (for Zähigkeit) in place of μ, they became literally the first to apply zn/p. Test data were replotted against zn/p, including those of Stribeck (1902) and Biel (1920).

Sommerfeld had shown in his third paper (1921) that these two investigations confirmed his conclusion that "the physical factors of the lubrication problem appear only in the combination $\mu U/W$." This is equal to π times $\mu N/P$ since W is the load per unit length. It may be conceded that U/W is more vivid and tangible than N/P. Sommerfeld goes on to say that this conclusion will be called "the similarity law of lubrication."

First to question the validity of zn/p was A. E. Becker (1926). He went back to Tower's experiments, where an apparent disagreement was found. The discrepancies were soon explained by Barnard in terms of the difficulty of measuring true film temperatures.

About this time, at the National Bureau of Standards, S. A. McKee began a series of experiments which, he found, could best be interpreted in the light of ZN/P. See references from 1926 to 1952 in Chapters II and VII. McKee, also, had the happy thought of promoting the notation from lower case to full capitals. The lubrication number ZN/P had now come of age.[1] Its use spread across the Atlantic, where it was taken up by Boswall in his second chapter (II: 1928), Goodman (VII: 1929), Hanocq (VII: 1929), Boswall &

[1] A convincing demonstration of ZN/P can be shown with the McKee "lubrication indicator" (Hersey I: 1964 and pages 66–67 of I: 1936). Adjust the load and speed so that the lamp flashes on, indicating metallic contact at the high spots. Double the speed, the light goes out. Double the load, light comes on again.

Brierly (VII: 1932), Jakeman & Fogg (VII: 1937), and others. In Germany, Gümbel's number $\omega\eta/\bar{p}$, based on classical hydrodynamic theory (1914), is frequently used as a correlating variable. Here η is the viscosity, ω the angular velocity of the shaft, and $\bar{p}$ the load per unit area. It remained for Vogelpohl (1949) to show that Gümbel's number can be derived by dimensions, thereby releasing it from the restrictions of classical theory. The symbol *Gü* was introduced by ten Bosch about 1932. A review was published by the writer in 1934.

Applications continued in the United States; see, for example, Tichvinsky (VII: 1938). Everyday use of the ZN/P diagram had become practically universal, when its validity was challenged by Linn & Irons (VII: 1941). These investigators could not represent their results by plotting f against ZN/P. Instead, they found that the friction coefficient increased with the rate of oil flow, Q, in such a manner that the data required plotting f against ZNQ/P. In the discussion C. S. L. Robinson, citing our paper of 1915, showed that a partial explanation could be found by retaining ZN/P for the main variable, while introducing Q/D^3N as an independent parameter.

5. THE SOMMERFELD NUMBER

Multiply ZN/P by the square of D/C and we have the *Sommerfeld number* S, a parameter applicable to infinite-length journal bearings of the clearance type, satisfying the classical assumptions of hydrodynamic theory. When the product of f and D/C is chosen for one coordinate, and S for the other, the graphs for all clearances are brought together into a single theoretical curve. The Sommerfeld number should therefore be used in preference to ZN/P, where applicable. The *Ocvirk number* C_n, described in Chapter II, is a further step in the same direction. With the aid of the Ocvirk number all curves for low L/D ratios, as well as the different clearances, can be brought together. It should be used in preference to the Sommerfeld number, where applicable.

Assumptions. The Sommerfeld number is derived by assuming geometrical perfection, neglecting side leakage, and endowing the lubricant with the ability to support negative pressure, or tensile stress. Sommerfeld pointed out that these assumptions are not realistic, but his aim was to see what could be learned from hydrodynamic theory in its conventional form. Dimensional theory is not limited by the classical assumptions. It applies to "fitted" as well as "clearance" bearings, to thrust as well as journal bearings; and even to the "Warlop bush" (Clayton, 1939)—a bearing shell so full of holes that it is practically a sieve.

Origin of Sommerfeld Number. The variable that we now call the Sommerfeld number can be traced back to the expression quoted on the first page of this chapter. Sommerfeld did not use it, or even propose it, as a parameter for plotting experimental data. He thought of it as an expression that came up in the theory, and had an interesting interpretation.

The first published application of the Sommerfeld variable seems to have been made not by Sommerfeld himself, but by H. A. S. Howarth—see discussion of McKee's paper on "Running In" (page 528 of VII: 1928). Howarth does not mention Sommerfeld. He employed Harrison's notation, and possibly acquired the idea from his study of Harrison's paper (IX: 1913). The same expression was later used by S. J. Needs (VII: 1934), but still with no name or symbol. It occurred to the writer that some credit should be given to Sommerfeld. Accordingly, when discussing Needs' paper, we introduced the name "Sommerfeld variable." This designation was repeated in our book (1936). It was then adopted by Norton in his book (II: 1942), and soon gained general acceptance.

Alternative Forms. It is optional whether one applies the Sommerfeld variable right-side up or in its reciprocal form. To the experimentalist watching the shaft turn around and thinking intently about the physics of the oil film, it seems natural to give speed and viscosity a place of honor in the numerator. To the designer, load belongs in the numerator.

Vogelpohl (II: 1943) proposed the symbol So for the Sommerfeld number, which he defined as ψ^2 times the reciprocal of Gümbel's number, where ψ denotes the clearance ratio C/D. By comparing the two forms, we can see that So $= 1/2\pi S$. Leloup (III: 1954) and Pestel (VIII: 1954) used the American form of Sommerfeld number—speed in numerator.

Barwell (1957) strongly favored the designer's form after reviewing the pros and cons. He called attention to Sommerfeld's Eq. (46) of 1904, where the reciprocal form is set equal to a function of the eccentricity ratio; thus, in our notation,

$$\frac{W}{\mu U}\left(\frac{C}{D}\right)^2 = f(\epsilon). \tag{24}$$

This may indeed be the first authentic statement of the Sommerfeld number; but Sommerfeld wrote it the other way up at the conclusion of his paper. Equation (24) is the form adopted by Barwell in his book (I: 1956). Our view is that this form with load in numerator might appropriately be called the "Sommerfeld load number" or "Sommerfeld reciprocal."

6. BEARING LUBRICATION UNDER VARIOUS CONDITIONS

We have seen that the coefficient of friction, f, and relative film thickness h_0/c for geometrically similar bearings are functions of a single variable ZN/P, when the bearings operate under simplified conditions that have been stated (Fig. 2). In particular, the conditions must be such that the ambient pressure p_a has no appreciable influence on the performance of the bearing. This assumption is one of those commonly made in classical hydrodynamic theory. It will be more closely examined later.

The friction coefficient and film thickness ratio are not the only performance characteristics that can be treated as functions of ZN/P. Other examples are the film pressure ratio p/P at any point, and the rate of oil flow Q/D^3N. The ZN/P relations apply to any such characteristic that can be expressed in dimensionless form.

The practice of combining viscosity values in cp with speed in rpm and load in psi, introduced by Wilson & Barnard (1922), has proved convenient. To distinguish ZN/P values so obtained from those corresponding to "normal" units, we have adopted the abbreviation "cu"

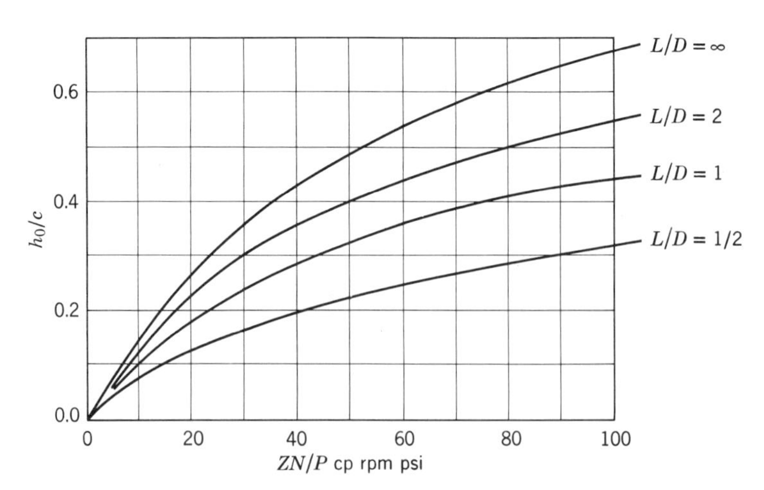

Fig. 2. Relative film thickness in a partial bearing, from Needs' observations on electrolytic models, ASME 1934.

to indicate the "customary" unit, cp rpm/psi. When units are not stated, normal or "consistent" units are understood. A ZN/P value of 1 cu is equal to 2.42×10^{-9} normal units, a pure number. In practice this would be a low value, corresponding, say, to the product of 1 cp by 300 rpm divided by 300 psi.

It is hardly correct to imagine that a change from normal to customary units destroys the nondimensional quality of ZN/P. Any function of x will be some function of $2x$, or even of $2.42 \times 10^{-9}x$.

Influence of Ambient Pressure. When the oil supply to a journal bearing is ample, and the inlet is at the ambient or surrounding pressure and located near the point of maximum film thickness or clearance, the simplest conditions apply. The film pressure is everywhere positive. This condition cannot always be realized since the angular position of maximum film thickness varies with ZN/P. If the inlet is not at the point of lowest film pressure, "negative" pressures are developed by hydrodynamic action. If such pressures are greater than ambient, cavitation occurs and the bearing performance will be different even at the same ZN/P. This can be corrected by raising the ambient pressure p_a; so that in general the performance characteristics depend not only on ZN/P, but also on p_a, and therefore on the ratio p_a/P.

Ambient pressure may be neglected when p_a/P is so small that cavitation is practically unopposed, as when operating in a partial vacuum under heavy load; or when p_a/P is so large that no negative pressures exist, as in deeply submerged bearings, or in lightly loaded bearings under an atmospheric ambient. At these two extremes, p_a/P may be treated as constants, equal to zero and infinity. But p_a cannot be disregarded merely because it remains constant at 14.7 psi, unless P is also held constant.

Cavitation was dealt with in our first papers on journal bearings (II: 1914, III: 1915) by including, as an independent variable, the fraction of the clearance volume occupied by oil. Use of the ambient pressure ratio is preferable, and may help to explain some of the small differences between theory and experiment. Other factors having a minor influence when cavitation occurs are the vapor pressure of the lubricant and its surface tension.

Forced Lubrication. When the inlet gage pressure p_1 is an appreciable fraction of the load per unit area, P, it is optional whether to use that ratio or its equivalent, the flow parameter D^3N/Q. Usually, p_1/P is easier to control and should be considered the independent variable. If the flow rate is sufficiently high for dynamic effects, the friction coefficient and journal running position will be influenced by the density of the oil, ρ. In any such case a new parameter $\rho N^2D^2/P$ comes into play. The dimensions thereof may readily be checked on the F, L, T system with the aid of Table 1. See also Table 2.

When p_1/P is taken as an independent variable, the rate of oil flow, Q, in volume units per unit time may be expressed by

$$Q/D^3N = \phi(ZN/P, p_1/P). \tag{25}$$

Thus in geometrically similar bearings unaffected by ambient pressure, the flow rate is proportional to the cube of the diameter, other factors being held constant.

Starved Lubrication. It is usually assumed in present-day bearing discussions that there will be an ample supply of the lubricant. If, however, the supply is restricted to a flow rate Q_1 less than the amount Q that the bearing could otherwise deliver, this must be taken into account by including Q_1/Q or an equivalent parameter—see Connors (VII: 1962). Starved lubrication is further discussed in Chapter VII.

Lightly Loaded Bearings. Plotting f against ZN/P tends to obscure the fact that the friction torque itself is frequently almost independent of the load, as noted by Barwell on page 99 of his book (I: 1956). It may be better at times to plot Dennison's coefficient as ordinate, in the form M/ZND^3 or equivalent, instead of the ordinary coefficient. Here M is the moment of friction on the journal—see Chapter I. The same purpose might be served by plotting M/M_0 or f/f_0 as ordinate, where the subscript refers to a concentric or Petroff bearing.

Oil Film Whirl. Translational whirl performance in geometrically similar systems may be defined by the shaft speed N_1 at which whirling occurs, together with the frequency and amplitude of the whirl, n and a (Chapter XII). The speed N_1 may be taken either as the "whirl-impending speed" or that at which the whirl amplitude has a peak value. These characteristics depend upon the usual operating factors Z, N, P, p_a, and p_1 as well as on the journal diameter D, shaft stiffness k, and rotor mass m. Here p_a is the ambient and p_1 the oil inlet pressure. The film viscosity Z is taken to be uniform and all factors are assumed constant in time. It follows from dimensional analysis that at fixed values of p_a/P and p_1/P,

$$ZN_1/P = f_1(k/DP, kZ^2/mP^2) \tag{26}$$

and

$$n(m/k)^{1/2} = f_2(k/DP, ZN/P), \tag{27}$$

with a similar expression f_3 for a/c, where c is the radial clearance.

The left-hand side of the last equation can equally well be written n/n_0, where n_0 is the natural frequency of the rotor. Any physical quantity appearing in either equation may be replaced by its dimensional equivalent from the remaining quantities. The functions f_1 to f_3 are to be found experimentally or by detailed calculations. They are unknown at the moment, and may all be identified by the same symbol f. The equations are intended to represent the general case of resonant whirl, or whip, as observed by Newkirk and Taylor; although somewhat restricted by the assumptions mentioned, and by neglecting the inertia of the lubricant. The oil density may have a secondary influence on whirl owing to acceleration, as well as turbulence in the film. When these effects are included, a dimensionless product $\rho N^2 D^2/P$, or its equivalent, must be added to the list of independent variables.

Equations (26) and (27) may be simplified in the case of an unloaded bearing, where $P = 0$; in that of a rigid shaft, where $k = \infty$; and in that of a rigid shaft without mass, where $k = \infty$ and $m = 0$. The corresponding expressions for N_1, n/n_0, and a/c are given in Table 3. The four cases represent resonant whirl with zero load; the Stodola condition; the Boeker & Sternlicht vertical shaft; and the Harrison problem, respectively (Chapter VIII).

The dimensional treatment is offered as a tentative guide in the study and coordination of the theories and test data now available, and in planning new experiments.

Compressible and Non-Newtonian Lubricants. The action of a gaseous lubricant depends on its elasticity as well as on its viscosity. Since the elasticity or bulk modulus of a gas is equal to its pressure, its value at every point in the film is fixed by the ambient pressure p_a in conjunction with the remaining independent variables. A compression exponent n should be included in the most general case. Under the usual isothermal conditions, $n = 1$; although it may ap-

TABLE 3. Dimensional Analysis of Fluid Whirl
(Special cases under the general equations)

Case	k	m	P	N_1	n/n_0 and a/c
1	k	m	0	$(k/DZ)f(D^2Z^2/km)$	$f(k/DNZ,\ D^2Z^2/km)$
2	∞	m	P	$(P/Z)f(DZ^2/Pm)$	$f(ZN/P,\ DZ^2/Pm)$
3	∞	m	0	const DZ/m	$f(DZ/Nm)$
4	∞	0	P	const P/Z	$f(ZN/P)$

proach γ, the specific heat ratio, when the flow is adiabatic. The ϕ-function for a gas-lubricated bearing therefore includes three dimensionless variables, ZN/P, p_a/P, and n.

Common lubricants of the non-Newtonian variety are (1) grease, and (2) polymer solutions. We have seen that the properties of a grease can be approximated by two constants, the yield shear stress τ_0 and mobility m, or its reciprocal, the "plastic viscosity" Z_0. The ϕ-function for bearing performance with a grease of this type would require Z_0N/P in lieu of ZN/P; and the ratio τ_0/P in addition.

The dimensionless variables needed for describing the performance of other non-Newtonian lubricants are similarly constructed from the constants of their shear stress equations, each constant being treated as one of the original n quantities.

Other Conditions. When there is a transverse couple M_t tending to create edge-loading, such as might result from belt pull, gearing, or overhung weights, the corresponding parameter will be M_t/PD^3. If there is elastic deformation of bearing and shaft, new parameters E/P and ν may be introduced for each material, where E denotes Young's modulus and ν Poisson's ratio. Alternatively, these symbols may be introduced for one material, with ratios E'/E and ν'/ν for the other. When the loading is high enough to bring into play the increase of viscosity under pressure, this factor may be approximated by the parameter bP, where b is the pressure coefficient of viscosity.

MHD Bearings. Application of a transverse magnetic field creates additional pressures in an electrically conducting fluid when in motion. The load capacity of a magnetohydrodynamic journal bearing, following Kuzma (VII: 1963), can be doubled by raising the Hartmann number M from 0 to 8, when the bearing is operating at an eccentricity ratio of 0.8. The performance of geometrically similar bearings when lubricated by liquid metals therefore depends not only on ZN/P, but also upon the new parameter M, a dimensionless number defined by

$$M = c\eta H(\sigma/Z)^{1/2}\,. \tag{28}$$

Here c is the radial clearance, η the permeability of the lubricant, σ its electrical conductivity, and H the applied magnetic field.

Engine Bearings. The film thickness ratio in geometrically similar engine bearings will be a function of the crank angle, the mean or maximum ZN/P_m, and the shape of the polar diagram for the instantaneous load P against crank angle. See Chapter VIII; as well as Chapter 3 of Shaw & Macks (1949).

Thrust Bearings. The ZN/P relations are equally applicable to thrust bearings, where the purely mechanical problems are usually simpler, not involving cavitation. Consider, first, the hydrodynamic types. A start can be made from Eq. (16), defining the coefficient of friction by M/WR. Here M is the moment of friction under a total load W, and R is the mean radius of the supporting surfaces. The result is expressed by Eq. (17). Film thickness h_0 at the trailing edge is

$$h_0/D = \psi(ZN/P) \, , \tag{29}$$

where D is the outside diameter of the thrust collar or runner. Although the ZN/P relations take the same general form for the tapered-land, pivoted-shoe, and other types of rigid thrust bearings, an additional parameter E/P is needed for the spring-supported bearings. Preloading and other design factors can be represented by ratios of like quantities. Geometrical similarity is presupposed.

The Kingsbury variable K, denoting ZU/PB, was introduced by Raimondi & Boyd (VI: 1955). Here B is breadth of shoe and U the surface speed at mean radius, so that K is proportional to ZN/P.

Next consider the hydrostatic bearing with parallel surfaces, whose operation is dependent on external pressure. Since the friction is inversely proportional to film thickness with uniform viscosity, and the film thickness depends on the flow rate, we can see that f is a function of D, N, P, Z, and Q. Two independent parameters are required, the simplest being ZN/P and D^3N/Q. Thus the problem can be explored by varying only two quantities instead of the original five. If it should be found by experiment that f is proportional, say, to the two-thirds power of the mean film viscosity and the first power of the speed, dimensional reasoning further shows the influence of P, D, and Q; thus

$$f = \text{const}\ (ZN/P)^{2/3}(D^3N/Q)^{1/3} \, ; \tag{30}$$

the constant being the same for geometrically similar bearings. The constant is evaluated in Chapter VI. Relative film thickness h_0/D is a function of the same two parameters and may be treated in a similar manner.

7. APPLICATION TO OTHER MACHINE ELEMENTS

Oil ring performance had been investigated by Karelitz (III: 1930), Baildon (1937), and by Baudry & Tichvinsky (1937), when our dimensional treatment was applied in discussing the last reference. Later investigations have been published by Blok (1946), Özdas (III: 1955), Lemmon (1960), and co-authors. Suppose the conditions are such that for geometrically similar systems, the rate of oil delivery Q depends only on D, N, P, Z, g, ρ, and τ. Here D and N refer to the diameter and speed of the journal and P is the ring weight per unit of projected bearing area and g the acceleration of gravity. The remaining quantities are properties of the lubricant; Z and ρ the viscosity and density, τ the surface tension. Then Q/D^3N will be some function of ZN/P, $\rho N^2D^2/P$, $\rho gD/P$, and τ/PD. Suppose it were found by varying the temperature, speed, and ring weight, while holding D, ρ, and g constant, that Q is proportional to Z^a, N^b, and P^c over a stated range. Neglecting the surface tension as a first approximation, it can be inferred that Q will vary with D^d where

$$d = \tfrac{5}{2} - \tfrac{3}{2}a + \tfrac{1}{2}b - c\,. \tag{31}$$

We expect to find $c = 0$ when ZN/P is low enough to preclude slip between ring and journal. Depth of ring immersion in bath should be included under geometrical similarity.

Although it often happens that some performance characteristic is jointly proportional to the variables with constant exponents, this does not follow from dimensions. All that was shown above is that if it be known that certain variables appear in this form, the remaining variables must appear with predictable exponents.

Rolling Contact. A survey of rolling contact with slippage was published by H. Blok (X: 1962), including many applications of dimensionless variables. See Chapter X.

Gear tooth friction is often represented by the power-transmitting efficiency η of the meshing gear and pinion, by the mean coefficient of friction f, or by the power loss ratio $1 - \eta$. Dimensionally, for rigid teeth separated by fluid films, η and f will be functions of ZN/P, where P is now the tooth load per unit of face width and pitch diameter. These characteristics will be the same for all pairs of gears that are geometrically similar and operating at the same ZN/P, assuming the viscosity uniform. The relative film thickness h_0/D at the point of nearest approach, D being the pitch diameter, is also a function of ZN/P (H & H, 1945, Fig. 2). The effects of pressure-viscosity, and of elastic deformation, are considered in Chapter XI.

Dynamically Similar Engines. Michell points out (I: 1950, pages 65–7) that the stresses will be the same in geometrically similar machines if the surface speeds are the same and the viscosity proportional to the size, in which case the angular speeds are inversely as the size.

This requires holding ZN/P constant. Suppose one machine is four times as big as another; it should then be run at one quarter of the rpm of the little machine. It will carry sixteen times as great a load, and require four times the viscosity. Meeting these conditions would provide an oil film four times as thick in the larger machine, which is hardly necessary. Again we see that the conditions for dynamical similarity are not necessarily those for optimum design.

These relations have been independently investigated by Professor C. F. Taylor (1950). His experiments were conducted on three dynamically similar engines of bore diameters $2\frac{1}{2}$, 4, and 6 in., respectively. Volumetric efficiency and indicated mean effective pressure were the same for all three engines when plotted against mean piston speed. Professor Taylor further applied dimensional criteria to the data on engine friction and wear in the literature. He concluded that for similar engines running at the same piston speed, both wear-life and weight per unit of power output are proportional to the cylinder bore, or absolute size.

8. THERMAL APPLICATIONS

Analysis of thermal effects requires additional physical dimensions for which the temperature unit Θ and a heat unit Φ are sufficient; but Φ is not required if we are concerned only with frictional heat and express it in work units. It will pay to run through a few examples.

(1) ***Friction and Film Thickness.*** We have seen that the performance of geometrically similar thrust and journal bearings and gearing depends only on a single dimensionless variable, ZN/P, when operating under the simplest conditions. The film viscosity is assumed uniform. If now we extend the picture to allow for local variations in viscosity caused by frictional heating, at least two additional variables must be included in the original list of n quantities. These are a, the temperature coefficient of viscosity, and q, the heat capacity per unit volume. The viscosity Z may be defined by its value at the inlet. To measure a and q or to specify their units requires one additional fundamental unit, making $i = 7 - 4 = 3$. The performance characteristic serves for Π_1 and ZN/P is Π_2. Inspection confirms the choice of aP/q for Π_3. A plot of test data against ZN/P with aP/q as the constant parameter, or vice-versa, should give consistent results under favorable operating conditions. Two such bearings are said to be "thermally similar" if they are dynamically similar and operating at the same value of aP/q.

(2) ***Thermal Equilibrium.*** Let it be required to determine the mean elevation T of the oil film above inlet oil temperature in a lightly loaded, high-speed journal bearing under forced lubrication. Disregard eccentricity and loading. Assume that some of the frictional heat is transferred externally from the housing, and that most of it is carried off by the oil flow Q. Then if the inlet temperature is somewhere near the ambient, the rate of heat transfer is given by $H' = KT$, where K denotes $C_1A + Qq$. Here C_1 is defined by Eq. (6) of Chapter III, in which A is the effective housing area. The equilibrium temperature elevation T will be governed by the inlet viscosity Z_1, speed N, the overall heat transfer constant K, the temperature coefficient of viscosity a, heat capacity q of the lubricant per unit volume, and the size and shape of the bearing as fixed by the journal diameter D, with the usual requirement of geometrical similarity. The list includes $n = 7$ quantities so we should expect $i = 7 - 4 = 3$ dimensionless products. By trial and checking of dimensions, $\Pi_1 = aT$, $\Pi_2 = Z_1N^2D^3/KT$, $\Pi_3 = aZ_1N/q$. All 7 quantities are now accounted for; but it is awkward to have the unknown T on both sides of the prospective equation. Therefore, we deftly multiply Π_2 by Π_1 to get rid of T in Π_2. The new Π_2 is $aZ_1N^2D^3/K$; therefore

$$aT = \phi\left(\frac{aZ_1N^2D^3}{K}, \frac{aZ_1N}{q}\right). \tag{32}$$

In the limiting case of an air-cooled bearing, where no appreciable fraction of the heat is carried away by the oil, $Q = 0$ and $K = C_1A$. The property q has no longer any part to play. The product containing q may be dropped from Eq. (32). This equation treats the viscosity-temperature curve as if it followed Reynolds' formula (Chapter IV) with a constant temperature coefficient. Equation (32) is therefore limited to a moderate temperature rise T. To cope with a wider range would require a second viscosity-temperature constant.

It may be shown that $a = -1/T_0$ when the viscosity can be represented by an hyperbola, as in Eq. (14) of Chapter III. Here T_0 is the elevation of the ambient above the apparent solidifying temperature of the oil. Dropping the q product, and substituting $-1/T_0$ for a in Eq. (32) reduces it to the simpler form

$$T/T_0 = \phi(Z_1N^2D^3/KT_0). \tag{33}$$

Keeping in mind that constant values of L/D and C/D are implied by geometrical similarity, we see that our analytical solution, Eq. (15) of Chapter III, is a particular example of Eq. (33).

(3) *Temperature Drop.* Under steady-state conditions the drop in temperature, $t_j - t_0$, from journal to bearing when running concentrically may depend on the film thickness h, surface speed U, mean viscosity Z or viscosity at the bearing surface, thermal conductivity of the lubricant, k, and its temperature coefficient of viscosity, a. The conductivity of the metals will be taken as infinite compared to that of the oil. Disregard any axial flow of heat. The journal would run pretty hot, yet to simplify the problem we shall assume there is no heat carried off by the oil. We have named $n = 6$ quantities. Therefore, we expect $i = 6 - 4 = 2$ dimensionless products and can set $\Pi_1 = k(\theta_j - \theta_0)/ZU^2$, $\Pi_2 = aZU^2/k$. What became of h? It fell out, leaving

$$\theta_j - \theta_0 = \frac{ZU^2}{k} \phi\left(\frac{aZU^2}{k}\right). \tag{34}$$

If the viscosity is but slightly dependent on temperature within the short range $\theta_j - \theta_0$, we can try $a = 0$ as a first approximation. The function ϕ in (34) then reduces to a constant, in accord with Eq. (30) of Chapter III. We thus have dimensional support for the well-known paradox that $\theta_j - \theta_0$ is independent of film thickness. If, however, D is added to the original list, $n = 7$, $i = 7 - 4 = 3$, and $\Pi_3 = h/D$. We had tacitly assumed h/D to be vanishingly small. But, when h is an appreciable fraction of D, Π_3 is by no means superfluous.

(4) *Metallic Conduction.* Buckingham treated the example of a bearing held at a fixed temperature on the outside (perhaps by water-jacketing), the journal being concentric, or running at a fixed eccentricity. He showed that the temperature elevation at the inside will be given by

$$\Delta\theta = \text{const}\left(\frac{ZN^2D^2}{\lambda}\right); \tag{35}$$

where λ is the effective conductivity of the metal. We start from $n = 5$ quantities that require 4 fundamental units, so that $i = 1$. In every such case, Π_1 is a function of no other variable, and must therefore be a constant. Here, Π_1 is $\lambda\,\Delta\theta$ divided by ZN^2D^2, leading to Eq. (35).

Let it be required, however, to express $\Delta\theta$ as a function of power loss H without reference to Z and N. This is a reasonable problem, but involves only $n = 4$ instead of 5 quantities. The usual rule, $k = 4$, makes $i = 0$. Where is the catch? By analogy with Eq. (35), the desired relation must be

$$\Delta\theta = \text{const}\,(H/D\lambda). \tag{36}$$

This equation is both complete and dimensionally homogeneous. The explanation can only be that the *minimum* number of fundamental units needed is *less* than 4. The dimensions are $\Delta\theta$: θ, H: FL/T, D: L, and λ: $F/T\theta$ in terms of the usual 4; but 3 will suffice, namely θ, L, and F' where F' denotes F/T. Actually then $i = 1$.

This latter problem is a variant of one offered by Hunsaker & Rightmire (page 109 of 1947) in discussing how to determine the number of independent dimensionless products.

(5) *Nonsteady Heat Conduction.* To express the temperature elevation as a function of the time t before equilibrium has been reached, we can start with $n = 6$ quantities, $\Delta\theta$, H, D, λ, q_m, and t; where q_m is the heat capacity of the metal per unit volume. As usual $k = 4$ so that $i = 6 - 4 = 2$ and

$$\Delta\theta = \frac{H}{D\lambda}\,\phi\left(\frac{\lambda t}{D^2 q_m}\right). \tag{37}$$

Observations on one bearing, plotted in terms of these dimensionless variables, might enable us to predict the temperature-time curve for geometrically and "generally" similar bearings of a different size and constructed from different materials; provided allowance is made for the fact that $\Delta\theta$ is the temperature rise above the bearing exterior, not above the ambient air at a distance. Two such bearings are "thermally similar" if observed when $\lambda t/D^2 q_m$ is the same for model and original. The temperature rise should then be inversely proportional to the size, or four times as great for a quarter-sized model. Possibly a dummy bearing would serve as the model, with heat provided by an electrical heating coil or ribbon instead of by friction.

(6) *Microhydrodynamic Lubrication.* Since the surfaces of solid bodies are rough and wavy, it follows that when two such bodies, nominally parallel and smooth, are brought close enough together, hydrodynamic pressures may be developed between high spots approaching tangentially. Hydrodynamic lubrication must be expected on a microscopic scale. Friction and film thickness will depend on such factors as the elastic constants of the bearing materials; the pressure and temperature coefficients of viscosity; and the heat capacity of the lubricant per unit volume and its thermal conductivity, as well as the mean or inlet viscosity, the speed, the load per unit of projected area, the shapes of the rubbing surfaces, and a reference linear dimension. A dimensional treatment was outlined in the writer's paper on thin-film lubrication (XIII: 1933). Thus for geometrically similar systems the coefficient f may be influenced by Young's modulus E,

Poisson's ratio ν, the ratios E'/E and ν'/ν for the two bodies, b, a, q, k, Z, N, P, and D. If the systems are "generally similar," E'/E and ν'/ν are constant, leaving a relation to be expressed between $n = 11$ physical quantities. With $k = 4$, $i = 7$, from which

$$f = \phi\left(\frac{ZN}{P}, \frac{aP}{q}, bP, \frac{aND^2P}{k}, \frac{E}{P}, \nu\right). \tag{38}$$

Optionally aP/q may be replaced by bq/a, which depends solely upon the lubricant; and bP may be replaced by its dimensional equivalent bE, which depends only on the materials chosen. Thus for a given set of materials, aP/q and bP may be dropped from Eq. (38). When the film is thick enough to serve as a perfect insulator, the result is the same as if $k = 0$ so that aND^2/k may be dropped. Under these conditions the coefficient of friction is independent of the absolute size D, and (38) reduces to Eq. (17).

The last few problems, as stated, have been limited to the simplest operating conditions. In any particular application the conditions should be reexamined, to see, for example, if ambient or inlet pressures should be included, of if turbulence may be expected.

Practically every step in the derivation of Eq. (38) applies to the relative film thickness h_0/D as well as to the coefficient of friction and may be extended to other machine elements as well as bearings.

The dimensional treatment of thin-film and elastohydrodynamic lubrication has been further advanced by H. Blok in a number of publications (1951, 1958, 1962), as shown in Chapters X and XI.

9. GENERAL CONSIDERATIONS

Geometrical Similarity. After so many repetitions of the limitation to geometrical similarity, the question may well come up: what good is it to have a lot of fancy formulas, if they can be applied only to bearings that happen to be geometically similar? The limitation can be avoided by explicitly writing down, as additional Π's, all the length ratios like C/D and L/D that may be needed to specify the shape of the bearing. Or we can introduce the more important ratios and reserve the similarity requirement for any that remain. These are sometimes denoted collectively by $r_1, r_2, \ldots$ lest we forget their existence.

Derivatives. At times it may be of interest to get rid of the unknown function ϕ accompanying the result of a dimensional analysis. This can be accomplished by differentiating (Hersey, 1919). A known relation between derivatives is thus obtained in place of an unknown relation among the primary quantities. Take the formula $f = \phi(ZN/P)$ for any bearing or other element in which the remaining parameters can be held constant. Differentiate partially with respect to the load. This gives an expression containing ϕ', the total derivative of ϕ with respect to its argument. Differentiating by speed leads to a similar expression. Eliminate ϕ' between the two equations. The result is

$$P(\partial f/\partial P) = -N(\partial f/\partial N). \tag{39}$$

Thus when it is more convenient to vary N than P over a short interval, we can predict the effect of a change in load from that observed experimentally under the same percentage change in speed or viscosity. This method is not limited to lubrication. It was applied to a problem in hydraulics by A. E. Roach and co-author (1951).

Note on Fundamental Units. Not more than $k = 4$ fundamental units are required in the treatment of thermal problems where mechanical energy is converted into heat by friction, as can be seen from the six examples given. The same results are obtained by including heat as an independent dimension provided the mechanical equivalent J is added to the list of n quantities, whereby i remains unchanged.

Thermal problems not involving the conversion of mechanical energy into heat require the recognition of heat as a fundamental unit, or independent dimension, without the addition of J to the original list. Examples are (1) forced convection of heat, as in a water-jacket, and (2) free convection, as in cooling the oil bath in the housing of a vertical thrust bearing by submerged copper tubing, chilled internally by water circulation. Heat transfer problems of this kind are dealt with like the foregoing except that in general $k = 5$ instead of 4.

It is optional in friction problems whether to include both Φ and J, or to omit them both as done in the examples above, since i will come out the same either way. But it is mandatory to include Φ among the k's and omit J from the n's if the heat involved does not come from mechanical energy in the phenomenon described.

Paradoxically, the fewer Π's we have the more informative the result, provided their number was not erroneously reduced by using too large a value of k. No harm comes from a superfluous Q among the original n; but it is essential to check carefully to insure that the k chosen is truly a minimum—watching especially for "associated units," a kind of bird-watching game. Various rules have been proposed in the effort to make dimensional analysis foolproof. No rule is as good as familiarity with physical facts, and cumulative experience in applying them.

10. SUMMARY

Dimensional technique has a variety of useful applications to lubrication, including:

1. Checking, memorizing, and deriving formulas, and converting units.
2. Reducing the number of independent variables needed for representing experimental data, as when plotting f or h_0/c against ZN/P.
3. Generalizing incomplete empirical equations.
4. Indicating the requisite conditions for model experiments.
5. Planning future investigations in the most economical manner.
6. Compulsory elimination of fuzzy thinking about physical factors.

A surprising number of books have been written on "dimensional analysis," as can be seen from Table 4. The term was coined by Professor Bridgman in choosing a title for his book. The first and fourth columns in our table give the dates of publication of the first editions. The third and sixth columns show the number of pages in the latest edition of each book. For details, see the reference list at end of this chapter.

Many books in the engineering field include one chapter expressly on dimensions. Countless articles have been published on dynamical similarity and dimensional analysis. Professor Bridgman's last article on the subject (1945) goes to the root of the matter by emphasizing the "absolute significance of relative magnitudes." But for everyday use, it would be difficult to improve on Edgar Buckingham's ASME paper (1915) in conjunction with his articles of 1921 and 1924.

TABLE 4. Books on Dimensional Analysis

Year	Author	Pages	Year	Author	Pages
1922	Bridgman	117	1953	Duncan	156
1933	Porter	80	1953	Focken	224
1936	Lanchester	314	1953	Kirpechev	92
1941	Hill	107	1958	Comolet	116
1944	Sedov	363	1960	Ipsen	236
1945	Esnault-Peltérie	236	1963	Baker	78
1948	Stubbings	107	1963	Gukhman	256
1950	Murphy	302	1964	Pankhurst	151
1951	Langhaar	166	1965	Kline	229
1952	Huntley	158			

REFERENCES

American Machinist (1915), "Laws of Lubrication of Journal Bearings." (editorial) **43**, 81–2 (July 8).

Baildon, E. (1937), "The Performance of Oil Rings." *GDLL* **1**, 1–7.

Baker, W. E. (editor) (1963), *Use of Models and Scaling in Shock and Vibrations* (Colloquium of 8 papers); *ASME*, New York, 78 pp.

Barwell, F. T. (1957), "The Sommerfeld Variable, Its Origin and Further Generalization." Paper I-205, *Ninth ICAM*, Brussels, 1956; *Actes* **4**, 272–83. (a) NEL, Glasgow, L.D.R. 29/56; 14 pp.

Baudry, R. A. & Tichvinsky, L. M. (1937), "Performance of Oil Rings." *ME* **59**, 89–97, 291–2.

Becker, A. E. (1926), "Does the *Zn/p* Function Fit the Facts?" *IEC* **16**, 856. (a) Reply by D. P. Barnard, 975.

Biel, C. (1920), "Die Reibung in Gleitlagern bei Zusatz von Voltolöl zu Mineralöl und bei Veränderung der Umlaufzahl und die Temperatur." *ZVDI* **64**, 447–83.

Blok, H. (1946), "Considérations sur le graissage par les bagues." *Soc. Belge de l'Étude du Petrôle,* Conf. Apr. 27; 20 pp.

Bridgman, P. W. (1922), *Dimensional Analysis.* Yale University Press, New Haven; 112 pp. (a) 2nd ed. (1931); 113 pp. (b) German ed. with H. Holl, *Theorie der physikalischen Dimensionen,* Teubner, Leipzig and Berlin, 1932; 117 pp.

Bridgman, P. W. (1945), "Dimensional Analysis." *Encyclopedia Britannica,* 14th ed., **7**, "W" and later printings, 387–387 J (11 pp.).

Buckingham, Edgar (1914) [1], "Physically Similar Systems." *JWAS* **4**, 347–53.

Buckingham, Edgar (1914) [2], "On Physically Similar Systems; Illustrations of the Use of Dimensional Equations." *Phys. Rev.* **4**, 347–76 (Oct.).

Buckingham, Edgar (1915), "Model Experiments and the Forms of Empirical Equations." *T* **37**, 263-96.

Buckingham, Edgar (1921), "Notes on the Method of Dimensions." *Phil. Mag.* (7) **42**, 696–719.

Buckingham, Edgar (1924), "Dimensional Analysis." *Phil. Mag.* (7) **48**, 141–5.

Clayton, D. & Wilkie, M. J. (1939), "An Investigation of the Warlop Bush." *The Engineer* **167**, 761–3.

Comolet, R. (1958), *Introduction à l'analyse dimensionelle et aux problèmes de similitude en mécanique des fluides.* Masson, Paris; 116 pp.

Duncan, W. J. (1953), *Physical Similarity and Dimensional Analysis.* Arnold, London; 156 pp.

Esnault-Peltérie, Robert (1945), *L'Analyse dimensionelle.* F. Rouge & Cie., S. A., Lausanne. (a) 2nd ed., Gauthier-Villars, Paris, 1946; 232 pp. (b) Ed. of 1948; 236 pp.

Focken, C. M. (1953), *Dimensional Methods and Their Applications.* Foreword by H. Dingle. Arnold, London; 224 pp.

Fourier, J. B. J. (1822), *Théorie analytique de la chaleur,* Didot, Paris (see Chap. II, 9). (a) Transl. by A. Freeman, London, 1878; 466 pp. (b) Reprint, Dover, New York, 1955; 466 pp.

Gukhman, A. A. (1963), *Introduction to the Theory of Similarity* (Russian). Vysshaya Shkola, Moscow. (a) Transl. by Scripta Technica, Inc., edited by R. D. Cess, 1965; 256 pp.

Gümbel, L. (1914), "Das Problem der Lagerreibung." *Monatsblätter des Berliner Bezirksvereines Deutscher Ing.* **5** (May–June) 87–104, 109–120.

Hersey, M. D. (1909), *Journal Friction and Carrying Power under Varied Conditions of Lubrication. Thesis, Dept. Mech. Eng.,* Massachusetts Institute of Technology; 156 pp. (a) *JWAS* **4,** 542–52 (1914).

Hersey, M. D. (1919), "A Relation Connecting the Derivatives of Physical Quantities." *Sci. Papers NBS* No. 331; 21–9. (a) *JWAS* **6,** 620–9 (Nov. 4, 1916).

Hersey, M. D. (1934), "Present Status of Dimensional Theory in Lubrication." *World Petrol. Cong.* (1933); *Proc.,* Inst. of Petroleum, London, 389–90.

Hersey, M. D. (1950), Discussion on "Turbulence in High Speed Journal Bearings." *T* **72,** 833.

Hersey, M. D. & Hopkins, R. F. (1945), "Viscosity of Lubricants under High Pressure." *ME* **67,** 820–4; 68, 576–83. (a) *T* **67,** 187.

Hill, W. S. (1941), *Teoria general de las Magnitudes fizicas.* Universidad de la Republica oriental del Uruguay, Facultad de Ingeneria, Montevideo; 107 pp.

Hunsaker, J. C. & Rightmire, B. G. (1947), "Dimensional Analysis and Similitude." Pages 98–121 in *Engineering Applications of Fluid Mechanics.* McGraw-Hill, New York.

Huntley, H. (1952), *Dimensional Analysis.* Macdonald, London; 158 pp.

Ipsen, D. C. (1960), *Units, Dimensions, and Dimensionless Numbers.* McGraw-Hill, New York; 236 pp.

Kirpechev, M. V. (1953), *Theory of Similarity* (Russian). Izd-vo Akademii Nauk SSSR, Moscow; 92 pp.

Kline, S. J. (1965), *Similitude and Approximation Theory.* McGraw-Hill, New York; 229 pp.

Lanchester, F. W. (1936), *The Theory of Dimensions and Its Application for Engineers.* Taylor & Francis, London; 314 pp.

Langhaar, H. L. (1951), *Dimensional Analysis and the Theory of Models.* Wiley, New York; 166 pp. (a) French transl. by C. Charcosset, Dunod, Paris, 1956; 230 pp.

Lemmon, D. C. & Booser, E. R. (1960), "Bearing Oil-Ring Performance." *JBE, T* **82,** 327–34.

Murphy, Glenn (1950), *Similitude in Engineering.* Ronald, New York; 302 pp.

Pankhurst, R. C. (1964), *Dimensional Analysis and Scale Factors*. Chapman & Hall, London; Reinhold, New York; 151 pp.

Porter, A. W. (1933), *The Method of Dimensions*. Methuen, London; 80 pp. (a) "Units, Dimensions of," *Encyclopedia Britannica* **22,** 1944, 853–60.

Rayleigh, Lord (1915), "The Principle of Similitude." *Nature* **95,** 66–8 (Mar. 18).

Riabouchinsky, D. (1911), "Variables de dimension zero . . . ." *L'Aérophile* **19,** 407–8 (Sept. 19). (a) *Bull. de l'Institut Aerodynamique de Koutchino* No. 4, 50–55, 90 (1912).

Roach, A. E. & Kehrl, H. H. (1951), "Performance of Centrifugal Pumps from Affinity Laws at Non-Corresponding Operating Points." *Indust. Math.* **2,** 32–8.

Sedov, L. I. (1944), *Similarity and Dimensional Methods in Mechanics* (Russian). Gosnia Izd-vo Tekhnika Teoret Literatury, Moscow, 1944. (a) 3rd ed., 1954; 328 pp. (b) 4th ed., 1957. (c) Transl. by M. Friedman, edited by M. Holt, Academic Press, New York, 1959; 363 pp.

Shaw, M. C. & Macks, E. F. (1949), *Analysis and Lubrication of Bearings*. McGraw-Hill, New York; 618 pp.

Smith, L. P. (1944), "Logical Methods of Investigation." *J. Am. Soc. Naval Eng.* **56,** 1–27.

Sommerfeld, A. (1921), "Zur Theorie der Schmiermittelreibung." *Z. Techn. Phys.* **2,** 58–62, 89–93. (a) *GDLL* **1,** 297–301.

Stribeck, R. (1902), "Die wesentlichen Eigenschaften der Gleit- und Rollenlager." *ZVDI* **46,** 1341–8, 1432–8, 1463–70. (a) *Forschungsarbeit* **7,** 1903; 47 pp.

Stubbings, G. W. (1948), *Dimensions in Engineering Theory,* Lockwood, London; 107 pp.

Taylor, C. F. (1950), "Effect of Size on the Design and Performance of Internal Combustion Engines." *T* **72,** 633–45. (a) *Technology Review,* **52,** 142–3 (1950). (b) *Proc. Engine Wear Symposium,* Southwest Research Institute, San Antonio, 3–21 (1958).

Vaschy, A. (1892), "Sur les lois de similitude en physique." *Ann. Télegraphiques*. Paris, 3rd Ser. **19,** 22–9. (a) *CR* **114,** 1416–9.

Vogelpohl, G. (1949), "Ähnlichkeitsbeziehungen der Gleitlagerreibung und untere Reibungsgrenze." *ZVDI* **91,** 379–84.

Wilson, R. E. & Barnard, D. P., 4th (1922), "The Mechanism of Lubrication." *TSAE* **17,** pt. 1, 203–38.

chapter VI Theory and Testing of Thrust Bearings

1. Characteristics of rectangular surfaces. 2. Tapered-land bearings. 3. Pivoted-shoe bearings. 4. Other forms of thrust bearing. 5. Effect of fluid properties. 6. Deformation and vibration. 7. Hydrostatic lubrication. 8. Thermal lubrication. 9. Water-lubricated thrust bearings. 10. Summary.

Thrust bearings support axial loading. Noteworthy applications are found in large vertical hydroelectric generators, in marine propeller shafts, steam and gas turbines, centrifugal pumps, and a variety of other equipment. Sliding surfaces with fluid film lubrication, rather than rolling contacts, are usually necessary in installations where no statistical risk of failure can be tolerated. The physical principles outlined in earlier chapters are applied here to the various types of thrust bearing in which the surfaces can be completely separated by a lubricating film.

1. CHARACTERISTICS OF RECTANGULAR SURFACES

From the review of Michell's theory in Chapter II it will be recalled that the load capacity of a finite rectangular plane surface is less, per unit area, than that of the surface in Reynolds' solution having an infinite length L at right angles to the motion. Michell found that the load capacity of a square bearing, per unit area, is little more than two-fifths that of Reynolds' bearing. The "square bearing," or its sector-shaped equivalent, is very commonly met in practice.

Let P denote the load capacity per unit area of a rectangular bearing shoe for any breadth-length ratio B/L, and P_0 the value of P for

$B/L = 0$, as in Reynolds' solution. The load ratio or relative load capacity, P/P_0, equals unity when there is no side leakage, and drops off sharply with increasing values of B/L. The load ratio curve used by Kingsbury (Fig. 10 of VII: 1932) was limited to near-optimum conditions.

When B/L is large, we have a *ski-like* shoe, to borrow a phrase from Professor Fuller. When B/L is small, the shoe is *oar-like.* The feathered blade of an oar glides like the Reynolds inclined plane bearing. The upper left-hand part of a load ratio curve represents the oar-like, and the lower right-hand part the ski-like shoe. The square bearing comes near the middle at $B/L = 1$.

Load Ratio Curves. There will be a family of curves for different values of the relative film thickness Y, defined as the ratio of the minimum film thickness h_0 to the taper-height h_t. Here h_t denotes the difference between film thickness h_1 at the entering edge and the value h_0 at the trailing edge. Note that Y is the reciprocal of $a - 1$, where a is Norton's film thickness ratio, h_1/h_0 in the present notation. Michell's factors, Table 2 of Chapter II, apply only to the curve for $Y = 1$. This value is often taken as a standard for reference, since it is close to the optimum value 0.85 for which Reynolds found a maximum load capacity when $B/L = 0$. The optimum is somewhat less when B/L is finite. From the table cited, $P/P_0 = C_P/0.159$.

Michell not only determined load capacities for $B/L = 1$ and 3 but showed how solutions might be obtained for B/L approaching infinity. His solution was extended by *Engineering* (1920), by Boswall (II: 1928), and by Kobayashi (1934). Kingsbury determined the relative load capacity for the square bearing with $Y = 1$ by his electrical method. Measurements reported in 1931 led to a mean value 1.6 per cent above Michell's, or $P/P_0 = 0.430$, now an accepted figure.

Many other theoretical solutions have been tried or proposed. The first, by Muskat and co-authors (1940), gave results from $B/L = 0$ to 4, at Y values up to 2.8. The second, by Frössel, was published in 1941. His solution was given in graphical form from $B/L = 0$ to $\frac{5}{4}$ and from $Y = \frac{1}{8}$ to 1. A third solution was proposed by Mrs. W. L. Wood (1949) but not applied. A solution was published by Hays (1958) and another by Jakobsson and co-author (1958). Hays' solution was presented graphically from $B/L = 0$ to 4, and for Y from $\frac{1}{4}$ to 4; Jakobsson's in numerical form over the range from $B/L = 0$ to 2 and from $Y = \frac{1}{5}$ to 2. The last two investigations made use of digital computers. See also the values by Wang (1950).

The results of nine completed investigations have now been averaged and plotted in Fig. 1 as a family of curves for P/P_0 against B/L at

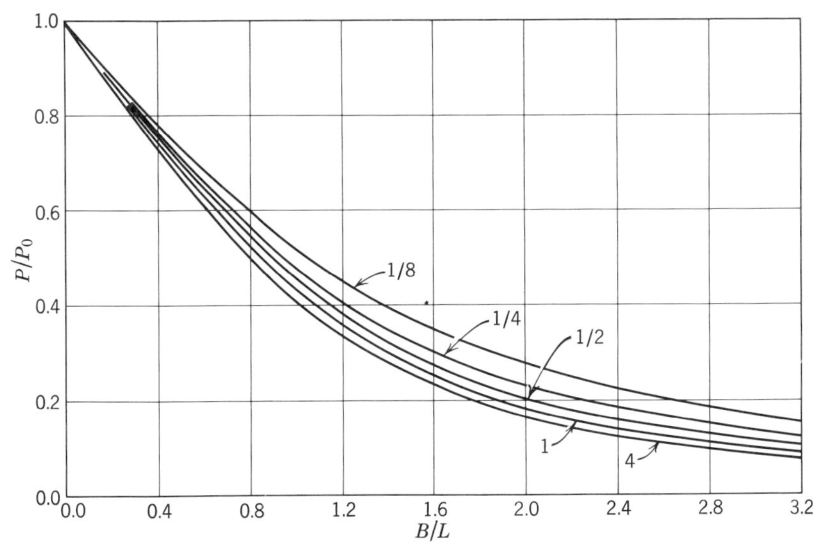

Fig. 1. Load ratios for plane rectangular surfaces (Y values on curves).

constant Y. Values read from Fig. 1 are assembled in Table 1 for convenient use. For example, what fraction, P/P_0, of the Reynolds load capacity can be expected from a bearing shoe whose breadth B in the direction of motion is 1.25 times its length L at right angles to the motion, (1) when the trailing-edge film thickness h_0 is half the taper-

TABLE 1. Load Ratios for Rectangular Plane Surfaces
(Mean P/P_0 from nine investigations)

B/L	$Y = \frac{1}{8}$	$Y = \frac{1}{4}$	$Y = \frac{1}{2}$	$Y = 1$	$Y = 2$	$Y = 4$
$\frac{1}{4}$	.88	.87	.85	.84	.84	.83
$\frac{1}{2}$	.74	.72	.71	.69	.69	.68
$\frac{3}{4}$	.62	.60	.58	.56	.55	.54
1	.52	.48	.46	.43	.42	.41
$\frac{3}{2}$	.38	.32	.30	.28	.27	.26
2	.28	.23	.20	.18	.17	.16
3	.17	.14	.12	.10	.090	.085
4	.10	.085	.070	.060	.055	.050

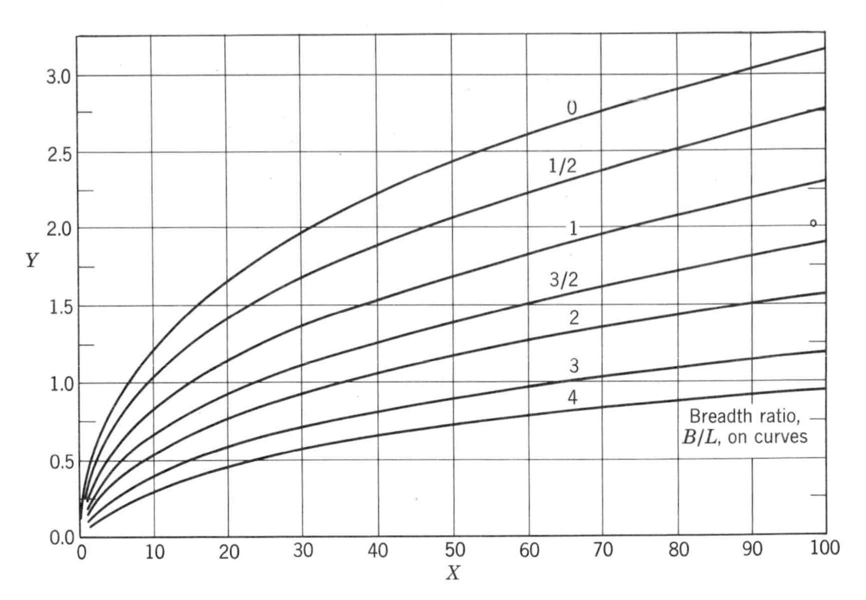

Fig. 2. Relative film thicknesses for rectangular plane surfaces.

height h_t? (2) when h_0 is twice h_t? In the first case, $B/L = \frac{5}{4}$ and $Y = \frac{1}{2}$; interpolating in Table 1 gives $P/P_0 = 0.38$. In the second case, $Y = 2$, making $P/P_0 = 0.345$, say 0.35.

Apparently P/P_0 approaches unity as Y approaches zero at any B/L, and approaches zero as Y approaches ∞. Side leakage is a more important factor the greater the relative film thickness.

Alternative Representations. The results of side leakage theory are presented in a different form in Fig. 2. Here, the relative film thickness Y is plotted against the thrust variable X for constant values of B/L, where X is defined by K/c^2; in which K denotes the Kingsbury number, ZU/PB (Chapter V), and c is the "taper angle," or inclination of bearing surface.

The linear speed U at mean radius may be set equal to $2\pi N$ times B/A, where N is the shaft speed, B the breadth of thrust pad or shoe at mean radius, and A the sector angle. Therefore,

$$X = \frac{ZU}{PBc^2} = \frac{2\pi}{Ac^2}\left(\frac{ZN}{P}\right). \tag{1}$$

Now from Reynolds' integration, Eq. (24) of Chapter II, putting $h_0 = Yh_t$ and $h_t = Bc$, it is seen that $X = Y^2/C_P$. And by substituting

$1 + (1/Y)$ for a in Eq. (25), an expression is obtained for C_P as a function of Y. Hence under the condition of no side leakage, B/L approaching zero,

$$\frac{1}{X} = 6\left[\ln\left(1 + \frac{1}{Y}\right) - \frac{2}{1 + 2Y}\right]. \tag{2}$$

The top curve in Fig. 2 is plotted from Eq. (2) for values of the thrust variable up to $X = 100$. See also Hersey II: 1954, 60–61 or (a) 55.

The remaining curves in Fig. 2 can be derived from the top curve by reference to the load ratios in Table 1, since X is inversely proportional to P at any fixed value of Y. Figure 2 is in good accord with similar charts constructed by Muskat (1940) and by Raimondi (1955) [1], and their co-authors, using different methods.

A still different representation is offered by Michell in his book (I: 1950), where load capacity contour lines are given in two folding charts. In both, the scale of ordinates corresponds to Y and the scale of abscissas to L/B. His Chart IV F is drawn for a constant inclination c between the two surfaces; Chart IV G for a constant minimum film thickness. See Charnes & Saibel (1952).

Formulas for Curves. Many investigators have tried to represent the P/P_0 curves by theoretical and empirical formulas. A reasonably good fit is obtained by an expression like Schiebel's of 1933

$$\frac{P}{P_0} = \frac{1}{1 + C(B/L)n^n}, \tag{3}$$

where $n = \frac{5}{3}$ and

$$C = 1.50 - 1.22/(1 + 6Y). \tag{4}$$

Equation (2) can be replaced by simpler expressions over stated portions of the X range. For relative film thicknesses not too far from Reynolds' optimum of 0.85, Y may be taken equal to 0.4 times the square root of X. Above the optimum, Y increases less rapidly than the square root; below, more rapidly. There is a point of inflexion near $X = 0.1$, $Y = 0.03$. Below that point, Y increases faster than the first power of X. The slope dY/dX is zero at the origin.

Optimum Values. Reynolds showed that the maximum load capacity P_0 occurs when $a = h_1/h_0 = 2.18$ and $Y = 0.85$, provided $B/L = 0$. But the optimum film ratio may be different when B/L is finite. From the definitions of X and Y it follows that

$$P = \frac{ZUB}{h_0^2}\left(\frac{Y^2}{X}\right). \tag{5}$$

TABLE 2. Optimum Film Ratios for Load Capacity

B/L	Y	h_1/h_0	B/L	Y	h_1/h_0
0	0.85	2.2	2	0.6	2.7
$\frac{1}{2}$	0.75	2.4	4	0.55	2.8
1	0.65	2.6	∞	0.0	∞

In the general case, P will be a maximum for constant values of h_0 and B/L when Y^2/X is a maximum. Turning to Fig. 2, we can read off the X and Y values for a series of points along any one curve and tabulate Y^2/X. These quotients are then plotted as ordinates against Y in a new family of curves. The optimum Y values corresponding to the maxima of Y^2/X are collected in Table 2. The optimum film thickness ratios h_1/h_0, computed from $a - 1 = 1/Y$, have been included. It appears that Y drops from 0.85 to 0.6 as B/L goes from 0 to 2, while h_1/h_0 rises from 2.2 to 2.7. These values, however, are not precise, since the maxima utilized were not at all sharp.

Another optimum problem is to determine the best rectangular proportions. Lord Rayleigh (1918) noted that if it were not for side leakage, the maximum load per unit area would be realized with the greatest possible breadth B in direction of motion. Let B/L be denoted by β. Then from Reynolds' solution, Eq. (24) of Chapter II, putting $P = P_0$ and $B = \beta L$, we find that $P_0 = \beta C_0$, where C_0 is a constant with respect to β. Substituting this expression for P_0 into Eq. (3) makes

$$P = \frac{\beta C_0}{1 + C\beta^n}. \tag{6}$$

Here C is a function of Y given by Eq. (4).

Differentiating Eq. (6) with respect to β and setting the first derivative equal to zero gives an expression for the optimum value of β, namely,

$$\beta_0 = \frac{1.275}{C^{0.6}}, \tag{7}$$

By trial and error we find the optimum Y equal to 0.65 from Table 2; hence by Eq. (14), $C = 1.251$, and by (7), $\beta_0 = 1.12$. This lends support to the frequent practice of designing with "square" pads or shoes, in which $\beta = 1$.

TABLE 3. Friction Ratios, Rectangular Plane Surfaces
(Mean values of F/F_0)

B/L	$Y = \frac{1}{4}$	$Y = \frac{1}{2}$	$Y = 1$	$Y = 2$
$\frac{1}{2}$	.90	.94	.97	.99
1	.83	.88	.94	.98
2	.74	.83	.92	.97
4	.69	.81	.90	.96

Friction Relations. Thus far we have been interested only in load capacity. The study of frictional resistance offers analogous problems. The solution by Reynolds' theory for $B/L = 0$ may be expressed in terms of the friction per unit area,

$$F_0 = C_F(ZU/h_0) . \tag{8}$$

The friction factor C_F was given by Norton (II: 1942) in the form

$$C_F = \frac{4}{a-1}\ln a - \frac{6}{a+1} \cdot \tag{9}$$

The coefficient of friction, F_0/P_0, approaches the limit $(\frac{2}{3})c$ as Y approaches zero. This follows by letting a approach infinity in the expression for C_F/C_P while keeping in mind the definition of Y, the relation $a - 1 = 1/Y$, and the fact that $h_t = Bc$.

In general, the friction F per unit area is slightly less than F_0, as shown by Table 3, the analog of Table 1. The values are taken from Hays (1958) and from Jakobsson & Floberg (1958). According to Floberg (1960), the optimum B/L for minimum friction is $\frac{4}{3}$; not far from the best shape for maximum load capacity. Kingsbury showed optimum characteristics of plane rectangular bearings in Fig. 4 and Plate X of VII: 1932.

2. TAPERED-LAND BEARINGS

Thrust bearings are of two kinds, the solid type and tilting shoe. Solid bearings are mainly of the tapered-land design.

The tapered-land idea may be traced back to Barr's discussion of Tower's report on the friction of a pivot bearing (IME, 1891). A patent was granted to de Ferranti in 1911. The construction and use

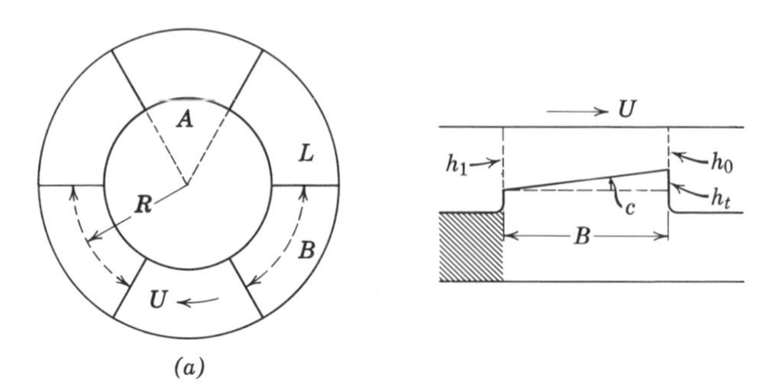

Fig. 3. Simple tapered-land bearing. (*a*) Axial view, (*b*) profile at mean radius.

of tapered-land bearings were reported by Gibbs in 1918. Later types are discussed by Linn & Sheppard (1937) and Wilcock & Booser (I: 1957). Small tapered-land bearings are described with test data by Levinsohn & Reynolds (1953).

Up to this point our calculations have been limited to rectangular plane surfaces. Tapered-land thrust bearings differ in two respects: (1) the sector shape, and (2) the composite profile. These features are sketched in Figs. 3 and 4. Three fixed pads are shown in Fig. 3, each having a central angle A, breadth B at mean radius R, radial length L, taper angle c, and taper height h_t. Note that $h_t = Bc = h_1 - h_0$. Figure 4 shows that the "composite" tapered-land has the additional feature of a flat-top to support the runner when it is standing still. Here r denotes the taper-fraction, or fraction of breadth occupied by the tapered surface.

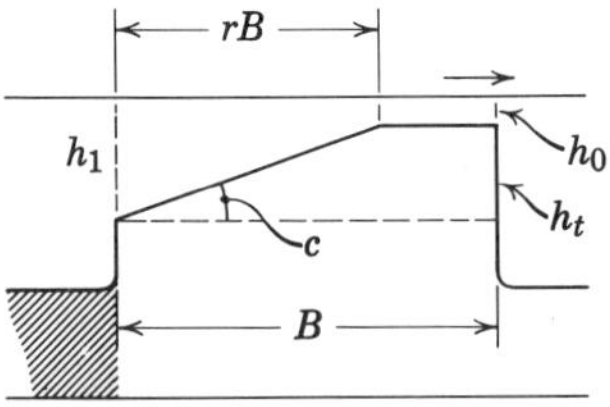

Fig. 4. Composite tapered-land (r denotes the taper fraction).

The Sector Effect. It is customary in most practical work to disregard any special influence of the sector shape. The sector-shaped pad can then be treated as a rectangular plane surface having the same B and L dimensions, as was done in Eq. (1). Scattered data are available in the literature from which a study could be made to determine the sector effect, P_s/P, where P_s is the load capacity of a sector-shaped pad equivalent in other respects to a rectangular one.

Several investigators have given theoretical solutions taking the sector shape expressly into account, notably Boswall in his Chapter VIII (II: 1928). See also Kingsbury (II: 1931), Skinner (1938), Christopherson (1941), Brand (1951), Charnes and co-authors (1953) [2], Kunin (1957), Sternlicht & Sneck (1957), and Tao (1959). Values obtained for P_s/P ranged from 0.85 to 1.44 under diversified operating and design conditions. Now the outer filaments are sheared over a longer distance, and at a higher speed than those near the inside radius. Intuitively we should expect P_s/P to average greater than unity. See Kettleborough (1956) [3], Pinkus (1958).

The Composite Profile. A simple composite bearing is the flat surface with $B/L = 0$ and a short chamfered approach of slope c. A solution for the case of a relatively thin film was offered by Harrison (1919). It was prompted by his study of Tower's report on the collar bearing (IME, 1888). In the limit, as r and Y both approach zero, Harrison's expression for load capacity per unit area reduces to

$$P_c = (\tfrac{3}{2})ZU/ch_0 \,. \tag{10}$$

Dividing the right-hand side by Reynolds' expression for P_0, and noting that h_0/Bc is equal to rY, we see that

$$P_c/P = (\tfrac{3}{2})rY/C_P \,; \tag{11}$$

where C_P is the load coefficient for the relative film thickness Y. The load factor P/P_0 is nearly unity for very low values of Y. Thus Eq. (11) may be taken as an approximation for the chamfered-edge effect regardless of B/L.

Equation (10) may be illustrated by calculating the load supported on an oil film $\frac{1}{100}$ of a mil thick, when a chambered-edge block is moved by hand at the speed of 1 in./sec. Consider a chamfer slope $c = \frac{1}{2}$ and viscosity 3.3 n or 23 cp. Substitution gives $P_c = 1.0$ psi. If $B = 1$ in. and the chamfer ratio $r = \frac{1}{100}$, $Y = \frac{1}{500}$; and C_P, by Eq. (19) of Chapter II, is 1.02×10^{-4}. Substituting in Eq. (11) makes P_c/P about 0.30. It seems surprising that such a short chamfer would create a load capacity nearly one-third that of a fully tapered-land—one having the same taper-height, $h_t = 5$ mils, in a distance of 1 in.

Gümbel derived an expression for the pressure distribution in a composite bearing of $B/L = 0$ without restricting the values of r and Y. He then calculated load capacities numerically (I: 1925). In the case $Y = 1$ he found maximum load capacity when r is about $\frac{3}{4}$.

It remained for G. S. Bower to complete the integration for $B/L = 0$ and tabulate load capacity as a function of r and Y both—see his discussion of Fogg's paper (1946). He found optimum values

$r = Y = \frac{4}{5}$. Under these conditions, P_c/P_0 is nearly equal to 1.20. Thus the composite bearing has a greater load capacity than a full taper, as shown earlier by Gümbel. Bower's formula is complicated, but it reduces to Harrison's when r and Y approach zero, and to Reynolds' when $r = 1$. Since the formulas by Harrison, Gümbel, and Bower disregard the sector effect, they are strictly applicable only to straight-line motion.

Tapered-Land Design. Commercial tapered-land bearings are usually designed with a definite radial taper at the leading edge of each pad, diminishing to zero at the trailing edge (Linn & Sheppard, 1938). Let δ_1 denote the taper-height at the inside and δ_2 that at the outside diameter; then h_t may be taken as the mean, while the effective slope c is h_t/rB. See Kettleborough (1956) [2].

In practice the outside diameter is about twice the inside, and the taper-fraction r is usually equal to $\frac{4}{5}$. Design data for composite tapered-land bearings are presented by Wilcock & Booser in Chapter II of their book (I: 1957). Following rules for determining the number of pads and load per unit area, charts and equations are given for estimating film thickness at the trailing edge. The safe thickness depends on cleanliness of oil and deformations to be expected. The authors advise a minimum oil-film thickness of 1.0 mil for small bearings and 2.0 mils for large, high-speed bearings.

The design example on pages 327–329 of Wilcock and Booser serves to illustrate their method. A steam-turbine thrust load of 70,000 lb. is carried on six pads at 3,600 rpm. The design required pad dimensions of $B = 5.25$ and $L = 4$ in., from which $B/L = 1.31$. Conventional practice for a pad of this size calls for δ_1 and δ_2 of 7 and 4 mils, respectively. Taking $h_0 = 1.5$ mils as a safe film thickness, we find P_t/P approximately 1.11. Here P_t is the load per unit area on a commercial-type tapered-land bearing, while P is the value for a plane rectangular surface of the same B/L.

Load capacities for tapered-land bearings operating with an assumed uniform viscosity have also been published by Sternlicht & Maginnis (1957), Sternlicht & Sneck (1957). They were determined for an 8-shoe bearing with $A = 0.69$ rad, inside diameter about $9\frac{3}{4}$ in., outside $22\frac{1}{4}$ in., running at 3600 rpm with a mean film viscosity estimated to be about 2.12 n or micro-reyns. Taper-heights δ_1 and δ_2 were taken equal to 9 and 6 mils, respectively. Three cases were solved using a digital computer. In Case 1, a total load of 450,000 lb was carried on a minimum film thickness $h_0 = 1$ mil. In Case 2, a load of 172,000 lb was carried on 2 mils; in Case 3, 27,000 lb on 5 mils. The corresponding Y values are approximately $\frac{1}{8}$, $\frac{1}{4}$ and $\frac{2}{3}$. In every case B/L is about $\frac{7}{8}$. From these data, P_0, P/P_0, and P are easily found. Dividing the published P_t values by P gives $P_t/P = 1.4$, 1.2, and 0.8 in the respective cases, with a mean of 1.20; slightly better than the book value cited, for which $B/L = 1.31$. The taper-fraction r may be assumed equal to $\frac{4}{5}$ as before.

Other Performance Characteristics. Similar methods have been used for calculating friction, oil flow, and temperature rise. See the references cited on load capacity; in particular Raimondi & Boyd (1955), whose charts are computed only for rectangular plane surfaces, but presented in a convenient form. Consider, for example, a square pad of 3×3 in. size having a taper-height of 3.0 mils, lubricated with an oil of 6.0 n or micro-reyns viscosity and heat capacity per unit volume $q = 110$ psi/dF. When running at 1200 in./sec under a load $P = 400$ psi, these charts indicate a film thickness $h_0 = 1.98$ mils, friction coefficient $f = 0.0064$, power loss $H = 4.24$ hp, total oil flow $Q = 7.0$ cu in./sec, temperature rise 43 deg F.

Experimental Results. Test data from at least eight investigations on tapered-land bearings are readily available: Linn & Sheppard (1938), Morgan and co-authors (1940), Levinsohn & Reynolds (1953), Drescher (1956), de Guerin & Hall, Sternlicht (both 1957), Brandon (1959), and Bahr (1961).

Linn and Sheppard reported friction tests on three bearings about 2, 11, and 14 in. in diameter. They were run at 20,000; 3600; and 1800 rpm, respectively. Power loss curves are convex upward until failure. The most favorable results were obtained with a mean taper-height of 3 mils in the small bearing, 7.5 in the medium size, and 5 in the largest.

Morgan, Muskat, and Reed measured friction coefficients for steel and brass sliders 1.0 cm sq, having fixed tapers from 1 to 7.5 min arc. They explored the X-range of Fig. 2 from 0.04 to 290. Good agreement was found between calculated and observed friction curves, considering the difficulty of measuring small taper slopes. The results by Levinsohn & Reynolds are reported in this chapter under water-lubricated bearings.

Drescher made friction and shaft-displacement measurements on a tapered-land bearing of 210 mm diameter, having fifteen square bronze pads with $B = 28$ mm and a taper-height of 26 microns (about 1 mil). These pads had plane surfaces with no flat-top. Load was carried to 6000 kg at speeds of 500 and 1500 rpm. The position of the shaft collar was measured both in operation and at rest under a load of 500 kg. The displacement, or difference in the two positions, was denoted by Δh_0. Contact occurred ($\Delta h_0 = 0$) when the load parameter PL/ZU

reached a critical value of 6×10^5, regardless of speed. A theoretical thickness $h_0 = 9$ microns was calculated under these conditions. The measured coefficients of friction agree well enough with theoretical values, but see discussion by Vogelpohl (1958).

An improved design was tested by de Guerin and Hall, oil grooves being closed at both inner and outer circumference and fed by a positive pressure. Thus a transition could easily be made to hydrostatic operation. Both a 26-in. and a 10-in. diameter were tested. Best results were obtained with a taper-height of only 2 mils in the 10-in. bearing. Inlet edge is well rounded and taper-fraction less than 0.8. These bearings performed favorably and were said to show a self-aligning tendency.

Sternlicht reported film thickness measurements on a $22\frac{1}{4}$ in. bearing of conventional design, with eight pads of nearly 40 deg central angle. The outer and inner taper-heights are, respectively, 6 and 9 mils. Taper-fraction may be assumed about $\frac{4}{5}$. Oil was supplied at 115 F prior to mixing with warmer oil in the radial grooves. Thus the mean operating viscosity could not be stated offhand. Curves show that the bearing carried loads from 230,000 down to 25,000 lb with trailing-edge film thicknesses from 1 to 5 mils at 3600 rpm. Sternlicht showed that his digital computer values are in reasonable agreement with the experimental results when variable viscosity and mixing in the groove are taken into account.

Brandon & Bahr tested both tapered-land and pivoted-shoe types. Bahr found means for improving heat conduction.

All these factual reports are worthy of closer examination for comparison with each other and with calculations. Tapered-land thrust bearings are in common use in small and medium sizes, where the requisite accuracy is available for machining double-tapered or warped surfaces. They are more sensitive to misalignment than pivoted-shoe types; and the preferred minimum film thickness can be realized only near one value of ZN/P. Such bearings with a flat supporting area at the trailing edge can have a high load capacity when designed to the optimum proportions, and with the correct taper for predetermined, constant operating conditions.

3. PIVOTED-SHOE BEARINGS

The tapered-land is equivalent to a pivoted shoe frozen in one particular position. This position can be the optimum for only one value of ZN/P, whereas the pivoted shoe is free to adjust itself to the optimum angle for any operating condition. This feature is especially valuable when starting up from rest, and when operating well above or below the designed load and speed.

Historical Development. The pivoted-shoe thrust bearing was invented independently by Kingsbury and Michell. The first such bearing was built and tested by Albert Kingsbury at the New Hampshire State College in 1898 (Baker, 1899). A British patent was granted to A. G. M. Michell of Melbourne, Australia (Michell, 1905); the U. S. patent to Kingsbury (1910). The history of the origin and development of the bearing is a fascinating story (Kingsbury, 1950, Archibald, I: 1955). Kingsbury tells us that, after reading the section on flat surfaces in Reynolds' paper,

> It occurred to me that here was a possible solution of the troublesome problem of thrust bearings . . . . If the block were supported from below on a pivot, at about the theoretical center of pressure, the oil pressures would automatically take the theoretical form, with a resulting small bearing friction and absence of wear of the metal parts . . . .

Kingsbury's expectations were fulfilled. The low friction, high load capacity, and freedom from wear in these bearings revolutionized shipbuilding and the hydroelectric industry. Other applications followed. The diversity and extent of thrust bearing applications are amazing. Appreciating the source of his inspiration, Kingsbury proposed and contributed toward an "Osborne Reynolds Fellowship" at the University of Manchester.

Early descriptions of the new bearing were published by W. W. Smith (1912), H. T. Newbigin (1914), and H. A. S. Howarth (1915). Textbook chapters and technical papers followed. Countless improvement patents have been issued since the original inventions.

Hydroelectric bearings installed in 1912 showed no significant wear after twenty-five years' operation.

Optimum Pivot Location. Consider here the ideal case of a lubricating film of uniform viscosity between rigid, flat surfaces. Let h_0 as before denote film thickness at trailing edge, x_p the pivot distance measured aft from the leading edge, and c the angle of tilt, or its tangent. Further, let h_1 be the film thickness at the leading edge, a denoting h_1/h_0. The film form may be identified by h_0 and c, by h_0 and h_1, or by h_0 and a. How may the film form be calculated in relation to the pivot location and operating conditions? This question is treated by Norton (II: 1942).

Let B as before stand for the breadth of the shoe in the direction of motion, L its length at right angles. For the moment take $B/L = 0$.

In order for the pivot to be located at the center of pressure, a requisite for equilibrium, it can be shown that

$$\frac{x_p}{B} = \frac{a(a+2)\ln a - 2.5(a-1)^2 - 3(a-1)}{(a-1)[(a = 1)\ln a - 2(a-1)]}\cdot \qquad (12)$$

Conversely for any chosen pivot location, a is determined by Eq. (12). This follows from Fig. 5, where $x_p \cdot W$ must equal the integral of $x \cdot Lp\, dx$. But that fixes only one element of the film form. To determine h_0, we go back to the equation for load capacity, Eq. (24) of Chapter II. This brings in the operating conditions together with C_P, a known function of a. In principle, the film ratio a can be eliminated between the Chapter II equation and (12) above, leading to the desired relation between film thickness h_0, pivot location x_p/B, and operating conditions.

The optimum pivot location is found by setting $a = 2.18$ in Eq. (12). It is equal to about 0.58 under the stated condition of uniform viscosity. Since there can be no hydrodynamic pressure when the surfaces are parallel, it would appear that the most inefficient pivot location would be found by setting $a = h_1/h_0 = 1$ in Eq. (12), making $x_p/B = \frac{1}{2}$. Thus we should not immediately expect a bearing with central pivots to work. Guided by this reasoning, some manufacturers locate their pivots aft of the center. Yet, Kingsbury bearings work, and have usually been built with central pivots. The paradox is mainly explained by the nonuniform viscosity caused by frictional heat in

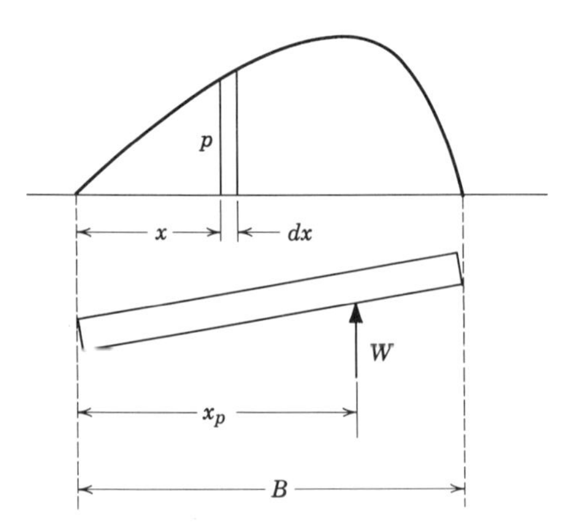

Fig. 5. Pivot position at center of pressure (p is film pressure at x).

actual bearings (*Engineering,* 1915); and partly by elastic deformation.

Theoretical Performance. Eliminating a leads to the film thickness relation

$$h_0/B = k_1 \sqrt{K}, \tag{13}$$

where k_1 is a coefficient depending only on the pivot location x_p/B; and K denotes the Kingsbury number ZU/PB. This dependence is shown by Table 4. The right-hand side of Eq. (13) is proportional to the square root of ZN/P. The constant of proportionality will be k_1 times the square root of $2\pi/A$, or $3k_1$ if the central angle is 40 deg (0.7 rad). If x_p/B is taken equal to 0.58, the optimum location, h_0/B will be 1.2 times the square root of ZN/P.

This rule agrees with the relation $Y = 0.4\sqrt{X}$ that was found to fit the curve for $B/L = 0$ in Fig. 2 at the optimum point. Only when optimum conditions are fulfilled can we expect the same film thickness curve for a fixed taper and a pivoted shoe. A comparison of the two types is especially interesting under starting conditions. We have seen that the Y, X curve for a fixed taper is tangent to the base line at $X = 0$, so that dh_0/dU is zero at the instant of starting. By contrast, in a pivoted bearing, dh_0/dU is infinite at the start. The first infinitesimal motion begins to build up a film.

We are now in a position to appreciate the statement frequently met, that a pivoted-shoe "automatically" takes up the correct angle of tilt. If the pivot is once set at the optimum location, every equilibrium film form will be the optimum one for the prevailing conditions. At a low speed under a heavy load, or more generally at a low value of ZN/P, the angle of tilt is small and h_0 is small; but the film ratio h_1/h_0 conforms to its optimum value. At high ZN/P, the angle of tilt is higher and h_0 greater but h_1/h_0 remains the same.

TABLE 4. Film Thickness Coefficient and Pivot Location

x_p/B	k_1	x_p/B	k_1
0.50	0.000	0.60	0.392
0.51	0.215	0.65	0.331
0.54	0.364	0.70	0.240
0.58	0.400	1.00	0.000

Presumably pivoted shoes would benefit by a composite profile—the trailing "flat" being machined into the shoe at an angle equal to the expected slope of the main surface, so that the flat would ride parallel to the moving surface at the intended ZN/P. New shoe designs are described by Diachkov, pages 5–103 in the collective treatise on hydrodynamic theory (1959). To streamline the radial flow between shoes, corners are removed from the sectors, leaving a kind of S-shaped channel.

Friction and Side Leakage. The classical friction law, Eq. (2) of Chapter III, holds true regardless whether pads are fixed or pivoted, but should be expressed now in terms of load and pivot location, instead of h_0 and a. To eliminate h_0, combine with Eqs. (24) and (25) of Chapter II. To eliminate a, return to Eq. (12). In this manner, when $B/L = 0$, it is seen that the coefficient of friction may be written

$$f_0 = k_2 \sqrt{K}, \tag{13}$$

where k_2 depends only on x_p/B, as shown by Table 5. Again the right-hand member is proportional to the square root of ZN/P, with a constant $= k_2$ times the square root of $2\pi/A$. The minimum friction seems to come at $x_p/B = 0.61$. The k_2 values were checked and rounded off to $2\frac{1}{2}$ significant figures.

The rate of flow of oil into a plane rectangular bearing without side leakage, Q_0, is equal to half the linear velocity U multiplied by the cross-sectional area of the film at the location of maximum pressure, from which $Q_0/BLU = k_3 \sqrt{K}$. Here k_3 is a constant depending only on the pivot location. The maximum, 0.288, comes at $x_p/B = 0.61$. See Fig. 10, Muskat, Morgan & Meres (1940). From the above it is apparent that the flow ratio is proportional to the square root of ZN/P.

TABLE 5. Friction Coefficient and Pivot Location

x_p/B	k_2	x_p/B	k_2
0.50	∞	0.61	1.8
0.51	4.6	0.65	1.85
0.52	3.3	0.70	2.1
0.58	1.85	1.00	∞

TABLE 6. Side Leakage Ratios for Pivoted Shoes

(Pivot location $x_p/B = 0.60$)

B/L	P/P_0	F/F_0	Q/Q_0	Q_z/Q
$\frac{1}{2}$	0.72	1.2	1.03	0.24
1	0.46	1.4	0.88	0.36
2	0.19	2.4	0.53	0.39
4	0.18	5.0	0.23	0.33

Figures 1 and 2 and the corresponding tables apply equally to the fixed taper and pivoted shoe, but should now be expressed with x_p/B in place of Y as the constant parameter. This has been done in Table 6 for $x_p/B = 0.60$, the optimum location for maximum load capacity of a square shoe. Optimum locations for other shoe proportions run from 0.58 to 0.64 and 0.69 as B/L increases from 0 to 2 and 4. In Table 6, Q_z denotes the combined leakage from both sides or $Q_0 - Q$. These results were obtained from the curves of Muskat (1940) and Raimondi (1955), and co-authors, by averaging and rounding off.

The charts by Raimondi & Boyd (1955) offer quick approximate solutions to problems on fixed-taper and pivoted-shoe thrust bearings. They are based on the integration by Muskat and co-authors (1940) for plane rectangular surfaces assumed to be rigid and lubricated with a uniform viscosity. The charts show performance characteristics in dimensionless form against the Kingsbury number K. Temperature rise was calculated from the rate of oil flow on the adiabatic assumption. This implies that all frictional heat is carried off by the oil, and may be more nearly fulfilled in pivoted-shoe than in tapered-land bearings because of less metallic conduction.

Heat Conduction has been discussed by Guilinger & Saibel (1958), as well as in Chapter III. Temperature distribution in a thrust shoe was computed by Pargin (1958) and confirmed experimentally. See pages 104–15 in Diachkov's book (1959), as well as Bahr (1961).

Experimental. Kingsbury's tests of 1898 carried loads to 4000 psi, or more, at 285 rpm on centrally supported shoes. The tests were continued by Baker and Putney and reported in their theses the following year. The first published account of pivoted-shoe experiments is found in the paper by W. W. Smith (1912), twelve years after the tests in New Hampshire. Those years were occupied by commercial development and full-scale testing in the field. The practical conclusions reached are described by Lieutenant Smith.

During 1910–1911, a test of special interest was made on the pivoted-shoe bearing in a 3600 rpm steam turbine. The bearing area was reduced in successive tests until the babbitt metal was crushed, and flowed plastically, under a load of more than 7000 psi; yet without overheating or any failure of the oil film (Kingsbury, 1950).

Early tests detailed in the manufacturer's catalogs (1922, 1925) include:

1. Friction data on a six-shoe hydroelectric bearing of 48-in. diameter, showing a coefficient of 0.0008 at 94 rpm under a load of 410,000 lb.
2. Overload tests on a 29-in. vertical bearing to 760 psi at 514 rpm with perfect operation; similar tests on a horizontal centrifugal pump bearing to 1430 psi at 1500 rpm.
3. Service tests on a 61-in. hydroelectric bearing, showing a loss of 12.7 hp at 55.6 rpm under a total load of 530,000 lb, as determined by the temperature rise in the circulating oil, when corrected for external cooling.
4. Determination of air-cooling constants for hydroelectric and marine applications.

Operation of pivoted-shoe bearings in a billet mill, and in a plate-glass grinding machine, over a period of years, was described by Howarth (1919). These applications involved a very great number of stops and starts.

In the meantime thrust bearings were investigated in Switzerland (von Freudenreich, 1917). Temperature and pressure distributions were determined experimentally, and the calculated film thickness checked by electrical resistance observations. The effect of changing the pivot location was determined. Shoes were supported on steel balls arranged to equalize the loading.

Howarth (1919) made "slow-speed tests" to learn what happens when the speed is insufficient for hydrodynamic lubrication. A 12-in., four-shoe bearing was used. Loads were carried up to 1000 psi at 0.34 rpm. Curves show the coefficient of friction against speed up to 16 rpm. Most of these curves lie well to the left of the minimum point. Static friction and running-in effects were also observed. It would be interesting to plot the data against ZN/P, since very few data are available on friction of modern thrust bearings under thin-film conditions.

The friction loss in propeller thrust bearings was measured by

Howarth and Ogden (1922) in a classic series of tests. The multicollar, horseshoe type of marine bearing was compared with a six-shoe, centrally pivoted design of Kingsbury bearing. Both bearings were of 14-in. diameter. They were lubricated with the viscosity grades specified in service. The friction in the multicollar bearing was found to be from 10 to 12 times that in the pivoted-shoe type. Had the latter been tested in the smaller size customarily installed for the known service, the friction ratio would have been nearer 22 to 1 in favor of the pivoted shoe.

Newbigin (1922) experimented with three-shoe bearings pivoted at 0.60 of the way from leading to trailing edge. His friction measurements were in close agreement with Michell's theory. On changing to centrally pivoted shoes, the friction nearly doubled, with a corresponding increase of temperature. These tests were made on a 3-in. bearing at 1400 rpm under loads up to 1000 psi. Note Michell's discussion with additional data, as well as a summary by Howarth (1935).

A test stand for vertical thrust bearings of 128-cm diameter, having a load capacity of 300 metric tons or over 660,000 lb, was described by Feifel (1925). Three series of friction tests were reported, using twelve shoes in the first series and six in the other two, the shoes being approximately "square." Loads were carried from 26 to 52 kg/sq cm with speeds near 155 rpm, and coefficients of friction fell from 0.0037 to 0.0012. Professor Fuller (I: 1956) selects a representative test and compares it with calculations.

Four experimental papers on thrust bearings will be found in the Proceedings of the General Discussion (IME, 1937). They are by Dowson, Kraft, Soderberg, and Tenot.

Morgan and co-authors extended their friction tests of 1940 to include square sliders with three pivot locations. The Kingsbury number was explored over a range of more than 260-fold, without significant deviations from hydrodynamic theory except where thin-film conditions intervened. The number of shoes in a thrust bearing was varied from one to ten in von Freudenreich's experiments (1941). He concluded that six or seven shoes would be an optimum number in practice.

Lafoon and associates (1947) investigated starting conditions. Good correlation was found between laboratory and field tests. Surface finish, and additives in the oil, were among the variables. Starting friction increased with idle time, as found by Coulomb. Tests on bearings up to 105-in. diameter showed that a partial hydrodynamic film is established in $\frac{1}{10}$ of a revolution. Arthur Lakey (1952) found that from 300 to 400 starts improved the surfaces and reduced starting torque. The starting friction met in large hydroelectric bearings would be less than in laboratory tests owing to vibrations induced hydraulically.

With the increasing size of modern thrust bearings, more elaborate research techniques are required. This is evident from two reports by Cassacci and associates (1951). Film thickness was measured by a capacitance method. An air-gaging technique was used by Barwell (1951) in testing a Michell bearing. Although Michell shoes are designed to tilt on the edge of a radial step, radial tilting was discovered. Kingsbury shoes, pivoted on spherical supports, are free to tilt in any direction. See Langlois-Berthelot (1951); Ambler (1955).

Kettleborough and co-authors (1955) experimented with two opposed Michell bearings. Film thickness was determined mechanically from the relative movement of opposed shoes. Pivot location ranged from 0.40 to 0.60, and ZN/P from 100 to 2600 cp rpm/psi. Plotting against ZN/P gave performance curves to which simple equations could be fitted. In Part II, Kettleborough and co-authors compared theory with experiment. It was concluded that fluid film lubrication exists for pivot positions on both sides of the center, owing to variable viscosity. Trailing-edge film thickness h_0 varied with ZN/P to the 0.71 power.

Twenty years after the historic "General Discussion" a comparable symposium was again arranged (IME, I: 1957). Three papers gave experimental data on pivoted-shoe bearings with oil lubrication.

Cole tested such bearings up to surface speeds over 350 ft/sec, far beyond the range of previous work. Outward flow of oil tended to starve the surfaces and create excessive shoe temperatures. A solution was found by supplying oil to each shoe through grooves at the inner radius and leading edge.

An eight-shoe bearing with radial supports was tested by de Guerin and Hall. Removal of six shoes reduced the loss 30 per cent. Friction was increased 15 per cent at the heaviest loads by changing the step location from 0.60 to 0.50.

In a companion paper the authors gave further results on bearings of the conventional Michell design. It was confirmed by measurements of film pressure that the shoes were not pivoting freely, and concluded that bearings with load-equalizing mechanisms might do better.

Several investigations published since the London Conference are of interest here:

1. Papers by Baudry, Kuhn, and co-authors on hydroelectric tests in the field (1958, 1959, 1960). Tests were conducted on an eight-shoe

bearing of 96-in. diameter, loaded up to 1,110,000 lb. Measurements of temperature distribution and shoe distortion were included. Best prediction of performance with centrally pivoted shoes could be made by combining the assumptions of (1) flat shoes with variable viscosity and density, and (2) crowned shoes with uniform viscosity.

2. Measurements of film pressure and film thickness in thrust bearings by Snegovskii and Medvinskii (1958).

3. Extension of Brandon & Bahr's tests to pivoted shoes (1959).

4. Experiments by Trifonev (pages 116–31) and Letkov (132–41) showing effects of load on temperature in different size thrust bearings, reported in Diachkov's book (1959).

5. Study of propeller-shaft bearing failures by Sternlicht and co-authors (1959). The photograph of a typical shoe indicates maximum wear near the center, "wiping" at the inner circumference. Such failures seem to be caused by lack of oil, dirty oil, operating overload, and flexibility of the ship's structure, as well as by elastic and thermal distortion of the shoe. Overload thrusts occur during rapid acceleration, and in hard rudder turns under full power. Thrust meter readings were taken aboard ship. It was noted that Kingsbury bearings equipped with leveling plates can tolerate misalignment as great as $\frac{1}{3}$ of a degree.

6. Experiments by El-Sisi on thrust bearings with radiused pads (1959).

7. Sea trials on the main propulsion bearings of the destroyer USS Barry are reported by Elwell and co-authors (1964). They were based on steady runs in both rough and smooth water. From a large number of readings of film thickness, bearing and oil temperatures, and propeller thrust, it was seen that the published methods of calculation are reasonably accurate. Film thickness measurements by the mutual inductance and capacitance systems were consistent. Observed film thicknesses were slightly greater in rough water than smooth, but appreciably less than calculated. Values ranged from about $\frac{1}{2}$ mil to 2 mils at middle of trailing edge. Actual temperatures were lower than calculated, apparently owing to neglect of heat conduction; making the analysis conservative in that respect.

Spherical Pivoted Shoes. Power (1930), Growdon (1946), Lakey (1952), and Needs (1954) have described the Kingsbury spherical—a combined thrust and journal bearing. The spherical runner is supported by concave pivoted shoes. When used as the bottom bearing to carry the weight of a vertical rotor, it eliminates the need for a lower guide bearing. It is particularly useful when the shaft must be positioned closely.

4. OTHER FORMS OF THRUST BEARING

We have discussed the theory and testing of tapered-land and pivoted-shoe bearings as leading examples of the solid type and the tilting-pad type. Several other forms may be considered. These include (1) plain thrust bearings, (2) the Rayleigh bearing, or stepped surface, (3) thrust plates with spiral and other grooving, (4) solid spherical bearings, (5) floating ring or shoe type, and (6) the bearing with spring supports.

Plain Thrust Bearings. Flat thrust bearings were in common use before the advent of tapered-land or pivoted-shoe bearings. Horseshoe-type, multicollar marine bearings operated after a fashion; better, it is said, in reciprocating engine than in turbine service, owing to replenishment of oil films under fluctuating load. Test data on collar and pivot bearings are found in Tower's reports to the IME Research Committee on Friction (1888, 1891). Lasche (1906) gives friction coefficients of a 50-ton, 4-collar bearing for the S. S. Kaiser.

Lewicke (1955) offered a theory showing the effect of pressures developed at the leading edge in parallel sliding. Wilcock & Booser (I: 1957) offer a chart for the friction in flat bearings lubricated by wide, radial oil grooves. Under loading of 250 psi, at a mean surface speed of 1 mile/min, a loss of $\frac{1}{2}$ hp/sq in. is indicated. Under heavier loads, and at lower speeds, imperfect lubrication is to be expected. Waviness, rounded edges, nonparallelism, and thermal effects may then be beneficial.

Good practice requires rounding off the entrance edges of thrust bearing shoes. The hydrodynamic effect was calculated by Gümbel (page 98 of I: 1925) and by Boswall (II: 1928, Fig. 14). The same principle applies to oil grooves, as described with new designs by Brillié (1928), and proved experimentally by Leloup (1949). See above under "The Composite Profile" for effect of chamfering.

Gümbel showed that an entrance curvature of radius r would develop a maximum film pressure between otherwise parallel surfaces,

$$p_m = \frac{5}{3} \cdot \frac{ZU}{h_0} \left(\frac{r}{h_0}\right)^{1/2}. \tag{16}$$

Here h_0 is the film thickness, and p_m comes at the point of tangency, close to the leading edge. Since the film is assumed thin compared to the dimensions of the bearing surface, side leakage may be neglected. Hence the load capacity per unit area, P, may be taken nearly equal to $p_m/2$. For application to a thrust bearing with a mean or pitch diameter $\frac{3}{4}$ of the outside diameter D, Gümbel's equation may be solved for the relative film thickness, and written

$$\frac{h_0}{D} = \frac{5}{3}\left(\frac{r}{D}\right)^{1/3}\left(\frac{ZN}{P}\right)^{2/3}. \tag{17}$$

Here the viscosity Z, speed N, and load P are to be expressed in the same system of units. As an example, take $D = 12$ in., $r = \frac{1}{4}$ in., $Z = 3$ n, $N = 60$ rps, and $P = 500$ psi; then the film thickness h_0 will be 0.26×10^{-3} in., about $\frac{1}{4}$ of a mil.

Boswall solved the problem by graphical and numerical methods applicable to pivoted shoes with variable viscosity. He showed that the effect of a rounded entrance is especially beneficial when the shoe is centrally pivoted. Boswall credits de Ferranti (1911) with the discovery that a surface with rounded edge can sustain loads even when the flat portion of the bearing is parallel to the moving surface.

Thrust bearing theory for fixed and pivoted shoes of a convex profile has been developed by Nahme (III: 1940), Frössel (1941), Abramovitz (1955), and Raimondi & Boyd (1955). The last two references describe experimental comparisons of flat and crowned shoes. A slight degree of crowning is recommended as beneficial, with exceptions noted. Crowning by deformation helps to explain the operation of bearings with centrally pivoted shoes. The effects of transverse curvature are discussed by Professor Charnes and co-authors (1953). See also Chapter IX.

The Stepped-Surface. It will be recalled that Lord Rayleigh foresaw as the requirement for a load-carrying film, that it "should be thicker on the ingoing than on the outgoing side." A tapered or wedge film is not the only effective shape. He found by calculus of variations that the optimum pad profile is that of two flat surfaces, parallel to the runner, separated by an abrupt step comparable in height to the expected film thickness (1918). The step acts as a dam. It restricts the flow of the lubricant dragged along by the runner (Fig. 6). The result is a hydrodynamic pressure, having its peak value at the edge of the step. Rayleigh confirmed his idea by spinning two English pennies one upon the other with the aid of an inertia bar. They had been ground to a fit and one of them etched to the required form. Oil was used as the lubricant.

The theory of the Rayleigh or stepped thrust bearing has been furthered developed by Archibald (1950), Kettleborough (1954, 1955 [1]), Osterle and co-authors (1955), and Neal (1957). The bearing has been tested experimentally by Johnston and Kettleborough (1956,

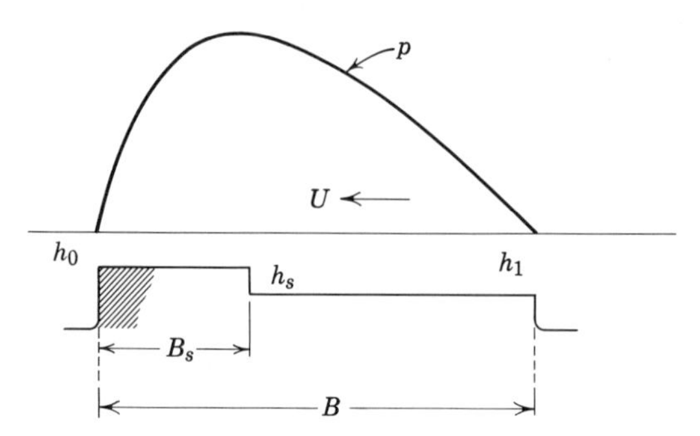

Fig. 6. The Rayleigh step (p is film pressure at runner speed U; h_s, step height; B_s, breadth of step).

1957), Cole (1957), and Neal (1957). These trials brought out the advantages and limitations of the Rayleigh type compared with pivoted-shoe bearings. The effects of density variation are described by Kettleborough (1959).

A variant of Lord Rayleigh's bearing is the pocket thrust developed by Wilcock (1955) and investigated by Kettleborough (1956 [1]). This combines the advantages of a stepped bearing and a hydrostatic bearing. The pressure developed by each step is retained in a pocket of considerable area. The removal of bearing metal to form pockets reduces friction. See also Brillié's description (1937[a]) of a thrust collar with oil pockets.

Thrust Plates with Grooving. Several forms of spiral and herringbone grooving have been found effective in maintaining a hydrodynamic film. Reference may be made to Gümbel for basic mathematical theory (I: 1925) and to Stone (1936) and Underwood (1937) for descriptions. See Chapter IX for gas film applications.

Solid Spherical. A novel design called the "hydrosphere" is described by Shaw & Strang (1946, 1948), in which the inertia of the lubricant plays an essential part. The runner is spherical and forms the bottom of a vertical shaft. It is supported hydrodynamically in a solid spherical bearing of 180-deg arc, having the same radius as the runner. The bearing surface, or cup, is unbroken except for an oil inlet at the lowest point. Here, oil is admitted at the ambient pressure. As soon as the rotor begins to lift, a crescent-shaped clearance is created through which the oil is drawn upward and outward by the motion of the runner. Curves obtained experimentally show that the coefficient of friction depends on two parameters, ZN/P and $\rho ND^2/Z$, as expected from dimensional requirements.

Floating Ring or Shoes. The principle of an intermediate floating ring of tapered shoes is seen in the Bostock and Moore patent (1922). The design was improved by Michell and Seggel to reduce adhesion by static friction (a). When the individual shoes float freely, they tilt as if pivoted. In either design the relative speed is cut in half, which reduces the friction loss.

Bearings with Elastic Support. Shoes mounted on springs were patented by de Ferranti in 1911 and Kingsbury in 1914. A spring-supported, flexible plate bearing was described by Reist (1918). This bearing and later forms are built for large, vertical machines. A friction loss formula is found on page 336 of Wilcock & Booser (I: 1957). A theoretical study was presented by Osterle & Saibel (1957) and discussed by S. E. Weidler. Experience dictated alterations evolving toward the equivalent of a pivoted shoe, as noted by S. J. Needs in his discussion of Lewis & Gordon's paper (1947).

Flexible shoes are described by Kingsbury (1914), Fulpius (1929), Leloup (III: 1954), and others.

MHD Bearings. Magnetohydrodynamic lubrication is described by Snyder (II: 1962) and others—see Chapter V.

5. EFFECT OF FLUID PROPERTIES

The influence of the density or inertia of the lubricant is discussed by Shaw & Strang (1946). It has usually been assumed negligible, and the viscosity treated as uniform throughout the film.

General Effects of Density. Several investigators have considered the influence of inertia under laminar flow, constant-speed conditions. Brillié (1928) introduced a term into Reynolds' theory proportional to the product of density by the square of the local velocity. Kingsbury (II: 1931) made use of his electrical analogy to investigate the effects of centrifugal force in the film, and the impact or Pitot pressure at the leading edge. The centrifugal effect developed only a small negative pressure. The leading edge or impact pressure could, however, be as great as 240 psi at a speed of 200 ft/min. Both effects are proportional to the density of the oil.

Slezkin and Targ (1946) derived a modified equation in which the inertia effect is averaged across the film. Kahlert used an iteration method (1948). The subject was further advanced by Brand (1955), Osterle & Saibel (1955), Milne (1959), and Bodoia & Osterle (1960).

Brand took the mean film thickness h for the linear dimension in

Reynolds' number, and the peripheral speed of the runner for V. He showed that inertia forces will be of the same order as viscous forces when $hV\rho/\mu$ has the magnitude R/h, in which R is the radius of the runner. The solution by Osterle & Saibel is based on the averaging method. The results compare closely with those of Kahlert. The treatment by Milne (1959) illustrates his stream-function theory. It leads to a general solution that reduces to Kahlert's for small inclinations. Although the effect of inertia is slight under laminar flow at constant speed, all solutions indicate that the load capacity is greater, the greater the density of the lubricant.

Acceleration Effects. Ladanyi (1948) extended Reynolds' equation to bring out the effect of a change in speed. Integration showed that the load capacity is reduced by a positive acceleration. The percentage reduction is proportional to the product of the density by the ratio of the acceleration to the speed. Evidently the effect is a maximum at each end of the stroke in a reciprocating motion, but is small at ordinary speeds.

Turbulence. The transition from laminar or streamline motion to turbulent motion occurs when Reynolds' number has a critical value. These facts were confirmed by Abramovitz in experiments on water-lubricated thrust bearings (1955). Performance characteristics were derived in more detail by Chou & Saibel (1959). This derivation is based upon Prandtl's "mixing-length," and thus retains an empirical constant. See also Tao (1960). Theory and experiment agree that the load capacity and friction loss of a bearing are both augmented by turbulence; and are therefore greater, the greater the density of the lubricant. Turbulent flow was discussed by Kettleborough (1965).

Viscosity-Temperature Effects. Fortunately the characteristics of a thrust bearing are not greatly different from the values calculated for a uniform viscosity equal to the mean of the inlet and outlet values.

The temperature effect on load capacity is shown by Norton (II: 1942) in his Fig. 38, where the load coefficient C_P is plotted against a for constant values of b. Here a is the ratio of film thickness at inlet to thickness at outlet, and b is the corresponding viscosity ratio μ_1/μ_0. Taking $a = 2.5$ for square bearings, we see that C_P is nearly twice as great for $b = 3$ as for $b = 1$, the value for a uniform viscosity. Norton's chart is to be used with our Eq. (24) of Chapter II after substituting the outlet viscosity μ_0 for the mean value μ. When $b = 3$, the mean viscosity is twice the outlet value, as seen from Fig. 7.

Norton's chart may be approximated by taking the percentage increase of C_P equal to 45 $(b - 1)$, a simple rule to remember. We are familiar with the round number $C_P = 0.16$ for the case of uniform

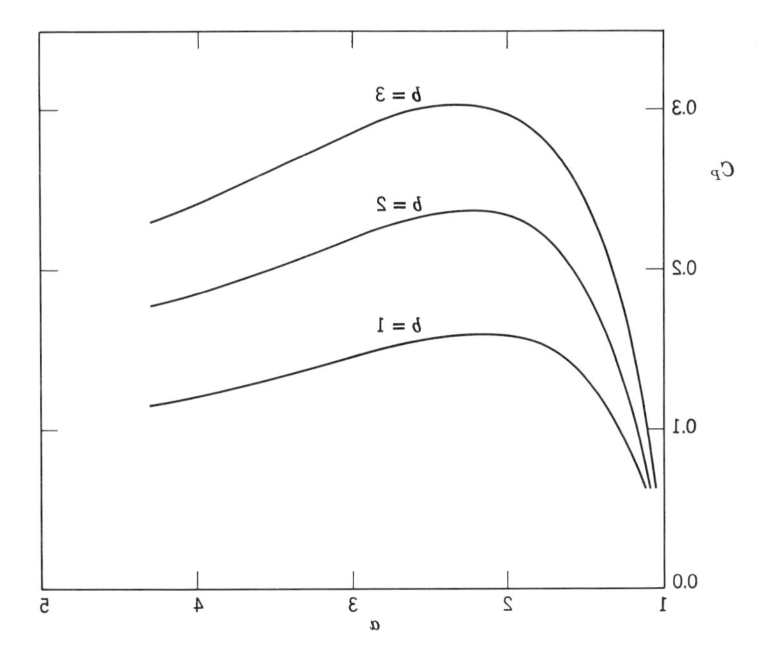

Fig. 7. Load factor with variable viscosity, plane rectangular shoes (a denotes h_1/h_0; b denotes viscosity ratio μ_1/μ_0). Based on Norton (1942), by permission; copyrighted © 1942, McGraw-Hill, Inc.

viscosity. From the above rule, C_P is 90 per cent greater for $b = 3$ than for $b = 1$, and therefore equal to 0.30. Although Norton's solution was derived for bearings without side leakage, it is a close enough for actual bearings, since P/P_0 is but slightly affected by variable viscosity; as shown by Duffing's solution (1931) when $b = 2$.

Similar charts are given by Norton for pivot location, friction, and temperature rise. He shows that the pivot location shifts toward the leading edge with increasing values of b, so that operation is possible with centrally pivoted shoes. The viscosity-temperature effect has only a slight influence on the temperature rise.

Stodola (1924) and Boswall (II: 1928) introduced graphical methods of integrating Reynolds' equation, which can be applied to variable viscosity, as shown neatly by Shaw and Macks (V: 1949). However, it is necessary to assume the viscosity, or the temperature, as a function of distance along the shoe. To predict the variation mathematically requires a detailed consideration of the mechanical work dissipated into heat, as described in Chapter III. This can be done, when necessary, with the aid of an energy equation. Christopherson (1941) offered a numerical solution for the sector-shaped pad. Cope (III: 1949) solved the problem analytically for surfaces of infinite width; Kunin (1957) for the sector shape, aided by an approximation. Other solutions will be reviewed here in conjunction with viscosity-pressure effects.

Sternlicht and co-authors, in papers cited previously (1957), investigated temperature effects by the use of a digital computer. Pressure and temperature distributions, and load capacity, were computed for tapered-land and pivoted-shoe bearings of eight pads each and about 22-in. diameter. Solutions were obtained under the realistic condition of oil mixing in the radial grooves—the fresh oil mixing with the discharge from a trailing edge, before entering the next shoe at its leading edge. Load capacities computed for this condition were roughly two-thirds of those corresponding to a uniform viscosity at the same oil supply temperature, although slightly greater than found experimentally. Possibly the difference is explained by the neglect of heat conduction. See also Gubolev (1958).

Viscosity-Pressure Effects. How the change in viscosity of lubricants under pressure affects bearing performance was studied analytically by Weibull (1925) and by Duffing (1931); Bradford and co-authors (1937), Needs (1937), Skinner (1938), Muskat & Evinger (1940), and others; assuming thrust surfaces of infinite width and isothermal operation. Charnes et al. (1953, 1955), Kochanowski (1953), and Tipei (1957) gave further solutions.

The combined effects of pressure and temperature on the oil film in thrust bearings have been studied by Christopherson (1941), Osterle and co-authors (1953, 1955), and Sternlicht (1957, 1958). Digital computer solutions were obtained by Sternlicht for the eight-pad tapered-land bearing mentioned previously. In the case of a high VI oil and $h_0 = 5$ mils, his calculations showed 4.5 per cent more load capacity, and 5.5 more friction than found without the pressure effect. With a low VI oil and $h_0 = 3$ mils, the new calculations showed 13 per cent more load capacity and 21 per cent greater friction than the conventional method. The operating speed was 3600 rpm and the load from 600 to 1200 psi in the two cases. At lower speeds or heavier loads, the differences would be greater.

Rheodynamic Lubrication. The rheological properties of non-Newtonian lubricants have been mentioned in Chapter IV. Many such materials have plastic or visco-elastic characteristics. Their behavior in thrust bearings cannot be predicted by conventional calculations.

Peek (II: 1932) derived equations for the rate of approach of flat plates under load when separated by a non-Newtonian layer. The formulas agreed reasonably well with observations and contained Stefan's law as a special case. Non-Newtonian lubrication of thrust bearings was described by Milne in 1954. At that time he introduced the term "rheodynamic."

The squeeze-film was further investigated by Osterle and co-authors (II: 1956), and followed by a study of the "rheostatic" thrust bearing (1955). Milne came up with two more papers at the London Conference of 1957, one on the Maxwell or visco-elastic liquid; the other on grease lubrication. See also to the theory of grease-lubricated thrust bearings by Slibar & Paslay (1957).

6. DEFORMATION AND VIBRATIONS

The Kingsbury shoe deforms into a slightly convex, nearly spherical or "crowned" surface under the fluid pressure load, whereas the Michell pad bends more like a beam. Calculations are given by Michell (I: 1950).

Elastic and Thermal Deformation. Shoe distortions are more apt to be detrimental the larger the bearing, hence the attention to this question by Osterle & Saibel (1957), Baudry et al. (1958), and Sternlicht and co-authors (1959).

In comparing geometrically similar shoes, note that their *stiffness* is proportional to the size. This follows from dimensions (Hersey, 1916). But when two such bearings of different size are deformed under equal loads per unit area, the *deflection* is proportional to size. If geometrically similar before loading, they remain similar under load, provided the load distributions are similar. See also Wright (1959).

Pad deformation by varying depth of babbitt was relied on by Schein (1928) to induce film formation.

Thermal expansion by frictional heating at the surface has a like effect in crowning the shoe. These effects are beneficial, in moderation. Michell points out that the load capacity of nominally flat bearings must be favored by thermal expansion of surface layers.

The elastohydrodynamic problem was solved for rectangular surfaces by Osterle & Saibel (1957) assuming isothermal flow, with no side leakage. This required integrating Reynolds' equation simultaneously with those for the runner and shoe deformation. The net effect can be increased load capacity. Computer solutions for sector pads have been obtained by Sternlicht and co-authors (1961), both on the isothermal and the adiabatic assumption.

Vibrations in Thrust Bearings. While rarely troublesome in thrust bearings, vibrations can be caused by torsional motion of the shaft, since any fluctuation of speed is accompanied by fluctuations of fluid pressure. Vibrations have been reported on starting up hydraulic turbines (Lafoon et al., 1947; Lakey, 1952). Hill (1958) describes how vibrations in aircraft gas turbines were remedied by an improved design of thrust bearing.

Osterle and co-authors (1954) extended Reynolds' equation to the nonsteady operation of pivoted-shoe bearings. They treated the case of a sinusoidal axial vibration. It was found that load capacity increases with amplitude. See also Nahavandi (1960).

7. HYDROSTATIC LUBRICATION

To initiate hydrodynamic action with minimum wear, motion may be started either without load or by "hydrostatic lift." Professor Fuller (1947) has modernized and applied the principles of hydrostatic lubrication, supporting the bearing loads continuously by pressure from an outside source.

The theory of hydrostatic, or externally pressurized lubrication was initiated by Lord Rayleigh in 1917 and advanced by Gümbel (I: 1925). The development of the subject led to an extensive literature summarized by Professor Fuller in his book.

Load capacity of a hydrostatic thrust bearing of radius R, supplied with an inlet pressure p_0 from an external source, is given by $P = Cp_0$, where

$$C = \frac{1 - (r/R)^2}{2 \ln (R/r)} \cdot \tag{18}$$

Here r is the radius of a central cavity, or recess, at the uniform pressure p_0, and P denotes the load divided by the total area πR^2. This may be proved by applying the slot formula of Chapter II to the radial flow of lubricant through the clearance space h from the inner radius $x = r$ to the outer radius $x = R$, thereby determining the pressure p as a function of x, and integrating to find the load.

The rate of flow Q needed for maintaining a chosen clearance or film thickness h is found proportional to Ph^3/Z, where Z is the viscosity of the lubricant, assumed uniform. The constant of proportionality is

$$C' = \frac{\pi/3}{1 - (r/R)^2} \cdot \tag{19}$$

The friction torque T may be calculated by applying Newton's law to an element of thickness h and area $2\pi x\, dx$, and integrating from r to R. The equivalent friction force F acting at some radius r_f is equal to

T/r_f. The friction radius r_f is customarily taken as the mean of r and R. The load W may be written $\pi R^2 P$. The coefficient is then found by dividing T/r_f by $\pi R^2 P$. Since by Newton's law the expression for T will be inversely proportional to h, an unknown quantity, it is necessary to eliminate h by the relation $h^3 = QZ/PC'$ implied above. It will then be seen that f is proportional to the two-thirds power of ZN/P multiplied by the one-third power of D^3N/Q, as in Eq. (28) of Chapter V, with a constant of proportionality

$$C'' = \frac{3.2(1-\rho)(1+\rho^2)}{(1-\rho^2)^{1/3}} . \tag{20}$$

Here ρ stands for r/R. Alternatively, if h is considered known and Q unknown, the latter may be replaced by its equivalent in terms of h, making the coefficient proportional to ZN/P divided by h/D. The new constant of proportionality will be

$$C''' = \pi(1-\rho)(1+\rho^2) . \tag{21}$$

We need go no further here since the subject is primarily one of design rather than lubrication. It has been carried on by many able investigators, leading into problems of stability and optimum proportions as described by Rippel in his book (1963). See also Boyd et al., Hahn, and Levesque (all 1965). Applications to gas-lubricated bearings are mentioned in Chapter IX.

8. THERMAL LUBRICATION

The experiments by Fogg (1946) and by Boussages & Casacci (1948) have been discussed in Chapter III. Reethof et al. (1958) ran tests on a 4-in. thrust bearing fitted with three shoes rigidly held to prevent tilting. Film thicknesses agreed remarkably well with Fogg effect calculations. Shaw (1946); Osterle et al. (1953) [2].

Kettleborough (1955) studied the effect of different numbers of radial grooves in annular bearing surfaces. The four-groove design gave the greatest film thickness, and least friction, in tests ranging up to ZN/P values of 7000 cu. See tests by Ulukan (1956).

Cole tested parallel-surface bearings of $7\frac{3}{8}$ in. mean diameter at speeds up to 9000 rpm, and found they would carry loads from 100 to 600 psi, although with greater friction than tilting pads. A design of more rigid construction led to excessive temperature rise and surface damage at loads above 90 psi (Cole, 1957).

Further research would be of interest to determine what proportion of the load capacity in parallel-surface bearings is due to the Fogg effect, the Cameron effect, geometrical factors, and elastic deformation; see Cameron & Wood (1946); Cameron (1958).

9. WATER-LUBRICATED THRUST BEARINGS

Footstep bearings were commonly lubricated by water in the early days of hydroelectric power. The water was supplied under pressure by an external pump. Model experiments and calculations leading to new types of water-lubricated, hydrostatic bearings are described by Gottwald & Vieweg (1950, 1954).

Experiments on hydrodynamic thrust bearings, using water as a lubricant, are reported by Abramovitz (1953). The most favorable performance, especially on starting, was obtained when the pivoted shoes were slightly crowned. The initial action of the shoe is described showing how the film is automatically formed. In a later investigation the author gives data on the friction losses in water-lubricated bearings before and after the beginning of turbulent motion (1955).

Levinsohn & Reynolds (1953) experimented with a series of nine tapered-land carbon bearings of $2\frac{3}{4}$ in. diameter in which the design factors were systematically varied. A good design is one having six radial grooves; rounded or beveled entrance; composite profile with taper-fraction $\frac{2}{3}$ to $\frac{3}{4}$; mean taper-height 0.6 to 0.8 mil. A radial taper of 0.3 mil was found beneficial at high speeds. Such bearings were tested from about 900 to 3500 rpm under 110-psi load, and the coefficients of friction plotted against ZN/P. Wepfer & Cattabiana (1955) report on a program applicable to reactor development. Thrust bearings of different materials having a high degree of hardness and corrosion resistance were compared. Tests were conducted on an 8-in. diameter, pivoted-shoe bearing in oxygenated distilled water at 200 F, normally at 3600 rpm, under a loading of 21 psi. These tests were run for 200 hr, followed by 500 start-and-stop cycles. Some materials gave high friction coefficients necessitating a reduction of speed. The best results were obtained using very hard materials; the shoe surfaces being crowned to avoid concavity at the higher temperature.

Newman (1958) reported favorably on water-lubricated bearings for marine use. An eight-shoe thrust bearing of 7.3-in. mean diameter (possibly 10-in. OD), in which the white-metal had been replaced by fabric-reinforced phenolic resin, was tested up to 10,000 rpm at 200 psi with no sign of trouble. By removing four shoes, the load was taken up to 1400 psi at 6000 rpm. The water-lubricated bearing was found to be in perfect condition. The same design of oil-lubricated bearing with eight shoes failed at 700 psi. The chief advantage of water as a lubricant was credited to its high specific heat compared to oil. Other advantages were nonflammability, and the fact that water is expendable (Hayes, 1962). To carry Newman's thought a step farther, it may be asked if the time is not at hand to study the possibilities of water lubrication more systematically. Surely the research laboratories would be at work on this question if water were a proprietary material.

10. SUMMARY

The theory of plane rectangular-surface bearings has been reviewed with special consideration of side leakage and optimum conditions. We have then gone on to see how the commercial forms of fixed-taper and pivoted-shoe thrust bearings perform in actual service. Other forms of thrust bearing are described. General agreement is found between theory and experiment, but some differences are noted. Secondary effects introduced by the inertia of the lubricant, the change in viscosity with temperature and pressure, and by elastic and thermal deformation of the metal have been studied. References are given to tests on parallel-surface bearings, a subject discussed theoretically in Chapter III. Favorable reports are noted concerning water-lubricated bearings.

Experiments on large thrust bearings are costly. It appears that much can still be learned by collecting and coordinating the published test data, and plotting curves with the aid of dimensionless variables. A closer comparison can then be made between experiment and calculation.

REFERENCES

Abramovitz, Stanley (1953), "Experiments on Hydrodynamic Lubrication Using Water as a Lubricant." *JAP* **24,** 1521–2.

Abramovitz, Stanley (1955), "Turbulence in a Tilting Pad Thrust Bearing." *JFI* **259,** 61–4. (a) *T* **78,** 7–11 (1956).

Abramovitz, Stanley (1955), "Theory for a Slider Bearing with a Convex Surface; Side Flow Neglected." *JFI* **259,** 221–38.

Ambler, G. C. (1955), "Contamination in Lube Oil Systems—and How to Keep It Out." *Bu Ships J* Feb., 13–6; May, 9–13.

Archibald, F. R. (1950), "A Simple Hydrodynamic Thrust Bearing." *T* **72,** 393–400.

Bahr, H. C. (1961), "Recent Improvements in Load Capacity of Large-Steam-Turbine Thrust Bearings." *JEP, T* **83,** 130–6.

Baker, H. C. (1899), *The Friction of a Step Bearing.* Thesis, New Hampshire State College, Durham; 21 pp. +. (a) Putney, H. N., 34 pp. +.

Barwell, F. T. (1951), "Some Aspects of Research on Friction and Wear." *Eng.* **174,** 649–51, 697–9. (a) *T Inst. Eng. and Shipbuilders in Scotland* **95,** 64–100 (1952).

Baudry, R. A. & Peterson, G. E. (1958), "Some Mechanical Considerations in the Design of Large Thrust Bearings." *TASLE* **1,** 225–31.

Baudry, R. A., Kuhn, E. C., & Cooper, G. D. (1959), "Performance of Large Water-Wheel Generator Pivoted-Pad Thrust Bearings Determined by Test under Normal Conditions." *TAIEE* **78,** pt. III, 1300–15.

Baudry, R. A., Kuhn, E. C., & Wise, W. W. (1958), "Influence of Load and Thermal Distortion on Design of Large Thrust Bearings." *T* **80,** 807–18.

Bodoia, J. R. & Osterle, J. F. (1960), "On the Attainment of Fully Developed Flow in Lubrication Film." *Wear* **3,** 165–9.

Bostock, F. P. & Bramley-Moore, S. (1922), *Improvements in Film Lubricated Bearings.* Brit. pat. 187,497. (a) Michell, A. G. M. & Seggel, A. J., Australian 101,857 (1937); (b) British 482, 868 and 582,967 (1946); (c) U.S. 2, 250,546 (1941).

Boussages, P. & Casacci, S. (1948), "Étude sur les pivots à graines parallèles." *La Houille Blanche,* July–Aug., 1–9.

Boyd, J., Kaufman, H. N., & Raimondi, A. A. (1965), "Basic Hydrostatic Pad Design." *LE* **21,** 391–4, 439–42.

Bradford, L. J. & Vandegrift, C. G. (1937), "Relationship of the Pressure-Viscosity Effect to Bearing Performance." *GDLL* **1,** 23–9. (a) with R. S. Wetmiller, *Machine Des.* **9,** 36–9.

Brand, R. S. (1951), "The Hydrodynamic Lubrication of Sector-Shaped Pads." *T* **73,** 1061–3.

Brand, R. S. (1955), "Inertia Forces in Lubricating Films." *JAM* **22,** *T* **77,** 363–4.

Brandon, R. E. & Bahr, H. C. (1959), "Load Capacity Tests on Tapered-Land and Pivoted-Shoe Thrust Bearings for Large Steam Turbine Application." *JEP, T* **81,** 208–14.

Brillié, H. (1928), "Théorie du graissage rationelle." *Bull. Techn. du Bureau Veritas* **10,** 105–10, 130–3, 151–3, 170–3, 198–201, 215–7, 237–42. (a) *GDLL* **1,** 38–51 (1937).

Cameron, A. (1958), "The Viscosity Wedge." *TASLE* **1,** 248–53.

Cameron, A. & Wood, W. L., Mrs. (1946), "Parallel Surface Thrust Bearing." *Sixth ICAM,* Paris. (a) *TASLE* **1,** 254–8.

Casacci, S. X. & Peuchmaur, A. (1951), "Études expérimentales sur les pivotéries industrielles." *La Houille Blanche,* Jan.–Feb., **6,** 23–43, (a) *Bull. soc. française des électriciens* (7) Oct., 1–18.

Charnes, A., Osterle, F., & Saibel, E. (1955), "On the Solution . . . VIII.

The Optimum Slider Profile for Viscosity a Function of the Pressure." *T* **77,** 33–6.

Charnes, A. & Saibel, E. (1952), "On the Solution of Reynolds' Equations for Slider Bearing Lubrication, I." *T* **74,** 867–73.

Charnes, A. & Saibel, E. (1953), "On the Solution . . . II. The Viscosity a Function of the Pressure." *T* **75,** 269–72.

Charnes, A., Saibel, E., & Ying, A. (1953) [1], "On the Solution . . . III. Effect of Transverse Curvature." *T* **75,** 507–13.

Charnes, A., Saibel, E., & Ying, A. (1953) [2], "On the Solution . . . V. The Sector Thrust Bearing." *T* **75,** 1125–32.

Chou, Y. T. & Saibel, E. (1959), "The Effect of Turbulence on Slider Bearing Lubrication." *JAM* **26,** *T* **81,** 122–6, 689–90.

Christopherson, D. G. (1941), "A New Mathematical Method for the Solution of Film Lubrication Problems." *Proc. IME* **146,** 126–35.

Cole, J. A. (1957), "An Experimental Investigation of Power Loss in High-Speed Plain Thrust Bearings." Paper 94, *CLW,* 158–63; 13, 748, 756, 825, 835.

Diachkov, A. K., editor (1959), *Development of the Hydrodynamic Lubrication Theory for Thrust Bearings* (Russian) AN, SSSR, Moscow; 153 pp.

Dowson, R. (1937), "Thrust Bearings for Steam Turbine Machinery." *GDLL* **1,** 72–83.

Drescher, H. (1956), "Zur Berechnung von Axial-Gleitlagern mit hydrodynamischer Schmierung." *Konstruktion* **8,** No. 3, 94–104; 228–31.

Duffing, G. (1931), "Die Schmiermittelreibung bei Gleitflächen von endlicher Breite." *Handbuch des Physik. u. Technischen Mechanik,* Auerbach-Hort, Barth, Leipzig, **5,** 839–50. (a) *ZVDI* **72** (1), 495–9 (1928).

El-Sisi, S. I. (1959), The Lubrication of Radiused-Pad Thrust Bearings (Italian). *Technica Italiana* **24,** 349–64.

Elwell, R. C., Gustafson, R. E., & Reid, J. C., Jr. (1964), "Performance of Centrally Pivoted Sector Thrust-Bearing Pads—Sea Trials . . . . " *JBE, T* **86,** 483–97.

Engineering (1920), "The Theory of the Michell Thrust Bearing." **109**; 233–6. (a) "The Theory of Lubrication," **100,** 101–3, 154–5, 196–7, 207–8 (1915).

Feifel, E. (1925), "Ein Versuchsstand für grosse axial Drucklager." *ZVDI* **69,** 679–82.

Ferranti, de, S. J. (1911), *Improvements in and Relating to Thrust Bearings and the Like.* Brit. Pat. 5035.

Floberg, Lief (1960), "The Optimum Thrust Tilting Pad Bearing." *T CUT,* No. 231; 23 pp.

Fogg, A. (1946), "Film Lubrication of Parallel Thrust Surfaces." *Proc. IME* **155,** 49–67.

Freudenreich, J. von (1917), "Zur Theorie moderner Drucklager." *Z. f. die gesamte Turbinenwesen* **14,** 293–5. (a) *Brown-Boveri Mitteilungen* **4,** 3–9, 35–44, 58–65, 80–5; **5,** 9–16, 27–35, 46–54, 71–5 (1918).

Freudenreich, J. von (1941), "Some Recent Investigations into Segmental Bearings." *Brown Boveri Rev.*, 366–7.

Frössel, W. (1941), "Berechnung der Reibung und Tragkraft eines endlich breiten Gleitschues auf ebener Gleitbahn." *ZAMM* **21,** 321–40; **22,** 176 (1942).

Fuller, D. D. (1947), "Hydrostatic Lubrication." *Machine Des.* **19,** I, "Oil-Pad Bearings," June, 110–6; II, "Oil Lifts," July, 117–22; III, "Step Bearings," Aug., 115–20, IV, "Oil Cushions," Sept., 127–31, 188, 190.

Fulpius, Edmond (1929), *Thrust Bearing.* U. S. Pat. 1,735,315.

Gibbs, E. U. (1918), "Thrust Bearings. Their Development for Hydraulic Turbines . . . ." *The Canadian Engineer* **36,** 553–4.

Golubev, A. J. (1958), Stationary Plane Flow of a Viscous Incompressible Liquid with Varying Viscosity in a Bearing (Russian). *TIM* **12,** 1958, 205–223. (a) *FWM* **12,** 197–214 (1960).

Gottwald, F. & Vieweg, R. (1950), "Berechnungen und Modellversuche an Wasser-und Luftlagern." *Z. angew. Physik* **2,** 437–43. (a) *ZVDI* **96,** 1005–7 (1954).

Gross, W. A. (1959), *Film Lubrication, VII. Finite Incompressible Lubricating Films.*" IBM Research Lab., San Jose, Calif.; 131 pp.

Growden, J. P., Terry, R. V., & Gnuse, H. H., Jr. (1946), "Nantahala Turbine." *T* **68,** 687–700.

de Guerin, D. & Hall, L. F. (1957), "An Experimental Comparison Between Three Types of Heavy Duty Thrust Bearings." Paper 79, *CLW,* 128–34; 12, 744, 828, 833; Plate 13.

de Guerin, D. & Hall, L. F. (1957), "Some Characteristics of Conventional Tilting-Pad Thrust Bearings." Paper 82 *CLW,* 142–6; 12, 200, 744, 828, 835, 848; Plate 15.

Guilinger, W. H. & Saibel, E. (1958), "The Effect of Heat Conductance on Slider-Bearing Characteristics." *T* **80,** 800–6.

Hahn, R. S. (1965), "Some Advantages of Hydrostatic Bearings." *LE* **21,** 89–96.

Harrison, W. J. (1919), "The Hydrodynamical Theory of the Lubrication of a Cylindrical Bearing under a Variable Load, and of a Pivot Bearing." *T. Cambridge Phil. Soc.,* **22,** 373–88.

Hayes, T. J. (1962), "A Steam Turbine Driven, Water Lubricated, High Speed Centrifugal Compressor for Oxygen Service." *ME* **84** (Oct.), 72.

Hays, D. F. (1958), "Plane Sliders of Finite Width." *TASLE* **1,** 233–40.

Hersey, M. D. (1916), "The Theory of the Stiffness of Elastic Systems." *JWAS* **6,** 569–75.

Hill, H. C. (1958), "Slipper Bearings, and Vibration Control in Small Gas Turbines." *T* **80,** 1756–64.

Howarth, H. A. S. (1915), "The Kingsbury Thrust Bearing." *Elec. J.* **12,** 351–5.

Howarth, H. A. S. (1919), "Slow Speed and Other Tests of Kingsbury Thrust Bearings." *T* **41,** 685–707.

Howarth, H. A. S. (1935), "The Loading and Friction of Thrust and Journal Bearings with Perfect Lubrication." *T* **57,** 169–91.

Howarth, H. A. S. & Ogden, Nelson (1922), "Friction Tests of Propeller Thrust Bearings." *JASNE* **34,** 1–15.

Institution of Mechanical Engineers (1888), "Third Report on Friction Experiments. Experiments on the Friction of a Collar Bearing." *Proc. IME,* 173–205. (a) "Fourth Report of the Research Committee on Friction. Experiments on the Friction of a Pivot Bearing," 111–40 (1891).

Institution of Mechanical Engineers (1937), *General Discussion of Lubrication and Lubricants* (*GDLL*). London, **1,** 648 pp.; **2,** 507 pp. (a) Reprinted, ASME, New York (1938).

Jakobsson, Bengt & Floberg, Leif (1958), *The Rectangular Plane Pad Bearing. T CUT,* Göteborg, No. 203; 44 pp.

Johnston, C. R. & Kettleborough, C. F. (1956), "An Experimental Investigation into Stepped Thrust Bearings." *Proc. IME* **170,** 511–22, 524–25, 532–33.

Kahlert, W. (1948), "Der Einfluss der Trägheitskräfte bei der hydrodynamischen Schmiermittel Theorie." *Ingenieur—Archiv.* **16,** 341–2.

Kettleborough, C. F. (1954), "The Stepped Thrust Bearing—A Solution by Relaxation Methods." *JAM* **21**; *T* **76,** 19–24.

Kettleborough, C. F. (1955) [1], "An Electrolytic Tank Investigation into Stepped Thrust Bearings." *Proc. IME* **169,** 679–88, 707–15. (a) "Oil Stream-lines . . . ," *JAM* **22,** *T* **77,** 1955; 8–10, 600.

Kettleborough, C. F. (1955) [2], "Tests on Parallel Surface Thrust Bearings." *Eng.* **180,** 174–5.

Kettleborough, C. F. (1956) [1], "The Hydrodynamic Pocket Bearing." *Proc. IME* **170,** 535–44.

Kettleborough, C. F. (1956) [2], "Tapered-Land Thrust Bearing." *JAM* **23,** *T* **78,** 581–3.

Kettleborough, C. F. (1956) [3], "The Sector-Shaped Pad." *JAM* **23,** *T* **78,** 584–6.

Kettleborough, C. F. (1957), "Stepped Thrust Bearings." Paper 37, *CLW,* 59–65; 13, 832.

Kettleborough, C. F. (1959), "Density Variation Effects in Step-Thrust Bearings." *JAM* **26,** *T* **81,** 337–40.

Kettleborough, C. F. (1965), "Turbulent and Inertia Flow in Slider Bearings." *TASLE* **8,** 286–95.

Kettleborough, C. F., Dudley, B. R., & Baildon, E. (1955), "Michell Bearing Lubrication. Part I. Experimental Results." *Proc. IME* **169,** 746–56, 759–66. "Part II. Correlation Between Theory and Experiment," 756–66.

Kingsbury, Albert (1910), *Thrust Bearings.* U.S. Pat. 947,242. (a) U.S. Pat 1,117,499 (1914).

Kingsbury, Albert (1950), "Development of the Kingsbury Thrust Bearing." *ME* **72,** 957–62. (a) *JASNE* **63,** 433–43, 513–4 (1951).

Kobayashi, Torao (1934), "A Development of Michell's Theory of Lubrication" (Japanese). Report 107, *Aeronautical Research Inst.* Tokyo Imperial University, **8,** No. 11, 385–414.

Kochanowski, W. (1953), "Die Druckabhängigkeit der Viskosität und ihre Auswirkung auf die Schmierung von Gleitflächen." *Kolloid-Z.* **131,** No. 2, 74–83.

Kraft, E. A. (1937), "Thrust Bearings: Theory, Experimental Work, and Performance." *GDLL* **1,** 146–54.

Kucharski, W. (1916), "Zur Theorie moderne Drucklager." *Z. f. die gesamte Turbinenwesen* **13,** 297–300. (a) "Der Einfluss der endlichen Lagerslange . . ." **15,** 53–6, 68–71, 75–8, 81–5 (1918).

Kuhn, E. C. & Cooper, G. D. (1959), "Some Factors Affecting Performance of Large Pivoted Pad Thrust Bearings." *LE* **15,** 250–5.

Kunin, I. A. (1957), On the Hydrodynamic Theory of Pad-Type Bearings (Russian). *AN, SSSR* (Eastern Branches), No. 4–5, 128–37. (a) *Wear* **2,** 9–20 (1958).

Ladanyi, D. J. (1948), *Effects of Temporal Tangential Acceleration on Performance Characteristics of Slider and Journal Bearings. TN* 1730, NACA; 34 pp.

Laffoon, C. M., Baudry, R. A., & Heller, P. R. (1947), "Performance of Vertical Water-Wheel Generator Thrust Bearings During the Starting Period." *T* **69,** 371–9.

Lakey, A. B. (1952), "Recent Developments in Hydro-Electric Thrust Bearings." *ASME Annual Mtg. Paper;* 19 pp. plus figs.

Langlois-Berthelot, R., Sauvage, P., Leroy, G., & Casacci, S. X. (1951), "Quelques methodes, récemment mis au point, pour le réglage et le contrôle du gros matériel électrique, notamment la mesure des petites épaisseurs. Application aux pivotéries d'alternateurs à axe vertical." *Bull. soc. française des électriciens* (7), **1** (Oct.) 1–18.

Lasche, O. (1906), "Kammlager für 900 Uml/min bei rd. 50 t Belastung für die Turbinen des Dampfers Kaiser." *ZVDI* **50,** 1355–61.

Leloup, L. (1949), "Relation d'une série d'essais sur les paliers de butée et les conclusions que l'on a pu en tirer au point de vue pratique." *RUM* (9) **5,** 258–72.

Levesque, G. N. (1965), "The Error Correcting Action of Hydrostatic Bearings." *ASME* 65-LUBS-12; 5 pp. (a) *ME* **87** (Aug.); 58.

Levinsohn, M. & Reynolds, N. E. (1953), "Experiments with Water-Lubricated Tapered-Land Thrust Bearings." *T* **75,** 1137–45.

Lewicke, W. (1955), "Theory of Hydrodynamic Lubrication in Parallel Sliding." *The Engineer* **200** (Dec. 30), 939–41.

Lewis, F. M. & Gordon, T. W. (1947), "The Design and Performance of the Vertical Generator Thrust Bearings at the Bonneville Plant of the Corps of Engineers, War Department." Paper 47–184, *TAIEE* **66,** 1231–9.

Linn, F. C. & Sheppard, R. (1937), "The Tapered-Land Thrust Bearing." *GDLL* **1,** 171–8. (a) *T* **60,** 245–52 (1938).

Michell, A. G. M. (1905), *Improvements in Thrust and Like Bearings.* Brit. Pat. 875.

Milne, A. A. (1957), "A Theory of Rheodynamic Lubrication for a Maxwell Liquid." Paper 41, *CLW,* 66–71; 14, 202, 743, 749, 751, 820. (a) *Kolloid-Z.* **139,** 96 (1954).

Milne, A. A. (1957), "On Grease Lubrication of a Slider Bearing." Paper 102, *CLW,* 171–5; 14, 201, 599, 751, 835.

Milne, A. A. (1959), "On the Effect of Lubricant Inertia in the Theory of Hydrodynamic Lubrication." *JBE, T* **81,** 239–44. (a) *HS* 423–527.

Morgan, F., Muskat, M., & Reed, D. W. (1940), "Studies in Lubrication, VIII. Lubrication of Plane Sliders." *JAP* **4,** 541–8.

Muskat, M. & Evinger, H. H. (1940), "Studies in Lubrication, IX. The Effect of Pressure Variation of Viscosity on the Lubrication of Plane Sliders." *JAP* **11,** 739–48.

Muskat, M., Morgan, F., & Meres, M. W. (1940), "The Lubrication of Sliders of Finite Width." *JAP* **11,** 208–19.

Nahavandi, A. (1960), "The Effect of Vibration on the Load-Carrying Capacity of Parallel Surface Thrust Bearings." *ASME* **60** LubS-3. (a) *ME* **82** (June), 99.

Neal, P. B. (1957), "Re-Examination of the Stepped Thrust Bearing." Paper 42, *CLW,* 72–81; 13, 835.

Needs, S. J. (1937), "Influence of Pressure on Film Viscosity in Heavily Loaded Bearings." *GDLL* **1,** 216–22. (a) *T* **60** (1938) 347–58 (1938); **61,** 160–2 (1939).

Needs, S. J. (1954), "Vertical Pivoted Shoe Thrust Bearings." Pages 82–91 in *Fundamentals of Friction and Lubrication in Engineering,* First ASLE Nat. Symposium (1952); ASLE, Chicago, 1954.

Newbigin, H. T. (1914), "The Problem of the Thrust Bearing." *Proc. ICE.* **196,** 223–65.

Newbigin, H. T. (1922), "Lubrication Tests." *Eng.* **114,** 260–1.

Newman, A. D. (1958), "Water-Lubricated Bearings for Marine Use." *Trans. N.E. Coast Inst. Eng. and Shipbuilders* **74,** pt. 7, 357–70. (a) *JASNE* **70,** 756–60.

Osterle, F., Charnes, A., & Saibel, E. (1953) [1], "On the Solution . . . IV. Effect of Temperature on the Viscosity." *T* **75,** 11–17.

Osterle, F., Charnes, A., & Saibel, E. (1953) [2], "On the Solution . . . VI. The Parallel-Surface Slider-Bearing without Side Leakage." *T* **75,** 1133–6.

Osterle, F., Charnes, A., & Saibel, E. (1954), "On the Solution . . . VII. The Non-Steady State Operation of Tilting-Pad Slider-Bearings." *T* **76,** 327–30.

Osterle, F., Charnes, A., & Saibel, E. (1955), "On the Solution . . . IX. The Stepped Slider with Adiabatic Lubricant Flow." *T* **77,** 1185–7.

Osterle, F. & Saibel, E. (1955) [1], "On the Effect of Lubricant Inertia in Hydrodynamic Lubrication." *ZAMP* **6,** No. 4, 334–9. (a) "Recent Advances . . . ," *LE* **11,** 187–92 (1955). (b) Saibel, E., *Actes, Ninth ICAM,* 1956; Brussels, **4** 302 (1957).

Osterle, F. & Saibel, E. (1955) [2], "The Rheostatic Thrust Bearing." *ASME Paper* 55-LUB-6; 9 pp.

Osterle, F. & Saibel, E. (1957), "The Spring-Supported Thrust Bearing." *T* **79,** 351–5.

Osterle, F. & Saibel, E. (1957), "Surface Deformations in the Hydrodynamic Slider-Bearing Problem and Their Effect on the Pressure Development." Paper 35, *CLW,* 53–8; 13, 200. (a) *TASLE* **1,** 213–6 (1958).

Pargin, D. P. (1958), Analysis of Temperatures in the Pad of a Thrust Bearing in a Hydroelectric Generator (Russian). *TIM* **12,** 224–41. (a) *FWM* **12,** 215–31 (1960).

Pinkus, O. (1958), "Solution of the Tapered-Land Sector Thrust Bearing." *T* **80,** 1510–6.

Power (1930), "Spherical Thrust Bearings . . . ." Vol. **71** (June 17), 944–5.

Raimondi, A. A. & Boyd, J. (1955) [1], "Applying Bearing Theory to the Analysis and Design of Pad-Type Bearings." *T* **77,** 287–309.

Raimondi, A. A. & Boyd, J. (1955 [2]), "The Influence of Surface Profile on the Load Capacity of Thrust Bearings with Centrally Pivoted Pads." *T* **77,** 321–30.

Rayleigh, Lord (1917), "A Simple Problem in Forced Lubrication." *Eng.* **104,** 617, 697.

Rayleigh, Lord (1918), "Notes on the Theory of Lubrication." *Phil. Mag.* (6) **35,** 1–12. (a) *Collected Papers,* 523–33.

Reethof, C., Goth, C., & Koro, J. (1958), "Thermal Effects in the Flow Between Two Parallel Flat Plates in Relative Motion." *ASLE Paper* 58-AM-4A-1, 11 pp. +.

Reist, H. G. (1918), "Self-Adjusting Spring Thrust Bearing." *T* **40,** 191–7.

Rippel, H. C. (1963), *Hydrostatic Bearing Design Manual.* Cast Bronze Bearing Inst., Cleveland; 75 pp.

Schein, A. E. (1928), *Thrust Bearing.* U. S. Pat. 1,682,190.

Schiebel, A. (ed. by Körner, K.) (1933), *Die Gleitlager.* Springer, Berlin; 70 pp. +.

Shaw, M. C. (1947), "An Analysis of the Parallel-Surface Thrust Bearing." *T* **69,** 381–387.

Shaw, M. C. & Strang, C. D., Jr. (1946), "Role of Inertia in Hydrodynamic Lubrication." *Nature* **158,** 452.

Shaw, M. C. & Strang, C. D., Jr. (1948), "The Hydrosphere—A New Hydrodynamic Bearing." *JAM* **15,** *T* **70,** 137–45.

Skinner, S. M. (1938), "Film Lubrication of Finite Curved Surfaces." *JAP* **9,** 409–21.

Slezkin, N. A. & Targ, S. M. (1946), The Generalized Equations of Reynolds (Russian). *Doklady, AN, SSSR* **54,** 205–8.

Slibar, A. & Paslay, P. R. (1957), "On the Theory of Grease-Lubricated Thrust Bearings." *T* **79,** 1229–34.

Smith, W. W. (1912), "The Kingsbury Thrust Bearing." *JASNE* **24,** 1161–5.

Snegovskii, F. P. (1958), Experimental Determination of Hydrodynamic Pressures and Thickness of Lubricating Films in Sliding Bearings (Russian). *Trudy, Tsentral'nyi Nauchnois sledov a tel'skii, Institut Teknologii i*

Mashinostroeniia, Moscow **90,** 48; Thermal Calculation . . . 76. (a) Medvinskii, M. D., Thickness of Oil Films, 109.

Soderberg, C. R. (1937), "Bearing Problems of Large Steam Turbines and Generators." *GDLL,* **1,** 285–96.

Sternlicht, B. (1957), "Energy and Reynolds Considerations in Thrust Bearing Analysis." Paper 21, *CLW,* 29–38; 13, 828, 832.

Sternlicht, B. (1958), "Influence of Pressure and Temperature on Oil Viscosity in Thrust Bearings." *T* **80,** 1108–12.

Sternlicht, B., Carter, G. K., & Arwas, E. B. (1961), "Adiabatic Analysis of Elastic, Centrally Pivoted, Sector Thrust Bearing Pads." *JAM* **28,** *T* **83,** 179–87.

Sternlicht, B. & Maginnis, F. J. (1957), "Application of Digital Computers to Bearing Design." *T* **79,** 1483–93.

Sternlicht, B., Reid, J. C., Jr., & Arwas, E. B. (1959), "Review of Propeller Shaft Thrust Bearing." *JASNE* **71,** 277–89.

Sternlicht, B., Reid, J. C., & Arwas, E. B. (1961), "Performance of Elastic, Centrally Pivoted, Sector, Thrust-Bearing Pads—Part I." *JBE, T* **83,** 169–78.

Sternlicht, B. & Sneck, H. J. (1957), "Numerical Solution of the Reynolds Equation for Sector Thrust Bearings." *LE* **13,** 459–63.

Stodola, A. (1924), *Die Dampf-und Gas Turbinen.* Springer, Berlin, 6th ed. (a) Transl. by Loewenstein, McGraw-Hill, New York, 1927; see pp. 1027–34. (b) 1945 ed., 2 volumes, Peter Smith, New York.

Stone, D. D. (1936), *Bearing Lubrication.* U. S. Pat. 2,030,232.

Tao, L. N. (1959), "The Hydrodynamic Lubrication of Sector Thrust Bearings." *Sixth Midwestern Conf. on Fluid Mechanics,* Texas University, 406–16.

Tao, L. N. (1960), "A Theory of Lubrication with Turbulent Flow and its Application to Slider Bearings." *JAM* **27,** *T* **82,** 1–15; 598.

Tenot, A. (1937), "Tests of Bearings: Methods and Results." *GDLL* **1,** 317–22.

Tipei, N. (1957), *Hidro-Aerodinamica Lubrificatiei* (Hydro-Aerodynamics of Lubrication). Editura Academiei Republicii Populare Romini, 1957; 695 pp. (a) Transl., *Theory of Lubrication,* ed. by W. A. Gross, Stanford University Press, 1962; 566 pp.

Ulukan, Von Lütfullah (1956), "Thermische Schmierkeilbildung." *Bull. Tech. Univ. Istanbul* **9,** 77–101. (a) *Actes, Ninth ICAM,* 1956; Brussels, **4,** 303 (1957).

Underwood, A. F. (1937), "Automotive Bearings—Effect of Design and Composition on Lubrication." *Symposium on Lubricants, ASTM,* Philadelphia, 29–52.

Vogelpohl, Georg (1958), *Betriebsichere Gleitlager. Berechnungsverfahren für Konstruktion und Betrieb.* Springer, Berlin; 315 pp.

Wang, T. T. (1950), Theory of Hydrodynamic Lubrication of Plane Slider and Full Journal Bearing with Side Leakage (Chinese). *Engineering Re-*

ports, Nat. Tsing Hua University **4,** No. 2, 96–106. (a) *AMR,* Dec. 1951, p. 679. (b) Chieng, W.-Z., *Chinese J. Phys.* **7,** 278–99 (1949).

Weibull, Waloddi (1925), "Glidlagerteorie med Variabel Viskosität" (Slider-Bearing Theory with Variable Viscosity). *Teknisk Tidskrift, Mekanik* **55,** 164–7.

Wepfer, W. M. & Cattabiani, E. J. (1955), "Water-Lubricated-Bearing-Development Program." *ME* **77,** 413–8; 923.

Wilcock, D. F. (1955), "The Hydrodynamic Pocket Bearing." *T* **77,** 311–9.

Wood, W. L. (1949), "Note on a New Form of the Solution of Reynolds' Equation for Michell Rectangular and Sector-Shaped Pads." *Phil. Mag.* (7) **40,** 220–6.

Wright, J. H. (1959), "Longitudinal Stiffness of Marine Propulsion Thrust Bearing Foundations." *JASNE* **71,** 103–6.

chapter VII Steady-Load Journal Bearings

1. Types of journal bearings. 2. Early investigations. 3. Load capacity of partial and full bearings. 4. Multiple films, and others forms. 5. Friction and temperature rise. 6. Effects of variable viscosity. 7. Inertia and turbulence. 8. Steady-state experiments. 9. Performance with special lubricants. 10. Bearing design. 11. Summary.

The journal bearing supports a radial or transverse load, as distinguished from the axial load on a thrust bearing. It ranges in size from the tiniest instrument bearings (Jaquerod, 1922) to rolling mill bearings loaded up to 5 or 6 million pounds (Dahlstrom, 1933; Hitchcock, 1954). It carries a steady load in the steam turbine, a fluctuating or dynamic load in the automobile and diesel engine. The journal bearing offers some difficult problems for calculation.

1. TYPES OF JOURNAL BEARINGS

A full 360-deg journal or "sleeve" bearing is required when the direction of the load is variable or unknown. Partial bearings of 180-deg arc, or less, are used where the load direction is fairly constant, as in railroad cars. It will be recalled that a bearing of 157-deg arc was used by Tower in his historic experiments. A shorter arc, 120 deg, is in common use today (Needs, 1946). Partial bearings are normally of the "clearance" type—journal radius less than bearing radius—but for some purposes a "fitted" bearing is preferred, in which the journal and bearing have the same radius (Howarth, 1924). Partial bearings are described as "centrally loaded" or "offset," according as the load line bisects the bearing arc or divides it unequally. The offset type is the more efficient when rotation is in one direction.

A split construction is common in which the bearing surface is a part of the lower member, while a cap of large clearance forms the upper half. Side reliefs are provided at the split to effect entrance of cool oil on one side, and its outflow on the other side after some of it passes through the load zone. Another split construction is the "elliptical" bearing, each arc less than 180 degrees (Rumpf, III: 1938; Wilcock, I: 1957).

Full journal bearings are often provided with an oil hole and short distributing grooves. Fancy grooving is obsolete (Karelitz, 1929). Multiple film bearings are formed by several longitudinal grooves (Frössel, 1953). The active areas may be machined on a tilt to insure convergent films, or they may be cut in the form of Rayleigh steps (Archibald, 1952). Length-diameter ratios are now rarely greater than 1.25. In the earlier days of the bearing art, L/D ran from 3 to 5, which explains the term "sleeve bearing." Clearance ratios range from $\frac{1}{1000}$ in the larger bearings to $\frac{1}{500}$, or more, in bearings under an inch in diameter (Howarth, 1934). Circumferential grooving is often applied as an aid to better cooling, although it divides the bearing into two parts, reducing its load capacity.

Besides solid bearings, there are tilting-shoe and floating-shell constructions (Michell, 1912). Tilting shoes or pads are useful when the eccentricity of the journal must be limited or when greater stability is required. Spring-supported and elastically deformable shoes (Kingsbury, VI: 1914, Michell, 1919) have been tried as a substitute for the pivoted type. References on hydrostatic bearings are given under "Bearing Design." Pictures of modern journal bearings are found in the books by Michell (I: 1950), Vogelpohl (VI: 1958), and Radzimovsky (1959), as well as in manufacturers' catalogs.

2. EARLY INVESTIGATIONS

The best-known experiments among the earlier investigations of journal bearings fall into three periods: (1) those by Morin, Hirn, and Thurston from 1832 to 1880, (2) by Petroff, Tower, and Goodman in the eighties, and (3) by Kingsbury, Lasche, Stribeck, and Moore from 1897 to 1906.

First Period. Morin (I: 1832) extended Coulomb's ideas to practical bearings. Hirn (I: 1884) found that the friction of well-lubricated surfaces was more nearly proportional to the square root of the load. Thurston (1878) experimented over a wide enough range of speeds and of temperatures to observe the now-familiar minimum point in the coefficient of friction curve. He speaks of a "critical temperature" at every speed and load.

Second Period. Petroff's classical work (I: 1883) is the theory of the concentric bearing. He confirmed it experimentally and found that the effective film thickness diminishes with increasing load. See also an authentic publication by the Soviet Academy of Sciences (Petroff, 1948) and the biographical note by Archibald (I: 1956, g).

Tower's first report came out the same year as Petroff's. Tower compared the oil bath with other methods of supply. He introduced a new measurement, that of the pressure distribution in the film (II: 1883). Goodman was the first to measure film thickness. He used a micrometer method (1886).

Third Period. Kingsbury's experiments with an air-lubricated journal are cited in Chapter IX. Published in 1897, they provide an early confirmation of Reynolds' theory, supplementing Goodman's observations and antedating Sommerfeld's theory. Kingsbury measured friction, pressure distribution, and film thickness.

Later came the investigations by Lasche (III: 1902), Stribeck (V: 1902), and Moore (1903). These brought into good focus the heat transfer characteristics of the bearing housing; the shape of the coefficient of friction curves, with their minimum points; and the effect of speed on load capacity, respectively. The period closed with Kingsbury's tests on large shaft bearings (III: 1906).

A New Study of Journal Bearings. Our own experiments (Hersey, V: 1909) were undertaken after the foregoing period. They began with friction tests on different oils, using a machine that had been set up for testing partial bearings with an oil bath (1923). The test bearing was soon modified by attaching a cap to the unloaded side. This converted the partial bearing into a full bearing. Oil was fed to the cap at rates varying from 2 to 60 drops per minute for comparison with bath lubrication. The friction results proved to be of unexpected interest because they led to the discovery of "ZN/P," as described in Chapter V. However, they were secondary to our main purpose, which was to be the study of "carrying power" (load capacity).

Little had been done on this problem, since it was commonly assumed that load capacity tests would be destructive. Designs were based on textbook formulas according to which the load per unit of area must be reduced when the speed in increased. The writer distrusted this rule in the light of Reynolds' theory, and believed that a well-lubricated bearing should be able to carry a heavier load the greater the speed. He was especially intrigued by Professor Moore's experiments with a partial bearing (1903). Moore had found that the electrical resistance of the oil film in his bearing would vanish, indicating metallic contact at any given speed when the load is sufficiently great. The critical load increased nearly as the square root of the speed in the experiments cited.

Upon connecting our full bearing and journal to a direct current source, similar results were obtained except that no amount of loading available would completely break down the electrical resistance. We could, however, plot the relative resistance R/R_0 against load and speed; where R_0 is the no-load resistance, or value of R for a concentric journal and bearing. This procedure minimized the effects of any change in the resistivity of the oil, since a check reading on R_0 could be taken just before and after each reading under load. A formula was then derived for the eccentricity ratio in terms of R/R_0, assuming the clearance space full of oil,

$$(e/c)^2 = 1 - (R/R_0)^2 . \tag{1}$$

Here e is the journal eccentricity and c the radial clearance. The observations showed a gradual increase in the minimum film thickness with increasing speed, up to a maximum value nearly equal to the radial clearance. At higher speeds the film thickness fell off sharply.

The results supported the belief that load capacity increases with speed. This belief was not yet generally accepted. The writer, however, had an opportunity to visit Professor A. Stodola at Zürich. This experienced engineer, a steam turbine authority, expressed a keen interest in the experiments. He went out of his way to encourage the visitor to offer his data for publication upon returning home. This advice was followed, resulting in a short paper the following year (II: 1914), and a life-long interest in lubrication research.

The resistance formula, Eq. (1), published in 1910, is equivalent to the 90-deg arc of a circle with R/R_0 and e/c as coordinates. It was shown that a similar formula holds for capacitance. These formulas were rediscovered by Schering and Vieweg some years later (1926).

3. LOAD CAPACITY OF PARTIAL AND FULL BEARINGS

For the purpose of reviewing load capacity calculations, journal bearings may be considered in four groups—partial, full, multiple film, and other forms, in the order named. Partial bearings of short arc are the simplest, and approximate the characteristics of thrust bearings.

Partial Bearings of Infinite Length. Reynolds applied hydrodynamic theory to the 157-deg bearing of Tower. Sommerfeld solved the 180-deg or half-bearing analytically, along with the better known full bearing. An editorial in *Engineering* (VI: 1915) extended the solutions to various bearings arcs with eccentricity ratios from 0.4 to 0.8. Arcs and eccentricities were so chosen as to avoid or minimize negative pressures. Howarth applied the theory to all bearings arcs by graphical integration, notwithstanding negative pressure. He included both fitted and clearance bearings, with offset as well as central loading (1924, 1925, 1926).

Negative Pressures. Owing to the small eccentricities, no negative pressure occurred in Reynolds' application. His solution was based on the idea that the circumferential pressure gradient $dp/d\theta$ must equal zero at the end point of the film where $p = 0$. This requirement is called the "Reynolds boundary condition." Sommerfeld accepted the existence of negative pressures in divergent film areas, although remarking that lubricants cannot actually support tension under practical conditions. The coexistence of symmetrical positive and negative pressure curves is known as the "Sommerfeld boundary condition." Howarth calculated load capacities by Sommerfeld's method, although Gümbel's objections had been published in 1922. One of Gümbel's proposals was to disregard the negative pressure loop, assuming positive pressures would begin at the maximum film thickness, and end abruptly at the minimum. This is known as the "Gümbel boundary condition." Later, Gümbel advocated the Reynolds condition.

Karelitz, in his paper of 1925, determined load capacity for infinite length at each eccentricity by a vectorial graphic integration, disregarding negative pressures. His solutions apply to all bearing arcs from 60 to 180 degrees at moderate eccentricities. They were extended, later (III: 1930), to an eccentricity of 0.9. This study led to a family of curves for the eccentricity ratio against a parameter A at constant values of the bearing arc. The Karelitz parameter is $1/\pi$ times the Sommerfeld reciprocal. End leakage disregarded, $A = 132P_0/ZN$ times $(1000\ m)^2$, expressed in cp, rpm, and psi. Here m is the clearance ratio, and P_0 = the load per unit of projected area for infinite length. This expression agrees with Professor Fuller's on page 201 of his book (I: 1956), where η has been used for the load ratio P/P_0. The eccentricity curves for different bearing arcs are reproduced in Fuller's Fig. 131.

Boswall (V: 1928) was canny enough to avoid negative pressures by limiting his solutions to cases where they could not occur. Strong support for the Reynolds boundary condition was offered by Stieber and by Swift (both in 1933). Stieber showed that this condition was necessary for continuity of flow in bearings of infinite length. He calculated the load capacities for a series of partial bearings. Swift derived it as a condition for stability of the journal position. He applied it in calculating bearings with arcs of 90 to 180 deg, centrally loaded. His paper gives a family of eccentricity curves for these bearings plotted against a load criterion Δ equal to the Karelitz A.

End Leakage. Gümbel was the first to propose corrections for end leakage (1917). Taking 126 deg as a typical bearing arc in order to come out with round numbers, he showed from a formula for plane surfaces that P/P_0 is approximately the reciprocal of $1 + (D/L)^2$. This reduces to $\frac{1}{2}$ when L/D is unity. Further approximations were offered by Karelitz (1925) and by Boswall (V: 1928). Karelitz (1925) assumed the longitudinal pressure distribution parabolic. This led to a fixed P/P_0 ratio of $\frac{2}{3}$.

Boswall's solutions are based on the Stodola approximation, whereby the film pressure is taken as the product of a function of x and a function of z. Application is made to a 90-deg fitted bearing of 6-in. diameter. The computations for clearance-type and fitted bearings of finite length are collected in Tables XIX and XX of his book, and in Table 11 of Norton's. These tables disclose the load angles necessary to insure positive pressures in arcs from 35 to 120 deg.

Kingsbury's electrical integration soon became available (II: 1931). Applications were made to the 180-deg offset and 120-deg centrally loaded bearings. Both investigations were carried over a wide range of B/L, though limited to an eccentricity of 0.4. In his "optimum" paper the following year, Kingsbury combined these results with previous determinations on plane surfaces, and on finite journal bearings calculated for optimum conditions, and plotted them in a master curve for P/P_0 against B/L—see Fig. 10 of the reference (1932). This famous curve served usefully for design problems, but is limited to optimum conditions. Less restricted conditions would require a family of curves marked with different eccentricities, and redrawn for each different bearing arc.

The Finite 120-Deg Bearing. Such a diagram resulted from Needs' investigation of the centrally loaded 120-deg clearance bearing (1934). He used the electrical integration method, neglecting negative pressures by the Gümbel condition. Howarth (VI: 1935), Dennison (I: 1936), and Fuller (I: 1956) applied the results. Needs evaluated

P/P_0 for eccentricities up to 0.8 at B/L ratios from 1 to 4. Here $B = \pi D/3$; thus $L/D = 1$ when $B/L = 1.047$. Needs' values of P/P_0 have been extended to an eccentricity of 0.99 in Professor Fuller's Fig. 130 (I: 1956).

Following Needs' investigation some characteristics of 120-deg bearings were calculated by Boswall (1937) from $L/D = 1$ to 2, both for central and offset loading. It was assumed that the film viscosity at outlet is half that at inlet. Tabulated results are limited to conditions that will not show negative pressure. The eccentricity ratio is found inversely proportional to the square root of the Sommerfeld number, when ϵ is small.

Other Bearing Arcs. The RCL now invited Professor Waters to make a supplementary study (1939, 1942). His investigation included arcs from 80 to 140 deg; and B/L from 0 to 5, at eccentricities from 0 to 0.9. These results were interpolated from a large number of combinations, including offset bearings. The investigation was conducted by iteration methods, excluding negative pressures by Reynolds' condition.

The Sommerfeld numbers are shown in Table 1 for the four bearing arcs investigated. Values were interpolated from Waters' Table 4.

The P/P_0 ratios required for converting load capacities into finite length values are shown in Table 2 for two of the bearing arcs studied by Professor Waters. These values were interpolated from his Figs. 4 and 5, except for the zero eccentricity column, which comes from a formula that he derived (page 8 of 1942).

A comparison of load ratios for the 120-deg bearing indicates that Needs' values averaged slightly lower. This may be due to his exclusion of negative pressures without the aid of Reynolds' boundary

TABLE 1. Sommerfeld Number for Infinite Length
(From Waters on "Centrally Supported Bearings")

Arc, deg	80	100	120	140
ϵ	Values of S			
0.4	0.46	0.26	0.19	0.16
0.6	0.169	0.112	0.085	0.070
0.8	0.050	0.038	0.033	0.0295
0.9	0.018	0.016	0.0145	0.014

When $\epsilon = 0$, $S = \infty$; when $\epsilon = 1$, $S = 0$.

TABLE 2. Load Ratios for Finite Length **(Waters)**

Bearing arc, deg	$\epsilon =$	0.9	0.8	0.6	0.4	0.0
	B/L	Values of P/P_0				
80	$\frac{1}{2}$	.81	.78	.72	.69	.67
	1	.65	.59	.485	.44	.41
	2	.39	.315	.205	.185	.17
	4	.165	.100	.061	.055	.048
120	$\frac{1}{2}$	.85	.825	.755	.72	.64
	1	.71	.65	.565	.51	.41
	2	.48	.375	.29	.25	.17
	4	.21	.155	.11	.075	.048

$P/P_0 = 1$ when $B/L = 0$; $P/P_0 = 0$ when $B/L = \infty$.

condition. Professor Water' load ratios for the 120-deg bearing are plotted in Fig. 1, making a diagram analogous to Fig. 1 of Chapter VI.

The Half-Bearing. Sommerfeld's solution for the infinite-length bearing of 180-deg arc had been challenged by Gümbel, Swift, and others since it involved negative pressure, or tension in the lubricant. Not only the pressure distribution, but the running position of the journal is affected by the choice of boundary conditions.

Draw a circle with radius representing the radial clearance of the bearing on a magnified scale, as in Fig. 2. Draw horizontal and vertical lines through the center O. Let these intersect the circle at A on the left and at B below. Say the load on the journal acts vertically downward, and the journal rotates clockwise. The path of the journal axis starts from O when journal and bearing are concentric. With increasing load or decreasing speed the path goes toward the left, dipping downward. It represents the locus of all steady-state positions and lies wholly in the third quadrant. Under the Sommerfeld boundary condition, the path begins to curve upward after reaching an eccentricity of about $\frac{3}{5}$, and terminates at A. But under the Gümbel condition it continues curving in the original direction and terminates at B. The Gümbel curve is nearly a semicircle.

Boundary Conditions, Full Bearing. Vogelpohl (1937) describes the situation in more detail with an argument in favor of the Reynolds condition. Imagine the pressure curve plotted in rectangular coordinates against the developed bearing surface as a base line, as in Fig.

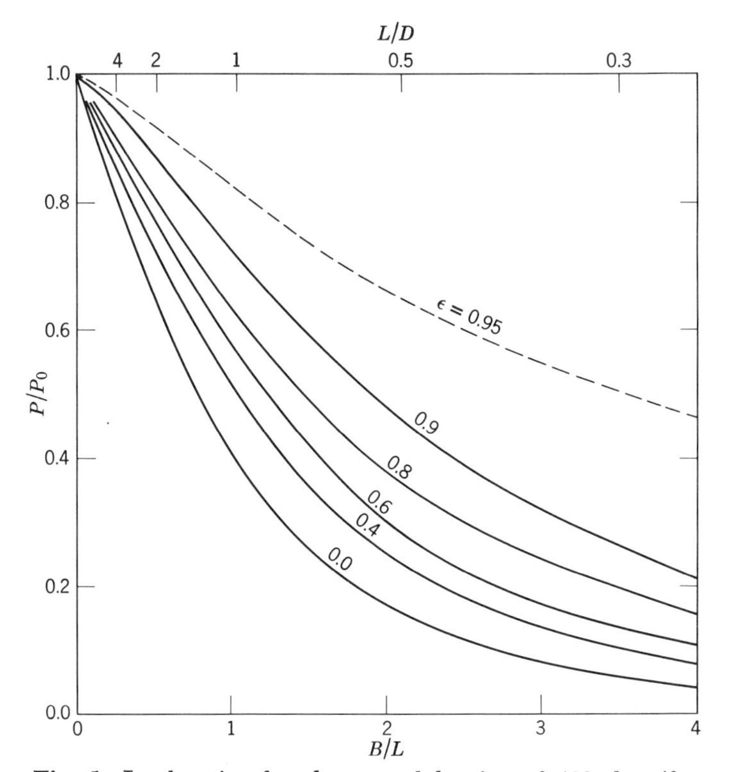

Fig. 1. Load ratios for the central bearing of 120 deg (from E. O. Waters, *Trans. ASME* 1942).

3. The curve for Gümbel's boundary condition drops abruptly to zero at the point of minimum film thickness. The curve for Reynolds' condition is swept past this point in the direction of motion. The value $p = 0$ is reached in a gentle manner, with $dp/d\theta = 0$, some distance down stream. The curve crosses above the point of minimum film thickness at an ordinate just half the maximum. Thus additional load capacity is provided by the Reynolds condition. The film behavior is similar to that in a "composite" thrust bearing. The Reynolds path of the journal axis is identical with the Sommerfeld path until ϵ exceeds 0.4. It is nearly the same as the Gümbel path from 0.9 to 1.0. The Reynolds path offers a smooth transition from one to the other.

Approximations. Schiebel's approximation for side leakage was applied to the half-bearing of finite length (VI: 1933). Vogelpohl, in

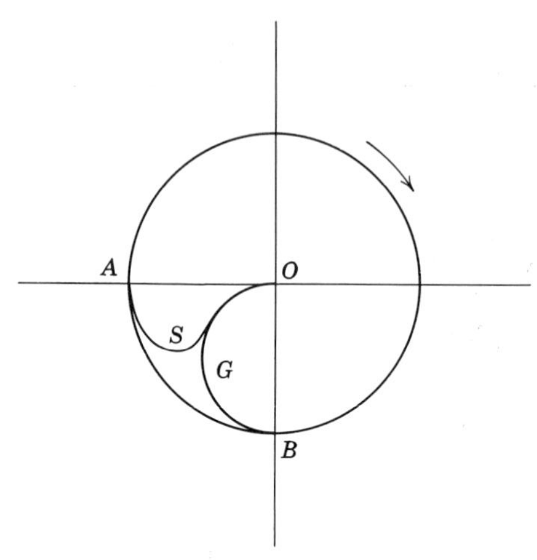

Fig. 2. Path of journal axis in 180-deg bearing of infinite length.

his paper of 1943, integrated Reynolds' equation on the assumption that the circumferential pressure is geometrically similar, at every cross section, to that of an infinite-length bearing. This approximation, he notes, agrees with Michell's solution for plane surfaces. Film pressures and load capacities were evaluated for $L/D = 0, \frac{1}{4}, \frac{1}{2}, 1$, and 2;

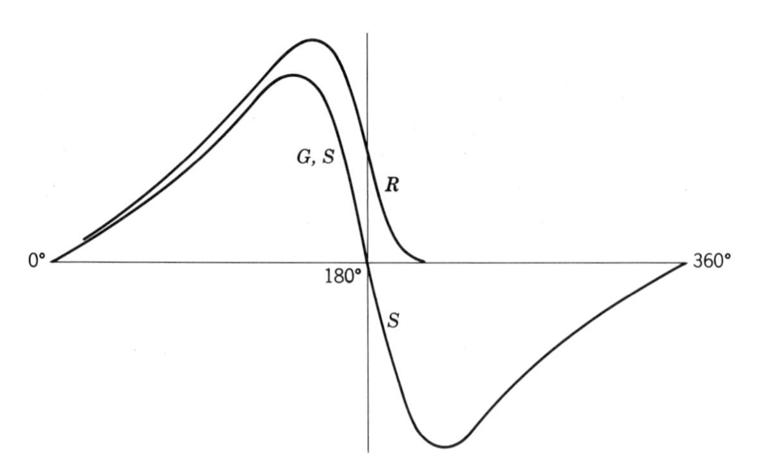

Fig. 3. Film pressure in full bearing for different boundary conditions (R, Reynolds' condition; S, Sommerfeld's; G, Gümbel's).

including P/P_0 ratios, and using Reynolds' boundary condition. Vogelpohl's values are close to those given by Swift (1933), averaging perhaps 2 per cent lower. They are in good agreement with the most recent computer results.

New Charts and Solutions. A new series of journal bearing charts began with a paper by Boyd & Raimondi in 1951. Their calculations are based on the Sommerfeld condition except for the 105-deg arc. Charts are for infinite length, accompanied by a load ratio diagram. A solution for the pressure distribution in fitted bearings of finite length was published by N. Tipei in 1951; see pages 328–366 of his book (VI: 1957). A new solution for partial bearings of infinite length was offered by J. C. Lee (1955). Eccentricities were carried up to 0.999 for railroad application. At such eccentricities the performance would be little different for finite length. Negative pressures were excluded by the Reynolds condition.

Sassenfeld & Walther (1954) came up with new calculations for the centrally loaded half-bearing and the full bearing. Eccentricities are carried to 0.95, with L/D's as low as $\frac{1}{8}$. An iteration method was employed, using the Reynolds boundary condition. A numerical solution for the half-bearing was given by Philipzik (II: 1956) using a relaxation method. The results came close to those of Sassenfeld & Walther, although obtained independently. Jakobsson & Floberg (1957) investigated the half-bearing of finite length, both offset and centrally loaded. This involves cavitation. The authors assumed atmospheric pressure in the cavity and determined the beginning of this region by the Reynolds condition. The solutions were obtained by numerical relaxation methods.

Computer solutions for centrally loaded, finite partial bearings were given by Pinkus in 1956 and 1958. Negative pressures are eliminated by the Gümbel condition. Eccentricities from 0.1 to 0.9 are plotted against the Sommerfeld number for four L/D ratios from $\frac{3}{2}$ down to $\frac{1}{4}$. A paper followed with solutions by Jakobsson & Floberg (1958) for offset partial bearings of finite length, having their minimum film thickness at the trailing edge; another (1959) with central loading and arcs from 60 deg up, considering cavitation. Offset bearings of 120 deg with $L/D = 1$ were studied by Floberg (1961), including cavitation. Pinkus, also, investigated offset bearings (1961).

Three papers were published by Raimondi & Boyd (1958) on centrally loaded bearings, and one by Raimondi on offset bearings (1959), all based on computer solutions. The papers on centrally loaded bearings covered arcs from 60 to 180 degrees. Performance characteristics are plotted against S for L/D's of 1, $\frac{1}{2}$, and $\frac{1}{4}$. In the first two papers, negative pressures are included for application to bearings under high ambient pressures. An example is given of a boiler-water circulating pump, where the bearing operates in water at 210 F and 200 psig. The negative pressures are merely subambient, not actually creating tensile stress. In the third paper, from which our Tables 3 and 4 were obtained, negative pressures are excluded by the Reynolds or "Type 2" condition. The charts are accompanied by tabulated values up to eccentricities of 0.97. The results are in good accord with Professor Waters' report of 1942, but more complete. Raimondi's paper on offset bearings is limited to the 120-deg arc with $L/D = 1$. Negative pressures are excluded as before.

An analytical solution for the finite-length fitted bearing was obtained by Jacobson & Saibel (1958, 1960). Load capacities were then evaluated for bearing arcs of 60 and 120 deg with the aid of a digital computer. This was repeated for different load angles.

The symbol δ has been introduced for simplicity to denote the relative film thickness h_0/c. Hitherto, journal positions have usually been indicated by the eccentricity ratio ϵ. Note that $\delta + \epsilon = 1$.

Resume on Partial Bearings. The reader might like to unscramble our narrative by assembling the information in tabular form. Entries would be required for the bearing geometry, showing whether clearance or fitted; centrally loaded or offset, finite or infinite length, and the range of bearing arcs; also, a column for the boundary condition used, whether Sommerfeld, Gümbel, or Reynolds. Imagining such a table at hand, arranged in chronological order, we could look back to see what has been accomplished. It appears that the ground is open

TABLE 3. Central Partial Bearings of Infinite Length
(Raimondi & Boyd's data by Reynolds' condition, *TASLE*, 1958)

δ $(= h_0/c)$	0.03	0.1	0.2	0.4	0.6	0.8
Arc, deg	Values of 100S					
60	0.50	2.41	7.55	32.2	93.1	266
120	0.41	1.47	3.28	8.45	18.1	43.1
180	0.39[a]	1.28	2.53	5.23	9.0	17.9

[a] As δ approaches zero it may be taken $= 7.69S$ (Kochanowsky). It is 0 for $S = 0$, and 1.0 for $S = \infty$.

TABLE 4. Load Ratios for Central Partial Bearings
(From the tables by Raimondi & Boyd)

δ (= h_0/c)		0.03	0.1	0.2	0.4	0.6	0.8	1.0[a]
Arc	L/D	Values of P/P_0						
60 deg	1	.85	.78	.75	.72	.70	.68[b]	.67
	$\frac{1}{2}$	.70	.57	.51	.46	.44	.41[b]	.40
120 deg	1	.82	.71	.62	.57	.47	.43	.40
	$\frac{1}{2}$	.64	.45	.34	.24	.20	.17	.15
	$\frac{1}{4}$	.39	.20	.10	.08	.06	.05	.04
180 deg	1	.80	.66	.55	.41	.32	.27	.24
	$\frac{1}{2}$	.61	.41	.27	.16	.11	.09	.08

Note that $P/P_0 = 1$ when L/D is infinite, and zero when L/D is zero. [a] From Waters' formula when $h_0/c = 1.0$. [b] Mean $\delta = 0.81$ (line of centers beyond trailing edge).

for more complete calculations on the clearance and fitted types with offset loading.

Although only the major investigations are reviewed above, other studies are not without interest. Kochanowsky (1955) and Vogelpohl (VI: 1958) showed that the initial slope of the relative film thickness curve δ or h_0/c against S is the same for the 180-deg arc at all L/D ratios under the Reynolds boundary condition, and equal to 7.69 (Table 3). It would be zero for the Sommerfeld condition. Further studies of cavitation were published in support of the Reynolds condition—see, for example, Robb (1933), Klimov (1947), Vogelpohl (1950), and Cole (1951). Dowson's (1957) experiments led him to suggest a modification of Reynolds' condition. See also Birkhoff & Hays (1963).

With the aid of Tables 1 through 4 it should be possible to draw up load capacity charts for centrally loaded clearance bearings of any L/D ratio, having arcs from 40 to 180 deg, all based on the Reynolds boundary condition. Figure 4, from Table 3, shows the relative film thickness for arcs of 60, 120, and 180 deg against the Sommerfeld number for bearings of infinite length. Figure 5 shows load ratios applicable to the central clearance bearing of 120-deg arc. It is constructed from the data of Tables 2 and 4, extrapolated

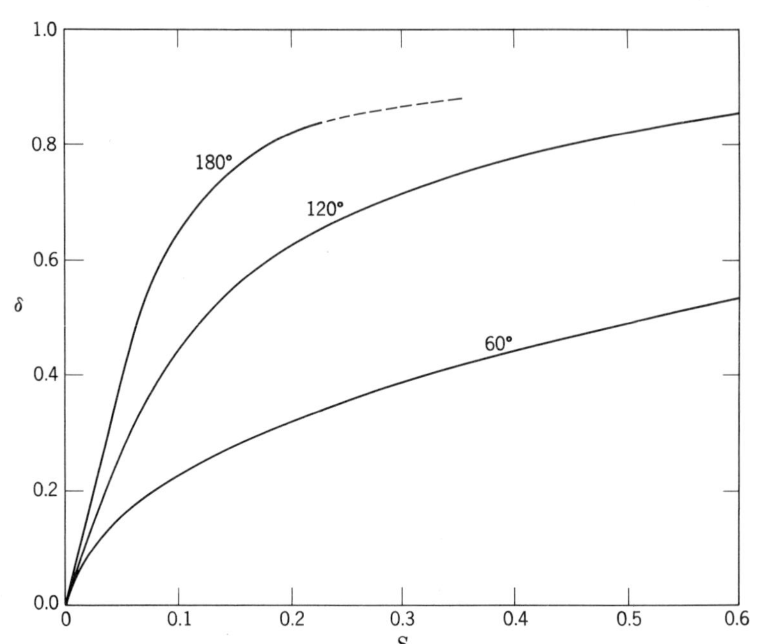

Fig. 4. Central partial bearings of infinite length (relative film thickness, δ, from Table 3).

to a film thickness $\delta = 0.01$ ($\epsilon = 0.99$), and terminated by the curve from Professor Waters' formula for $\delta = 1.0$ ($\epsilon = 0$),

$$\frac{P}{P_0} = 1 - \frac{\tanh x}{x} \cdot \tag{2}$$

Here x denotes $\pi/2$ divided by B/L, or $\frac{3}{2}$ divided by D/L, and the formula applies to central partials of all arcs. The curves are in reasonable agreement with Professor Fuller's Fig. 130, though derived from more complete data utilizing Reynolds' boundary condition. These load ratios should not be applied to bearing arcs greater or less than 120 deg except for the highest eccentricities. Thus from Table 4 it is seen that for $L/D = \frac{1}{2}$ and $\delta = 0.8$, the load ratio drops from 0.41 to 0.09 with an increase of bearing arc from 60 to 180 deg. Table 4 should be used in preference to Fig. 5 for bearing arcs very much different from 120 deg.

The charts and tables by Raimondi & Boyd now provide direct solutions to practically all journal bearing problems, making it un-

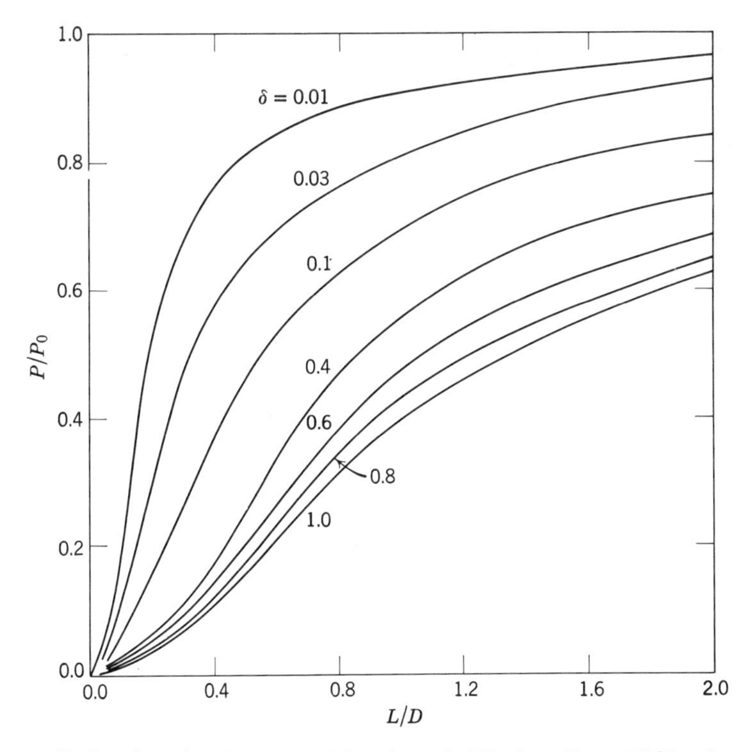

Fig. 5. Load ratios for central bearing of 120 deg (from Tables 2 and 4).

necessary to resort to the indirect load ratio method. These ratios, however, help to visualize the physical factors and to present the relations in a compact form.

For example, let it be required to estimate the minimum film thickness in a bearing of 120 deg, centrally loaded on a journal of 6-in. diameter. Suppose $L/D = \frac{2}{3}$, $m = C/D = 10^{-3}$, mean viscosity $Z = 3$ n or micro-reyns, speed 3000 rpm ($N = 50$ rps), and load per unit area, P, 150 psi. Then $ZN/P = (\frac{1}{2})10^{-6}$ and $S = ZN/Pm^2 = \frac{1}{2}$. From Fig. 4 it is seen that δ would be 0.83 for a bearing of infinite length at $S = \frac{1}{2}$ ($100S = 50$). The film must be considerably thinner for $L/D = \frac{2}{3}$; take $\delta = 0.60$ as a trial value to start from. Entering Fig. 5 at $L/D = 0.67$, we read $P/P_0 = 0.31$ from the curve for $\delta = 0.60$. The film thickness for a finite length therefore corresponds to a trial Sommerfeld number $S' = 0.31\ S = 0.155$. Enter Fig. 4 at $100S = 15.5$. This leads to a second approximation, $\delta = 0.56$. Figure 5 now gives $P/P_0 = 0.325$. For the next Sommerfeld number, then, $S'' = 0.325S = 0.163$. Figure 4 gives 0.572 for the new δ. We could settle for 0.57, making $h_0 = 0.57mD/2 = 0.0017$ in., or 1.7 mils.

The preceding methods are substantially equivalent to the use of Fuller's Fig. 131 when the bearing arc is near enough to 120 deg. It might be worthwhile to repeat Fig. 5 for other arcs.

Full Journal Bearings, Infinite Length. In the same year with Sommerfeld's paper Zhukovskii & Chaplygin (1904) offered a more general treatment, which reduces to Sommerfeld's when the clearance is small. Three investigations of a similar character followed by Duffing (1924), Reissner (1935), and Wannier (1950). These derivations start from the Navier-Stokes relations in order to avoid the limitations of Reynolds' equation. Duffing alone offers a remedy for negative pressures—namely, the G boundary condition. Wannier illustrates his theory by an example with a large radial clearance. He finds a closed whirlpool of laminar flow trapped near the maximum film thickness, confirmed experimentally by Cole. All four papers will interest theoreticians.

Many computations have been published on the full bearing of infinite length. Among the few to employ the Reynolds boundary condition are those by Cameron & Wood (1949), Sassenfeld & Walther (1954), Floberg (1957), Raimondi & Boyd (1958), Jakobsson & Floberg (1959), and Floberg (1961). Table 5 records film thickness ratios computed by Raimondi & Boyd. It is interesting to compare these ratios with the corresponding values for partial bearings in Table 3, the Sommerfeld curve, Fig. 6 of Chapter II, and Hays' solution using the Gümbel condition (II: 1959).

The treatment of cavitation is more critical for the full bearing than for partials. It is clearly explained by Floberg. Figure 2 applies

TABLE 5. Film Thickness in Full Bearings of Infinite Length
(From Raimondi & Boyd, 1958)

$\delta = 0.01$ $\epsilon = 0.99$	0.03 0.97	0.1 0.9	0.2 0.8	0.4 0.6	0.6 0.4	0.8 0.2	0.9 0.1
Values of 100S							
0.135	0.385	1.145	2.10	3.89	6.26	12.3	24.0

Note: $S = 0$ when $\delta = 0$, $\epsilon = 1$; and S approaches infinity as δ approaches unity. At $\delta = 0.95$, $100S = 39.3$

to the full bearing except that the Sommerfeld path is now a straight line from O to A at right angles to the load line. The Reynolds path again approximates a semicircle. Figure 3 extends the pressure distribution to a full bearing.

Ambient pressure is taken equal to 1 standard atm in the usual solution by Reynolds' condition. Load capacities would be less at an altitude of 20,000 ft, where the ambient is about $\frac{1}{2}$ atm; and still less on the moon. A useful varient on Raimondi & Boyd's solution might therefore be to include the ambient pressure ratio p_a/P as an additional parameter. It is customary in the ideal case of a sleeve bearing of infinite length to interpret p_a as the film pressure in concentric operation.

Full Bearings of Finite Length. Nearly all of the contributors to journal bearing theory who have been cited here extended their calculations to the full bearing of finite length. Some others are listed in Table 6, and referenced in this chapter or in Chapter II. It would be instructive to compile a complete tabulation of full bearing investigations, recording in some way the geometrical factors and range of variables, boundary conditions used, and the special assumptions and approximations involved. This done, we could coordinate past work. Are there any gaps in our knowledge? Should the effect of the oil supply and its location be further explored? At least, the new theories should be carried through to some tangible result.

For practical design purposes the computer solutions by Raimondi & Boyd (1958) and by Connors (1962) cover the ground very well. Load ratio curves are similar to those for the 120-deg bearing. Fig. 5, but with lower P/P_0 values. The difference, $\Delta P/P_0$, ranges from 0 for the thinest films to 0.27 for $\delta = 1.00$ at $L/D = 1.0$.

TABLE 6. Full Bearings of Finite Length; Additional Theory

Authors	Dates	Authors	Dates	Authors	Dates
Reissner	1935–6	Baranov	1947	Hirano & S.	1958
Muskat & M.	1938–9	Pinkus	1947–61	Hays	1959
Gutyar	1939	Weber	1950	Tao	1959
Kino	1940	Steller	1954	Fedor	1960–1–2
Morgan, M. & R.	1940	Tipei	1954–6		
Bauer	1943	Nishihara & S.	1956	Ramachandra	1961
Fränkel	1943	Sasaki & M.	1956	Connors	1962
Klemencic	1943	Tipei & N.	1957	Ishii	1962

The Raimondi & Boyd full bearing tables assume oil supply at ambient pressure through an imaginary axial groove at the variable location of maximum film thickness. Similar tables based on computer solutions were published by Jakobsson & Floberg (1957) as well as by Pinkus & Sternlicht (II: 1961). The earlier reference also gives tables for the more realistic case of a groove fixed in position in the converging space 90 deg ahead of the load line. This reduces the load capacity for an infinite length bearing, but augments the load ratios for finite lengths, so that the resulting film thicknesses are nearly the same.

Floberg discussed the full bearing with two grooves at ambient pressure, each 90 deg from load (1960). His tables and curves were based on Raimondi & Boyd values for 180 deg.

It is further assumed in such tables that the oil supply is just adequate to provide for the calculated hydrodynamic outflow. An extension of the computations to allow for a more limited supply was offered in the study by Connors. His results are charted only for $L/D = 1$. At $S = 0.50$ the classical flow rate makes $\delta = 0.76$. This drops to 0.61 at half rate, and to 0.52 at one-quarter. Conversely if the bearing is submerged, or the inlet pressure otherwise raised enough to neutralize negative pressures, the Sommerfeld boundary condition can be applied, leading to $\delta = 0.86$ for $L/D = 1$ in the example chosen. This follows from the R & B tables for the Type 1 or Sommerfeld condition.

The effects of forced lubrication on load capacity were discussed by Muskat & Morgan (III: 1939), but in general this has been considered an experimental problem. That completes our review of load capacity theory for journal bearings of the simpler geometric forms, with rigid surfaces. The lubricants are assumed Newtonian, incompressible, without inertia, and of uniform viscosity.

4. MULTIPLE FILMS, AND OTHER FORMS

Multiple-film journal bearings may be classed as rigid or nonrigid. The rigid class includes axially grooved bearings and the multilobe types. The nonrigid class includes pivoted-shoe and floating-shell types as well as journal bearings with flexible elements.

Rigid Bearings. Axially grooved sleeve bearings with three or more straight grooves were advocated by Schein for the gyrostabilizer as early as 1933; and by Frössel (1942, 1953) for steel mill and engine use to preserve stability and to resist shock loads from any direction. The theory has been worked out by Sasaki and Mori (1956). The theory of the two-groove design has been given by Pinkus (1958) and by Floberg (1960). Thus Floberg finds that for $L/D = 1$, at say $\epsilon = 0.8$, $P/P_0 = 0.545$ as compared with 0.472 for the case of ambient pressure at maximum film thickness. But the load capacity for infinite length is reduced to 0.84 of the classical load, a net loss of 8 per cent, compensated in practice by more efficient cooling.

The active surfaces in a mutigroove bearing are not necessarily parts of one circular cylinder. They may be scraped to a slight convexity (Waring, 1917), set off on a tilt (Frössel, 1942), or stepped in the Rayleigh manner (Archibald, 1952). A typical four-groove bearing is shown in Vogelpohl's book (VI: 1958, page 190). Each film may be analyzed as in partial bearing theory, the vector eccentricity of the journal being such that the resultant fluid pressure loads form a system in equilibrium with the external load.

Multilobe bearings of "elliptical" or "lemon-shaped" cross section have been analyzed by Ott (1954) and by Pinkus (1956). Theories for the three-lobed bearing have also been derived by the same authors—Pinkus (1956, 1959), Ott (1959). All have been confirmed experimentally. Other bearings with interrrupted surfaces are described by Brillié and Robb (1929, 1933). The sleeve bearing with side reliefs was investigated by Wilcock & Rosenblatt (1953; III: 1952).

Nonrigid Bearings. Pivoted-shoe journal bearings have been described by Michell (1912; II: 1929, 1937), Odqvist (1935), Haydock (1939), and others. Performance calculations are given by Michell in his paper of 1937; by Hentsch with application to high-speed bearings (1946); and more fully by Boyd & Raimondi (1953, 1962) in the form of charts for practical use. In certain charts two curves are plotted against the Sommerfeld variable, one for a plain bearing, the other for the pivoted shoe. It is shown that the pivoted-shoe bearing is not invariably superior, but only under particular conditions.

Floating-pad and ring bearings are similar to those described in Chapter VI. An illustration is found in Michell's book (I: 1950, Fig. VI, 26). Spring-supported shoes and flexible necks were described by de Ferranti (VI: 1911), Kingsbury (VI: 1914), and Fulpius (VI: 1929); flexible tips by Michell (1919). See Fig. VI, 22 of Michell's book for an illustration of a large journal bearing of the flexible-tip variety. A variant having interconnected tips is described with test data by G. Fast (1941). Such constructions, though simpler, are less efficient than freely pivoted shoes—their operation is like playing a piano with mittens on.

Other examples of deformable bearings are the fluted rubber bearings for outboard marine use (Busse & Denton, 1932); Schein's bearing with soft babbit of varying depth (VI: 1933); and Higginson's model of an elastohydrodynamic bearing (1962). See also Korovchinskii (1962) on elastorheology.

Floating sleeve bearings were investigated by Shaw and co-authors (1947, V: 1949), as well as by Kettleborough (1954) and Pinkus (1962). A completely flexible type, the "foil bearing," is described by Blok & Van Rossum (1953). It was further studied by Patel & Cameron (1957), Baumeister (1964), Eshel & Elrod (1965), and Ma (1965).

Special Types. Tanner (1958) analyzed the lubrication of ball and socket joints, treating the case of a constant speed with socket completely surrounding the ball, both for half-cavitation and no cavitation—the Gümbel and Sommerfeld conditions, respectively. See also Wannier (1950) and Peeken (1960) on the spherical journal bearing.

Counterrotating journal bearings were investigated by Pinkus (1962). It had often been assumed that no load capacity can be developed when journal and bearing rotate with equal speeds in opposite directions. Pinkus showed that this applies only to perfectly circular, ungrooved bearings. His calculations were confirmed by tests on four-groove, floating-pad, floating-ring, and tapered-land journal bearings. Camella bearings with "restricted clearance" are described by *The Marine Engineer and Naval Architect* (1963). The journal is supported by three or more eccentric bores to restrict the movement of the shaft axis, while allowing normal clearance and film thickness in each bearing element. The theory of porous metal bearings is given in papers by Morgan (1957, 1964); Morgan & Cameron (1957), Tipei (1959); Cameron, Korovchinskii, and Rouleau (all 1962).

5. FRICTION AND TEMPERATURE RISE

The theory of load capacity as reviewed in the preceding sections requires a knowledge of the mean film viscosity, which in practice is usually unknown. Trial-and-error methods are possible for estimating the steady-state film viscosity. More rational methods are described in Chapter III. In any case, we begin by calculating the friction and temperature rise corresponding to the inlet viscosity.

Numerical Examples. Such calculations will be illustrated by three examples, based on journal bearings of 4-in. diameter with $L/D = \frac{1}{2}$, and m or $C/D = \frac{1}{500}$, operating at 50 rps with an eccentricity ratio 0.8. The lubricant is assumed to have a constant viscosity of 3 n or micro-reyns. The first example is for an end-lubricated Ocvirk bearing; the second for a full bearing with two axial grooves 90 deg from the load line; the third for a partial bearing of 120-deg arc, centrally loaded. It is required to determine the load capacity, oil flow, friction coefficient, and temperature rise.

Consider Ocvirk's solution for the full bearing. Substitution of the given data in Eq. (29) of Chapter II gives for the load capacity $P =$ 588 psi. This may be checked by curve 3, Fig. 6 of that chapter since for a given eccentricity ratio, P varies with the square of L/D. From Eq. (35), $f = 3.1 \times 10^{-3}$. This may be checked by curve 2 of Fig. 9, since f is inversely proportional to $(L/D)^2$. From Ocvirk's Eq. (22) of 1952 it is seen that Q is equal to the clearance volume $\pi DLC/2$ multiplied by $N\epsilon$; from which $Q = 4.04$ cu in./sec.

The heat generated by friction per unit time will be $H = fWU$, where $W = PDL$ and $U = DN$. Then $\Delta t = H/Qq$, where q, the heat capacity of the oil, as noted in Chapters III and IV, may be taken equal to 140 psi/dF. It follows that $\Delta t = 16$ dF.

The solution for the second example is taken from Floberg (1960). That for the third comes from Jakobsson & Floberg (1959). These authors utilized the computations of Raimondi & Boyd (1958) with a minor correction to allow for the fact that the cavitation space is not solid full of oil. All solutions are assembled for comparison in Table 7. Note that if a smaller clearance ratio $m = \frac{1}{1000}$ had been used, P and Δt would be four times as great, Q and f half as great.

Alternatively it would be possible, and was formerly customary, to calculate friction with the aid of a friction ratio F/F_0 analogous to

TABLE 7. Friction and Temperature Rise
($D = 4$ in., $L/D = 1/2$, $m = 1/500$, $N = 50$ rps for a constant viscosity, $Z = 3$ n; with $\epsilon = 0.8$)

Result		Full bearing		Partial bearing
		Ocvirk	2-groove	120 deg central
P	psi	588	408	386
Q	cis[a]	4.04	4.32	1.74
f	10^{-3}	3.1	5.4	4.2
Δt	dF	16	18	3.4

For clearance ratio $m = \frac{1}{1000}$, new values are $P' = 4P$, $Q' = Q/2$, $f' = f/2$, $\Delta t' = 4\Delta t$.

[a] Cu in./sec.

the load ratio P/P_0. The frictional resistance per unit area, F_0, for the bearing of infinite length would be multiplied by the appropriate F/F_0 to give the required value F for a finite length. Or the coefficient ratio f/f_0 could be taken equal to F/F_0 divided by P/P_0. It was found that L/D has much less influence on F/F_0 than on P/P_0. The friction ratio is of the order of 0.86 for $L/D = \frac{1}{2}$ at $\epsilon = 0.8$.

The effect of the temperature rise Δt is to reduce the film viscosity by an amount depending on the viscosity-temperature curve of the oil used. Consider example 3, where $\Delta t = 34$ dF. Suppose the inlet temperature, at which the oil has its stated viscosity of 3 n, is 130 F. Then if this oil conforms to the R & B diagram (Fig. 2 of 1958), its viscosity will be 1.45 n at 164 F. The true temperature rise must be less than 34 dF; we can therefore be sure that the true viscosity lies between 1.5 and 3.0 n. It can be found by successive approximations or by the graphical method of Chapter III for determining equilibrium conditions. Tables might well be constructed like Table 7 based on a constant load instead of a constant eccentricity.

The solutions for many different bearing designs are assembled by Wilcock in his paper of 1957; as well as in the books by Wilcock & Booser and Vogelpohl.

6. EFFECTS OF VARIABLE VISCOSITY

At least four approaches have been made to the Newtonian variable viscosity problem: (1) by considering radial heat conduction, (2) by specifying circumferential temperature distributions, (3) by adiabatic flow calculations, and (4) by isothermal pressure-viscosity solutions. The effects of variable viscosity on bearing performance have been summarized by Professor Saibel (1960).

Radial Heat Conduction. It will be recalled from Chapter III how Kingsbury and others determined the radial temperature distribution and viscosity variation in the unloaded journal bearing, or film between concentric cylinders. The mathematical treatment was confirmed experimentally. A theory worked out by Hummel (III: 1926) led to a formula for circumferential distribution, Eq. (31) of Chapter III. These studies were based upon radial conduction, disregarding heat transfer by oil flow.

Specified Temperature Distribution. The variation of film viscosity with temperature was further discussed by Fränkel (1944), who noted that the effective viscosity is less than the mean of inlet and outlet viscosities. He concurred with others in taking viscosity proportional to film thickness as a first approximation. Nishihara and

Sugimoto (1956) calculate the pressure distribution in the full bearing of finite length when the temperature distribution, and the viscosity-temperature relation, are given by empirical formulas. See also Tipei (VI: 1957, pages 257–266 of translation), where it is shown that the effective film viscosity is fixed by the outlet temperature.

Adiabatic Solutions. Boswall applied adiabatic flow to journal bearings, including the full bearing of infinite length. Here there is no end flow. So what becomes of the heated oil? Either it disappears miraculously after traveling 360 degrees; or it mixes with fresh oil in the supply groove. In this event the true temperature is that of a mixture, rather than the inlet temperature of the fresh oil, measured externally. We have to choose between a miracle or an unknown quantity. Boswall's solution is regarded by Norton as "artificial." Wilcock and Rosenblatt considered the mixing problem in detail, with special reference to the larger sleeve bearings of finite length, having 30-deg side reliefs.

Christopherson (VI: 1941) was the first to solve for the combined effects of pressure and temperature. This he did by assuming adiabatic flow with Reynolds' boundary condition. The bearing chosen was of 120-deg arc, with $L/D = 0.7$. Line of centers was placed 90 degrees beyond leading edge, with ϵ fixed at $\frac{2}{3}$. The operating variable Z_1UL/h_0^2 was set equal to 7500 psi, where Z_1 is the entering viscosity. Under these conditions the load capacity for a uniform viscosity is 250 psi, regardless of bearing size. The temperature rise T, with $q = 119$ psi/dF, still keeping a uniform viscosity, was found to be $= 38.5$ dF.

Solutions for variable viscosity require the use of a Z, p, t diagram showing the viscosity Z at any pressure p and temperature t; and a knowledge of the temperature t_1 at the leading edge, say 100 F. The diagram chosen shows an oil of 8.5 n at atmospheric pressure and 100 F. The pressure coefficient b appears to be 1.9×10^{-4} (psi)$^{-1}$ at that temperature. The corresponding temperature coefficient a is about 3.4 per cent per deg F drop in temperature. Christopherson plots the contour lines of pressure and temperature using Southwell's relaxation method, but finds little change in performance compared to the uniform viscosity solution. It would be instructive to recompute the problem with a more complete energy equation as suggested by Charnes, Osterle, and Saibel (VI: 1955). These authors and Cope (III: 1949) called attention to the omission of terms representing "flow-work" done by the fluid pressures. Only the work of the shearing stresses was considered in calculating the heat generated.

It remained for M. A. Oksal to apply Christopherson's method to the full journal bearing. His results were reported in a thesis (1946) at Cornell University under Professor P. H. Black. Oksal chose a bearing 7.64-in. in diameter by 12 in. long, operating at 700 rpm with an eccentricity ratio of 0.4. The oil inlet was kept at the maximum film thickness. These data were chosen to duplicate a problem on the 180-deg bearing already solved by Kingsbury for uniform viscosity. Satisfactory agreement was obtained. Oksal then chose an oil of variable viscosity having a viscosity of 10 n at the inlet temperature 90 F. The Z, p, t chart gives $Z_1 = 8.16$ n, $b = 3.0 \times 10^{-4}$ (psi)$^{-1}$, and $a = -2.0\%$/dF at 100 F. The value $q = 119$ psi/dF was retained. Results show a temperature rise of 6.0 dF; with an increase of 29 per cent in load capacity compared to the case of uniform viscosity. It might be of value to continue Dr. Oksal's investigation using a digital computer with the more exact energy equation, and some variation of lubricants. Fixed locations for the inlet groove are suggested, as well as higher eccentricity ratios.

Two adiabatic solutions appearing in 1957, one by Pinkus and Sternlicht, the other by Purvis, Meyer, & Benton are cited in Chapter III. Both solutions give the temperature distribution around the circumference of a full bearing, assuming the viscosity independent of pressure. The first mentioned solution is an approximation derived from Sommerfeld's equation for the pressure distribution. The energy equation used is based solely on the work done by the shear stress. The location of the oil inlet, θ_0, is retained as a parameter. The sleeve bearing with axial supply grooves on both sides is treated by setting $p = p_0$ at θ_0 and $\theta_0 + \pi$. Reasonable agreement is found with test results on bearings of 9-in. and 13 in.-diameter over a wide range of loads and speeds, although the calculated temperatures tend to run higher than the observed values.

The solution by Purvis and co-authors applied to both infinitely long and infinitely short bearings. Longitudinal as well as circumferential distributions of temperature are worked out for the short bearing. The calculations are based upon an energy equation including pressure work as well as shear work. This equation is then combined with the Reynolds equation, while treating both density and viscosity as functions of temperature. The film is shown to reach its maximum temperature where the positive pressure vanishes, shortly beyond the point of nearest approach. A comparison is made with the high-speed, heavy-load experiments of Clayton & Wilkie (III: 1948) on their 2-in. diameter journal bearing. The calculated temperature rise is greater than observed.

An adiabatic solution was published by Hughes & Osterle (III:

1958) for partial bearings of infinite length, in which the viscosity of the oil is treated as a function of pressure and temperature both. The theory is based upon a complete energy equation combined with the Reynolds equation, using Reynolds' boundary condition. Application is made to a bearing of 180-deg arc with the aid of a digital computer. The peak pressure works out to be 778 psi with a temperature rise of 17 dF, as compared to 750 psi in the case of uniform viscosity. A method for determining the film pressure and temperature at any point was derived by Motosh (1964) for application to cylindrical journal bearings. It is based on minimum dissipation, the energy equation, and the temperature-pressure-viscosity law.

The scope of adiabatic solutions is indicated by Table 8.

Isothermal Pressure-Viscosity. Professor Waters contributed a numerical solution in 1928. He took the problem of a 4-in. full bearing with $m = \frac{1}{500}$, L/D infinite, operating at 1000 rpm with a fixed eccentricity ratio of 0.7. The oil viscosity was 2.7 n (about 100 SUS) at atmospheric pressure. Recomputing with due allowance for pressure-viscosity, he found an increase of 6 per cent in load capacity. Higher eccentricities would show greater effects.

A full-bearing problem was considered by Burwell (1942), who expressed the mean viscosity of the film as a function of pressure and temperature, disregarding heat removal by oil flow. He assumed that the effective pressure would be equal to the load per unit of projected area, and undertook to show how friction and load capacity at equilibrium temperature would depend on the operating conditions, properties of the oil, and geometry of the bearing. Closer clearances were recommended.

Reynolds' equation was integrated by Korovchinskii (1954) as well

TABLE 8. Adiabatic Solutions, Variable Viscosity

Authors	Viscosity function		Bearing type		L/D	
	$Z(t)$	$Z(p,t)$	full	partial	∞	finite
Christopherson	–	x	–	x	–	x
Oksal	–	x	x	–	–	x
Pinkus & S.	x	–	x	–	x	–
Purvis, M. & B.	x	–	x	–	x	x
Hughes & O.	–	x	–	x	x	–
Motosh, N.	–	x	x	–	x	–

as by Nishihara and co-author (1956) for the full bearing with viscosity a function of pressure.

In discussing Needs' paper of 1958, Hartung showed how to find the approximate ratio of friction with and without the pressure effect for a given operating temperature. Hartung calculated the ratios in this way for the 120-deg bearing tested by Needs. The agreement between calculated and observed results at 130 and 210 F is surprising. It augurs well for more elaborate and precise methods of prediction. The near agreement provides convincing evidence that the pressure effect exists in real bearings, not just in viscometers.

Tao (1959) derived an isothermal solution of the Reynolds equation for a full bearing of infinite length, taking the pressure effect into account. The film is assumed continuous around the bearing, without cavitation. Although the expressions obtained are too complicated for immediate application, the author proves that the line of centers will no longer be at right angles to the load when the effect of pressure on viscosity is considered.

Korovchinskii extended his solution in 1962 to include elastohydrodynamic conditions, as, for example, water-lubricated rubber bearings in rolling mills.

7. INERTIA AND TURBULENCE

The effects of the inertia of the lubricant, noted in thrust bearings, apply also to journal bearings. These conditions are described in Chapter VI; also by Fishman (1950), Fuller (pages 67–71 of I: 1956), Osterle, Chou, & Saibel (1958), Tipei, Constantinescu, Nica, & Bita in Chapter VI of their book (1961), and by Milne in pages 423–527 of the volume cited (1965).

Critical Speed. It was early recognized that in bearings "with exceptionally high speeds and wide clearances the effects of centrifugal force and of turbulent motion would begin to be felt, thus involving the density of the oil in addition to its viscosity" (quoted from our paper of 1914).

Turbulence between concentric cylinders was discovered by Couette (II: 1890), who, however, rotated only the outer cylinder. G. I. Taylor (1923) showed that the critical speed is lower when the inner cylinder rotates. In this case it follows that streamline or laminar motion between cylinders of infinite length will change to turbulent motion when Reynolds' number, based on the radial clearance, c, exceeds $41/m^{1/2}$, where m denotes the ratio of radial clearance to radius. Reynolds' number may be written cU/ν, where U is the surface speed and ν the kinematic viscosity. This criterion was confirmed by Lewis (1928) and again by Taylor in 1936. The critical speed in rpm is given conveniently by

$$\frac{N}{1000} = \frac{77}{m^{3/2}}\left(\frac{\nu}{D^2}\right), \tag{3}$$

where m is now the clearance ratio in mils per inch, ν is in centistokes, and D, the journal diameter, is in inches.

To apply Eq. (3) consider a journal bearing of 1-in. diameter with a clearance of 1 mil/in., lubricated with water for which ν is 1 cs; then m, ν, and D are all equal to unity, and the critical speed will be 77,000 rpm. Double m and D and the speed drops to a critical of 7000. Then for an oil of 1.0 n or 6.9 cp viscosity, specific gravity near 0.9, $\nu = 7.7$ cs, making $N = 7.7 \times 7000$ or 54,000 rpm. Thus oil-lubricated bearings are not likely to develop turbulence except in larger sizes than a 2-in. diameter; or, in smaller bearings, at speeds above 50,000 rpm. The effect of axial flow on the critical speed has been described by Buckingham (II: 1923) and Goldstein (1937).

Careful experiments on the frictional torque of water in turbulent motion between concentric cylinders were reported by E. M. Wagner (1932). His results should be compared with the others. Tests have been conducted by J. A. Cole (1957) to study the influence of eccentricity. A converging clearance tends to suppress turbulence, so that an eccentricity ratio of $\frac{1}{2}$ raises the critical speed about 22 per cent, while a ratio of $\frac{4}{5}$ will raise it 70 per cent in bearings of moderate or high L/D. See Di Prima's calculations (1959, 1963) supporting the Cole effect; as well as the studies by Schultz-Grunow (1959) and by Burton (1964).

Bearing Performance. The best-known experiments on turbulent friction in journal bearings are those of Wilcock (1950) and of Smith & Fuller (1956). Wilcock's bearings were of 4-in. and 8-in. diameter. They were conventional split bearings with oil grooves along each side, closed near the ends. Taylor's observations were confirmed, with a marked rise in frictional torque after passing the critical speed.

Smith and Fuller investigated pressure distribution, load capacity, and journal positions as well as friction loss. The report of experimental data is preceded by a tentative theoretical study. Their reasoning indicated that we may expect to find greater friction and greater load capacity above the critical speed. These expectations were confirmed by a series of tests on a 3 in.-diameter water-lubricated bronze bearing, having a clearance ratio of about 2.9 mils/in. Speeds were varied up to 7450 rpm, more than five times the critical. See also the high-speed tests by Neale & Love (1957).

Bearings lubricated with hot liquid potassium, and with water, were tested by Stahlhuth & Trippett (1962). Critical speeds were higher than expected from Taylor's criterion, owing to the effects of eccentricity (Cole). Friction torques were as much as forty times greater in the turbulent region than would be calculated for laminar flow. Two partial bearings were tested by Orcutt over a wide range of Reynolds' number (1965). Results compared well with eccentric bearing theory by Di Prima.

Theory. Following the approximate calculations by Smith & Fuller, Tao (1958) set up a theory of short-bearing load capacity. This led to a family of curves for ϵ against the Ocvirk number $S(L/D)^2$ at constant values of (L/D)Re. Here Re denotes Reynolds' number in the form cU/ν. The derivation is based upon a recognized law of turbulent friction giving an expression that reduces to the "slot formula" in laminar flow, and to the "one-seventh power law" when empirical constants are introduced.

The theory was further advanced by Constantinescu (1959, 1962); by Tipei (1960) who offered a three-dimensional approximation; and by Arwas and co-authors (1964). The last study was applied to both 180- and 360-deg bearings. It showed that the ratio of load capacity of a turbulent film to that of a laminar film is approximately 0.415 times the three-fifths power of Reynolds' number for ϵ from 0.1 to 0.6. And from a comparison of the present theory with Smith & Fuller's experiments, it can be inferred that the load ratio for side leakage, P/P_0, is only three-eighths as great for turbulent motion as for the laminar film. See also Constantinescu's lecture in the Houston volume, pages 153-213 (1965) as well as Ng & Pan (both 1965). Derivations support the general conclusion that the load capacity and friction of a turbulent film are greater than those of a laminar film.

8. STEADY-STATE EXPERIMENTS

We come now to the most interesting part of lubrication research—steady-load experiments on journal bearings—a field reviewed thoroughly by Professor Fuller in his survey of the literature (1958). Some of the earlier experiments are mentioned at the beginning of this chapter, and in publications on the history of lubrication (Chapter I). Representative investigations of later years are listed in Tables 9 and 10, covering a middle period of three decades.

American Investigations. Karelitz (III: 1930) experimented with waste-packed and ring-oiled bearings, and determined the side leakage from 120-deg bearings (1937). McKee explored the laws of lub-

TABLE 9. Journal Bearing Experiments, USA, 1926–55

1926	Karelitz		1939	Morgan & M.	
1927	McKee	II[a]	1941	Linn & I.	
1927	McKee	VII	1943	Muskat & M.	III
1928	Karelitz		1944	McKee, W., B., & S.	
1929	McKee & M.	II	1946	Needs	VII
1930	Karelitz	III	1947	Shaw & N.	
1931	Bradford		1948	Boyd & R.	III
1932	Atwater[b]		1948	McKee, W. & S.	
1932	Busse & Denton[b]		1949	Levinsohn	VII
1932	McKee & M.		1950	Hitchcock	
1932	Newkirk[b]		1950	Roach	
1935	Karelitz	VII	1951	DuBois, M. & O.	
1937	Bradford & W.		1952	DuBois & O.	
1937	Karelitz		1952	McKee	III
1937	McKee & M.		1952	Wilcock & R.	III
1937	Needs		1955	DuBois, O. & W.	VII
1938	Tichvinsky				

[a] Roman numerals indicate chapter where reference is given; for others, see Fuller's *Survey of Journal Bearing Literature*, ASLE (1958). [b] Water lubrication

TABLE 10. Journal Bearing Experiments Abroad, 1926–55

1926	Vieweg & V.	VII[a]	1937	Vogelpohl	II
1926	Schering & V.		1938	Frössel	
1929	Goodman		1938	Rumpf	III
1929	Hanocq	VII	1942	Frössel	VII
1930	Schneider		1946	Clayton	
1932	Boswall & B.		1947	Hanocq	III
1932	Nücker	III	1948	Clayton & W.	III
1933	Goodman		1948	Fogg	
1934	von Schroeter[b]		1948	Thomson	
1936	Clayton & J.		1949	Leloup	XII
1937	Baildon		1951	Cameron	III
1937	Clayton		1953	Blok & V.	
1937	Fogg & H.[c]		1953	Stephan	
1937	Guy & S.		1954	Artemiev	VII
1937	Jakeman & F.		1954	Kettleborough	VII
1937	Samuelson[c]		1954	Ott	VII
1937	Swift & H.		1955	Özdas & F.	III

[a] See footnote, Table 9. [b] Grease lubrication. [c] Water lubrication.

rication of full journal bearings with respect to the effect of abrasives; running-in and the influence of clearance and length; pressure distribution with misalignment, and the friction curves for different bearings metals; friction and temperature relations under forced feed, and oil flow as affected by the geometrical and operating variables.

Professor Bradford and his students made good use of a machine designed by Kingsbury in tests on a full journal bearing. Friction, film pressure, film thickness, and temperature distribution were studied in relation to the design factors and operating conditions. Bradford & Wetmiller (1937) compared three oils under a mean load of 20,000 psi using a small partial bearing. Needs (1937) tested 120-deg bearings on a 4-in. journal in a study of the pressure-viscosity effect, comparing seven different lubricants. Loads were carried to 7000 psi at a speed low enough to avoid heat effects. Pronounced differences in friction were found that could be correlated with the pressure coefficients. More advanced experiments with the same pair of bearings have been mentioned under "Variable Viscosity" (1958). He also reported tests on railroad bearings of improved design (1946).

Tichvinsky (1938) experimented with a 7 by $10\frac{1}{2}$ in. bearing having 90-deg side reliefs. Tests were run at 3600 rpm under loads controlled by a magnet. The data appeared to be consistent with theory and were cited later in a thermal equilibrium problem (Chapter III). Morgan & Muskat investigated the effects of forced lubrication, as well as temperature conditions in journal bearings. Linn & Irons determined power loss in commercial turbine bearings up to 8 in. in diameter, at speeds to 12,000 rpm. Bearings were insulated to eliminate heat loss to the atmosphere. Power loss was computed from the difference between inlet and outlet oil temperature, and viscosity was estimated at the mean of these temperatures. It was found that the coefficient of friction was a function of ZNQ/P (Chapter V). The data were shown to be in good agreement with tests on larger bearings by other manufacturers. Floating sleeve bearings were tested by Shaw & Nussdorfer.

Boyd & Robertson investigated oil flow and temperature relations. Their results confirmed earlier calculations and served to illustrate the graphical method for predicting equilibrium temperatures (III: 1948). Hitchcock (1950) measured operating temperatures in roll-neck bearings of $17\frac{1}{4}$ in. diameter at loads up to 1700 psi. The results appeared concordant with tests on smaller bearings by Levinsohn (1949). Roach studied pumping action of the 180-deg bearing (1950). The effects of speed and flow control were observed and then generalized by dimensional analysis.

Professor DuBois and associates (1951, 1952) determined the influence of operating and design factors on friction, oil flow, pressure distribution, load capacity, and journal positions in full bearings. Wilcock & Rosenblatt (III: 1952) experimented with commercially designed bearings up to 8-in. diameter and 9000 rpm. Oil flow and temperature rise were especially investigated. The "operating line" was used for determining equilibrium values. It was concluded that large bearings with axial grooves act as "mixers" so that the effective film temperature is approximated by the outlet oil.

British Investigations. The steady-load experiments by Goodman, Boswall & Brierly, Clayton, Guy, Jakeman, Swift, and Fogg; also by Thomson, Kettleborough, Cameron, Özdas & Ford, are among the best known in Britain and the Commonwealth. Goodman (1929, 1933) determined friction, temperature rise, pressure distribution, load capacity, and journal positions in partial bearings from 90 to 180-deg arc. Boswall & Brierly (1932) followed with tests on bearings from 30- to 120-deg arc.

Clayton & Jakeman (1936) determined friction and running positions in a full bearing. Temperature was raised to the point of impending seizure in each test. Distortion and wear measurements were included. Guy & Smith (1937) tested commercial turbine bearings of 15-in. diameter. Results were compared with available theories. Jakeman & Fogg (1937) studied the effects of many variables, with clearance ratios from $\frac{1}{2}$ to 8 mils/in. Swift & Haslegrave (1937) experimented with an 8-in. sleeve bearing. They measured journal and bearing friction independently, as well as journal running positions. Clayton (1946) compared twenty-five styles of oil grooving in a small bearing. Clearance, speed, and inlet pressure were varied. Clayton & Wilkie (III: 1948) explored the temperature distribution in a full bearing under a wide range of oil feed pressures.

Fogg (1948) determined friction in small bearings for L/D ratios from $\frac{1}{4}$ to $\frac{9}{8}$. He obtained ZN/P graphs very much like those in the McKee investigation (II: 1929). Frictional heat diminished the viscosity so much that ZN remained practically constant at the higher speeds. Professor Thomson (1948) investigated partial bearings under varied conditions to establish design data. Cameron tested a 6-in. full bearing with side reliefs up to 6000 rpm, measuring friction, oil flow, and temperature rise (III: 1951). Mechanically measured friction values came close to the Cameron-Wood theory. He concluded that so much heat is usually carried away by conduction in bearing tests that friction cannot be determined accurately from the temperature rise in the outgoing oil. Kettleborough (1954) experimented with floating bearings. Dr. Özdaş and Professor Ford (III: 1955) measured oil delivery and cooling in oil-ring bearings.

Investigations in Germany. Among the better known experiments are those of the Vieweg brothers, Schneider, Nücker, Vogelpohl, Frössel, Rumpf, and Stephan. Richard and Volkmar Vieweg (1926) devised a novel type of frictionless support for use in the dynamometer measurement of journal friction. H. Schering and R. Vieweg (1926) determined film thickness in full bearings by an electrical capacitance method. Schneider (1930) experimented with full bearings. He found the transitions from perfect to imperfect lubrication by electrical resistance observations.

Nücker's experiments (III: 1932) under Professor Heidebroek set a high record for accuracy and completeness. His tests were conducted with fluid films at medium to high speeds. The bearings were 220-mm (8.7-in.) in diameter. Full bearings of conventional design, and one with three-fourths of the unloaded side relieved were included in the tests. Bearing friction was measured mechanically, film pressure by thirty-two pressure taps, film thickness by twelve small capacitance probes, and bearing temperatures by well distributed thermocouples. A mineral turbine oil was used. Its dielectric constant was determined as a function of temperature. Good agreement was found between calculated and measured values. Vogelpohl (II: 1937) confirmed Reissner's streamline calculations by glass-bearing observations. The experiments led to a better understanding of cavitation. Frössel (1938) investigated pivoted-shoe and multiple-film bearings. The results confirmed hydrodynamic theory with respect to film thickness and frictional resistance (see Chapter VI). Frössel also made experiments on glass bearings.

Nücker's tests were supplemented by Rumpf's investigation (III: 1938), using the same basic apparatus. Rumpf added the lemon-shaped or elliptical bearing, and found it good for suppressing oil film whirl. He confirmed the doctrine that the coefficient of friction and relative film thickness are functions of ZN/P, which he attributed to Gümbel's publications. He gave a linear equation for the coefficient having a constant term equal to 0.00165. Rumpf's investigation includes various comparisons between theory and experiment.

Stephan (1953) tested journal bearings up to 25,000 rpm. Performance was observed for a variety of designs, including 60-deg side reliefs. Results agree well with theory and with data from earlier investigations. Practical formulas are derived for the mean operating temperature, friction loss, oil flow, journal eccentricity, and optimum clearance.

Other Experiments. Hanocq (III: 1947) investigated the laws of lubrication of full and partial bearings. He found general agreement with hydrodynamic theory and certain published tests. Professor Hanocq measured journal friction by deceleration, bearing friction by the restoring couple, and plotted their ratio against ZN/P. In further experiments he determined heat transfer from the housing as a function of the temperature elevation. Leloup (XIII: 1949) extended Hanocq's program into the field of imperfect lubrication. Lubricants of widely different composition were compared over a 10 to 1 load range and a 200 to 1 speed range. Different loads gave separate curves when f was plotted against ZN/P at left of minimum.

Blok and Van Rossum (1953) experimented with their "foil bearing" to good effect. Artemiev (1954) studied the influence of clearance and of oil-inlet location on load capacity. Ott (1954) experimented with elliptical and three-lobed bearings. His results seem consistent with hydrodynamic theory and with data from other tests.

This completes our review for the 1926–1955 period. It would be useful to greatly amplify Tables 9 and 10. They could be extended back to an earlier date. Columns could be added to show the performance characteristics measured (film thickness, film pressure, friction, and the like); also the independent variables explored (shape and size of bearing; viscosity, speed, load, and other factors). Such tables might serve as a first step in coordinating or comparing the test data, and checking theory against experiment.

Steady-Load Experiments, 1956–1965. It remains to review certain investigations of the decade just past, on (1) film extent and cavitation, (2) high-speed and heavily loaded bearings, and (3) test methods.

1. Oil flow and film extent were studied by Cole & Hughes as noted in Chapter III. Photographs confirmed Stieber's filaments ("Stromfäden"). Busby & Pigman tested railroad bearings (1957). Friction data agreed with theory when cavitation was considered. Dowson (1957) experimented on cavitation by sliding a lubricated plane surface under a stationary, lightly loaded, transparent lens or cap. Film pressures were measured with the aid of a geometrically similar brass cap.

Floberg investigated cavitation under several conditions:

a. By friction-loss measurements on a full bearing with $L/D = 1$, showing the effect of oil strips and air spaces in the cavitation region (1959).

b. By the use of end seals and external pressure to realize the

Sommerfeld bearing of infinite length (1961), thus confirming the theory.

c. By visual observations in a transparent full bearing of $L/D = 1$. Film pressures and eccentricities were measured in a thick-walled bearing with and without an axial groove (1961). Tests showed good agreement with theory. Paper contains a valuable review of cavitation research.

A transparent full bearing was used by Auksmann (1961) to investigate pressure distribution, eccentricity, and cavitation. Results agreed with short-bearing theory when cavitation was suppressed by pressurizing.

2. The development of high-speed, heavily loaded marine bearings was described by Brown & Newman (1957). Kolarev (1959) measured the pressure distribution in full bearings. His observations show the effects of deformation and running-in. Tests were carried up to seizure-impending loads. Full bearings were tested to 60,000 rpm by Neale & Love (1957). Temperature-rise data were plotted in a dimensionless form against Reynolds' number. The graph showed a break at the critical speed, above which there is a greater rate of temperature rise. Dimensionless oil flow shows a similar break, above which there is a reduced rate of relative flow.

New tests on heavily loaded partial bearings were reported by Needs, showing distinct differences between lubricants depending on their pressure coefficients of viscosity (1958). Dowson & Longfield also experimented on partial bearings under heavy load (1964). Pressure and temperature distributions were observed and correlated with elastohydrodynamic and pressure-viscosity effects.

3. Methods of testing were featured in five investigations. Thus Drescher (1956) devised a pressurized dynamometer to measure journal torque. Davis & Krok (1957) described a new testing machine and its use in determining running-in characteristics. Sternlicht & Fuller (1957) compared dial-gage contact, microscope, and air-gaging or pneumatic methods for measuring eccentricity. Axis of rotating shaft was held fixed, so that the measurements could be more conveniently made on a nonrotating member. The pneumatic method was found to be the best of the three but electromagnetic indicators were being considered. Kaufman reports favorable experience with an air-gaging system (1961) for steady-state measurements. Cook (1964) describes an ingenious technique for measuring journal friction as distinguished from bearing friction.

These last references bring to mind the need for a logical classification of bearing test methods. The study must not be limited to instrumentation, but should include the methods used for applying the load, and varying the shaft speed, oil supply, and temperature.

9. PERFORMANCE WITH SPECIAL LUBRICANTS

Journal bearings, like thrust bearings, have been lubricated with materials ranging in consistency from grease to water and air. Molasses has been used in a sugar mill, liquid sodium in reactor pumps. Sulfuric acid and other "process fluids" served well as lubricants (Lialin, 1959; Fuller, I: 1965). Most liquid lubricants are Newtonian, all giving the same performance when operating at the same ZN/P. The action of grease and other non-Newtonian lubricants is not so simple.

Non-Newtonian Lubrication. A general theory of non-Newtonian or "rheodynamic" lubrication has been given by Milne (VI: 1954, 1957). He shows that a "core" of undeformed lubricant adheres to the bearing surface in the vicinity of maximum film thickness, while a smaller one is continually forming at the surface of the journal near the point of minimum film thickness. This behavior recalls Buckingham's treatment of "plug flow" in capillaries (II: 1921). A theory of the full journal bearing lubricated with polymer-thickened oils has been worked out by Horowitz and Steidler (1960, 1961). Digital computer solutions were obtained both for isothermal and adiabatic operation. They found that such oils had different viscosities in the axial and circumferential directions; with less load capacity and lower friction than mineral oils. Similar investigations have been reported by DuBois and co-authors (1960). Tanner (1963) showed how friction can be reduced in short bearings by the use of non-Newtonian lubricants.

Grease-Lubricated Bearings. Westcott's experiments (1913) showed that cup greases compared well with oils in the lubrication of journal bearings, when the bearing geometry favored admission of the grease. Michell (II: 1929) theorized that journal bearings would give higher friction when lubricated with grease than with oil of equal high-shear viscosity, yet the load capacity would be unchanged. Investigations by von Schroeter confirmed these conclusions (1934). He studied the effects of yield stress and mobility, and improved grease inlet grooving. So comprehensive was von Schroeter's research that it merits further analysis and application.

Cohn & Oren (1949) measured film pressures in a full bearing with grease lubrication. They observed practically hydrodynamic conditions except for the reduced end leakage, and found that the pressure components integrated to give total load. Pressure distributions were obtained mathematically by Lawrence (1950). He used Christopherson's finite difference form of the Reynolds equation. Slibar and Paslay (1959) calculated the effect of "retarded flow" in Bingham plastics between concentric cylinders.

Friction experiments on a full journal bearing with grease lubrication are reported by Chakrabarti & Harker (1960). An empirical equation is set up in terms of dimensionless variables. The bearing was operated under zero load to evaluate the yield stress τ_0 and the "plastic viscosity" Z_0 of the different compositions. The variation of each property with percentage of soap base is plotted for the two oils used in blending. The empirical equation chosen may be expressed, for geometrically similar bearings, by

$$f = \phi\left(\frac{\tau_0}{P}, \frac{Z_0N}{P}\right). \tag{4}$$

The specific form of ϕ is developed in the paper by a graphical treatment of the data. Equation (4) agrees with the symbolic solutions offered in Chapter V for the Bingham plastic, and reduces to the fluid form when $\tau_0 = 0$ and $Z_0 = Z$.

Bradford, Barber, & Muenger described a long program of research (1961). A cast bronze half-bearing, used in most of the tests, was replaced by a transparent bearing for visual observations. With proper application of the grease it was not difficult to insure thick-film, rheodynamic lubrication. Friction curves are plotted against ZN/P, where Z is the viscosity, at the bearing temperature, of the oil used in compounding the grease. Friction coefficient on the bearing member was a linear function of ZN/P from 10 to 350 cu. Variation of feed rate and bearing clearance, and the effects of thixotropy were studied. It was found that grease-lubricated bearings operate well under heavy loads at low speeds. They offer little starting friction after a period of rest, a characteristic attributed in part to good boundary lubrication, but also to the ability of plastic lubricants to resist being squeezed out. See also Horowitz & Steidler (1963).

Water as a Lubricant. As early as 1879 Professor Thurston mentioned bearings in which the journal is floated by water pressure from an external pump (Chapter I). Naylor (1906) described shaft bearings that continually ran hot with oil as a lubricant but were operated for eleven years with water. They were oiled only when standing idle overnight, to prevent rust. Gümbel pointed out the hydrodynamic requirement in 1914—that the product of viscosity and speed shall be the same for water as for oil in a given bearing (Chapter V). When this requirement is met, the minimum film thickness will be the same at a given load.

Water has been used as a lubricant with nonmetallic bearing materials such as lignum vitae (Atwater, 1932), fabric-plastic compositions (Newkirk, 1932), and soft rubber (Busse & Denton, 1932; Fogg & Hunwicks, 1937; Raimondi, 1959). The ready conformity of such materials is equivalent to a reduction in clearance. Samuelson (1937) lubricated conventional steam-turbine bearings with a water-base lubricant, 1 part of soluble oil to 50 parts water. Water-lubricated bearings were tested by Wepfer & Cattabiani (VI: 1935), and others, using carbon and other hard materials for the bushing. Wear properties of materials in high-temperature water were investigated by Westphal & Glatter (1954). Smith & Fuller (1956) tested bronze bearings with water as a lubricant. Twenty combinations of materials were tested by Hother-Lushington & Sellors (1963) for journal bearing applications in the power generation industry.

To aid in finding references see Tables 9 and 10.

10. BEARING DESIGN

Journal bearing design makes full use, today, of lubrication theory and test data. This is evident in treatises like those of Barwell and Fuller (both I: 1956); Wilcock & Booser (I: 1957); Vogelpohl (VI: 1958), and Rippel (XII: 1959). Vogelpohl's book contains fifty fully worked examples of practical design problems, most of them on journal bearings. See also Korovchinskii (1954). Good chapters on the journal bearing are found, also, in many of the textbooks on machine design, as for example Black (1955) and Shigley (XI: 1963).

Self-Acting Bearings. Modern bearing design requires the calculation of optimum values for the factors at our disposal, such as clearance and length. Kingsbury determined optimum conditions for maximum load capacity and minimum power loss (1932) by making reasonable assumptions as to film extent, and correcting for end leakage by electrical integration. Optimum solutions have been obtained by Jakobsson and Floberg, as well as by Raimondi and Boyd in the references cited, by means of the digital computer. The upgrading of machine design by Kingsbury was reflected in the title of a paper by Barber & Davenport, "Machine Design for Lubrication" (1933). It led to the phrase "designing an oil film" used by Sydney Needs in a lecture that we recall.

Many authors have contributed articles expressly on bearing design; see, for example, McKee (III: 1937), Klemencic (1943), Wilcock & Rosenblatt (1953), Thomson (1954), Barwell (1956, 1959), and Fuller (1965).

Most bearing designs aim for a fixed minimum film thickness. McKee introduced a combination criterion for load capacity, based on the maximum film temperature and minimum coefficient of friction. The location of the minimum point is found from tests on similar bearings. Leloup and Vogelpohl have proposed formulas for predicting the location of this critical point, as described in Chapter XIII.

Bearings with Hydrostatic Load Support. Externally pressurized journal bearings are described by Professor Fuller in the articles of 1947 (Chapter VI) and in his books (1956, 1958). Compensated hydrostatic journal bearings were investigated by Raimondi and Boyd (1957). A study of optimum proportions was published by Loeb & Rippel (1958); see also Rippel's book (VI: 1963), and Gross (1965). These few references out of a large number will indicate the state of the art, which has undergone a remarkable development. The use of hydrostatic load support eliminates the problems of lubrication as we ordinarily think of them. It is analogous to magnetic support. However, the power consumed in operating an externally pressurized bearing includes the pumping requirement as well as the bearing loss.

Hydrostatic support bearings have often been used in test machines. An interesting application was made by H. T. Grandin, Jr. (1960) in the design of a dynamometer-type machine for testing small bearings under exceptionally heavy loads. The test bearings are placed in the positions normally occupied by the support bearings. Load is applied hydrostatically through a relatively thick film at the middle of the shaft. Friction loss in the test bearings can then be read from the dynamometer with little or no correction.

Other Design Problems. Magnetohydrodynamic journal bearings are described by Hughes (VI: 1963) and Kuzma (1963). See the note on MHD bearings in Chapter V, leading to Eq. (26) for the action of an electrically conducting lubricant in a magnetic field.

Bearing designers have been very much concerned with vibration problems. The earliest forms of vibration causing trouble in rotating machines were due to unbalance, shaft flexibility, and dry friction. Fluid film whirl came into the picture later, as described in Chapter XII.

11. SUMMARY

We have seen how the principles discussed in earlier chapters apply to actual bearings. A knowledge of the properties of lubricants has been found useful both in calculation and experiment. Hydrodynamic theory appears to be in general agreement with test results when allowance is made for practical conditions not always foreseen (Appendix D). The better known experiments on steady-state journal bearings are listed in chronological order, with brief indications of their scope and significance. Advantages of special types of lubricant are compared.

The chapter concludes with a quick look at bearing design. Nearly every aspect of lubrication theory, and every technique of research, has some application to the journal bearing.

REFERENCES

Artemiev, Y. N. (1954), Experimental Study of Influence of Oil Inlet Angle and Clearance on Load Capacity of a Bearing (Russian). *TIM* **9,** 91–113.

Arwas, E. B., Sternlicht, B., & Wernick, R. J. (1964), "Analysis of Plain Cylindrical Journal Bearings in Turbulent Regime." *JBE, T* **86,** 387–95.

Auksman, Boris (1961), "Experimental Investigation of Oil Film Behavior in Short Journal Bearings." *ASME* 61-LUB-11; 8 pp.

Barber, E. M. & Davenport, C. C. (1933), "Machine Design for Lubrication." *Tech. Bull.* **18,** School of Eng'g, Pennsylvania State College, 27–51.

Barwell, F. T. (1956), "Hydrodynamic Lubrication and its Application to Design." *JIP* **42,** 304–15. (a) *AMR* **12,** 149–53.

Baumeister, H. K. (1964), "A Note on the Nominal Clearance of the Foil Bearing." *LE* **19,** 377–80.

Pirkhoff, G. & Hays, D. F. (1963), "Free Boundaries in Partial Lubrication." *J. Math. & Phys.* **42,** 126–38.

Bisson, E. E. (1964), "Friction and Bearing Problems in the Vacuum and Radiation Environments of Space." Pages 259–87, *ABT*.

Black, P. H. (1955), *Machine Design.* McGraw-Hill, New York, second ed., 471 pp. (a) Third ed. (1966), in press.

Blok, H. & van Rossum, J. J. (1953), "The Foil Bearing—A New Departure in Hydrodynamic Lubrication." *LE* **9,** 316–20.

Boyd, J. & Raimondi, A. A. (1951), "Applying Bearing Theory to the Analysis and Design of Journal Bearings, I and II." *JAM* **18,** *T* **73,** 298–310.

Boyd, J. & Raimondi, A. A. (1962), "Clearance Considerations in Pivoted Pad Journal Bearings." *TASLE* **5,** 418–26.

Bradford, L. J. (1940), "Teaching Lubrication." *J. Eng. Educ.* **30,** 387–86.

Bradford, L. J., Barber, E. M., & Muenger, J. R. (1961), "Grease Lubrication Studies with Plain Bearings." *JBE, T* **83,** 153–61.

Brown, T. W. F. & Newman, A. D. (1957), "High Speed Highly Loaded Bearings and Their Development." Paper 16, *CLW,* 20–7; 10, 199, 361, 362, 744, 820; Plates 5–7.

Burton, R. A. (1964). "Turbulent Film Bearings under Small Displacements." *TASLE* **7,** 322–32.

Burwell, J. T. (1942), "The Effect of Diametral Clearance on the Load Capacity of a Journal Bearing." *T* **64,** 457–61.

Busby, A. L. & Pigman, G. L. (1957), "Experimental Investigation of Railroad Journal Bearing Operating Characteristics." *LE* **13,** 546–52.

Cameron, A. (1962), "Critical Conditions for Hydrodynamic Lubrication of Porous Metal Bearings." *Proc. IME* **176,** 761–70.

Chakrabarti, R. K. & Harker, R. J. (1960), "Frictional Resistance of a Radially Loaded Journal Bearing with Grease Lubrication." *LE* **16,** 274–80, 290.

Cohn, G. & Oren, J. W. (1949), "Film Pressures in Grease-Lubricated Sleeve Bearings." *T* **71,** 555–60.

Cole, J. A. (1957), "Experiments on the Flow in Rotating Annular Clearances." Paper 15, *CLW,* 16–9; 11; Plate 4.

Connors, H. J. (1962), "An Analysis of the Effect of Lubricant Supply Rate on the Performance of the 360° Journal Bearing." *TASLE* **5,** 404–17.

Constantinescu, V. N. (1959), "On Turbulent Lubrication." *Proc. IME* **173,** 881–900.

Constantinescu, V. N. (1962), "Analysis of Bearings Operating in Turbulent Regime." *JBE, T* **84,** 139–51.

Constantinescu, V. N. (1965), "Theory of Turbulent Lubrication." Pages 153–213 of *HS.*

Cook, N. (1964), "Hydrodynamic Bearing Test Apparatus." *LE* **20,** 178–9.

Dahlstrom, F. P. (1933), "The Morgoil Roll-Neck Bearing." *T* **55,** IS-55-2; 9–18.

Davis, A. J. & Krok, T. V. (1957), "Sleeve Bearings, Part I. Development of a Testing Machine." *Proc. IME* **171,** 943–50." Part II. Running-In Characteristics," 951–66.

Di Prima, R. C. (1959), "The Stability of Viscous Flow between Rotating Concentric Cylinders with a Pressure Gradient Acting Round the Cylinders." *J. Fluid Mech.* **6,** 462–8.

Di Prima, R. C. (1963), "A Note on the Stability of Flow in Loaded Journal Bearings." *TASLE* **6,** 249–53.

Dowson, D. (1957), "Investigation of Cavitation in Lubricating Films Supporting Small Loads." Paper 49, *CLW,* 93–9; 12, 757; Plates 10, 11.

Dowson, D. & Longfield, M. D. (1964), "Distribution of Pressure and Temperature in a Highly Loaded Contact." Paper 3, *LWG,* 24–30.

Drescher, H. (1956), "Versuchsstand für Querlager mit schwimmender Reibungswaage." *Konstruktion* **8,** 228–31.

DuBois, G. B., Ocvirk, F. W., & Wehe, R. L. (1955), *Experimental Investiga-*

tion of Eccentricity Ratio, Friction and Oil Flow of Long and Short Journal Bearings with Load Number Charts. NACA TN **3491**; 63 pp.

DuBois, G. B., Ocvirk, F. W., & Wehe, R. L. (1957), "Properties of Misaligned Journal Bearings." *T* **79**, 1205–12.

DuBois, G. B., Ocvirk, F. W., & Wehe, R. L. (1960), *Study of Effects of a Non-Newtonian Oil on Friction and Eccentricity Ratio of a Plain Journal Bearing.* NASA TN **D-427**; 43 pp.

Eshel, A. & Elrod, H. G. (1965), "The Theory of the Infinitely Wide, Perfectly Flexible Self-Acting Foil Bearing." *JBE, T* **87**, 831–6.

Fedor, J. V. (1961), "Journal Bearings with Arbitrary Position of Source—II." *JBE, T* **83**, 572–8.

Fishman, I. M. (1950), On the Motion of a High Viscosity Liquid between Journal and Bearing. *PMM* **14**, 593–610. (a) *AMR* 3471; 593 (1951).

Floberg, Leif (1957), *The Infinite Journal Bearing, Considering Vaporization. TCUT*, Göteborg, No. 189, 83 pp.

Floberg, Leif (1959), *Experimental Investigation of Power Loss in Journal Bearings, Considering Cavitation. TCUT,* Göteberg, No. 215; 16 pp.

Floberg, Leif (1960), *The Two-Groove Journal Bearing, Considering Cavitation. TCUT,* Göteberg, No. 232; 30 pp.

Floberg, Leif (1961), *Attitude-Eccentricity Curves and Stability Conditions of the Infinite Journal Bearing. TCUT,* Göteborg, No. 235; 43 pp.

Floberg, Leif (1961), *Boundary Conditions of Cavitation Regions in Journal Bearings. TASLE* **4**, 282–6.

Floberg, Leif (1961), *Experimental Investigation of Cavitation Regions in Journal Bearings. TCUT,* Göteborg, No. 238; 28 pp.

Frössel, W. (1953), "Stossunempfindliche Gleitlagern." *MTZ* **14**, 179–80.

Fuller, D. D. (1958), *A Survey of Journal Bearing Literature.* ASLE, Chicago; 260 pp.

Fuller, D. D. (1965), "Design of Fluid Film, Hydrodynamic and Hydrostatic Thrust and Journal Bearings." Pages 855–77 in *HS*.

Goldstein, S. (1937), *Modern Developments in Fluid Dynamics.* Oxford University Press, 2 volumes, 702 pp.

Goodman, John (1886), "Recent Researches in Friction." *Proc. ICE* **85**, 376–92; **89**, 421–54 (1887).

Grandin, H. T., Jr. (1960), *The Design and Analysis, Construction and Operation of a Journal Bearing Testing Machine.* M.S. Thesis, Dept. of Mech. Eng., Worcester Polytechnic Institute; 50 pp.

Gross, W. A. (1965), "Externally Pressurized Bearing Lubrication." Pages 307–42 of *HS*.

Gümbel, L. (1917), "Einfluss der Schmierung auf die Konstruktion." *Jahrb. Schiffbautechn. Ges.* **18**, 236–322.

Gümbel, L. (1922). "Zuschriften an die Schriftleitung zur Theorie der Schmiermittelbreibung." *Z. Tech. Phys.* **3**, 94–101.

Haydock, J. (1939). "The Wedge-Shaped Oil Film Does the Trick." *Am. Mach.* **83**, 823–5.

Hays, D. F. (1961), "Oil Flow in a Full Journal Bearing." *JBE, T* **83**, 312–4.

Hersey, M. D. (1910), "Resistance, Inductance, and Capacity of Eccentric Cylinders." *Elect. World* **56,** 434–6.

Hersey, M. D. (1923), "Problems of Lubrication Research." *JASNE* **35,** 648–74.

Higginson, G. R. (1962), "A Model Experiment in Elastohydrodynamic Lubrication." *Int. J. Mech. Sci.* **4,** 205–10.

Hitchcock, J. H. (1954), "Hydrodynamically Lubricated Roll Neck Bearings." *First ASLE Nat. Symp.,* Chicago, 1952; *Proc.,* 49–53. (a) *Oilways* **16** (Oct. 1949).

Hitchcock, J. H. (1960), "Lubrication of Roll Neck Bearings and Gear Drives in Continuous Rolling Mills." Pages 36–47 in *RV;* ASME, New York.

Horowitz, H. S. & Steidler, F. E. (1960), "The Calculated Journal Bearing Performance of Polymer-Thickened Lubricants." *TASLE* **3,** 124–33.

Horowitz, H. S. & Steidler, F. E. (1961), "Calculated Performance of Non-Newtonian Lubricants in Finite Width Journal Bearings." *TASLE* **4,** 275–81.

Horowitz, H. S. & Steidler, F. E. (1963), "Calculated Performance of Greases in Journal Bearings." *TASLE* **6,** 239–48.

Hother-Lushington, S. & Sellors, P. (1963), "Water-Lubricated Bearings: Initial Studies and Future Prospects in the Power Generation Industry." Paper 13, *LWG,* 131–8.

Ishii, Akira (1962), "On the Vaporization of the Oil Film in the Short Journal Bearing with a Circumferential Oil Groove." *Bull. JSME* **5,** No. 20; 745–52.

Jacobson, M. J. & Saibel, E. A. (1960), "The Finite Partial Fitted Journal Bearing." *TASLE* **2,** 242–7.

Jakobsson, B. & Floberg, L. (1958), *The Partial Journal Bearing. TCUT,* Göteborg, No. 200; 60 pp.

Jakobsson, B. & Floberg, L. (1959), *The Centrally Loaded Partial Journal Bearing. TCUT,* Göteborg, No. 214; 34 pp.

Jaqerod, A., Defossez, L., & Mügeli, H. (1922), "Recherches expérimentales sur le frottement de pivotement." *J. Suisse d'Horlogerie et de Bijouterie,* 269–74, 293–7; (1923) 8–10, 29–35, 53–7. Transl. by K. H. Beij, *NACA TM* **566** (1930); 54 pp.

Karelitz, G. B. (1937) "Oil Supply in Self-Contained Bearings." *GDLL* **1,** 151–6.

Kaufman, H. N. (1961), "A Pneumatic Gaging System for Measuring Oil Film Thickness in Journal Bearings." *LE* **17,** 342–5.

Kaufman, H. N. & Boyd, J. (1959), "The Conduction of Current in Bearings." *TASLE* **2,** 67–77.

Keller, W. M. (1955), "Effect of Viscosity of Car-Journal Oils on Running Temperature and Other Characteristics of Journal Bearing Performance." *T* **77,** 385–91.

Kettleborough, C. F. (1954), "Frictional Experiments on Lightly Loaded, Fully Floating Journal Bearings." *Australian J. Appl. Sci.* **5,** 211–20.

Khrisanova, L. B. (1959), Pressure Measurements in the Oil Film of a Sleeve Bearing (Russian). *TIM* **13,** 197–213. (a) Transl. by I. Malkin, pp. 183–208 in *FWM* **13;** ASME, New York (1961).

Kingsbury, Albert (1932), "Optimum Conditions in Journal Bearings." *T* **54,** RP-54-7, 123–48.

Klemencic, A. (1943), "Bemessung und Gestaltung von Gleitlagern." *Z. VDI* **87,** 409–18. (a) *J. ASNE* **64,** 104–19 (1952).

Klimov, V. Ia. (1947), Theory of the Motion of an Oil Film in the Non-Working Space of a Bearing (Russian). *Trudy II Vsesoinznoi Konferentzii po Trenii i Iznos v Mashinak,* **V;** AN SSSR.

Kochanowsky, W. (1955), "Grenzwerte von Tragfähigkeit und Reibung des Gleitlagers." *Schmiertechnik* **2,** 152–4.

Korovchinskii, M. V. (1954), *The Applied Theory of Fluid Film Bearings* (Russian). Mashgiz, Moscow.

Korovchinskii, M. V. (1962), Certain Problems of Elastorheology Applied in the Theory of Friction (Russian). *TIM* **15,** 332–74. (a) Transl., *FWM* **15,** 301–39; ASME, New York (1964).

Korovchinskii, M. V. (1962), Theory of Hydrodynamic Lubrication of Porous Bearings (Russian). *TIM* **16,** 151–218. (a) Transl., *FWM* **16,** 131–91; ASME, New York (1964).

Kuzma, D. C. (1963), "The Magnetohydrodynamic Journal Bearing." *JBE, T* **85,** 424–8.

Lawrence, K. B. (1950), "A Mathematical Evaluation of Pressures in a Grease-Lubricated Bearing." *T* **72,** 429–43.

Lee, J. C. (1955), "Analysis of Partial Journal Bearings under Steady Loads." *ASME* 55-LUB-1; 11 pp. (a) *ME* **77,** 1007.

Legett, W. D., Jr., Neely, G. L., & Ritch, J. B., Jr. (1949), "Some Case Histories of Shipboard Lubricating Problems During World War II." *TSNAME* **50,** 352–73.

Levinsohn, M. (1949), *Sixth Report on Test of Journal Bearings to Determine Performance at High Speeds and High Loads.* EES Rep. C-3041-F, USN Eng. Expt. Sta., Annapolis, Md.

Lewis, J. W. (1928), "An Experimental Study of the Motion of a Viscous Liquid Contained between Two Coaxial Cylinders." *PRS* **117,** 388–407.

Lewis, Paul (1965), "Lubrication in the Environment of Space." Pages 905–39 of *HS.*

Lialin, E. V. (1959), Investigation of Friction and Wear of Sleeve Bearings with Sulphuric Acid Lubrication (Russian). *TIM* **13,** 59–83. (a) Transl. by Th. Ranov, pp. 54–78 in *FWM* **13,** ASME, New York (1961).

Loeb, A. M. & Rippel, H. C. (1958), "Determination of the Optimum Proportions for Hydrostatic Bearings." *TASLE* **1,** 241–7.

Ma, J. T. (1965), "An Investigation of Self-Acting Foil Bearings." *ASME* 65 LUBS-4; 10 pp. (a) *ME* **87,** 58 (Aug.).

Marine Engineer and Naval Architect, The (1963), "Restricted Clearance Bearings." **86,** 462–4. (a) *JASNE* **76,** 219–22.

McKee, S. A. (1937), "Journal Bearing Design as Related to Maximum Loads, Speeds, and Operating Temperatures." *GDLL* **1,** 179–86. (a) RP 1037, *NBSJR* **19,** 457–65.

Michell, A. G. M. (1912), *Improvements in Journal Bearings.* Brit. Pat. 23, 496/1911; issued 1912.

Michell, A. G. M. (1919), *Bearing Element.* U. S. Pat. 1,315,735.

Milne, A. A. (1965), "Inertia Effects in Self-Acting Bearing Lubrication Theory." Pages 423–527 of *HS.*

Moore, H. F. (1903), "Experiments, Constants and Formulas for the Design of Bearings." *Am. Mach.* **26,** 1281–3, 1316–9, 1350–4.

Morgan, V. T. (1957), "Study of Design Critera for Porous Metal Bearings," Paper 88, *CLW;* 405–8, 362, 767, 775, 824.

Morgan, V. T. (1964), "Hydrodynamic Porous Metal Bearings." *LE* **20,** 448–55; **21,** 33 (1965).

Morgan, V. T. & Cameron, A. (1957), "Mechanism of Lubrication in Porous Metal Bearings." Paper 89, *CLW,* 151–7; 12, 362, 744, 767, 775; Plates 20, 21.

Motosh, N. (1964), "Cylindrical Journal Bearings Under Constant Load, the Influence of Temperature and Pressure on Viscosity." *LWG.* (a) *Sci. Lubn.* **16,** 19.

Naylor, C. W. (1906), "A New Valve Gear for Gas, Steam, and Air Engines." *T* **27,** 432–3.

Neale, M. J. & Love, P. P. (1957), "High Speed Journal Bearings." Paper 78, *CLW,* 123–7; 11, 758; Plate 13.

Needs, S. J. (1946), "Tests of Oil-Film Journal Bearings for Railroad Cars." *T* **68,** 337–53.

Needs, S. J. (1958), "Viscosity-Pressure Effect on Friction and Temperature in a Journal Bearing." *T* **80,** 1099–103.

Nishihara, T. & Sugimoto, Y. (1956), "On the Theory of Lubrication in Journal Bearings." *Kyoto University, Eng. Res. Inst. Tech. Rep.* **6,** No. 5, 75–97.

Ocvirk, F. W. & DuBois, G. B. (1959), "Surface Finish and Clearance Effects on Journal Bearing Load Capacity and Friction." *JBE, T* **81,** 245–53.

Odqvist, F. K. G. (1937), "The Nomy Pad Bearing: Development and Design." *GDLL* **1,** 227–33. (a) *Eng.* **140,** 577–9 (1935).

Oksal, M. A. (1946), *The Introduction of Variable Viscosity in the Analysis of Full Journal Bearings.* Ph.D. Thesis, Cornell University, Ithaca; 74 pp.

Orcutt, F. K. (1965), "Investigation of a Partial Arc Pad Bearing in the Superlaminar Flow Regime." *JBE, T* **87,** 145–52.

Osterle, F., Chou, Y. T., & Saibel, E. A. (1958), "The Effect of Lubricant Inertia on Journal Bearing Lubrication." *JAM* **24,** *T* **79,** 494–6; **80,** 420.

Ott, H. H. (1954), "The Coefficient of Friction and Shaft Position in Turbine Bearings with Reduced Vertical Clearance." *Brown Boveri Rev.* **41,** 256–63.

Ott, H. H. (1959), "Position of the Shaft and Friction in a Triple Wedge Bearing." *Brown-Boveri Rev.* **46,** 395–406.

Patel, B. J. & Cameron, A. (1957), "The Foil Bearing." Paper 73, *CLW,* 219–23; 201, 743, 833.

Peeken, H. (1960), "Die Berechnung von Kugelgleitlagern." *Forschung* **26,** 117–28.

Petroff, N. P. (1948), *Hydrodynamic Theory of Lubrication* (Russian). Collected papers; biographical articles. AN, SSSR, Moscow; 551 pp.

Pinkus, O. (1956), "Power Loss in Elliptical and Three-Lobe Bearings." *T.* **78,** 899–904; 965–73. (a) *JBE, T* **81,** 49–55 (1959).

Pinkus, O. (1958), "Solution of Reynolds' Equation for Finite Journal Bearings." *T* **80,** 858–64. See also (a) *T* **79,** 1213–7 (1957); (b) *JBE, T* **83,** 145–52 (1961).

Pinkus, O. (1962), "Counterrotating Journal Bearings." *JBE, T* **84,** 110–8.

Radzimovsky, E. I. (1959), *Lubrication of Bearings.* Ronald, New York; 338 pp.

Raimondi, A. A. (1959), "A Theoretical Study of the Effect of Offset Loads on the Performance of a 120° Partial Journal Bearing." *TASLE* **2,** 147–57.

Raimondi, A. A. & Boyd, John (1958), "A Solution for the Finite Journal Bearing and its Application to Analysis and Design." *TASLE* **1,** I: 175–93; II: 175–93; III: 194–209.

Raimondi, A. A. & Boyd, J. (1957), "An Analysis of Orifice and Capillary Compensated Hydrostatic Journal Bearings." *LE* **13,** 28.

Ramachandra, S. (1961), "A Solution of Reynolds' Equation for a Finite Full Journal Bearing." *JBE, T* **83,** 589–94.

Robb, A. M. (1933), "A Contribution to the Theory of Film Lubrication." *PRS, (London),* **140,** 668–94.

Rouleau, W. T. (1963), "Hydrodynamic Lubrication of Narrow Press-Fitted Porous Metal Bearings." *JBE, T* **85,** 1235–8. (a) **84,** 205–6 (1962).

Rylander, H. G. (1952), "Effects of Solid Inclusions in Sleeve Bearing Oil Supply." *ME* **74,** 963–6; **75,** 416–8.

Saibel, Edward (1960), "The Effect of Temperature and Pressure on Viscosity as Related to Hydrodynamic Lubrication." Pages 105–7 in *RV,* ASME, New York.

Sasaki, T. & Mori, H. (1956), "Analytical Solution for Full Journal Bearings of Short Length by Successive Approximation." *Mem. Kyoto University Fac. Eng.* **18,** 341–53.

Sasaki, T. & Mori, H. (1956), "Fluid Lubrication Theory of Multiple-Film Journal Bearings." *Proc. Sixth Japan Nat. Cong. Appl. Mech.,* University of Kyoto, 319–22.

Sassenfeld, H. & Walther, A. (1954), *Gleitlagerberechnung.* Forschungsheft 441; 28 pp. (a) *Forschung* **20,** 36.

Schultz-Grunow, F. (1959), "Zur Stabilität der Couett-Strömung." *ZAMM* **39,** 101–10.

Shaw, M. C. & Nussdorfer, T. J. (1947), "An Analysis of the Full Floating Journal Bearing." *NACA Rep.* **866**; 95–107.

Sizer, Harold (1939), "The Care and Lubrication of Grinding Machine Spindles for Fine Finishing." *Machinery* **45,** 565–7.

Slibar, A. & Paslay, P. R. (1959), "Retarded Flow of Bingham Materials." *JAM* **26,** *T* **81,** 107–13.

Smith, M. I. & Fuller, D. D. (1956), "Journal Bearing Operation at Super-Laminar Speeds." *T* **78,** 469–74.

Stahlhuth, P. H. & Trippett, R. J. (1962), "Liquid Metal Bearing Performance in Laminar and Turbulent Regions." *TASLE* **5,** 427–36.

Sternlicht, B. & Fuller, D. D. (1957), "A New Method for the Determination of Attitude & Eccentricity in Journal Bearings." *LE* **13,** 208–14.

Sugimoto, Y. (1959), "A Theory of Lubrication in Journal Bearings with an Elastically Bent Shaft." *Wear* **2,** 329–34.

Tanner, R. I. (1958), "Hydrodynamic Lubrication of Ball and Socket Joints." *Appl. Sci. Res. A* **8,** 45–51.

Tanner, R. I. (1963), "Non-Newtonian Lubrication Theory and Its Application to the Short Journal Bearing." *Australian J. Appl. Sci.* **14,** 129–36. (a) *JAM* **31,** *T* **86,** 350–1 (1964). (b) *JAM* **32,** *T* **87,** 781–7 (1965).

Tao, L. N. (1958), "The Theory of Lubrication in Short Journal Bearings with Turbulent Flow." *T* **80,** 1734–40.

Tao, L. N. (1959), "General Solution of Reynolds' Equation for a Journal Bearing of Finite Width." *Quart. Appl. Math.* **17,** 129–36. (a) *JAM* **26,** *T* **81,** 170–83.

Taylor, G. I. (1923), "Stability of a Viscous Liquid Contained between Two Rotating Cylinders." *Phil. Trans. Roy. Soc. (London) A.* **223,** 289–343. (a) *Proc. First ICAM,* Delft (1924), 89–96.

Thomson, A. S. T. (1954), "Some Factors in Design and Lubrication of Journal Bearings." *Trans Inst. Engrs. and Shipbuilders in Scotland* **97,** Pt. 4, 257–312; Pt. 5, 313–28.

Thurston, R. H. (1878), "Friction and Its Laws." *Proc. Am. Assoc. Adv. of Sci.* **27,** 61–71. (a) *JFI* **95,** 83 (1873).

Tipei, N. (1959), "Le lubrification des corps permeables." *Rev. Mecanique Appl.* **4,** 63–71.

Tipei, N. (1960), Three-Dimensional Lubrication of Short Bearing Surfaces at High Speeds (Rumanian). *Studii si Cercetari de Mecanica Aplicata* **11,** No. 3, 595–601, ARPR. (a) Chap. 13 of Transl., *Theory of Lubrication* (1962).

Tipei, N., Constantinescu, V. N., Nica, Al., & Bica, O. (1961), *Sliding Bearings (Calculation, Design, Lubrication).* Academei RPR, Bucharest; 435 pp.

Vieweg, V. & Vieweg, R. (1926), "Über Lager Versuche." *Maschinenbau* **5, 201–6.**

Vogelpohl, G. (1965), "Thermal Effects and Elasto-Kinetics in Self-Acting Bearing Lubrication." *HS,* 763–822.

Wagner, E. M. (1932), *Frictional Resistance of a Cylinder Rotating in a*

Viscous Fluid Within a Coaxial Cylinder. Thesis, Stanford University (a) *ME* **55,** 39 (1933).

Wannier, G. H. (1950), "A Contribution to the Hydrodynamics of Lubrication." *Quarterly of Appl. Math.* **8,** 1–32 (April).

Waring, E. H. (1917), *Bearing Surface for Machinery*. U.S. Pat. 1,236,511.

Waters, E. O. (1939), "Theoretical Pressure Distribution in Journal Bearings." *Fifth ICAM,* Cambridge, Mass. 1938; *Proc.* Wiley, New York, pp. 631–7.

Waters, E. O. (1942), "Characteristics of Centrally Supported Journal Bearings." *T* **64,** 711–9.

Westcott, A. L. (1913), "The Lubricating Value of Cup Greases." *JASME* **35,** 1143–67.

Westphal, R. C. & Glatter, J. (1954), "The Wear of Bearings Lubricated by Water." *ASME* 54-SA-13. (a) *ME* **76,** 768.

Wilcock, D. F. & Rosenblatt, M. (1953), "Nomographic Method for Sleeve Bearing Design." *Mach. Design* **25** (Aug.), 143–57.

Zhukovskii, N. E. & Chaplygin, S. A. (1904), *On the Friction in the Lubricating Film Between Journal and Bearing* (Russian). Moscow: 31 pp. (a) *Fortschritte der Math.* **35,** 766–8.

NOTE: Certain investigations covered in the text are not listed above, but can readily be found.

chapter VIII Reciprocating Machines

1. Piston and ring lubrication. 2. Hydrodynamic action. 3. Experiments on piston ring friction. 4. Ring and cylinder wear. 5. Principles of dynamically loaded bearings. 6. Engine applications. 7. Classical hydrodynamic theory. 8. Hahn's solution. 9. Experimental investigations. 10. Summary.

The bearings in a reciprocating machine are subject to fluctuating, impulsive, or *dynamic* loads and therefore offer problems not met in the normal operation of a steady-load journal bearing. Reciprocating machines include the steam engine; automobile, diesel, and aircraft piston engines; pumps, compressors, stamping and forging machines, and various machine tools. We start with a review of piston and piston ring lubrication, and then discuss dynamic loading as seen, for example, in the connecting rod, crankpin, and main bearings of an engine.

1. PISTON AND RING LUBRICATION

A diesel engineer, wishing to minimize liner wear, asked, "Which carries the heavier load, a thick film or a thin one?" Paradoxically, the answer is a "thin one." This seems contrary to common sense because, obviously, the thicker the oil film the greater the protection offered. The answer given is correct for steady-state operation. The common sense view pictures the effect of a suddenly applied load.

Piston Lubrication. The piston of a reciprocating engine, pump, or compressor must not fit too tightly. A reasonable clearance should be allowed between the piston and the cylinder wall or liner. When there is no crosshead the piston acts very much like a thrust bearing. Visual demonstrations are described by Beaubien (1946), Shaw (1946), and co-authors.

Manufacturers have made careful studies of piston design to compensate for thermal expansion. Aluminum pistons are scientifically contoured. Little difficulty is met in lubricating the piston skirt compared to the rings in a reciprocating engine, though piston tightening and seizure have been reported by Hartung & Savin (1962). Piston temperature measurements leading to improvements in design are described by Cavileer (1963).

A form of piston-sticking due to instability in hydraulic cylinders, and known as "hydraulic lock" has been investigated by Manhajm, Mannam, Kettleborough, and others (1955, 1959, 1961). Remedies are described by Boyd and by Raimondi (both in 1964).

Operation of Piston Rings. Tight forms of packing were used in the earlier piston machines, with consequent high friction and frequent replacement. Albert Kingsbury (1908) showed how packing could be eliminated in hydraulic testing machines by accurate workmanship, closer fitting, and compensation for permissible leakage. Oil-free compressors were later commercialized for air, oxygen, and other applications. They depend on a close fit or a labyrinth design, as in the Sulzer type (1938); or on special ring materials, as described by Taber, Summers-Smith, Arnold, and co-authors (1957, 1963, 1964). See also Bush (IX: 1961) and Notaro (1966).

Customary practice in engine design depends on the familiar steel or cast-iron ring with a gap to allow for a spring fit in the piston groove. Piston ring nomenclature was standardized by the SAE (1943). Many different cross-sections have been tried. Grooves are provided near top of piston for "compression rings" to reduce "blow-by" with grooves for oil control or "scraper rings" near bottom to minimize oil consumption.

Piston rings operate on relatively thin films owing to concentrated loading. Stanton (1925) concluded that ring friction was of the boundary type. According to C. F. Taylor, "piston rings operate most of the time in the partial film region" (1937). Hunsaker & Rightmire (1947) described piston ring lubrication as "the most critical problem of boundary lubrication." The late A. G. M. Michell put these ideas into quantitative form (V: 1950, pages 246–50). A typical piston ring of rectangular section is shown in Fig. 1, where the axial width AB will be denoted by l. The ring groove in piston is shown at the right, the cylinder wall of radius r at the left. The ring is an imperfect seal, permitting a pressure drop Δp from A to B. The cylinder pressure p, above and behind the ring, keeps it in contact with the wall and with the bottom of the groove. Let p_e denote the elastic force per unit of face area needed to hold the ring in the groove; then the net force acting on the ring toward the right in Fig. 1, per unit of circumference, is l times the sum of $p_e + \Delta p/2$. But $\Delta p = (p_1 - p_0)n$, where p_1 is the cylinder pressure above the piston and p_0 the pressure below, and n the number of rings. It follows that the total frictional force on the ring assembly is

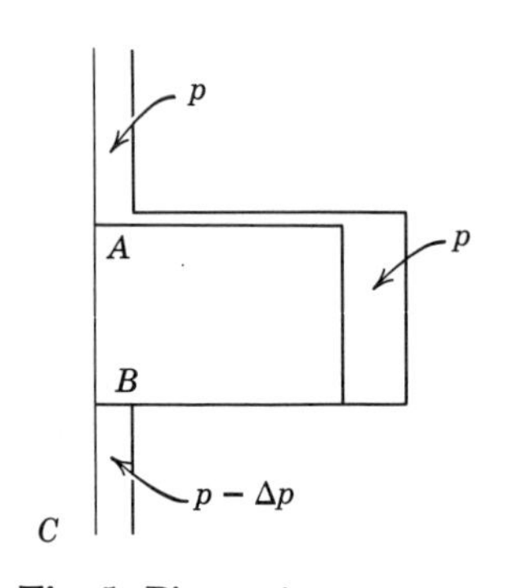

Fig. 1. Piston ring allows pressure drop Δp over axial width AB in Michell's theory; cylinder wall at C.

$$F = \pi rl(p_1 - p_0 + 2np_e)f\,, \qquad (1)$$

where f denotes a constant coefficient of friction.

For comparison with the friction experiments by Hawkes & Hardy (1936), Michell calculates the ratio of the friction on ten rings to that on four rings for $p_1 - p_0$ values from zero to 90 psi. The calculated ratios vary from 2.50 to 1.57; experimental ratios from 2.44 to 1.70, respectively, showing good agreement. He goes on to derive an expression for the loss ratio (1 minus efficiency) in a complete cycle; thus

$$1 - \eta = \frac{l}{r}\left(\frac{p_m + p'_m + 4np_e}{p_m - p'_m}\right)f\,, \qquad (2)$$

where p_m and p'_m are the mean effective pressures on the power stroke and the return stroke. Working back from the experiments cited, Michell finds f values from 0.034 to 0.047. He takes p_e equal to $\frac{1}{2}$ atm., or say 0.7 psi, in the light of published measurements.

2. HYDRODYNAMIC ACTION

Castleman showed what might be expected from hydrodynamic wedge formation (1936). He applied thrust bearing theory to piston rings worn to a double taper in accord with measurements by the National Bureau of Standards on slightly used Ford automobile rings (Fig. 2). In the case of a single tapered or conical surface, subject to

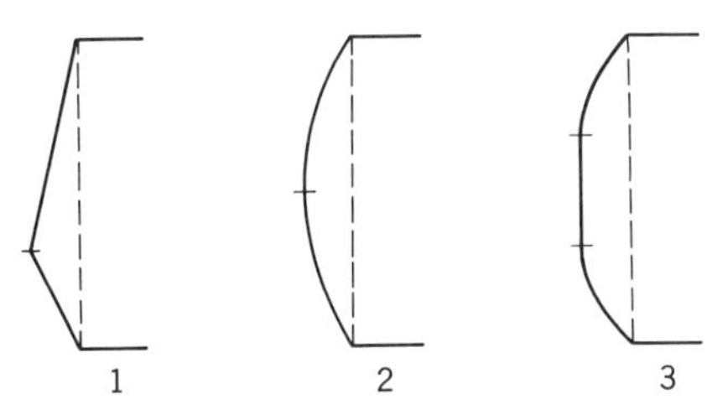

Fig. 2. Profiles of slightly worn rings: 1, Castleman's double taper, 2, Eilon & Saunder's parabola, 3, Furuhama's truncated parabola.

a load W_1 per unit of circumference, the minimum film thickness h_0 can be found from the load capacity equation

$$\frac{W_1 m^2}{6\mu U} = \ln (M + 1) - \frac{2M}{M + 2}. \tag{3}$$

Here m is the taper slope, μ the viscosity of the oil, U the piston speed, and M denotes $ml/2h_0$. Castleman applied this equation, and the film pressure equation, to the actual double taper by assuming ambient pressure at h_0 as well as at the two edges of the ring; and by disregarding negative pressures greater than ambient. From the measured slopes, averaging about $\frac{1}{200}$, he calculated that the minimum film thickness would be of the order of $\frac{1}{5}$ mil. Operating conditions were taken to be $U = 1200$ cm/sec, $\mu = 10$ cp, and ambient pressure 12 kg/sq cm.

This result cannot be far off. It started a new trend of thinking in lubrication research, whereby greater recognition is continually being given to full or partial hydrodynamic action. See, for example, the electrical resistance measurements of several investigators, reviewed by Bowden & Tabor (pages 248–50 of Part I, (I: 1950). Two refinements might be introduced: (1) the film pressure at h_0 can be treated as an unknown to be determined from simultaneous equations, since in fact the pressure development carries over the point of nearest approach part way into the divergent space; and (2) to agree with the measurements reported, the tapers need not be assumed equal, but may be taken to intersect at 0.72 of the distance from top to bottom of ring. It would be interesting to make these changes, and to extend Castleman's theory to a calculation of friction for comparison with Michell's findings.

Fluid friction was calculated by Eweis (1935) for the parallel-faced ring with rounded edges by applying Gümbel's formula. He also conducted experiments on friction and leakage, as noted in Section 3.

The next contributions to our understanding of hydrodynamic action are the investigations by Dykes (1957) and by Eilon & Saunders (1957). Dykes studied oil flow experimentally using a piston guided above and below. His observations are applied to the control of oil consumption. The others measured friction as described in the next section, and calculated the mean film thickness between ring and liner from the friction values. Noting that their flat rings wore parabolic (Fig. 3), they derived hydrodynamic formulas for the corresponding friction and load capacity in terms of film thickness. Both relations are complicated but reducible to the forms

$$\frac{F_0}{p_1} = \phi\left(\frac{\mu U}{p_1 b}, \frac{h_0}{b}, \frac{R}{b}\right) \tag{4}$$

and

$$\frac{P_0}{p_1} = \psi\left(\frac{\mu U}{p_1 b}, \frac{h_0}{b}, \frac{R}{b}\right). \tag{5}$$

Here F_0 and P_0 are the friction and load per unit of face area, and h_0 the minimum film thickness, in our notation; p_1 is the cylinder pressure above and behind the ring, b the axial half-breadth of the ring, R the parabolic radius of curvature, μ and U as before. The functions φ and ψ each take different expressions on the upstroke and downstroke of the piston. The equations refer only to the performance of a single ring on a piston without side thrust. Uniform viscosity is assumed, without consideration of inertia or nonsteady states. The pressure distribution is based on Gümbel's approximation of ambient pressure at the point of nearest approach, as in Castleman's solution.

Both papers were followed by discussions in which Barwell, Christopherson, and others took part. See also a paper by Lewicke (1957). The experimental results are reviewed later.

Squeeze-film action was included in the theory by Furuhama (1959) and checked experimentally in a following paper. The theory was developed both for a constant and a variable pressure behind the ring. Reynolds' equation was integrated numerically with the load per unit of circumference and the piston speed both functions of time, $W(t)$ and $U(t)$. It was shown, for example, that when the maximum pressure of 20 kg/cm^2 comes at top-dead-center, the film thickness will be 1.5 microns or 0.06 mil at 1000 rpm; rising hydrodynamically to 2.8 microns or 0.11 mil at 3000 rpm. Dynamical similarity relations were also noted.

3. EXPERIMENTS ON PISTON RING FRICTION

In the NPL experiments (Stanton 1925), two coupled pistons oscillated in horizontal cylinders, heated to simulate practical conditions. The pistons were driven by a long, heavy pendulum, whose damping served to measure the friction of the piston assembly. The coefficient of friction of the rings was found to be 0.023 at room temperature and 0.028 at 100 C. Castor oil was used as the lubricant. Since the friction was independent of amplitude and nearly proportional to the load, it was thought to be of the boundary type. These tests, however, were conducted at an abnormally low speed, 15 cycles/min, leaving open the possibility of fluid film lubrication at normal speeds. Sparrow & Thorne (I: 1927) concluded from their motoring and brake tests on aviation engines that fluid friction prevailed between piston and rings, and the cylinder walls. Friction increased with speed, but less than in direct proportion. It became nearly independent of speed at the lower speeds. Friction was found to vary with the viscosity of the oil film, and with changes in piston design. Eweis (1935) investigated the effect of the number and arrangement of rings, using a stationary piston with movable cylinder. The cylinder was heated electrically to provide operating temperatures to 120 C. Compressed air was supplied to establish the required pressure behind the rings. Fluid friction was indicated, although test speeds were not available above 2 m/sec. Eweis found that the friction of piston rings without gas loading is proportional to the number of rings.

Fluid friction was again found in the classical tests by Hawkes & Hardy (1936), using a reciprocated liner while the two connected pistons were controlled by calibrated springs. The pistons were of 5.87-in. diameter, each with six ring grooves. Air pressure was introduced in a central groove between the two crowns. Tests were made at piston speeds from 240 to 340 in./min, the piston temperatures ranging from 220 to 370 F. Ring friction was found proportional to the square root of the lubricant viscosity at the piston bulk temperature. With a single ring, frictional resistance was proportional to the square root of the net radial load. High friction was indicated at the end of each stroke. This effect was most pronounced at the lowest speeds. Experiments by M. P. Taylor (1936) consisted in motoring a six-cylinder engine at several constant gas pressures in the cylinders. Piston friction increased with gas pressure and speed, and was greatly reduced by running without rings. Changes in oil viscosity with change in jacket water temperature had a marked influence on piston friction.

Instantaneous values of piston ring friction were obtained by Tischbein (1939) with a spring-mounted liner. The stroke of the piston was 21 cm, liner diameter 20.5 cm. Piston rod was driven by a crosshead to avoid side thrust. From one to four rings could be fitted to the piston. Mean coefficients of friction are plotted against mean rubbing speed, with surface temperature and ring pressure against wall as constant parameters. These curves pass through a minimum point, indicating mixed lubrication at the lower speeds, hydrodynamic at the higher speeds. A fair correlation between friction coefficient and ZN/P or its equivalent was found at the right of the minimum point, as shown by Fig. 17 of the reference. Tischbein calculated the friction coefficients to be expected from the Gümbel-Eweis theory of parallel-surface rings with rounded edges. Surprisingly, the calculated curves fell very much higher than the observed values. This suggests some form of profile wear as in Fig. 2, a possibility consistent with the long period of running-in.

A method for measuring piston friction in a firing cylinder was developed through a series of investigations at M.I.T. These experiments were described by Forbes & Taylor (1943) and later authors, and are summarized by Professor Rogowski (1961). It was found that the rings may account for three-fourths of the friction of the piston assembly, and that the latter may be responsible for three-fourths of the mechanical friction of the whole engine. The special cylinder and crosshead arrangement are mounted on a CFR engine crankcase. Piston friction might be greater without crosshead. Measurement of instantaneous friction is accomplished by elastic mounting of the light combustion cylinder sleeve, so that small axial movements of the sleeve can be indicated electrically, and recorded by camera and oscilloscope. Ring friction increased with speed and inlet pressure and decreased with higher jacket temperatures. Discontinuities in friction records indicated metallic contact at top and bottom of the stroke. In the meantime a study of piston ring friction was reported by E. A. Bogdonov (1957).

For convenient application, the test data from Eilon & Saunders might well be charted as shown by Fig. 46 in Neale's discussion, page 488 of the reference. The lower part of the graph can be approximated by $y = cx^n$, where y denotes h_0/b, the ratio of minimum film thickness to the half-axial-width of the ring; and x, in our notation, stands for ZN/p_m, where p_m is the mean pressure between ring face and cylinder wall. If we take $n = 2$, then $c = 10^{-7}$, nearly. Above $x = 100$ cp rpm/psi the curve can be fitted by $Y = 0.21X^{1/2}$, where Y denotes $10^3y - 1$, and X has been written for $x - 100$. Taking h_0/b instead of h_0 as ordinate permits applying the data to geometrically similar systems of different absolute size.

Experiments were conducted with a fixed one-ring piston and reciprocating liner, under controlled conditions of gas pressure and oil viscosity. Cylinder bore was 5 in. with a 3.5-in. stroke at speeds from 800 to 1500 rpm. Aluminum alloy piston had a clearance of 0.010 in. Liner finish was from 0.8 to 1.5 micro-inches. Surface finish on ring seat was 18 to 20 micro-inches, leading to an estimated friction coefficient of 0.15 for use in calculations. Wall pressure of cast-iron ring $p_e = 16.5$ psi. The reported mean film thickness is the equivalent value for a uniform film corresponding to the measured friction. It was concluded that hydrodynamic lubrication prevails throughout the stroke at the lower temperatures, with a breakdown at the equivalent of top-dead-center leading to boundary lubrication when the local temperatures are sufficiently high. Adoption of a parabolic profile and wider top rings was suggested.

The experiments by Furuhama used a stationary piston with a reciprocating cylinder of 9.0-cm stroke at speeds up to 2000 rpm. A novel feature was the variable pressure applied by rubber tube behind the ring. The cast iron piston ring was of 7.6-cm diameter, 3 by 3-mm cross section, with a ring tension p_e of 1.0 kg/sq cm. Ring profile showed a crowning of 2.6 microns with a flat middle third (Fig. 3). Inside of cylinder was honed. Friction was measured by deflection of a phosphor-bronze diaphragm. It was concluded that fluid lubrication changed to the boundary type at a film thickness of the order of 1.5 microns, about 0.06 mil, for the particular degree of roughness in present surfaces. In general the experimental results were in agreement with calculations, which served thereby to throw new light on squeeze-film action at end of stroke.

Ten oils and a grease were compared in a study of piston movement by Volarovich (1959). Thixotropic as well as viscous properties were brought to bear in explaining the relations between force and motion from —60 to +20 C. Low-viscosity oils were compared by Barros & Dyson (1960) in the Eilon-Saunders rig. Reduction of friction at higher gas pressures was attributed to a possible twisting of rings in their grooves. Friction forces were greatest near dead-center positions. It was concluded that hydrodynamic lubrication can be maintained only during the last four-fifths of the power stroke. Using a motored engine, Pike & Stillman (1964) found that the friction of the piston assembly increased with peak pressure and was proportional to the square root of the ring-belt oil viscosity at constant speed. They emphasized that the establishment of a hydrodynamic film is strongly influenced by roughness of ring and liner surfaces.

4. RING AND CYLINDER WEAR

A comprehensive review of lubrication and wear data on piston rings and cylinder liners was published by R. Poppinga in his well-known book, which includes an extensive bibliography (1942).

Several more recent investigations are of special interest. H. V. Nutt and co-authors found that diesel engines wear less with oils having a higher viscosity at high temperatures; decreased volatility; and maximum thermal stability (1955). Watson and associates reported on abrasive wear of piston rings (1955). The Navy's long-continued development of chromium plating for steel liners was summarized by R. W. Reynolds (1957). Two types of porous chrome plate were found superior by wear-machine tests. Trends in piston ring wear in gasoline engines were determined from tests by Calow & Epton (1961). Cree & Thiery investigated factors affecting wear under starting conditions (1961). Cavileer used a radioactive tracer technique (1963). In the opposed-piston diesel of $6\frac{3}{4}$ in. bore, 8-in. stroke, 1335 hp at 1335 rpm, the rate of ring wear was very much greater on the lower pistons. A rapid increase occurred when operating above rated output. These facts were traced to local overheating, which led to the research on piston temperatures and design improvements previously cited. In their wear tests on a motored engine Pike & Spillman (1964) found mechanical wear most acute on the top ring. It was caused by increasing the peak cylinder pressure or the liner temperature but eventually became negligible with the liner surface highly polished. It was believed hydrodynamic lubrication would be attainable throughout the stroke, if roughening of surfaces by the effects of combustion products could be prevented. Attritional wear need be expected only when ZN/P falls below the value at which film thickness matches the surface roughness.

A selective bibliography of piston-ring lubrication, including friction and wear, was published by the NACA (Hersey, 1944). It goes back to the work of J. M. Denton, 1886, on the friction of piston rings in steam cylinders; of V. G. Karpenko, in 1913; of the late A. E. Flowers, who designed test apparatus in 1915; and other contributions of historical or scientific interest.

5. PRINCIPLES OF DYNAMICALLY LOADED BEARINGS

When the load applied to a bearing fluctuates in magnitude or direction, or both, the bearing is said to be "dynamically loaded." In the typical case of a reciprocating machine the fluctuations are periodic. The term is often extended to bearings subjected to a varying speed—oscillatory motion, for example—under a constant external load (Bidwell, 1954).

From a physical viewpoint the complicated movement of the journal under a dynamic load is the resultant of two elementary actions, the "squeeze-film" and the "equivalent-speed" effect.

The Squeeze-Film. The Stefan or Underwood squeeze-film has been described in Chapter II. Imagine a stationary journal, concentric in the bearing, suddenly subjected to a constant load in a fixed direction. It will move gradually closer to the bearing, cushioned by the remaining film of lubricant. The rate of approach is governed by Stefan's law in a modified form. Take the piston pin of a reciprocating engine as an example. Gümbel (I: 1925) and Fuller (I: 1956) solved the problem by disregarding the oscillation, and treating the bushing as a half-bearing of infinite length. An expression was first found for the velocity of approach. Integration gives the time Δt needed for the eccentricity ratio to change from ϵ_1 to ϵ_2 under a constant load. This integration was obtained graphically by Fuller and analytically by Licht (see Fuller, *op. cit.*, p. 143) and by Archibald (II: 1956). The latter reference includes a derivation for the full journal bearing, which had been approximated by Orlov in his book (1937).

From dimensions it is seen that Δt must be the product of Z/P and some function of m, ϵ_1, and ϵ_2; where Z is the viscosity, P the load per unit area, and m the clearance ratio. The formula can be simplified if the initial eccentricity is taken as zero, while the final value, close to unity, is denoted by ϵ. Under these conditions.

$$\Delta t = \frac{Z}{P}\left[\frac{13.3}{m^2(1-\epsilon)^{1/2}}\right]. \tag{6}$$

In the example chosen by Professor Fuller, a constant load of 950 psi is applied for a time Δt to a piston pin bushing with $m = 0.69$ mil/in., lubricated by oil of viscosity 1.56 n. How long must the load be on before the eccentricity reaches 0.9 of the radial clearance? From Eq. (6), $\Delta t = 0.15$ sec. This might fall to perhaps 0.10 sec if corrections for end leakage could be applied. The cushioning effect of the squeeze-film is made evident by the piston-pin example. The effect may be very similar when the journal is rotating and the load varying, as shown by Orlov.

Equivalent Speeds. The "equivalent-speed" idea is associated with rotating loads of constant magnitude, rotating at a constant angular velocity (Orlov, 1934). Suppose the direction of the load vector is changing at the rate of n revolutions per unit time, while the shaft is turning at the speed N. The simplest case is that of an unbalanced rotor exerting a constant centrifugal force on the supporting bearings. In this case, n and N are equal. A convenient model is shown by Orlov (1937, Fig. 82) to aid in visualizing the load vector.

We know that the load capacity of a statically loaded bearing, or the load that can be carried on a given film thickness at constant viscosity, is proportional to the journal speed N. The equivalent speed N_e is defined as the journal speed that will give the same load capacity when the load is rotating at a constant speed n. It can be shown that

$$N_e = N - 2n\,. \tag{7}$$

To see this, picture the bearing as a hole in a square block. Sketch the load vector W rotating with a clockwise speed n, while the journal has a clockwise speed N (Fig. 3A). At this point, to borrow a phrase from Professor Shaw's lectures, the observer should "sit on the load vector." He can then record the motions which he sees relative to that vector, and so create the equivalent static problem. Make another sketch showing the load vector stationary with respect to the observer

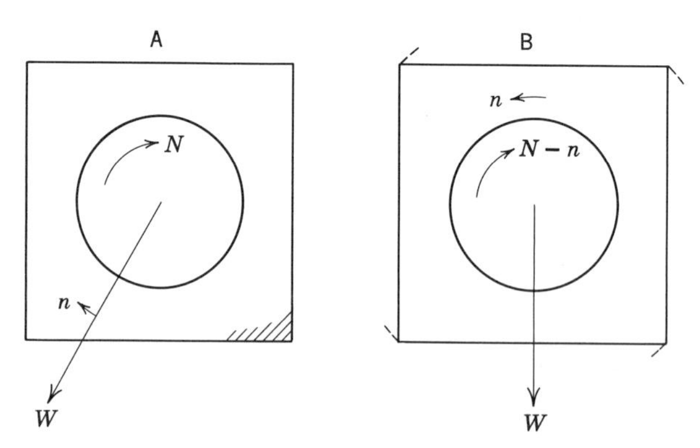

Fig. 3. Bearing with rotating load (A, bearing stationary, journal speed N, load speed n; B, load vector stationary, bearing speed n anticlockwise, journal speed $N - n$ clockwise, equivalent speed $N - 2n$).

(Fig. 3B). Here the journal rotates with a clockwise speed $N - n$, the bearing block with an anticlockwise speed n. While the motion of one surface drags oil into the convergent space on the loaded side, that of the other surface is dragging it out. The equivalent speed for a single moving surface is the algebraic sum $(N - n) - n$, or $N - 2n$ as in Eq. (7).

The equivalent-speed theorem was published independently by Orlov (1937) and by Swift (1937). It was rediscovered by Stone and Underwood (1947), experimentally demonstrated, and put to good use.

6. ENGINE APPLICATIONS

In applying the equivalent-speed theorem to reciprocating engines it is necessary to have a complete knowledge of the forces acting on the bearing. This is usually expressed by a polar diagram—a closed curve whose radii indicate the direction and magnitude of the resultant load at equal intervals of time or crank angle, say every 10 or 15 deg. In a connecting rod bearing, the load angle is customarily measured from the axis of the rod. The resultant at any instant depends on the gas load, reciprocating inertia, and centrifugal force of big end. Hence the diagrams for various bearings in a given engine may take on fantastic shapes. A heat classification of these shapes into categories having similar characteristics is much to be desired.

A well-known load diagram is that of the two-cycle Model 6-71 diesel engine crankpin bearing (Stone & Underwood, 1947; Burwell, 1947, II). Another is the "swan diagram" described by Professor Tichvinsky, Fig. 3 of 1948 or Fig. 10 of Hersey (II: 1949). This applies to a double-acting two-cycle diesel engine main bearing. It is the perfect outline of a swan with its proudly arched neck, sharp bill, and paddling feet (Fig. 4). A third example is the plain vertical ellipse, chosen as an average diagram by two companies building test machines. One such machine, of Dr. Ernst Lehr's design, produced a maximum vertical load of 100 metric tons up and down, 33 to the right and left; see Fig. 3 of Hersey & Snapp (1957). Another is the hydraulic machine described by Pigott & Walsh (1957).

The computation of load diagrams for reciprocating machines has been described by Root (1932); Tichvinsky (pages 38–41 of Hersey, II: 1949) and Vaughan (1963). It is an arduous task that must be repeated for each different speed. When the polar diagram is available, it is a simple matter to plot from it a rectangular coordinate diagram for the load angle against crank angle. The slope of the curve at any point is the instantaneous value of n/N. Hence it would appear from

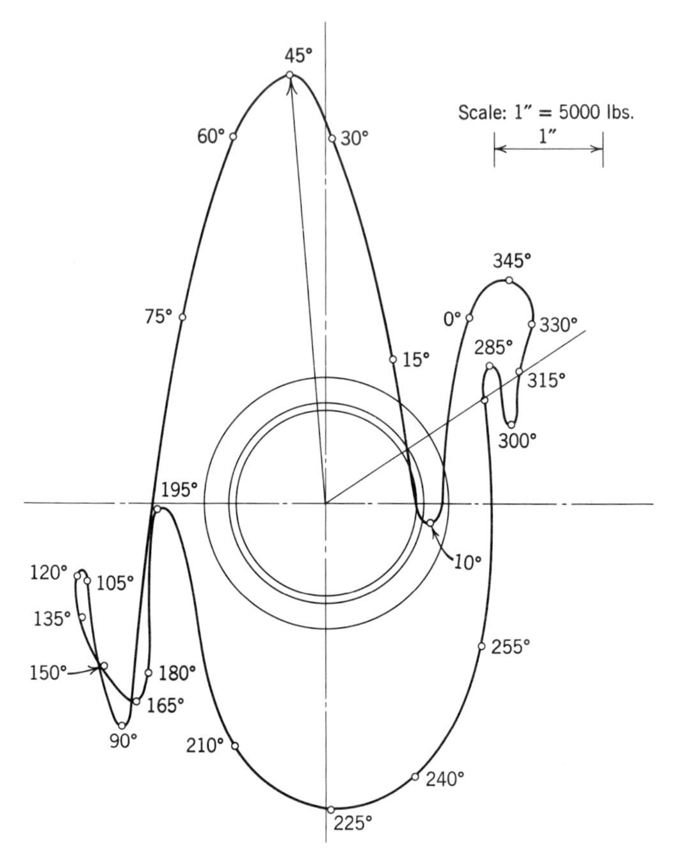

Fig. 4. Main bearing loads in a double-acting diesel engine (maximum load 900 psi at 45-deg crank angle); from Tichvinsky's discussion (Hersey, 1949).

Eq. (7) that N_e/N is equal to unity less twice the slope. A positive slope would indicate a deficiency in load capacity. A safe diagram would be one in which the deficient intervals do not coincide with heavy loads. Some designers prefer to plot N_e/N against crank angle. A line is then drawn horizontally at $N_e/N = 1$. The bearing is considered safe at all crank angles for which the curve lies above unity. A new criterion for engine design was thus recognized and zealously applied. Several types of failure were explained and many faults corrected. A striking example is an application of the equivalent-speed principle by Sahin (1957) to the diesel engine test data published by

Kollmann & Hockel (1953). This principle shows that the hydrodynamic load capacity for rotating loads can be very much greater than that for a static load. Possibly the most interesting result of the equivalent-speed theory is the prediction of film failure, or zero load capacity, when the load speed is half the journal speed, that is, when $n = N/2$ in Eq. (7).

The equivalent-speed equation applies accurately only when the load is rotating with a constant angular velocity. Hence, under practical conditions, only qualitative or approximate results are to be expected. It would be useful if some rule could be found indicating the permissible departure from constant load-speed while keeping within a stated margin of error.

Dynamic Load Literature. More than 400 references to the literature on dynamically loaded bearings will be found in a bibliography compiled by the U.S. Naval Engineering Experiment Station (Snapp & Dray, 1958). The references are identified as providing theoretical, experimental, or service information. Many such references are reviewed by Professor Fuller in Section 22 of his Survey (VII: 1958). See also Hersey & Snapp (1957) on bearing test machines and the treatise by Kamps & Perret (1957).

7. CLASSICAL HYDRODYNAMIC THEORY

Hydrodynamic theory was first applied to dynamically loaded bearings by Harrison (VI: 1919), and later by Orlov (1937) and by Swift (1937). Harrison modified Reynolds' equation by introducing a term for the velocity of the journal axis. He was able to calculate the motion of the axis both for a constant and moderately variable load. His solution led to orbital motion for a constant load. The solution for a variable load was accomplished by superposing a small fluctuating load w on the constant load W. It leads to a pair of simultaneous equations for the time rates of change of the eccentricity and the attitude angle. An exact solution is obtained when w is a simple periodic function of the time and remains fixed in direction.

Orlov, in an early paper (1934), pointed out the equivalence of a rotating and a constant load. Solutions for the squeeze-film under both a constant and a variable load are to be found in his book (1937). He concluded that the minimum film thickness under a load of variable magnitude, in the case of a revolving journal, is determined by the maximum value of the load—see Chapter I, Sections 26–27.

It remained for Professor Swift to work out the more complete solutions for special cases, and finally to derive a general equation for the pressure distribution (1937). Integration led to a pair of simultaneous equations for the rates of change of eccentricity, attitude angle, and other factors in terms of the Sommerfeld number, now a function of time. These equations are analogous to Harrison's, and reduce to Sommerfeld's in the static case. Appropriately interpreted, they include the squeeze-film and equivalent-speed relations. Swift's theory assumes freedom from end leakage and cavitation; hence, he points out, it must be considered an approximation.

He solved the problem of the nonrotating journal under a fluctuating load in some detail, both for the full and half-bearing. If the load per unit area, P, is expressed as a function of the time t and the journal is concentric at $t = 0$, the eccentricity ratio will be given by

$$\epsilon = k(1 + k^2)^{-1/2} \tag{8}$$

in which

$$k \equiv \frac{m^2}{6\pi Z} \int P \, dt\,. \tag{9}$$

Here m is the clearance ratio and Z the film viscosity as before. Equation (9) leads back to Eq. (6) when P is constant. If P is a periodic function $P_0 \sin \omega t$, the integral appearing in Eq. (9) may be replaced by P_0/ω, and the journal axis will oscillate with an eccentricity amplitude found by substituting the new value of k into Eq. (8). Note that k and ϵ approach zero as ω is indefinitely increased. Thus the minimum film thickness h_0 can be greatly increased by raising the frequency of the load impulses. Swift found that the optimum clearance ratio for a maximum h_0 would be 2.7 times the square root of $Z\omega/P_0$. He showed in his treatment of the rotating journal that it is simpler mathematically to calculate the load relation corresponding to a given motion of the journal axis than the other way around—a fact put to good use by Dick, Fränkel, and others.

Kutsayev (1943) calculated film thickness in aircraft engine bearings under a variable load with the aid of simplifying assumptions. Dick treats the case of the journal axis moving in an ellipse (1944). He finds that the load will be approximately sinusoidal, and drops to zero when the load frequency is half the speed of journal rotation. Fränkel considers, first, a circular motion of the shaft center (1944). This leads to the equivalent-speed theorem. He considers two cases in which the shaft center oscillates in a straight line. In one, the center of oscillation is at the bearing center; in the other, it is displaced to an eccentricity of 0.3. Polar diagrams of resultant load are plotted for a maximum eccentricity of 0.6. The two diagrams differ greatly, the first being symmetrical with reference to the load origin, while the second is offset and much smaller in area.

Burwell and Contemporaries. In a series of papers Professor Burwell begins with a general review of the dynamic load problem (1947), including Swift's theory, which he treats more fully than above. He then proceeds to an analytical solution for reciprocating loads in two cases—first the sinusoidal, and second the square-wave loading, at various ratios of load to journal speed. Although loads are given analytically, the journal-center paths must be found by trial-and-error. The calculated paths are shown graphically for both types of loading. Relative load capacities for frequency ratios n/N from 0 to 3 are compared in Table 1, where n is the load frequency in cycles per unit time, and N the journal speed. Values are rounded off from Burwell's table. Solutions are limited to a minimum eccentricity of 0.58. End leakage is considered negligible for comparative purposes. The inference is drawn that the more "peaked" the wave form, the greater the load capacity. Top row shows relative load capacities found by the equivalent-speed theorem.

Returning to the differential equations for eccentricity ϵ, and angle ϕ from load line to the point of minimum film thickness, Burwell shows how they may be applied to engine problems where the load cannot be expressed analytically, but must be read from a polar diagram. He takes for his first example the big-end connecting rod bearing of the diesel engine above, and for his second the master-rod bearing of a radial aircraft engine. Solutions were obtained by the M.I.T. differential analyzer. In one example the bearing diameter is 2.75 in., length 1.81 in. and clearance ratio $\frac{1}{1000}$. Maximum load on polar diagram is 9360 lb at a crankshaft speed of 2000 rpm. In his Paper II, corrections are made for end leakage and cavitation by the

TABLE 1. Relative Load Capacity
(From Burwell, 1947)

Type of Load	Frequency Ratio n/N				
	0	$\frac{1}{2}$	1	2	3
Constant rotating	1	0	1	3	5
Sinusoidal	1	0	1.7	4.7	7.9
Square-wave	1	0	1.3	3.3	5.2

Cameron & Wood method. A diagram is plotted showing path of crankpin center relative to the clearance circle and connecting rod axis, with crank angles marked on it. The eccentricity ratio fluctuates between 0.82 and 0.92, indicating oil film thicknesses from about $\frac{1}{4}$ to $\frac{1}{8}$ mil.

In Paper III Burwell applies dynamic load theory to Ocvirk's "short bearing," assuming, however, the real existence of Sommerfeld's negative pressures. This approach simplifies the mathematics and includes the L/D effect explicitly. It leads to results not greatly different from those derived for bearings of infinite length. The general case of a periodic load $P = P_0 p(\tau)$ is expressed by two simultaneous equations in ϕ and ϵ, the angle from load line to location of minimum film thickness, and eccentricity ratio. In a slightly changed notation these may be written

$$p(\tau) \sin \phi = \lambda d(\alpha - 2\beta - 2\phi)/d\tau \tag{10}$$

and

$$p(\tau) \cos \phi = 2d\lambda/d\tau\,, \tag{11}$$

where

$$\lambda \equiv 2\pi^2 S_0 \left(\frac{L}{D}\right)^2 \frac{n}{N} \cdot \frac{\epsilon}{(1 - \epsilon^2)^{3/2}}\,. \tag{12}$$

Here τ is the dimensionless time, $2\pi nt$, for a load frequency n; P_0 is the amplitude of P so that $p(\tau)$ gives the load ratio P/P_0. The angles α and β are the instantaneous shaft angle, and load angle, relative to some fixed line on the bearing. The Sommerfeld variable S_0 with subscript corresponds to the load amplitude P_0.

An analytical solution of Eqs. (10) and (11) is given for the square-wave load; a numerical solution for the sinusoidal load. Journal-center paths are shown graphically both for the rotating and non-rotating journal. Friction coefficients are calculated by the Muskat and Morgan formulas in terms of ϵ, ϕ, the Sommerfeld number, and clearance ratio. The treatment of the short bearing in Paper III runs parallel to that of the long bearing in Paper I. For this reason it throws new light on the concepts and methods described in Paper I and in Swift's paper, making them easier to follow.

Other studies appeared prior to the conclusion of the foregoing. Ott (1948) continued the investigation of infinite-length bearings started by Fränkel. A new feature was the exclusion of negative pressures by the Gümbel boundary condition, whereby the pressure remains ambient over an arc of 180 degrees. A differential equation is derived for the film pressure in terms of dimensionless variables. The first applications are to problems in which the motion of the journal axis is known, and the load variation must be calculated. Polar diagrams are determined for the load corresponding to ten different elliptical paths of the axis. Maximum eccentricity ratios are taken equal to 0.6 and 0.8, with ellipticities from zero to unity. The shapes and areas of the diagrams obtained are surprisingly diversified. In the second part of his treatise, Ott turns to the problem of calculating the path when the loading is known. He gives an exact solution for the effect of sinusoidal forces acting in a constant direction when the eccentricities are small, followed by an approximate solution for heavier loads producing greater eccentricities.

Publications by Professor Shaw helped to advance our understanding of dynamically loaded bearings (1949). An hydraulic analogy is offered whereby it is more easily seen that the load capacity of a journal bearing is proportional to the sum of the two surface velocities. Piston-pin lubrication is explained in terms of the load fluctuation and bearing oscillation. In their book, Shaw & Macks (V: 1949) show how any polar diagram for engine bearing loads may be generalized to apply to other bearings and other conditions with the aid of dimensional analysis.

Several investigations by Russian authors soon appeared. D'Iachov described load capacity failures (1949) and methods for calculating effective loads on engine bearings (1950). Fishman's theory (VII: 1950) was the first to include the effect of the inertia of the lubricant in dynamically loaded bearings. Solutions were obtained by iteration for motions under a constant load and a periodic load of fixed direction. Original studies were offered by Gut'iar and by Korovchinskii in 1953. Gut'iar criticizes Kutsayev for neglecting the inertia of reciprocating parts, and gives a new equation of the second order for sinusoidal loading. This can be integrated for moderate eccentricities to give ϵ and the attitude angle as simple functions of two dimensionless parameters. Korovchinskii reviews the work of D'Iachov and of Gut'iar with especial reference to the impulse, or time integral, of the applied load per cycle; and derives a solution for the bearing of infinite length under periodic loading.

Papers followed by Shawki, Constantinescu, and others. Shawki presents a formal solution based on exclusion of negative pressures (1956). Methods are given for calculating the extent of the positive film. General equations are derived for determining the applied load and the friction coefficient when the motion of the journal center is known, or the reverse. Friction is neglected in the cavitation area. Solutions are first obtained for a constant load and a rotating load of constant magnitude. In the static case, where Swift's formula reduces to that of Sommerfeld, Shawki's comes closer to Cameron and Wood's. In other cases, Shawki's theory agrees better with experiment than Swift's theory. Having established a measure of confidence in his theory by familiar problems, Shawki tabulates the numerical values of the principal functions needed in future applications.

Radial motions of the journal in a bearing under variable loading, and with variable speed, were investigated by Constantinescu (1956). Garski applied lubrication theory to the bearings of a large gas engine (1956). Dynamically loaded bearings are discussed by Tipei in Chapter VII of his book (VI: 1957).

8. HAHN'S SOLUTION

Hahn was the first to solve the dynamic load problem in detail for bearings of finite length (1957). Others had been content to apply end-leakage corrections derived for static loading or for the "short bearing." Hahn's treatment cannot be considered exact, since he excludes negative pressures by the Gümbel condition, making the film pressure p equal to zero at the points of maximum and minimum film thickness. His derivation starts from a form of Reynolds' equation in which V_1 and V_2 are the bearing and journal surface velocities, and the film thickness h is a function of the time t at any distance x along the bearing surface. The distance is measured in the direction of motion from a point vertically above the center of the bearing. The equation is put into dimensionless form by the substitutions $\phi = x/r$, where r is the radius of the journal; $\bar{z} = 2\, z/L$, where z is the axial coordinate from middle of bearing; $\beta = L/D$; $H = h/\Delta r$ (relative film thickness), Δr being the radial clearance; $\bar{\omega} = (V_1 + V_2)/r$; and $\Pi = pm^2/\mu\bar{\omega}$, in which m denotes the clearance ratio and μ the viscosity of the lubricant. The customary symbols L and m are used here in preference to the author's B and Ψ. Substitution of a new "dimensionless pressure" u for $\Pi H^{3/2}$ leads to the more convenient Reynolds' equation

$$\frac{\partial^2 u}{\partial \delta^2} + \frac{1}{\beta^2} \cdot \frac{\partial^2 u}{\partial \bar{z}^2} + au = b\,, \tag{13}$$

where δ represents $\phi - \gamma$, and γ is the value of ϕ at the minimum film thickness. In Eq. (13) a is a known function of δ and ϵ; and b is a known function of δ, ϵ, E, and G. The symbols E and G denote $(2/\bar{\omega})\, \partial\epsilon/\partial t$ and $(2/\bar{\omega})\, \partial\gamma/\partial t$.

The complete expressions for a and b are given in the paper, thus rendering Eq. (13) available for numerical integration. A method of differences was used similar to that of Sassenfeld and Walther. To illustrate the general solution, diagrams are shown for three special cases in which the film pressure p is plotted against the angular coordinate δ for bearings of $L/D = \frac{1}{2}$. The first case represents the circular motion of journal center with eccentricities from 0.2 to 0.8; the second a purely radial motion at eccentricities from 0.2 to 0.9. The third case represents both motions superposed. It is taken at an eccentricity of 0.6 with different combinations of E and G. These combinations are specified by a new parameter $q \equiv E/|G'|$, where $|G'|$ denotes the absolute value of $1 - G$, regardless of sign.

Load diagrams obtained by integrating the pressure distribution are drawn up for $L/D = \frac{1}{2}$. The first diagram shows the load characteristic $S_0|G'|$ as ordinate against the eccentricity ratio as abscissa. Here S_0 is a conventional form of the Sommerfeld reciprocal. The second diagram shows the load angle ϕ_P from location of minimum film thickness, plotted against ϵ as before. Both diagrams comprise a family of curves for constant values of q ranging from $+2$ to -2. We are told how the curves may be used in calculating the polar load diagram for any given path of the journal center. Special cases are worked out similar to those published by Ott for bearings of infinite length. Then, to compare theory with experiment, measurements were made on the gas pressures and main-bearing journal eccentricities in a diesel engine while running, the inertia forces being known. The agreement obtained was found to be within the tolerances of the measurements and calculations.

It is next shown in Hahn's paper how the journal displacement corresponding to a given polar load diagram may be calculated for a finite L/D ratio. After outlining the general solution, it is applied to sinusoidal loading in a fixed direction for $L/D = \frac{1}{2}$. In one case the load frequency is the same as the shaft speed. Journal paths are shown for a tenfold range of load amplitudes. In another case the load amplitude is held constant while the speed ratio n/N is stepped up from zero to unity. The journal paths and the phase-angle curves against n/N are very similar to those found by Shawki in his experiments for which $L/D = 1.5$. In conclusion, Hahn points to the assumptions made and to some of the factors requiring further investigation. He notes that the relations found provide means for calculating the spring and damping constants needed in problems of oscillating bearings and stability.

Later Theory. Kamps & Perret (1957) and Karatyshkin (1958) reviewed the state of the art. Karatyshkin offered a new theory for journal motions under a variable load. Holland (1959) integrated the Reynolds and Stefan film pressures separately and combined the results vectorially. Sternlicht recommended that pressure-viscosity be considered (VII: 1959). Zommer calculated the effects of a fluctuating load on a bearing with 60-deg side reliefs (1959). Hays calculated film thickness under sinusoidal load in bearings of finite length with a nonrotating journal (1961). Horsenell and co-author predicted journal bearing characteristics under static and dynamic loads (1963). Gupta & Phelan reported on load capacity of short bearings with oscillating speed (1964). Radermacher (1964) compared Holland's solution with the double Fourier series in a thesis by Someya. Booker (1965) introduced a "mobility method" for the solution of dynamically loaded bearings. The mobility vector is defined by a function of the eccentricity vector and L/D ratio. It is then applied to numerical examples in which the load vector is known as a function of time, and the path of the journal center is to be determined. See Hahn (1966).

9. EXPERIMENTAL INVESTIGATIONS

Journal bearing tests with dynamic loading can be described in three groups: (1) practical operating tests, (2) experiments on research machines, including motored engines, and (3) endurance tests.

Practical Operating Tests. Kollmann & Hockel (1953) measured oil-film thickness in the main bearings of two six-cylinder diesel engines. Bearings were 220 mm diameter, $L/D = \frac{1}{2}$. Three bearings were tested in each engine, using seventeen capacitance probes. Minimum film thickness decreased with higher speeds and temperatures. See Kollmann's discussion of Hersey & Snapp (page 1257 of 1957). Analyzing these data, Sahin (1957) saw no correlation with static calculations, but did find agreement with the equivalent-speed concept. Later tests on a single-cylinder engine confirmed Hahn's theory.

Tests on engines under their own power were also reported by Klusener (1957), Holland (1959), and Sims (1961). Klusener determined the journal-center path in two main bearings of a four-stroke, four-cylinder engine by an inductance method. Bearing diameters were 180 mm. He found eccentricities as high as 0.95 in one bearing. Using a small two-stroke diesel, Klusener determined the percentage oil flow to the main and crankpin bearings. He measured temperatures around the circumference of a main bearing for inlet pressures from $\frac{1}{2}$ to 4 atm. At an inlet temperature of 72 C the maximum circumferential temperature dropped from 92 to 76 C with increasing inlet pressure. Holland measured film thickness while varying the load, speed, clearance, and inlet conditions. He then compared the data with calculations. Sims measured film thickness in a V-8 main bearing.

Oil flow through the connecting rod and other bearings of an operating engine was investigated by Brooks & Sparrow (1927), Spiers (1941), and Withers & Wachal (1950). The effects of clearance, speed, and inlet pressure and temperature in raising the flow rate are clearly shown. See, also, the papers by Pigott (1941), Tichvinsky (1945), Bidwell (1954), and Klusener (1957).

Experiments on Research Machines. Oscillating bearings were investigated by Stanton as early as 1923; see Fogg and co-author (1937, 1938). Friction coefficients diminished with increasing frequency of oscillation. The Vieweg brothers showed that hydrodynamic conditions could be assured in oscillating bearings by imparting a continuous rotation to one member (1927). A long investigation was completed by Barwell and co-authors simulating piston-pin oscillation (1955). Friction, temperature, and wear rate were observed under a variety of conditions. A particular design of grooving was recommended. Friction coefficients were plotted against the product of cyclic speed and viscosity divided by load, showing more favorable conditions at the higher values. Cole and Hughes, however, using a transparent bearing (1957), report no hydrodynamic film visible with an oscillating shaft under constant load. A marked interest continues in oscillating bearings—see Blount (1957), Glaeser (1959), Brown (1963), and co-authors.

Investigations have been conducted on numerous test machines designed to simulate the centrifugal and reciprocating inertia loads, gas load, edge-loading, and static load either separately or in combination (Hersey & Snapp, 1957). These types of loading may be designated C, R, G, E, and S. Thirteen published investigations are described in the reference cited and discussions thereof. To these should be added the study by Manson & Morgan of a shaft supported on multiple bearings (1947), and tests by D'Iachov (1950).

Recent experiments on dynamically loaded bearings have been reported by Shawki (1956), Pigott & Walsh (1957), Karatyshkin (1958), and Özdas (1958); also by Snapp (1958), Zommer (1959), Phelan (1961), Carl (1964), and Radermacher (1964). Snapp reports an experimental demonstration of the gas-load unit in a proposed diesel bearing research machine of the multiple-load hydraulic type, based essentially on designs by P. G. Exline. Operation of the unit was successful at loads up to 153,000 lb (over 10,000 psi on a 5 in.-diameter bearing, $L/D = 0.6$); and at pulse frequencies to 500 cycles/min. Since these limits are imposed only by the driving motor, the feasibility of the design is confirmed. Zommer investigated journal positions under alternating loads in the bearing with 60-deg side reliefs. Phelan did the same for a full bearing with nonrotating journal. Carl determined pressure distribution and journal displacements under sinusoidal load, finding results in agreement with Hahn's theory. Radermacher observed the journal path for various loading cycles in a hydraulic bearing test machine. Results were compared with both methods of calculation cited above and with test data from operation of a single-cylinder diesel. Consistent results were obtained except for differences attributed to elastic deformation and viscosity variation.

Endurance Tests. Among the best-known machines for endurance testing of bearings are the Underwood reciprocating and Ryder centrifugal types. The Cornelius and Barten machine, CRGE type, is of historical interest. These machines and others are described in the review by Hersey & Snapp, and are listed in Table 1 thereof. See also a paper by B. C. Kroon (1947). Later additions are the Navy's diesel bearing test machine of CRG type developed by R. B. Snapp and others (1957, 1958), a machine for GE loading described by A. E. Russell (1958), and the Holfelder alternating load machine (1959).

The Underwood type mentioned is often called the "crossed-rod" machine. It employs two connecting rods at right angles with big ends mounted on the test journal. Small ends are pivoted at fixed points on the frame of the machine. Test shaft is loaded by eccentric weights so that the net effect is a reciprocating load on each test bearing. These are usually of $2\frac{1}{2}$ in. diameter. The Underwood machine is used extensively in the automotive and oil industries. Mr. Ryder's machine applies a constant rotating load to the test bearing. Short endurance runs can be made by observing the sudden rise of temperature due to incipient seizure. The Ryder machine gave timely aid in perfecting the master-rod bearing of radial aircraft engines. The Cornelius and Barten machine is one of ingenious design illustrated diagrammatically in Fig. 5, page 1259, of the review cited. It served to stimulate a wide interest in the design of multiple-load machines. Professor Kroon's tests revealed "cavitation erosion" at high speeds.

The Navy machine described by Snapp is a motored six-cylinder diesel with gas load produced by closing valves and compressing air. The load is intensified by adjusting initial pressure and decreasing the clearance volume, so that the bearings can be subjected to an accelerated life test. Bearing materials are judged primarily by inspection and wear measurements, aided by a novel rating system. Development was initiated by W. F. Joachim, Superintendent, Internal Combustion Engines Laboratory, assisted by P. G. Kestler and the writer. The Russell machine simulates impulsive gas loading by hydraulic means, and provides edge-loading by misalignment control. A misalignment of 1 mil/in. at a peak load of 3000 psi led to pitting and extensive fatigue cracks. Full-sized diesel bearings were tested in the Holfelder machine.

No sharp line need be drawn between endurance and research machines. An endurance machine can be used for research if instrumented for the purpose. Thus an investigation of film-thickness indicators for dynamically loaded bearings was conducted on the Navy machine. And a research machine can be used for endurance tests if it is ruggedly built, provided the design is allowed to stay put, so that the tests will be comparable. A given piece of apparatus might be called an "endurance machine" if the engineers in charge know nothing about bearing clearances and have no curiosity as to the mechanical action of the oil film. The same apparatus would be called a "research machine" if an oscilloscope is found in the wreckage.

Motor-driven machines are not essential for endurance tests, since the tests can be conducted in engines operating under their own power. These are often called "service" or "suitability" tests, and seem to be more convincing if run in the open so as to create an unbearable noise. They prove the suitability, or otherwise, of a bearing design, material, or lubricant for a particular type of engine, but the results are of limited application. Sooner or later the authorities realize that an engine bolted to a semi-infinite concrete foundation may not perform the same as it would in the hull of a ship weaving its way through heavy seas.

Other Interesting Results. A machine of the centrifugal or "C-type" was set up by E. J. Lourie in his thesis experiments (1932). Bearings were of automobile type. In addition to a study of heat effects, Lourie discovered that the McKee friction law for static loading applies equally to the C-load. This was confirmed within 6 per cent above $ZN/P = 32$ cu. Another early experiment was that of C. C. Davenport, who worked with a simulated gas load (1933). The test bearing was of 2-in. diameter by 2-in. length, with $m = 4.5$ per thousand. A static load could be superposed. Load capacity was determined by the flashing on of a light bulb in series with the oil film. Davenport found a substantial increase in load capacity upon raising the load-impulse frequency to 600 cycles/min.

Later R. M. Phelan took for a thesis (1950) the development of a research machine with reciprocating load. He devised a photoelectric method of film-thickness measurement. This work, too, seems to have escaped publication. The machine was used again by W. L. Smith in a thesis under Professor Phelan (1955). Smith describes experiments on film thickness and oil flow using a stationary journal at load frequencies up to 600 cycles/min. Maximum eccentricity ratios are plotted against a "dynamic Sommerfeld number" for constant inlet pressures. These eccentricities lie definitely higher than predicted, even by the "short-bearing" theory, as noted by Phelan (1961).

A manufacturer of diesel locomotives reassembled one of their four-cycle engines in the form of a C-machine very much like Lourie's, so that the center main bearing could be tested under a rotating load. Performance was judged by fatigue, and by signs of metallic contact. The tests led to an improved design. A builder of diesel and gas engines set up a large-scale replica of the Underwood machine, big enough for testing 6-in. diameter crankpin bearings. On a later visit we were told that it had shaken itself to pieces in the trial run. Evidently magnified models require something more than *geometrical* similarity.

During the visit to a factory pioneering in metal stamping we were shown instrumentation for recording instantaneous rod loads. Impulsive loads up to 3000 or 4000 psi are transmitted to the crankpin bearings at 150 to 600 strokes per minute. In practice, operations are limited by the frictional heat generated in these bearings. At the high loads and low speeds encountered, a squeeze-film cannot be maintained as thick as might be wished. It was thought that forced lubrication might prove to be the remedy.

10. SUMMARY

This chapter may be summarized in two parts: (1) piston and ring lubrication, and (2) dynamic loading. References are grouped accordingly.

1. We have seen that piston ring friction is largely hydrodynamic, although traditionally assumed to be of the boundary or imperfect-lubrication type. This conclusion is supported by calculation and experiment. Boundary friction occurs chiefly when starting down from top-dead-center.

2. Dynamically loaded bearings operate under the combined action of the wedge-film and the squeeze-film. Under favorable conditions they will have thicker films, and offer greater load capacity than a statically loaded bearing. However, under a constant load rotating at half the shaft speed the load capacity of such a bearing would vanish. Although the problem is complicated, hydrodynamic theory indicates at what point in the cycle the minimum film thickness will be least.

Theoretical expectations have been reasonably well confirmed by diesel engine experiments.

More research seems to be needed on the lubrication of reciprocating equipment used in manufacturing operations, including both sliding ways and dynamically loaded bearings.

REFERENCES

Part 1. Piston and Ring Lubrication

Arnold, W. C. & Eser, W. J. (1964), "Materials and Designs for Non-Lube Applications." ASME 64-OGP-3; 8 pp. *ME* **86,** 73 (June).

Barros, A. de F. & Dyson, A. (1960), "Piston Ring Friction—Rig Measurements with Low Viscosity Oils." *JIP* **46,** No. 433, 1–18.

Beaubien, S. J. & Cattaneo, A. G. (1946), "Piston Lubrication Phenomena in a Motored Glass Cylinder Engine." *JSAE* **54,** 60–7 (Oct.).

Bogdanov, E. A. (1957), Friction of Piston Rings (Russian). *Avtomob i Trakt. Prom.* No. 5, 39–71; No. 6, 11–15.

Boyd, John (1964), "The Influence of Fluid Forces on the Sticking and the Lateral Vibrations of Pistons." *JAM, T* **86,** 397–401.

Calow, J. R. B. & Epton, S. R. (1961), "Trends in Piston-Ring Wear in Automotive Gasoline Engines." *Proc. IME* **175,** No. 10, 506–12, 546–76.

Castleman, R. A., Jr. (1936), "A Hydrodynamical Theory of Piston-Ring Lubrication." *Physics* **7,** 364–7. (a) *Bull. Am. Phys. Soc.* **11,** 10 (Jan.); 35–6 (Apr.).

Cavileer, A. C. (1963), "Investigation of Piston Ring Wear Using Radioactive Tracer Techniques." *SAE Paper* S 345; 10 pp. (a) *JSAE* **71,** 104 (Sept.)

Cavileer, A. C. (1963), "Piston Design Improvement through Research Investigation." *SAE Paper* 636 B; 8 pp. *JSAE* **71,** 96–7 (May).

Cree, J. C. G. & Thiery, J. (1961), "An Investigation into the Factors which Affect Piston-Ring Wear under Starting Conditions." *Proc. IME Auto. Div.* No. 8, 298–316.

Dykes, P. de K. (1957), "An Investigation into the Mechanism of Oil Loss Past Pistons." *Proc. IME* **171,** No. 11, 413–26; 444–60.

Eilon, Samuel & Saunders, O. A. (1957), "A Study of Piston-Ring Lubrication." *Proc. IME* **171,** No. 11, 427–62.

Eweis, M. (1935), "Reibungs und Undichtigkeitsverluste an Kolbenringen." *VDI Forschungsheft* 371; 23 pp.

Forbes, J. E. & Taylor, E. S. (1943), "A Method for Studying Piston Friction." *NACA* W-37; 10 pp. + 23 figs. (Mar.).

Furuhama, Stoichi (1959), "A Dynamic Theory of Piston-Ring Lubrication." First report, "Calculation." *Bull. JSME* **2,** No. 7, 423–8. Second Report, "Experiment." **3,** No. 10, 291–7 (1960).

Hartung, H. A. & Savin, J. W. (1962), "Lubrication of Small Two-Cycle Engines." *ME* **84,** 52–5 (July); 92–3 (Dec.). (a) *LE* **18,** 78–82 (Feb.).

Hawkes, C. J. & Hardy, G. F. (1936), "Friction of Piston Rings." *Trans. NE Coast Inst. Engrs. & Shipbuilders* **52,** 143–78, D 49–58.

Hersey, M. D. (1944), "Bibliography on Piston Ring Lubrication." *NACA TN* 956; 34 pp.

Hunsaker, J. C. & Rightmire, B. G. (1947), "Lubrication of Piston Rings and Cylinders," Pages 338–9 in *Engineering Applications of Fluid Mechanics,* McGraw-Hill, New York.

Kettleborough, C. F. (1961), "The Effect of Pressure on Viscosity in Hydraulic Lock." *Int. J. Mech. Sci.* **3,** 137–42.

Kingsbury, Albert (1908), "Elimination of Friction in Hydraulic Testing Devices." *Eng. News.* **60,** No. 6, 154.

Lewicke, W. (1957), "Hydrodynamic Lubrication of Piston Rings and—Commutator Brushes." *The Engineer* **203,** 84–6, 122–4.

Manhajm, J. & Sweeney, D. C. (1955), "An Investigation of Hydraulic Lock." *Proc. IME* **169,** No. 42, 865–80.

Mannam, J. (1959), "Further Aspects of Hydraulic Lock." *Proc. IME* **173,** 699–716.

Notaro, J. (1966), "Non-Lubricated Piston Rings for High Pressure Applications." *LE* **22,** 104–8.

Nutt, H. V., Landen, E. W., & Edgar, J. A. (1955), "Diesels Wear Less with Oils of Increased High-Temperature Viscosity, Decreased Volatility, and Maximum Thermal Stability." *JSAE* **63,** 64–7 (Apr.). See (a) 694–703.

Pike, W. C. & Spillman, D. T. (1964), "The Use of a Motored Engine to Study Piston-Ring Wear and Engine Friction." Paper 11, *LWG;* 10 pp.

Poppinga, Reemt (1942), *Verschleiss und Schmierung, insbesondere von Kolbenringen und Zylindern.* VDI Verlag, Berlin; 183 pp. (a) Transl. by Edmund Kurz (D. D. Fuller, Techn. Adviser), ASLE, Chicago; 201 pp.

Raimondi, A. A. & Boyd, J. (1964), "Fluid Centering of Pistons." *JAM,* **31;** *T* **86,** 390–6.

Reynolds, R. W. (1957), "Chromium Plating Development at the U. S. Naval Engineering Experiment Station." *ASME* 57-OGP-6; 9 pp. (a) *ME* **79,** 677 (July).

Rogowski, A. R. (1961), "Measuring Instantaneous Friction of Piston Rings in a Firing Engine." *SAE* 379 F; 4 pp. *JSAE* **69,** 80–3 (Aug.). (a) *Sci. Lubn.* **14,** 29, 32 (Jan. 1962).

Shaw, M. C. & Nussdorfer, T. J. (1946), "A Visual Study of Cylinder Lubrication." *SAE* June 2–7; 8 pp. + figs.

Society of Automotive Engineers (1943), "SAE Standardizing Piston and Ring Nomenclature." *JSAE* **51,** 39 (Nov.) (a) *SAE Handbook,* 1944; 1957.

Stanton, T. E. (1925), "The Friction of Pistons and Piston Rings." *The Engineer* **139,** 70, 72. (a) *Aero. Res. Com.* R & M 931; 4 pp. + figs.

Sulzer Bros. (1938), "Oil-Free Compressed Air." *JFI* **226,** 243–4. (a) Walti, O., *Sulzer Tech. Rev.* No. 2 (1945). (b) *ME* **68,** 563–4 (1946).

Summers-Smith, D. (1963), "Performance of Unlubricated Piston Rings." Paper 24, *LWG* (May); 8 pp.

Taber, R. D. & Robbins, F. A. (1957), "Teflon-Based Piston Rings for Non-lubricated Applications." *ME* **79,** 838–41.

Taylor, C. F. (1937), "Problems in the Field of Internal Combustion Engine Lubrication." *GDLL* **1,** 577–82.

Taylor, M. P. (1936), "The Effect of Gas Pressure on Piston Friction." *TSAE* **31,** 200–5 (May).

Tischbein, H. W. (1939), "Reibung an Kolbenringen." *Kraftstoff* **15,** 83–7 (Dec.); **16,** 6–8, 39–42, 71–5 (Jan. 1940). (a) *NACA* Transl. by J. Vanier, *TM* 1069; 31 pp. + figs.

Volarovich, M. P. (1959), "Investigation of Piston-Cylinder and Shaft-Bearing Friction at Low Temperatures." *Wear* **2,** No. 3, 203–16.

Watson, C. E., Hanly, F. J., & Burchell, R. W. (1955), "Abrasive Wear of Piston Rings." *TSAE* **63,** 717–28.

Part 2. Dynamic Loading

Barwell, F. T., Milne, A. A., & Webber, J. E. (1955), "Some Experiments on Oscillating Bearings." *Trans. Inst. Eng. Shipbuilders in Scotland* **98,** 267–326.

Bidwell, J. B. (1954), "Engine Bearing Design Today." *LE* **10,** 272–81.

Blount, E. A. & de Guerin, D. (1957), "Importance of Surface Finish, Loaded Area Conformity, and Operating Temperature in Small End Plain Bearings for High Duty Two-Stroke Engines." Paper 80, *CLW,* 224–9; 202; Plate 14.

Booker, J. F. (1965), "Dynamically Loaded Journal Bearings Mobility Method of Solution." *JBE, T* **87,** 537–46.

Brooks, D. B. & Sparrow, S. W. (1927), "Oil Flow Through Crankshaft and Connecting Rod Bearings." *JSAE* **21,** 127–34.

Brown, R. D., Burton, R. A., & Ku, P. M. (1963), "Evaluation of Oscillating Bearings for High Temperature and High-Vacuum Operation." *TASLE* **6,** 12–9.

Burwell, J. T. (1947), "The Calculated Performance of Dynamically Loaded Sleeve Bearings." I, *JAM* **14,** *T* **69,** A 231–45; II, **16, 71,** 358–60 (1949); III, **18, 73,** 393–404 (1951); **19, 74,** 239–42 (1952). (a) *LE* **4,** 67–71, 74 (1948).

Carl, T. E. (1964), "An Experimental Investigation of a Cylindrical Journal Bearing under Constant and Sinusoidal Loading." Paper 19, *LWG;* 22 pp. (a) *Konstruktion* **15** (1963), No. 6, 209–17.

Cole, J. A. & Hughes, C. J. (1957), "Visual Study of Film Extent in Dynamically Loaded Complete Journal Bearings." Paper 87, *CLW,* 147–50; 12, 750 Plates 18, 19.

Constantinescu, V. N. (1956), Calculating the radial motion of a journal under variable load and speed (Rumanian). *Stiinte Teknice, Com. Acad. RPR* **6,** No. 11. (a) *Studii si Cercetari de Mecanica Aplicata, Acad. RPR* **8,** 789–804 (1957).

Davenport, C. C. (1933), *Breakdown Pressures of the Oil Film in a Bearing under Steady and Intermittent Loading*. Thesis, Pennsylvania State College, 1933; 44 pp.

D'Iachov, A. K. (1949), A Study of Dynamically Loaded Bearings (Russian). *TIM* **4**, 3–114.

D'Iachov, A. K. (1950), The Design of Bearings for Piston Engines. Application of the Theory of Effective Loads (Russian). *Izv. AN SSSR, Odtelenie Tekhnicheskikh Nauk,* No. 11, 1615–44. (a) *Eng. Digest* **12,** 119–22, 163–4 (1951).

Dick, J. (1944), "Alternating Loads on Sleeve Bearings." *Phil. Mag.* (7) **35,** 841–8.

Fogg, A. & Jakeman, C. (1938), *Friction in Oscillating Bearings*. Lubn. Res. Tech. Paper No. 3, DSIR, London; 28 pp. (a) *Eng.* **116,** 12–4. (b) *GDLL* **2,** 302–7; 1937.

Fränkel, A. (1944), *Berechnung von zylindrischen Gleitlagern*. Mitt. Inst. f. Thermodynamik u. Verbrennungsmotorenbau, ETH Zürich, No. 4; 134 pp.

Garski, W. (1956), "Lagerfragen an Grossgasmaschinen als Beispiel für die praktische Anwendung der Schmierungstheorie." *Stahl u. Eisen* **76,** 18–26 (Jan. 12).

Glaeser, W. A. & Allen, C. M. (1959), "A Study of Design Criteria for Oscillating Plain Bearings." *TASLE* **2,** 32–8.

Gupta, B. K. & Phelan, R. M. (1964), "The Load Capacity of Short Journal Bearings with Oscillating Effective Speed." *JBE, T* **86,** 348–54.

Gut'iar, E. M. (1953), On the Calculation of Dynamically Loaded Bearings (Russian). *Izvestiia AN SSSR, OTN* No. 5; 762–6.

Hahn, H. W. (1957), "Dynamically Loaded Journal Bearings of Finite Length." Paper 55, *CLW,* 100–10; 12, 749–50.

Hahn, H. W. (1966), "New Calculation Methods for Engine Bearings." *SAE* Paper No. 660033; 21 pp. (Jan.).

Hays, D. F. (1961), "Squeeze Films: A Finite Journal Bearing with a Fluctuating Load." *JBE, T* **83,** 579–88.

Hersey, M. D. & Snapp, R. B. (1957), "Testing Dynamically Loaded Bearings—I. A Short History of Bearing Test Machines." *T* **79,** 1247–59.

Holfelder, O. (1959), "Wechsellast-Lagerprüfmaschine und Erprobung hochbelasteter Gleitlager in Originalgrösse." *MTZ* **20,** 173–7.

Holland, J. (1959), *Beitrag zur Erfassung der Schmierverhältnisse im Verbrennungskraftmaschinen*. VDI Forschungsheft **25,** No. 475; 32 pp.

Horsenell, R. & McCallion, H. (1963), "Prediction of Some Journal-Bearing Characteristics under Static and Dynamic Loading." Paper 12, *LWG,* 123–30.

Kamps, R. & Perret, H. (1957), *Lager und Schmiertechnik insbesondere von Verbrennungsmotoren*. VDI-Verlag, Düsseldorf, 364 pp.

Karatyshkin, S. G. (1958), On the Theory of the Oil Film in a Dynamically Loaded Bearing (Russian). *TIM* **12,** 163–81. (a) Transl., *FWM* **12,** 155–70 (1960), ASME, New York.

Klusener, Otto (1957), "Beitrag zur Lagerschmierung bei nicht stationäres Belastung." *MTZ* **18,** 89–93. (a) *Sci. Lubn.* **9,** 28–30.

Kollmann, Karl & Hockel, H. L. (1953), "Ermittlung der Dicke des Schmierfilms in den Grundlagern eines stationären Diesel-Motors." *MTZ* **14,** 133–7.

Korovchinskii, M. V. (1953), On the Theory of Dynamically Loaded Journal Bearings (Russian). *Izvestia AN SSSR, OTN,* No. 5; 767–76.

Kroon, B. C. (1947), Some Conditions Leading to Bearing Failure in High-Speed Engines (Dutch). *De Ingenieur* **59,** pt. 1, No. 3, Jan. 17; Mk 1–8.

Kutsayev, S. N. (1943), *Determination of Thickness of Lubricating Film in Low-Clearance, Long Bearings under Known Variable Loads* (Russian). Oborongiz, Trudy TSIAM, No. 59. (a) Doctoral Thesis, 1950.

Lourie, E. J. (1932), *Study of Frictional Heat in Automotive Bearings.* Thesis, MIT; 103 pp.

Manson, S. S. & Morgan, W. C. (1947), *Distribution of Bearing Reactions* . . . . NACA TN 1280; 16 pp. +.

Orlov, P. I. (1934), Application of Hydrodynamic Theory to the Design of Aviation Engine Bearings (Russian). *Tekhnika Vozdushnogo Flota,* Moscow, No. 5; 39–66.

Orlov, P. I. (1937), *Lubrication of Light Internal Combustion Engines* (ed. by N. R. Brilling). Glavnaia Redaktsiia Mashinostroitel noi i Avtobraktornoi Literatury, Leningrad; 462 pp. (a) *JAM* **6,** *T* **61,** A 95 (1939).

Ott, H. H. (1948), *Zylindrische Gleitlager bei instantärer Belastung.* Mitt. Inst. f. Thermodynamik u. Verbrennungsmotorenbau, ETH Zürich, No. 7, Gebr. Leeman & Co.; 103 pp.

Özdaş, M. N. (1958), "The Behavior of the Lubricating Film and Side Leakage in Dynamically Loaded Bearings." *T* **80,** 826–32. (a) *Ninth ICAM,* Brussels, 1956; *Actes* **4,** 259–71 (1957).

Phelan, R. M. (1950), *The Design and Development of a Machine for the Experimental Investigation of Dynamically Loaded Sleeve Bearings.* Thesis, Cornell, University, Ithaca; 44 pp.

Phelan, R. M. (1961), *Non-Rotating Journal Bearings under Sinusoidal Loads.* ASME 61-Lub S-6; 16 pp. *ME* **83,** 76 (Aug.).

Pigott, R. J. S. (1941), "Engine Design Versus Engine Lubrication." *TSAE* **48,** (May), 165–76.

Pigott, R. J. S. & Walsh, B. R. (1957), "A Universal Bearing Tester." *T* **79,** 1267–74.

Radermacher, Karlheinz (1964), "Experimental Investigation into Cylindrical Plain Bearings under Loads Varying in Magnitude and Direction." Paper 23, *LWG;* 30 pp. (a) *MTZ* **24** (1963) No. 12, 427–38.

Root, R. E. (1932), *Dynamics of Engine and Shaft.* Wiley, New York; 184 pp.

Russell, A. E. (1958), "Some Bearing Tests Made with a Machine for Producing Impulsive Loading." *Proc. IME* **172,** 1047–64.

Sahin, A. M. (1957), "Ermittlung der Schmierfilmdicke in der Grundlagern eines Verbrennungsmotor" *MTZ* **18** (Feb.), 39–43. (a) *Diss.*, TH München, Nov. 1954; 46 pp.

Shaw, M. C. (1949), "Dynamically Loaded Bearings." *LE* **5,** 218–25.

Shawki, G. S. A. (1956), "Journal Bearing Performance for Combinations of Steady, Fundamental, and Law-Amplitude Harmonic Components of Load." *T* **78,** 449–55. (a) *Proc. IME* **171,** 795–804 (1957).

Shawki, G. S. A. (1956), "Analytical Study of Journal Bearing Performance under Variable Loads." *T* **78,** 457–64.

Simons, E. M. (1950), "The Hydrodynamic Lubrication of Cyclically Loaded Bearings." *T* **72,** 805–16.

Sims, W. D. (1961), "Measuring the Oil Film Thickness in Engine Bearings." *LE* **17,** 123–6.

Smith, W. L. (1955), *Experimental Investigation of Dynamically Loaded Sleeve Bearings with No Shaft Rotation.* Thesis, Cornell University, Ithaca, 1955; 82 pp.

Snapp, R. B. (1958), *Summary Report on the EES Diesel Bearing Test Machines.* RD Rep. 520009 C, USN Eng. Expt. Sta., Annapolis; 79 pp. +.

Snapp, R. B. (1958), *Development of a Hydraulic Pulse Loading Unit for a Diesel Bearing Research Machine.* RD Rep. 520010 A, USN Eng. Expt. Sta., Annapolis; 33 pp. +.

Snapp, R. B. & Dray, J. F. (1958), *Bibliography on Dynamically Loaded Journal Bearings.* RD Rep. 090009 D, USN Eng. Expt. Sta., Annapolis; 58 pp.

Snapp, R. B. & Hersey, M. D. (1957), "Testing Dynamically Loaded Bearings. II. A Diesel-Engine Bearing Test Machine." *T* **79,** 1260–6.

Spiers, J. (1941), "Oil Flow through Engine Bearings." *J. Inst. Auto Engrs.* **9,** 7–34.

Stone, J. M. & Underwood, A. F. (1947), "Load Carrying Capacity of Journal Bearings." *SAE Quarterly Trans.* **1** (Jan.) 56–70.

Swift, H. W. (1937), "Fluctuating Loads in Sleeve Bearings." *J. ICE* No. 4, 161–95.

Tichvinsky, L. M. (1945), "Diesel Engine Bearings—Discussion of Failure and Progressive Inspection Methods." *ME* **67,** 297–308; **68,** 71–3 (1946).

Tichvinsky, L. M. (1948), "Heavy Duty Diesel Engine Bearings." *LE* **4,** 170–4, 190.

Vaughan, P. S. (1963), "Factors Influencing Bearing Performance in High-Output Diesel Engines." *ASME* 63-WA-296; 9 pp.

Vieweg, R. & Vieweg, V. (1927), "Über die Bildung des Ölfilms in Schwinglagern." *Maschinenbau* **6,** 234–5.

Withers, J. G. & Wachal, A. L. (1950), "Flow of Lubricating Oil from Big-End Bearing." *Eng.* **169,** 247–9; 291–2.

Zommer, E. F. (1959), Investigation of the Position of the Journal in the 120 Deg. Bearing with Fluid Friction for a Load with Constant and Changing Direction (Russian). *TIM* **13,** 136–88. (a) Transl., *FWM* **13,** 132–82 (1961); ASME, New York.

chapter IX Gas-Lubricated Bearings

Gas-lubricated bearings are governed by the same basic principles as liquid-film bearings, but they differ in performance owing to the differences in the physical properties of the lubricant. Liquid lubricants are still required for the heavier loads per unit area. Gaseous lubricants offer advantages for operation under light loads, at high speeds, at high temperatures, and in special environments.

1. HISTORY

Hirn spoke of air as a possible lubricant (I: 1854). Stefan showed that when two flat surfaces like gage blocks are pressed together, a viscous film of air remains between them sufficient to explain the "apparent adhesion" (II: 1874). Klein and Sommerfeld (1903) thought a better understanding of dry friction might be attained by allowing for the lubricating effect of air between the rubbing surfaces. Charron came to the same conclusion after comparing the results of dry friction experiments conducted in air and in a vacuum (1910).

Albert Kingsbury found that a close-fitting horizontal piston was supported by an air film without metallic contact except when at rest (1896; VI: 1950). Pursuing this discovery, he built the first gas-lubricated bearing, a journal bearing of 6-in. diameter (1897; VII: 1932). It was demonstrated at a Navy Department conference in 1896. Professor Kingsbury tested this bearing both with air and with hydrogen as lubricants. The data indicated the practicability of gas-lubricated bearings and served as one of the first confirmations of Reynolds' theory. De Ferranti patented air-lubricated bearings for textile machinery (1909). In the same year Sawtelle described a traverse spindle grinder with air-lubricated bearings used as a lathe attachment. The writer saw this type in production at the Pratt & Whitney plant, West Hartford, years ago. A red tag attached before shipment assured us that the grinder had been tested "and runs cool at 10,000 rpm without any lubricant in the bearings."

The first theoretical study was that of W. J. Harrison (V: 1913), who revealed the effect of compressibility by comparing Kingsbury's data with hydrodynamic theory. An air-lubricated thrust bearing was set up by W. Stone in 1921 at the suggestion of A. G. M. Michell. It comprised a glass cellar supported on quartz shoes so as to permit film-thickness measurements by interference fringes. Stone's observations confirmed hydrodynamic theory and brought to light certain facts regarding instability not then understood. Kingsbury's air-lubricated thrust bearing was described in 1922—see Fig. 6 of the catalog (III: 1922); Fig. 4 in Norton's book (II: 1942), and Fig. 119 in Fuller's (I: 1956). It is a portable, three-shoe model arranged to close a flashlight circuit on slowing down to the point of metallic contact. This model has been widely used in lectures on lubrication.

The bearings cited above are all of the "self-acting," aerodynamic or hydrodynamic type. The earliest example of an "externally-pressurized," aerostatic or hydrostatic type known to us is that of George Westinghouse (1904). Here, the rotor of a vertical steam turbine is supported on an air cushion. A device using steam pressure for the same purpose was described by Barbezat (1912). Air-lubricated thrust and journal bearings of the pressurized type were patented by W. G. Abbott (1916). Extremely high speeds were reached by Henriot and Hugenard in 1925 by means of a conical rotor driven and supported by air. Modifications were applied by Lawrence, Beams, and co-authors in their ultracentrifuge designs (1928, 1937). A working model was shown in the lectures at Brown University in 1934. In later developments, before adopting magnetic controls, speeds beyond 1 million rpm were reached by Professor Beams using hydrogen to spin the rotor. See early thrust designs by Mueller (1951).

A helium liquefier with gas-libricated piston was described by Kapitza (1934); a spindle bearing by Wahlgren (1938).

2. PRINCIPLES OF SELF-ACTING AND EXTERNALLY PRESSURIZED BEARINGS

Self-acting bearings in a steady state operate on the principle of the "converging wedge"—the lubricating film must be thicker on the ingoing than on the outgoing side. As in liquid-lubricated bearings, the coefficient of friction, f and the eccentricity ratio ϵ or film thickness ratio h_0/D depend on the parameter ZN/P, among others; and are independent of absolute size in geometrically similar bearings. The performance of gas-lubricated bearings depends on the ambient pressure, because of the compressibility of the lubricant. This property may be measured by its reciprocal, E, the bulk modulus or elasticity of the gas.

The *isothermal elasticity* of an ideal gas is equal to its pressure p and is the same for all gases at a given pressure. At the entrance to the film it is therefore equal to the ambient pressure p_a. Its value at any interior point is fixed by p_a in conjunction with the remaining variables governing the pressure distribution. We can therefore assume that for geometrically similar bearings

$$f = F(D,Z,N,P,p_a) \,, \tag{1}$$

with like expressions for ϵ and h_0/D. Here F is an unknown function of the independent variables listed, and D is the bearing diameter; the remaining symbols have their usual meanings. Other reasons for including the ambient pressure are given by Professor Fuller, based on the pumping action of the bearing (1961). Dimensional analysis shows that Eq. (1) must reduce to

$$f = \phi(ZN/P,\, p_a/P) \,. \tag{2}$$

Here ϕ is another unknown function, but one depending on only two independent variables, the lubrication number ZN/P and the ambient pressure ratio p_a/P. Thus the diameter has dropped out, so the coefficient will be unaffected by size in any one series of bearings that are strictly geometrically similar. Expressions for ϵ and h_0/D, the attitude angle, and other dimensionless performance ratios take a like form, depending only on two parameters.

The adiabatic elasticity of a gas is equal to γp where γ is the specific heat ratio c_p/c_v. For the intermediate condition of "polytropic" compression we may take the elasticity equal to np, where $n = 1$ for isothermal and $n = \gamma$ for adiabatic compression. More generally, Eq. (2) could be written with n as an independent variable in addition to

p_a/P, but the facts are usually represented well enough by isothermal conditions.

The isothermal eccentricity of a journal bearing is given by

$$\epsilon = \psi(ZN/P, p_a/P) , \tag{3}$$

where ψ is a different function from ϕ. Equation (3) is the natural way of expressing the relation between an unknown quantity or dependent variable, ϵ, and the known quantities or independent variables. Designers, however, like to think of ϵ as known and P as unknown. Thus from Eq. (3)

$$P/p_a = \psi_1(ZN/p_a, \epsilon) , \tag{4}$$

where ψ_1 is different from ψ. The parameter ZN/p_a might be called the ambient lubrication number. Since P is now regarded as unknown, we avoid having it on both sides by substituting its dimensional equivalent p_a on the right. The same result could be arrived at by making a fresh start.

In liquid lubrication P is directly proportional to ZN for a fixed eccentricity ratio. We expect the same relation in gas-film bearings at the lower values of P/p_a, where there is little compression. But at high P and low p_a compressional effects will be prominent. The load ratio P/p_a must then increase less rapidly than the first power of ZN/p_a. Even so we can see that speed and viscosity will continue to have equal effects on the load capacity. At high enough values of the ambient lubrication number such that this parameter has little effect, load capacity will be nearly proportional to the ambient pressure. These indications from dimensional theory are confirmed by calculation and experiment.

The Gas Bearing Number. Fluid dynamic theory makes use of a parameter Λ whereby for journal bearings of constant L/D ratio, Eq. (4) takes the form

$$P/p_a = \psi_2(\Lambda,\epsilon) . \tag{5}$$

Here Λ, called the compressibility bearing number or simply the "bearing number," denotes 12π times the product of ZN/p_a by the inverse square of the clearance ratio. In terms of the surface speed U, radius r, and radial clearance c,

$$\Lambda = 6ZUr/p_ac^2 . \tag{6}$$

Equation (5) goes beyond Eq. (4) in showing how clearance enters the picture. A similar equation may be written for thrust bearings with a differently expressed bearing number. The steps leading to the derivation of these parameters will be reviewed under calculations for thrust and journal bearings.

The Differential Equations. Harrison confined his study to the two-dimensional problem under isothermal conditions. He showed that two modifications of Reynolds' theory are needed in order to extend it to compressible fluids: (1) substitution of mass flow for volumetric flow in the equation of continuity, and (2) use of Boyle's law, making density proportional to pressure at every point in the film. If we set $\rho = p/R\theta$, where R is a constant for the particular gas and θ the absolute temperature, $R\theta$ appears in every term and so drops out. Reynolds' equation then becomes

$$\frac{\partial}{\partial x}\left(ph^3\frac{\partial p}{\partial x}\right) + \frac{\partial}{\partial z}\left(ph^3\frac{\partial p}{\partial z}\right) = 6\mu U\frac{\partial}{\partial x}(ph)\ . \tag{7}$$

Ignoring the $\partial/\partial z$ term and integrating gives for infinite length

$$h^3\, dp/dx = 6\mu U(h - C/p)\ , \tag{8}$$

where C is a constant of integration. In fact C is $h'p'$, where p' is the maximum or minimum pressure, and h' the corresponding film thickness. This can be seen by setting $dp/dx = 0$ in Eq. (8). An approximate solution in three dimensions was offered by Sheinberg in 1953. The general equations for polytropic flow in laminar films were discussed by Tipei and Constantinescu in 1954 and 1955, and soon extended to turbulent films (1961).

Equations for polytropic and adiabatic flow were derived by Ausman in 1957. The latter is to be applied in conjunction with an energy equation. The Reynolds equation was expressed in "normalized" (dimensionless) form by W. A. Gross in 1958. Explicit introduction of the polytropic exponent avoids the need for an energy equation. The integration of Reynolds' equation in any of these forms with due regard to boundary conditions shows the pressure distribution in the film. Further integration gives the load capacity of a self-acting thrust or journal bearing.

Externally pressurized gas bearings operate on the same principle as the hydrostatic thrust and journal bearings described in Chapters VI and VII. The load is supported by fluid pressure from an external source without the need for a viscous wedge or any form of relative motion. This type need be discussed only briefly here, since the problems offered relate more to bearing design than to lubrication. See, however, Gross (1965).

3. CALCULATIONS FOR STEADY-STATE THRUST BEARINGS

Externally Pressurized. In a thrust bearing of disk area A the load to be supported is $A(P+p_a)$, where P is the load per unit area W/A. Let p_0 be the absolute pressure at exit from the supply orifice opening into a pocket or recess of area A_0. Then if A_0 is nearly equal to A and α denotes A/A_0, the necessary recess pressure is given by

$$p_0/P = (p_a/P) + 2\alpha/(1+\alpha) \,. \tag{9}$$

It is interesting that the equation does not involve h, the gap between upper disk and annular bearing surface. The same *load* can be supported with a given supply pressure under laminar conditions regardless of the gap height, but the *flow rate* will depend on h. From dimensions we can see that the isothermal flow rate Q_a, measured in volume units per unit time at the ambient pressure, necessary to maintain a chosen film thickness h is given by

$$Q_a = \frac{Ph^3}{Z}\,\phi\left(\frac{p_a}{P}\right), \tag{10}$$

where ϕ is the same for all geometrically bearings. Here Z is the viscosity of the gas and P the load per unit of disk area (Fig. 1).

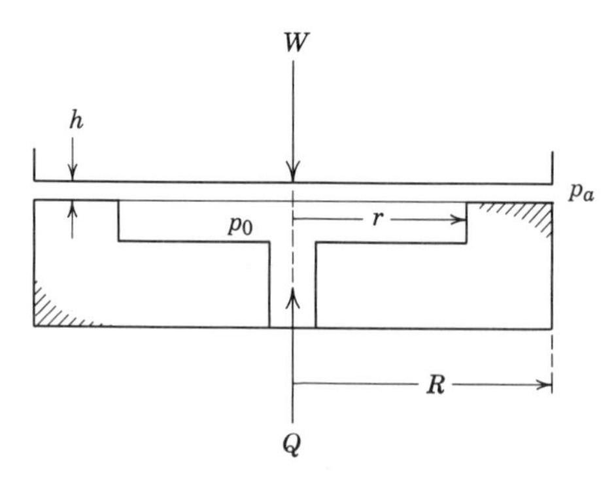

Fig. 1. Hydrostatic thrust bearing (h, film thickness; p_a, ambient pressure; p_0, inlet pressure; Q, flow rate).

When α is close to unity (narrow annular bearing), the effective supporting pressure p_m in the gap is approximately the arithmetic mean of p_0 and p_a. More generally it is given by

$$\frac{p_m}{p_0} = \frac{1}{\rho_0^2 - 1} \int_1^{\rho_0} (1 - k \ln q)^{\frac{1}{2}} q \, dq \,, \tag{11}$$

where q has been written for r/R_0 and ρ_0 for R/R_0. Here r varies from the radius of the recess, R_0, to that of the bearing, R. The constant k is defined by

$$k = 12\mu Q_0/\pi h^3 p_0 \,, \tag{12}$$

where the flow rate Q_0 is measured at the recess pressure p_0. Equation (11) was first integrated graphically by Gottwald & Vieweg (VI: 1950); then analytically by Licht & Fuller (1954) in terms of the "error function." When the flow rate is high enough to introduce the Bernouilli effect, the load capacity may be neutralized by suction. Higher rates are complicated owing to effects described by Comolet and others (1957). In the case of isothermal flow represented by the foregoing equations the load capacity has been computed by Laub (1960), and recorded in a family of curves.

Further calculations are required for evaluating "orifice compensation" and vibrational stability. When the equilibrium assumed in Eqs. (9) through (11) is disturbed, say by a transient overload, the radial outflow of gas will vary. This action can be utilized to *compensate* the overload by automatically raising the pressure p_0, if there is an orifice or capillary in the passage from a constant-pressure reservoir to the bearing recess (Richardson, 1958; Laub, 1960). The overload decreases the gap h and thereby reduces the rate of flow through the bearing and orifice. Immediately the pressure drop across the orifice is decreased; but since the reservoir pressure is constant, the result is an increase in the recess pressure p_0. This tends to balance the overload and restore equilibrium. In general the compensation is overdone, thus setting up vibrations. Their frequency and amplitude depend upon the stiffness of the system, hence upon the depth and diameter of the recess, the geometry of the orifice, and other parameters. Stability investigations were accordingly undertaken and are typified by the publications of Roudebush (1957), Licht (1958, 1961), and co-authors.

The coefficient of friction has the same formula either for a gas or a liquid film in pressurized thrust bearings. It is proportional to ZN/P divided by h/D, where D is the outside diameter; see Eq. (21), Chapter VI. The constant, C''', is the same for all geometrically similar bearings. Pumping loss must be added to bearing friction to obtain total power required for comparison with self-acting bearings.

Self-Acting Thrust Calculations. Harrison obtained an exact integration of Eq. (8) for infinitely wide plane surfaces. It cannot be solved directly for the pressure, but shows that the peak pressure lies nearer the trailing edge than in a liquid-lubricated bearing. With increasing speed the peak moves closer to the trailing edge, and gets there when the speed is infinite. The solution for low speeds is the same as found by Reynolds.

Spherical bearings were studied by Kowaski (1947). Solutions for thrust bearings of a sinusoidal profile were given by Artobolevski & Sheinberg (1950), and for the Rayleigh type by Drescher (1953). The Russian authors considered the pivoted shoe too complicated for gas bearings, and noted that journal bearing theory could be applied to the sinusoidal surface.

The Thrust Bearing Number. Ford, Harris, & Pantall (1957) plotted curves showing the pressure rise above ambient, Δp, for all points from the leading to the trailing edge of an inclined-plane bearing. They were obtained from computer solutions of the Harrison equation and are, therefore, limited to films of infinite width under isothermal compression. Each curve is plotted for a constant value of a thrust bearing parameter defined by

$$G = 6\mu UB/h_0^2 p_a . \tag{13}$$

Here B is the breadth in direction of motion and h_0 is the film thickness at trailing edge. The authors were led to this parameter by Professor Norton's expression for the maximum pressure rise in the oil-film bearing (II: 1942). The symbol Λ is now used indiscriminately for both thrust and journal bearings, but confusion may be avoided here by retaining G for the thrust bearing number.

Curves are shown for G values from 2 to 200 at a fixed ratio of h_1/h_0 (Norton's a) equal to 3, where h_1 is the film thickness at the leading edge. For low values of G the calculated pressure is practically the same as for a liquid film. At higher values it is less for a gas than for the liquid. The curves are initially concave upward. The peak value drops off 80 per cent when G is stepped up 100-fold, and moves toward the trailing edge. At G above 200 the film pressure conforms closely to the relation $ph = p_a h_1$, a form of Boyle's law. The load capacity can then be approximated by integrating p from $x = 0$ to $x = B$ after expressing h in terms of x. The authors point out that P/p_a may be taken as the performance criterion by expressing it in terms of G while holding Norton's film ratio constant. Conversely, dimensional analysis indicates that

$$h_0/B = \psi_0(ZN/P, pc/P) \tag{14}$$

for thrust bearings that are geometrically similar, including constancy of h_1/h_2. Equation (14) is the counterpart of the journal bearing equation (3).

Numerical Analysis. The research papers by W. A. Gross on gas-film lubrication, beginning in 1958, include new and even surprising results of interest for thrust bearing calculations. Side leakage becomes negligible in gas films at high values of the bearing number; the coefficient of friction is greater for a gas than for a liquid film at high G; and a region of instability is found at moderate values. Here the load capacity of a plane pivoted shoe, after reaching a maximum, falls off with further decrease in film thickness. Convergency is lost and the load capacity drops to zero. No such effect occurs in the gas bearing with a fixed taper; or in the liquid film with a pivoted shoe. In liquid lubrication the film ratio a remains constant regardless of load for a given pivot location, so parallelism cannot occur. This new effect, discovered by Gross, is seen in Fig. 2, where it can be shown that G or $\Lambda = 0.12$ divided by the square of $10^3 h_0/B$. See also Fig. 4.4.17 in Gross (1962), where the theory is compared with W. Stone's data. The effect exists only with a relatively thin film and can be avoided by a slight amount of crowning, or convexity, as de-

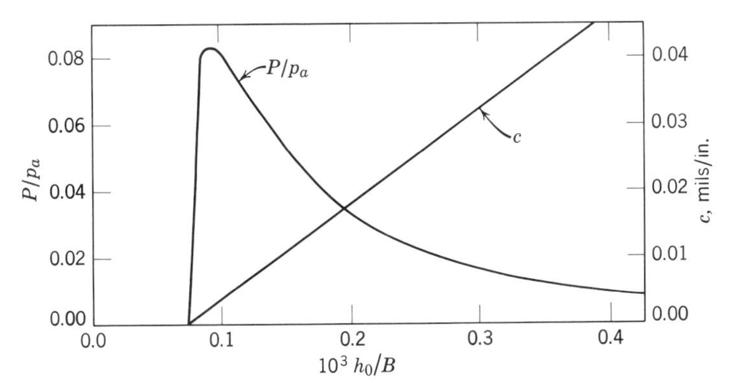

Fig. 2. The "Gross effect" in air-lubricated thrust bearings, a sudden loss of load capacity (from W. A. Gross, 1959). $L/D = 1$, standard pivot location, $c =$ angle of tilt; applicable to Stone's bearing with $B = 0.61$ in., $U = 69.3$ in./sec).

scribed by Brunner and co-authors (1959). Gross worked out the load capacity and friction characteristics of the various slider profiles—plane, curved, and composite—by integrating the normalized equation. Analytical solutions are obtained when G approaches zero or infinity, with computer solutions for the intermediate range (Michael, 1959). See also Sneck (1965).

Pivoted Shoes. Consider the isothermal lubrication of a plane slider. Let $\bar{x}_p$ denote the pivot location, or center of pressure, measured back from the leading edge and expressed as a fraction of the breadth B in direction of motion. The load ratio and bearing number for a given $\bar{x}_p$ with infinite L/B may be obtained from the tabular data of Gross (1958 and 1961), or by graphical interpolation from Figs. 9 and 10 (1959). The first figure shows a family of curves for pivot location against G at constant values of H_1 (in our notation, h_1/h_0). The second shows curves for the load ratio W', denoted here by P/p_a or P'. We can eliminate h_1/h_0 by cross-plotting and thus find the relation connecting G with P' at constant $\bar{x}_p$.

Film Thickness and Kingsbury Number. For practical use we should like to express h_0/B in terms of K and P', where K is the Kingsbury number $\mu U/PB$, as defined by Raimondi & Boyd (VI: 1955). It follows from the definitions of K and G that

$$h_0/B = k\sqrt{K}\,, \tag{15}$$

where k denotes the square root of $6P'/G$. Equation (15) is identical with Eq. (13) of Chapter VI, but in gas films k is a function of P' as well as of L/B and $\bar{x}_p$. A study of the P'/G values by graphical interpolation for different load ratios at $\bar{x}_p = 0.675$, in the figures cited, provides values of k from $P' = 0.07$ to 0.47. These values seem to approach 0.29 as P' approaches zero and they vary but little. For a liquid film, $k = 0.286$, as shown by Table 4 of Chapter VI.

The calculations by Gross have been supplemented by Brody (1960) in tables for C_W (our k^2) and C_s (our $\bar{x}_p$) at four values of G. Thus four pairs of k and G may be read from the tables with the help of graphical interpolation, corresponding to any constant value of $\bar{x}_p$. The load ratio P' is then taken equal to $k^2G/6$. Additional points can thus be obtained on a graph for $\bar{x}_p = 0.675$. No systematic trend away from the k value of 0.29 is apparent up to $P' = 48$.

Finite Width. It was seen in Chapter VI that K is $2\pi(ZN/P)$ divided by the sector angle A in radians. Since A is a constant for geometrically similar bearings, and since any function of P/p_a is equally a function of p_a/P, it will be seen that Eq. (15) is an example of the symbolic Eq. (14). That the effect of finite L/B is the same in a gas film as in a liquid film at low values of G, but diminishes at higher values, is shown by Fig. 13 of Gross (1959). Here is a family of curves for P' against G at $a = 2$, each labeled with a different value of L/B. The curves are widely separated at moderate values of G but converge toward a single value as G approaches infinity.

Values similar to those reported for $\bar{x}_p = 0.675$ at infinite L/B may be obtained from Fig. 7.8 of Gross (1961) for $\bar{x}_p = 0.558$ with $L/B = 1$. Inspection of the curves for film thickness against speed confirms the square-root law, Eq. (15). Computations based on data from the curves show that k varies only from about 0.271 to 0.277 as P' drops from 0.075 to 0.015. The corresponding k for a liquid film is 0.252. These calculations can be extended to other values of L/B and $\bar{x}_p$ with the aid of Brody's tables, which include both load and friction coefficients for L/B from $\frac{1}{4}$ to 5, and infinity.

Polytropic Flow. The characteristics of polytropic flow are given in the references by Gross, where curves are shown for P' against G with $L/B = 1$, $a = 2$ and 3. For example, three curves are given for $a = 2$ with exponent n ranging from 1 to 1.66. The curves coalesce as G approaches zero; but at $G = 50$ the P' values range from 0.25 to 0.38, the higher P' corresponding to the higher n. At infinite G the range of P' is from 0.39 to 0.77.

Crowning. The effect of crown height in cylindrical sliders is twofold. As shown by Gross, very slight crowning protects against instability. Further crowning augments the load capacity until after an optimum has been reached. An example is seen in Fig. 7.3 of 1961 for $L/B = 0.733$ with $\bar{x}_p = 0.60$, operating isothermally. The optimum crown height is about equal to the minimum film thickness. Solutions covering a wide range in L/B are given by Figs. 7.4 and 7.5 for $\bar{x}_p = 0.625$ at a stated speed and film thickness.

Solid Thrust Bearings. Consider next the calculations for fixed-taper and composite thrust bearings; the Rayleigh step; and other solid types in references below and in Gross' book (1962).

The Fixed Taper. Film conditions are different in solid bearings because the angle of inclination is fixed. The dimensional relation (14) still applies, but the square-root equation (15) does not. Curves equivalent to Eq. (14) for isothermal flow with $L/B = 1$ can be derived from Figs. 6.11 and 7.2 of Gross (1961). These figures show curves for P' against h_0 for stated values of B and the angle α (our c); but can be generalized with the aid of Eq. (15). For practical use, curves are desired showing the relative film thickness Y against a thrust variable X at constant P'. Here Y denotes h_0/h_t, as in Chapter VI, where h_t is the taper-height αB, a fixed element of the bearing; and

TABLE 1. Isothermal Flow in Fixed-Taper Bearings
(Relative film thickness Y for $L/B = 1$)

X	$P' = 0.15$	0.11	0.065	0.023	Liquid
5	0.55	0.57	–	–	0.59
10	0.78	0.81	0.85	–	0.84
20	–	–	1.20	1.25	1.15
40	–	–	–	1.75	1.57
80	–	–	–	2.60	2.08

X denotes K/α^2. Data processed from the reference were plotted in smooth curves leading to Table 1. The "liquid" column is from Fig. 2 of Chapter VI. It appears that the film thickness is greater in the gas- than in the liquid-lubricated bearing when P' is below say 0.08. With more complete data it should be possible to plot an X, P' graph for equal thicknesses in the gas and liquid films. Similar calculations can be made from Fig. 6.11 of Gross (1961) for $L/B = 2$. The results should be extended to higher values of P' and to infinite L/B. For conversion of X values to ZN/P, we recall that X is equal to the product of ZN/P and the geometrical constant $2\pi/A\alpha^2$, in which A is the sector angle.

The effect of *crowning* a fixed-taper bearing may be seen by comparing Figs. 13 and 14 of 1958. In practice the fixed taper is followed by a flat land at the trailing edge, parallel to the runner, making up a "composite" bearing. Solutions for such a bearing with an isothermal gas film, neglecting side leakage, have been given by Boeker and co-authors (1958), Gross (1958a), and Kochi (1959). Mow and Saibel (1959) and Gross (1961) dealt with finite length ratios.

Rayleigh and Other Types. The last-named authors and Yen (1960) published gas-film solutions for the Rayleigh bearing with infinite L/B. Kochi went into design problems. Yen derived optimal shape factors by a variation method. He saw that there was no single optimum, the most favorable design depending on the chosen operating conditions. He treated the general case of polytropic flow. A numerical solution for the Rayleigh step with $L/B = 1$ is given by Gross (1961). All these calculations were made by numerical methods except as noted, and are limited to rectangular surfaces. Analytical solutions for the Rayleigh-type sector shape have been derived by Ausman (1961) as well as by Toba & Saibel (1961). See Wildman *et al.* (1965).

Ford, Harris, & Pantall (1957) reviewed reports by Whipple and

Fortesque on the theory of spiral-grooved and herringbone thrust plates, with test data. See also Malinowski, Vohr, and their co-authors (both 1965).

Parallel and Near-Parallel Surfaces. The Fogg effect (Chapter III) was investigated by Hughes & Osterle (1957). Their study was reviewed by Gross (1958), who adds a polytropic solution. The Fogg effect is less pronounced in gas than in liquid films. A finite-width solution is given by Stein (1961).

Taylor & Saffman (1957) calculated the effects of compressibility in air films between nominally parallel surfaces. They thought the Reiner effect (1957, 1961) might be explained away by nonparallelism, nonflatness, or vibrations. It is true that there is a great difference between the action of strictly parallel and near-parallel surfaces when the films are very thin, without departing from the Navier-Stokes equations. Nahavandi & Osterle (1961) applied some of these facts in proposing a simple type of gas-lubricated thrust bearing—one of limited load capacity operating with flat but not quite parallel surfaces. See also Krzywblocki (1957).

Effect of Molecular Free Path. Burgdorfer (1959) modified the Reynolds equation to allow for slip at the wall. This occurs when the film thickness is not very large compared to the molecular mean free path. The slip effect reduces friction and load capacity. It can be disregarded when the mean free path λ is less than say $\frac{1}{50}$ of the minimum film thickness; but when λ is as large as $\frac{1}{10}$, the load capacity is reduced appreciably. Burgdorfer solves his equation for the fixed-taper and Rayleigh bearings with L/B infinite. At a bearing number $G = 5$ he shows that the load capacity is reduced to about two-thirds of its classical value when λ is raised from 0 to one-third of the minimum film thickness. At atmospheric pressure and temperatures, λ is about 2.5 micro-inches in air, and nearly three times as much in helium. The values are greater at low ambient pressures—that for helium goes up to 36 micro-inches at $\frac{1}{5}$ of an atmosphere. Further discussion of slip theory is given by Toba and Saibel.

4. STEADY-STATE JOURNAL BEARINGS

Externally pressurized bearings will be treated first, followed by more detail on the self-acting type. The calculations for externally pressurized gas bearings are the same as for hydrostatic liquid bearings up to a moderate pressure. Design calculations taking compressibility into account were published by Grinnel & Richardson (1957) and Richardson (1958). See also Adams, Fox and co-authors in

1961. Similar calculations have been reported by Allen, Stokes, & Whitley (1961), Laub (1961), and Lemmon (1962). All are found to agree with experiment. See also Shires & Pantall on aerostatic jacking (1963). Gas-film journal bearings are usually self-acting except when external pressure is needed for starting under load.

Beginning with the case of infinite length, the theory of gas-lubricated journal bearings advanced through finite-length approximations to the present-day computer solutions. We turn now to the *self-acting* type.

Self-acting Infinite Length. Pressure distribution in the self-acting bearing follows Eq. (8) when isothermal conditions may be assumed. Harrison was unable to integrate this equation analytically for the full journal bearing, although he had done so for plane surfaces. Instead, he used a numerical method aided by Kingsbury's data. The results are shown in a series of diagrams for different speeds. Each comprised three curves for the circumferential pressure distribution: (1) the theoretical pressure in a liquid, (2) that for a gas, and (3) one for the pressures observed by Kingsbury. A comparison of the first two curves reveals the effect of compressibility, and the contrast between the second and third curves on each diagram shows the effect of end leakage.

Harrison obtained similar results analytically from a model of the journal bearing in which the converging and diverging surfaces were simulated by two planes. The solution showed that at low speeds the load capacity is proportional to speed, as with liquid lubricants, whereas at higher speeds it approaches a constant value, owing to compressibility. Harrison was the first to note this important difference between the action of liquid and gaseous lubricants.

The bearing number $\mu Ur/p_a c$, now commonly denoted by $\Lambda/6$, seems to have been introduced by Sheinberg in his paper of 1947 (see his page 194). Further solutions using this parameter were published by Katto & Soda (1953, 1962) and by Sheinberg (1953).

Katto & Soda found new variables by means of which Eq. (8) could be integrated analytically. Following Harrison, they determined the constant of integration from the fact that the mass of gas in the clearance space must be the same at all speeds for bearings of infinite length. Their solution led to a series expansion useful at the smaller eccentricities, say for ϵ less than $\frac{1}{2}$. This result was based upon the parameters elsewhere designated P' and Λ. It will be recalled that Λ/P' is 12π times the Sommerfeld number S. In Katto and Soda's paper all performance characteristics, including $12\pi S$ and fr/c, are conveniently shown by a polar diagram, in one quadrant, having the eccentricity ratio and attitude angle as coordinates.

Equation (8) was integrated by Sheinberg using a vectorial graphic construction like that of Karelitz. This led to a numerical solution for high as well as low eccentricities. He determined the constant of integration from the variable mass content of finite-length bearings. Sheinberg's load capacities and friction coefficients are in reasonable agreement with later computations. The methods of Harrison and of Tipei (1954) were applied by Constantinescu (1955, 1956) to obtain solutions for infinite-length journal bearings, including optimum design factors.

Ausman (1957) expressed the gas film equation in forms applicable to polytropic flow and derived a perturbation solution, assuming a constant mass of gas per unit of axial length. The first three terms suffice for eccentricity ratios below $\frac{1}{2}$. Performance characteristics are given in terms of Λ/n, where n is the polytropic exponent. Solutions have since been obtained that are accurate to higher eccentricities, yet the foregoing is frequently cited for comparison. As ϵ approaches zero, the load ratio P' approaches $\pi\Lambda\epsilon/2$ for all values of n.

Elrod & Burgdorfer (1961) integrated the Katto and Soda equation by means of a digital computer. The result was an improved solution for the isothermal case good for high eccentricities. Diagrams show the load ratio, attitude angle, and journal friction against Λ at constant eccentricity. Friction is expressed by the ratio of its moment M to the Petroff or concentric moment M_0. These authors picture an endless bearing as being open to the ambient pressure at each end! The mass content would then be a function of eccentricity, as shown by Gross in his book (pages 126–8 of 1962). The same concept was adopted by Gross & Zachmanoglou in their analytical solution (1961) —a series built on Λ as the perturbation parameter for small bearing numbers, and on Λ^{-1} for large ones. Graphs show the load ratio from $\Lambda = 0$ to ∞ at nine values of ϵ between 0.20 and 0.95.

Finite-Length Approximations. Sheinberg's numerical solution (1953) was applied to finite bearings with the aid of Gut'iar's approximation and by assuming the same axial distribution of pressure at all points of the circumference. This led to a relation between mass content and eccentricity nearly the same as found by Elrod and Burgdorfer. Final results are shown by tables and graphs for load capacity, attitude angle, and friction coefficient at ϵ ratios from $\frac{1}{3}$ to $\frac{5}{7}$ with L/D's from $\frac{9}{8}$ up. Sheinberg's load capacities are slightly higher than found by later computations, except when Λ approaches zero and infinity. It was noted by Sheinberg that the finite gas bearing acts like one of infinite length at high values of Λ.

Constantinescu determined pressure distribution for polytropic flow by successive approximations. At each step part of the infinite length solution is fed back into the general equation (1955). Ausman (1957) derived end-flow corrections by perturbation. They are plotted against Λ/k for constant L/D. Although derived only for small eccentricities, the results served to extend the Katto & Soda solution, and were compared with experiment. Ausman improved the isothermal case by choosing ph instead of the pressure itself as a dependent variable (1961). The load ratios and attitude angles lie between those of his earlier solution and recent computer results.

Torque produced by misalignment was studied by Ausman, again using a perturbation method. Summarizing theory and design, Ausman offers a graphical comparison of the methods available (1961). He recommends the Elrod & Burgdorfer solution corrected by means of his end-flow factor (1961a). Tipei and Constantinescu also brought their work up to date at the Symposium (1961).

Katto & Soda (1962) derived a polytropic solution for bearings of finite length, applicable to high eccentricities. Satisfactory agreement was found with published experiments for $L/D = 2$. The authors concluded that gas bearings operate under a condition intermediate between isothermal and adiabatic at high speeds, and under isothermal conditions at low speeds. In this paper, too, they explain just how the mass content remains constant in bearings of infinite length, yet is a function of Λ, and hence of eccentricity, in finite bearings. Gas film lubrication was reviewed by Constantinescu in his book of 1963. Sneck calculated film pressures in the short journal bearing (1963).

Computer Solutions. Isothermal results were obtained by Sternlicht & Elwell (1958) for comparison with experiment. They plotted film pressure, Sommerfeld number, and attitude angle against ϵ up to 0.9 for three clearance ratios at $L/D = 1$ and 1.5. Solutions for $L/D = \pi$ were added by Gross (1960).

A wider range of variables was covered by Raimondi (1961). Tables and charts are given for the load ratio, attitude angle, friction ratio, Sommerfeld number, and film pressure characteristics against λ for constant ϵ and L/D ratios—the symbol λ being used for $\Lambda/6$. Eccentricity ratios are carried to 0.8 at $L/D = \frac{1}{2}$, 1, and 2; and up to 0.9 at $L/D = \infty$. His data for infinite length are taken from Elrod and Burgdorfer. Details of the computer method are given and results compared briefly with experiment. A condensed selection from Raimondi values is shown in Table 2. It is interesting that the load ratio

TABLE 2. Gas Lubrication of the Full Journal Bearing

(Isothermal data from Raimondi, 1961, λ denoting $\Lambda/6$)

L/D	ϵ	λ	P/p_a	F/F_0
$\frac{1}{2}$	0.4	0.1	0.035	1.11
		1.0	0.29	1.11
		∞	0.80	1.09
	0.8	0.1	0.20	1.84
		1.0	1.34	1.72
		∞	3.67	1.67
1	0.4	0.1	0.10	1.15
		1.0	0.55	1.11
		∞	0.80	1.09
	0.8	0.1	0.43	2.00
		1.0	2.28	1.73
		∞	3.67	1.67
2	0.4	0.1	0.32	1.24
		1.0	0.77	1.10
		∞	0.80	1.09
	0.8	0.1	0.91	2.34
		1.0	3.31	1.73
		∞	3.67	1.67

for $\epsilon = 0.4$ comes back to 0.80 whenever $\lambda \to \infty$, regardless of L/D. And for $\epsilon = 0.8$ it is always 3.67, because there is no end leakage at infinite λ. The table shows only slight variations in the ratio of friction F to concentric friction F_0 as affected by λ and L/D (Fig. 3).

Cooper (1961) solved Reynolds' equation in the isothermal case both by relaxation and computer methods. Results for various operating conditions agree well with experiment. He found that a molecular mean free path of 25 micro-inches drops the load capacity about 15 per cent. Later computations by Sternlicht (1961) led to charts and tables for the Sommerfeld reciprocal, attitude angle, and friction ratio in terms of Λ at constant ϵ. The load ratio may be found upon dividing Λ by $12\pi S$. Sternlicht's results are consistent with Raimondi's charts at the lower bearing numbers. See Castelli et al. (1964).

Inertia and Turbulence. Gas film turbulence was described by Constantinescu at an early date (1958). It was shown by Osterle & Hughes (1958) that the inertia of a gas can reduce the load capacity

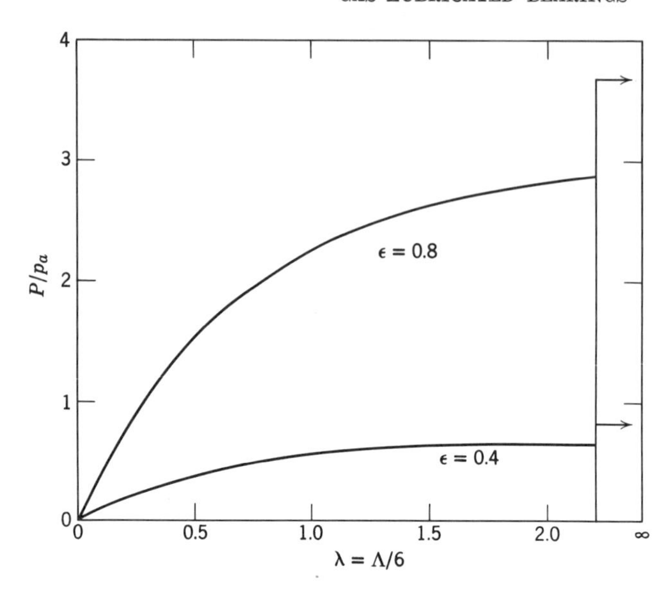

Fig. 3. Load ratios and bearing number for constant eccentricity (from Table 2).

of a bearing at high speeds while it is still operating with a laminar film. Equations are reviewed and values tabulated by Pinkus and Sternlicht in their Chapter 5 (II: 1961). The effect of turbulence in the film is to increase the load capacity as shown by Tipei & Constantinescu (1961) and more fully by Constantinescu in his book of 1964 and paper of same date.

Thermal Conditions. The isothermal assumption does not imply that there is no frictional temperature rise, but only that the temperature and viscosity remain uniform. Elrod & Burgdorfer (1961) showed that the radial temperature drop can never be very great. Ausman found only a few experiments tending to support adiabatic conditions (1957). Calculations based on isothermal operation are sufficiently accurate for most applications.

The question of equilibrium temperature came up some years ago (Hersey, 1938, III: 1939). It can be dealt with more simply here. By Petroff's equation, the power loss may be taken equal to aZ, where Z is the mean viscosity and a is a constant. As in Chapter III, the rate of heat transfer in power units may be written bT, where T is the temperature elevation above ambient and b a constant. When thermal equilibrium has been reached, $aZ = bT$, so the viscosity is equal to

bT/a. But the viscosity of a gas is also given by $c\theta^n$, where θ denotes the absolute temperature and c and n are practically constant for a moderate temperature rise. Now θ can be written as $T_a + T$, where T_a is the ambient temperature absolute. Treating T as a small fraction of T_a, we can equate the two expressions for the viscosity, finding that

$$T/T_a = K(1 + nK)\,. \tag{16}$$

Here K denotes ac/b divided by the $(1 - n)$ power of T_a.

As an example, consider Kingsbury's bearing of 1897 operating in air at 4400 rpm. At this speed and from the geometrical data, $a = 2.45 \times 10^{11}$ cu in./sec². For b, take βA, where A is the effective area of the housing, 410 sq in., and β the heat-transfer coefficient, say 0.13 in.-lb/sec per sq in. and deg C, making $b = 53.3$. The value assumed for β comes from the traditional 2 Btu/hr sq ft and deg F in "still air," allowing the customary factor of 2 for a moderate degree of air movement. Disregard a possible drop of $\frac{1}{2}$ deg C across the metal wall of the bearing. From the properties of air at 20 C, $n = 0.77$ and $c = 2.75 \times 10^{-11}$ in the units chosen. With $T_a = 293$ Kelvin, this makes $K = 0.0406$. Substituting into Eq. (16) gives $T = 12.2$ deg C as observed in Kingsbury's experiment, a coincidence that may be attributed to the uncertain heat-transfer coefficient. See also Kao (1963) and Powell (1963).

5. NONSTEADY STATES

Thrust bearings operate in an unsteady state when the load is supported on a Stefan or "squeeze" film; as well as when the rotor mass can vibrate. It is an odd fact that Stefan's law, although derived to explain the properties of an air film, is mathematically exact only for a liquid film. Gross (1958, 1961, 1962) extended Stefan's law to the compressible film. He gives first an isothermal solution by numerical methods, taking into account the time rate of change of the gas density. The solution with ρ constant follows as a special case. Load capacity of the gas film is greater than that of the corresponding liquid film when under compression but less when in tension; and it is reduced by superposing tangential motion. See also Salbu (1964).

Thrust bearing vibrations depend upon the stiffness of the lubricating film and the mass of the rotor as described by Gross in his book. Damping is provided by the squeeze-film. See Rood (1965).

Starting Journal Bearings. Both Wildmann (1960) and Drescher (1961) saw that there must be a condition of "mixed lubrication"

(Chapter XIII) between surfaces starting under load from the moment static friction is overcome until the rotor is fully supported by a fluid film. Drescher points out that the horizontal component of eccentricity shifts to the opposite direction when fluid friction gives way to mixed friction. If "boundary lubrication" predominates, the friction moment on the journal will be less than that on the bearing. Drescher's conclusions were confirmed experimentally.

Wildmann analyzed both the usual case of the journal rotating in a fixed bearing, and that of the bearing member rotating about a stationary journal. When the inner member rotates, solid friction and the fluid-supporting force oppose each other; when the outer member rotates, they act in the same direction. Consequently the energy input required to overcome solid friction, and accelerate the rotor up to its take-off speed, is greater when the journal rotates than when the outer member rotates, provided the weight and moment of inertia of the rotor are the same in each case. Formulas and curves are given for use in design.

Journal Bearing Whirl. High speed combined with low viscosity in gas bearings tends to accentuate any instability of the journal position (Chapter XII).

Instability of *externally pressurized* bearings led to lively discussions among designers. See, for example, Licht & Elrod (1960) and three publications by Gross in 1962; as well as one by Larson & Richardson the same year. Whirl calculations for *self-acting* bearings were discussed by Professor Boeker in an early report (pages 78–115 of 1958). It became evident that a theory of fluid-film whirl should be developed expressly for gas bearings. See Poritsky (1962).

Such a study was conducted by Constantinescu (1959) for polytropic conditions in the full bearing of finite length. He found that the journal position can be stable with respect to eccentricity and unstable with respect to the attitude angle. Snell (1959) concluded that the resonant frequencies in pivoted-shoe bearings would be closely proportional to the cube of the clearance. This investigation led to the successful operation of a gas-bearing machine with 22-lb rotor at 24,000 rpm. Cooper (1961) treated the case of an unloaded rigid shaft. Radial and tangential force components were plotted against the ratio of whirl speed to journal speed, showing that the graphs cross at 0.5. An open slot in the position leading to a minimum attitude angle is shown to be favorable for controlling whirl.

Sternlicht, Poritsky, & Arwas (1961) found theory and experiment in reasonable agreement but indicated directions for further study. Pan & Sternlicht (1962) introduced a coordinate transformation to eliminate the time-dependent term in Reynolds' equation, and thus calculated the translatory whirl of a vertical rotor. They find the threshold of "half-frequency whirl" to be at zero speed. Rentzepis & Sternlicht (1962) advanced the problem of a rigid shaft by "variational" equations of motion. Stability curves are plotted for dimensionless journal speed against a dimensionless clearance parameter, both for constant ϵ and constant load ratio. The latter curves exhibit a novel minimum point. Note the investigation of self-acting bearings by Reynolds & Gross (1962). See Whitley et al. (1962); Ausman (1963, 1965); Cheng et al. (1963, 1966); Sternlicht & Winn (1963); Wernick & Pan (1964); Castelli & Elrod (1965).

6. THRUST BEARING EXPERIMENTS

Self-Acting Bearings. Stone's observations on pivoted shoes (1921) are of special interest in exhibiting the type of instability explained by Gross. Representative data from the stable operating range in Stone's experiments are discussed by Shaw & Macks (V: 1949, page 330). An experimental report was published by Sheinberg (1950). Thrust plates with spiral grooving were tested by D. V. Wordsworth in 1952 (Sciulli, page 601 of 1961). Load capacity at the higher speeds fell below theoretical, apparently owing to thermal distortion. Further investigations are reported by Whitley (1961). Drescher (1961) gives film thickness and pressure distribution in Rayleigh-type thrust bearings.

Professor Fuller described the deceleration test on a Kingsbury three-shoe model (1953, I: 1956, pages 183–5). The coefficient of friction came to 0.002, including windage, at a mean speed of 350 rpm. In the same year, 1953, D. K. McKinley reported the failure of a pivoted-shoe bearing that he had constructed with great care for operating on air. Since the shoes were flat and centrally pivoted, such a failure confirms thrust bearing theory. Similar results are described by Raimondi & Boyd (VI: 1955)—flat shoes failed where convex shoes of crown height 50 micro-inches operated smoothly. Professor Reiner demonstrated a centripetal pumping effect in the air film between nominally flat, parallel disks when close together and in relative motion at high speed (1957, 1961, 1965). This effect is not yet well understood, but opposite in direction from the centrifugal action familiar with thicker films. The numerical solutions by Gross and Michael are strikingly borne out in experiments on tilting shoes by Brunner and co-authors (1959). Curves are given for load capacity and inclination against the minimum film thickness both for flat and crowned sliders. Thus a flat shoe, pivoted behind center, failed under a load of 1.6 psi at a film thickness over 0.3 mil, inclination 2×10^{-4}. The corresponding shoe crowned less than 12 micro-inches continued beyond 2.7 psi, with film thickness reduced below 0.1 mil at an inclination less than 10^{-4}, yet no failure occurred.

Experiments by the writer, mentioned in discussing Drescher's paper of 1961, led to a friction coefficient of 0.0010 for the Kingsbury three-shoe model at the speed of minimum friction, 54 rpm. The shoes in this model are centrally pivoted. Measurements revealed a crown of the order of 50 to 60 micro-inches near the middle of each shoe. This is enough to explain the results and to confirm the findings of Raimondi and Boyd. The tests were made at Brown University assisted by William Barlow, incident to an ASME study.

Pressurized Thrust Bearings. Experiments on a spherical bearing were reported by Kowaski in 1947. Gottwald & Vieweg (VI: 1950) tested flat and conical rotors, noting good correlation with theory. Comolet found radial outflow of gas between parallel plates isothermal in practice, and proportional to the cube of the gap when allowance is made for roughness (1957, 1961). Professor Fuller described the bearings of an 80,000-rpm centrifuge operating on air supplied at 5 psi. Results agreed well with calculation (1953, I: 1956). Pigott & Macks tested a 6-in. diameter nonrotating bearing with flat surfaces up to 1000 F. Load capacity increased with rising temperature (1954). Deviations from theory were attributed to geometrical changes. Spherical thrust bearings were tested by Corey and co-authors (1956). The test reported by Fuller on a flat thrust bearing with central recess illustrates the effect of compressibility (I: 1956, pages 87 and 301–4). The flow rate would be 1110 cu ft/min by incompressible theory and 2244 by gas film theory. Actual measurement gave 2240.

Grinnell (1956) experimented with bearings in which the acceleration of the fluid is a factor. Licht and co-authors (1954, 1960) found that vibrations could be suppressed by minimizing the recess depth and maintaining high recess pressures. Large nozzles were better than small ones. Experiments by Laub on orifice-regulated bearings agreed well with calculations (1960, 1961). Allen, Stokes, & Whitley (1961) tested bearings with carefully designed orifices. Vibrations were greater than expected. Fischer, Cherubim, & Decker investigated thrust bearing instability (1961). Laub & Norton (1961) determined the characteristics of spherical bearings. Nemeth & Anderson (1961) tested parallel surface, capillary-compensated bearings of 10-in. diameter containing sector-shaped recesses. Deviations from theory were accounted for by surface deformation. Sixsmith and co-authors (1961) measured the load capacity of circular plates with various recess and orifice combinations at supply pressures to 250 psi using nitrogen and helium. Demonstrations with an "air track" can be instructive (Stull, 1964).

7. JOURNAL BEARING EXPERIMENTS

Self-Acting Bearings. Kingsbury's experiments were accurate and are still referred to in support of present-day theories. Ausman (1957) plots the film thickness data as evidence for adiabatic flow, although by our interpretation of Kingsbury's values, they average only 2.4 per cent above the isothermal curve. His friction data are relatively high and not easily explained. Forty-six years passed before the publication of further journal bearing tests. These were by Sasaki (1943) and, we are told, included static as well as kinetic friction (Sciulli, 1961). Vitellozi reported tests on steam-lubricated bearings (1950). A silver-impregnated carbon bearing gave promise of satisfactory performance.

Film thickness and pressure measurements on full bearings of 48-mm diameter, with $L/D = \frac{9}{8}$, are given by Sheinberg in his papers of 1950 and 1953. Three clearances were chosen with C/D from 0.26 to 0.70 per thousand. Load ratio P/p_a was varied from about $\frac{1}{5}$ to $\frac{3}{2}$ at speeds from 3000 to 10,000 rpm. Eccentricities ranged from $\epsilon = \frac{1}{3}$ to $\frac{5}{8}$. Load capacities averaged 3 per cent below present isothermal theory. Sheinberg's friction measurements, like Kingsbury's, are higher than calculated. Drescher made friction and load capacity tests on air-lubricated journal bearings (1953). Low-speed and high-speed operating limits are described in a later publication (1961). Experiments were reported by Wildmann (1956) using air, helium, and neon as lubricants, mostly on a $\frac{3}{4}$-in. diameter. Geometrical factors and loads were suitably varied at speeds from 5000 to about 12,000 rpm. The resulting eccentricity ratios rarely exceeded $\frac{1}{2}$. These data led to empirical equations and curves for design use at moderate loads.

Cole & Kerr experimented on glass bearings of 1-in. diameter with $L/D = 1$ and C/D from 0.7 to 1.9 mils/in. (1957). Loads ranged to 26 psi and speeds from 3000 to 60,000 rpm. Film pressures agreed with Katto and Soda in the positive area but were less in the negative loop. Temperature rose 7 deg C at 53,000 rpm under 8 psi. Condensation of moisture could be seen in the bearing. Temperature was nearly uniform, indicating isothermal operation. Loads below 4 psi and eccentricity ratios less than 0.2 were accompanied by "half-speed whirl." Maximum safe loads were marked by bright flashes due to contact. It was shown that a convenient rule for estimating load capacities up to a moderate speed with clearance ratio close to 1 mil/in. is to allow 1 psi for every 1000 rpm. The authors suggest volatile liquids as lubricants for starting gas bearings under load. Ford and co-authors (1957) tested plain journal bearings from 1- to 2.5-in. diameter over a wide range of operating and design variables. Their chart for P' against Λ at constant ϵ shows plainly the effect of compressibility when the bearing number is greater than unity. Comparison of four gases with specific heat ratios from 1.11 to 1.67 reveals no appreciable departure from isothermal flow. A gradual change from synchronous to half-speed whirl was observed at the higher speeds. In a CO_2 circulator with 2-in. diameter bearings the threshold for half-speed whirl could be raised from below 5000 to above 14,000 rpm by venting the unloaded side of the film. See Schutten et al. (1958).

Experiments on air-lubricated bearings of 2.0- and 2.5-in. diameter, with $L/D = 1.5$ and clearance ratios below 1 mil/in., are described by Sternlicht & Elwell (1958). Pressure distributions were obtained for eccentricity ratios as high as 0.9. The volatile liquid method was found successful in starting from rest. Results agreed with isothermal calculations up to moderate values of Λ and ϵ, and confirmed the Cole & Kerr rule. Brix (1959) reported experiments on synchronous whirl in gas bearings due to unbalance. His tests were made on a stiff shaft in bearings of 2-in. diameter with $L/D = 3$ and a clearance of 1.5 mils/in. Curves of whirl amplitude against speed show what may be expected from plain bearings without modification of design. Whitley & Betts experimented on full journal bearings of 2-in. and 7-in. diameter, L/D from 1 to 4.5, at speeds up to 20,000 rpm and ϵ to 0.8, with a short axial groove on the unloaded side (1959). Results confirmed isothermal operation. Load capacities were less than expected from Ausman's leakage factors.

Drescher's investigation of operating limits was described at the Symposium (1961). Mixed friction prevails at speeds below the point of minimum friction, very much as in liquid-film bearings. He experimented with three bearings of 120-mm diameter, all having $L/D = 1$ and a clearance ratio of 0.9 per thousand. A constant load of 0.12 kg/sq cm was applied at speeds from 0 to 5000 rpm. The bearings chosen were a full bearing of hardened steel, one of bronze, and a half-bearing of bronze. Curves are plotted showing the friction number fD/C against $2\pi S$. All start out convex upward in the low-speed region, dropping off sharply as the minimum point is approached, then rising gradually with increase of speed. The three curves differ characteristically at the left of the minimum but coalesce at the right. These tests were accompanied by film pressure and eccentricity measurements, doubtless the first such to be made on gas bearings under mixed friction or "thin-film" conditions (Chapter XIII). Elastic deformations of asperities, together with their wedge-film action, were visualized just as for liquid films. The Sommerfeld number at minimum friction was 0.008 for the steel bearing, and greater for the bronze ones. It would be interesting to determine the separate effects of load and clearance on the location of the minimum point, as well as the effects of surface finish and running-in. The minimum point, with a factor of safety, is generally accepted as fixing a low-speed limit for safe operation. The value $S = 0.008$ corresponds to a speed of the order of 250 rpm under the stated conditions. Drescher found high-speed operation limited by half-speed whirl. It began at $S = 0.16$, or about 5000 rpm in the steel bearing. The eccentricity-attitude locus is shown both for a plain bearing, and for a more stable design with Rayleigh steps. Drescher concludes that load capacities up to at least 4 psi may be expected under atmospheric conditions when the clearance space is kept free from dust or other contamination.

Thirteen pairs of bearing materials were compared in a series of 1000 start-stop cycles by Block & Braithwaite (1962) at loads from 4 to 6 psi. Good performance was obtained with a chrome-plated steel journal running in soft Meehanite bearings.

Cooper determined pressure distribution, eccentricity, and attitude in bearings of 2-in. diameter with $L/D = 2$. Agreement was found with isothermal calculations (1961). Operating conditions were chosen to give eccentricity ratios from 0.4 to 0.6. Likely sources of error and necessary precautions are pointed out. The author concludes that bearing design can be more reliable when guided by theory and not based entirely on experiment. Whitley (1961) distinguishes four modes of half-speed whirl. The effects of ovality, of an axial slot, and of circumferential grooves are shown. Load capacity experiments by the U. K. Atomic Energy Authority are said to check steady-load theory to about 10 per cent, or within the probable accuracy of measurement. Best correlation is obtained with isothermal calculations.

Air-lubricated journal bearings with $L/D = 1$ were tested by Kerr at high compressibility numbers (1962). From $\Lambda = 6$ to 12 he found the minimum film thickness directly proportional to the ambient pressure ratio p_a/P regardless of speed or clearance. The constant of proportionality is 1.35×10^{-4} in. per inch of diameter.

Reynolds & Gross (1962) experimented on whirl conditions with clearance ratios from about 0.27 to 2.4 mils/in. and L/D from $\frac{1}{4}$ to 1. Many aspects of current theory were confirmed, although some conclusions are contradicted. Thus it was found that no-load operation can be free from whirl at favorable L/D and C/D ratios, and that the whirl threshold speed does not necessarily decrease with increasing L/D, but depends also on the clearance ratio. Some other results of special interest are: (1) large-amplitude self-excited whirl can be prevented by unbalancing the rotor, decreasing the clearance, or venting to the atmosphere, (2) the whirl path need not enclose the bearing axis, (3) speed can safely be taken up to ten times the threshold speed for light loads with appropriate length and clearance ratios, and (4) the threshold eccentricity ratio can be anywhere from 0 to 1 depending on L/D and C/D. The author's discussion leads to a better understanding of whirl phenomena with both compressible and incompressible lubricants. See Whitley (1962); Gunter (1963); McCann (1963), and others.

Experiments on the load capacity, friction, and temperature rise in self-acting bearings are reported by Powell (1963). Miniature-type gyro spin bearings were successfully tested up to speeds of 275,000 rpm at 1000 F by Trugman (1965).

Pressurized Journal Bearings. The use of an external source of pressure is less common with journal bearings than with thrust bearings, but offers a way to avoid surface damage when starting under load.

Many interesting experiments were conducted by Grinnell & Richardson (1957), Rieger (1958), Gross (1960), Adams and co-authors (1961), and Allen, Stokes, & Whitley (1961); also by Fischer, Cherubim, & Decker; Larson & Richardson; Laub, Licht, Sixsmith and co-authors, all in 1961. These investigations cover a variety of designs and usually indicate practical ranges for the design parameters, including means for controlling whirl. See Powell et al. (1963).

8. PROPERTIES OF GASES

Gases differ from liquids in their compressibility, their lower viscosity and density, and in the fact that their viscosity increases with rising temperature. The low viscosity favors low friction at high speeds, and the effect of temperature offers a marked advantage in high-temperature applications.

The viscosity of air is about 0.018 cp or 2.6×10^{-9} lb-sec/sq in. at 68 F (20 C)—say $\frac{1}{55}$ that of water at the same temperature, or $\frac{1}{4400}$ that of an SAE 10W oil. At 212 F (100 C) the viscosity of such an oil falls to one-sixteenth of its room temperature value, while the viscosity of air increases by more than a fifth. Table 3 shows some of the physical properties of air at atmospheric pressure that may be of

TABLE 3. Some Properties of Air

Temp. F	Temp. C	Viscosity 10^{-9} reyns	Temp. Coeff. %/dF	Press. Coeff. %/100 psi	$10^3 k$ lb/sec dF	$10^2 q$ psi/dF
68	20	2.63	0.146	0.47	3.2	9.8
100	38	2.75	0.133	0.42	3.4	9.2
212	100	3.14	0.108	0.23	4.0	7.7
1000	538	5.31	0.043	–	7.3	3.9

Note: One-millionth of a reyn or lb-sec/sq in. = 6.9 cp. At 20 C, ρ = 1.21 grams/liter = 4.35×10^{-5} lbm/cu in. = 1.13×10^{-7} lb-sec^2/in.^{4}; ν = 1.57×10^{-7} in.^{2}/sec; c_p = 0.24 cal/gram dC or Btu/lbm dF; $\gamma = c_p/c_v$ = 1.40.

interest in lubrication. As before, the abbreviations *dC* and *dF* are for differences in temperature; lb and lbm for pounds force and pounds mass. Here k is the thermal conductivity and q denotes the heat capacity per unit volume, or ρc_p; where ρ is the density and c_p the specific heat at constant pressure. In the footnote, ν is the kinematic viscosity and c_v the specific heat at constant volume. Density of air may be taken inversely as the absolute temperature. The temperature coefficient values indicate that it requires about 7 dF to raise the viscosity of air by 1 per cent at moderate temperatures and pressures. It is a curious fact, not shown in the table, that the temperature coefficients become negative at higher pressures. The tabulated pressure coefficients are computed from the slope, at 1 atm, of the isothermal curve for viscosity against pressure. The effect of pressure is seen to be negligible for the loads commonly met in air-lubricated bearings, and is less at high temperatures.

The properties of five other gases relative to air at room temperature and atmospheric pressure are shown in round numbers in Table 4. Here μ is the viscosity and *TC* its temperature coefficient, while the subscript a refers to values for air. The gases are listed in decreasing order of viscosity. The relative pressure coefficients, not tabulated, are small compared to air except for nitrogen—about the same—and ethylene, nearly twice that of air. Relative values for carbon dioxide and hydrogen are less than $\frac{1}{5}$, while the pressure coefficient of helium is negative and less than $\frac{1}{10}$ that of air. The data for both tables were averaged or interpolated from published values: Hersey (1923), Comings (1944), Iwasaki (1951), AIP Handbook

TABLE 4. Relative Properties of Gases
(Compared to air at 1 atm and 20 C)

Gas	μ/μ_a	TC/TC_a	ρ/ρ_a	k/k_a	q/q_a	γ/γ_a
He	1.08	0.88	0.137	6.0	0.71	1.17
N_2	0.97	0.98	0.97	1.01	0.99	1.00
CO_2	0.81	1.16	1.52	0.62	1.3	0.93
C_2H_4	0.55	1.08	0.98	0.72	1.6	0.89
H_2	0.49	0.85	0.070	7.1	0.99	1.07

(1957), and Marks (1958); Eckert, Golubev, Kaye, and Kestin (all 1959); Hilsenrath (1955); Andrussow (1962), and others.

Assuming that the viscosity of a gas varies as the nth power of its absolute temperature over a limited range (Hersey, 1938), its temperature coefficient TC in per cent/deg F at 20 C should equal $n/528$, from which $n = 5.28TC$. Accordingly, at room temperature, n ranges from 0.77 for air up to $\frac{9}{10}$ for carbon dioxide, and down to about $\frac{2}{3}$ for helium and hydrogen.

Note the contrast in viscosity between helium and hydrogen, and in volumetric heat capacity between the low value for helium and the high for ethylene. A value of γ as low as 1.11 has been reported by Ford and co-authors for dimethylether, a gas having 0.56 of the viscosity of air. Such contrasts may prove useful in selecting gases for

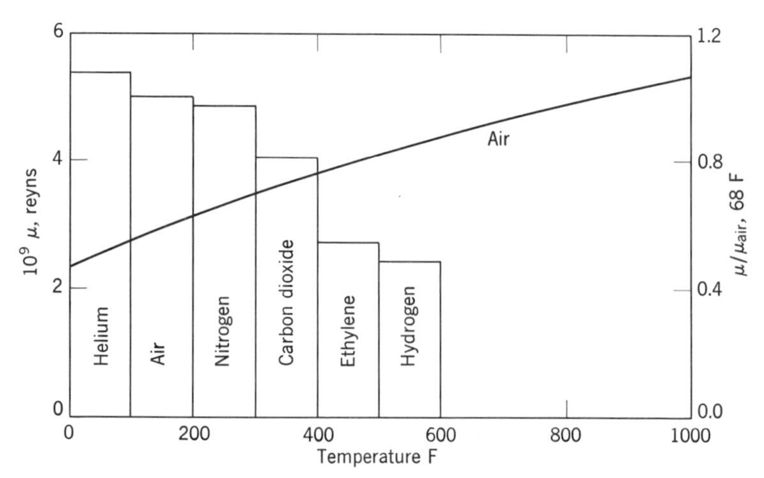

Fig. 4. Viscosity of five gases compared to air.

experimentation where a wide range of variables might be desired. See Fig. 4 for the viscosity of dry air against temperature, and for the viscosities of several gases compared to air.

9. APPLICATIONS

Modern applications followed with increasing frequency after the first historical developments. A panel meeting on gas bearing applications was conducted by the American Society of Mechanical Engineers (1963). A symposium on recent advances and applications of gas lubrication was held at the University of Southampton (1965).

Self-Acting Applications. Small centrifugal blowers and differential pressure gages were constructed by Brubach in 1947 (Fuller I: 1956). Centrifuges were designed by Artobolenskii & Sheinberg in 1950; polishing and boring machines by Hagen the following year. A vertical spinner was devised by Bild & Vial for testing small devices like clock mechanisms rotated at high speeds (1933). See discussion by Brewster, page 518.

Five new applications were described by Drescher (1953). These are (1) a squirrel-cage motor of 1.5-kw output, (2) a horizontal converter of 4 kw, (3) a vertical converter with external rotor, 12.5 kw, (4) a rotary optical filter, 3000 rpm, its accuracy increased ten-fold by replacement of the ball bearings that caused vibration, and (5) asynchronous motors, for horizontal or vertical operation, with recessed journal to eliminate whirl. Farrand applied air-lubricated thrust bearings to the magnetic memory of computers (1957).

Ford, Harris, & Pantall described the CO_2 circulating pump and motor for a gas-cooled nuclear reactor (1957). Journal bearings are of 2-in. and 5-in. diameter, speeds 2000 to 13,000 rpm. These authors report also on the motor bearings in a pump for radioactive molten lead and bismuth, hermetically sealed. The bearings operate on a vertical axis at 400 C with helium or argon as the lubricant. The chrome-plated steel shaft runs in Meehanite cast-iron bearings. Thrust bearings are made of copper-carbon to reduce starting torque and to avoid galling at starts and stops. Air-lubricated motor and grinding-head bearings were tested by Sheinberg and Kharitonov under both industrial and laboratory conditions (1958). Manufacturing methods and tolerances are reported showing how the most accurate alignment can be obtained. Hutton developed a tilting dead-weight piston gage (1959). Gordon designed the vertical air bearings of an instrument for measuring ball-bearing friction (1960); air-bearing torque only 100 dyne-cm at 10,000 rpm. See also Scott (1962).

Gas-bearing gyroscopes for inertial navigation are described by Slater (1957) and Savet (1961); helium-lubricated ceramic bearings for gyro-spin motors by Gettings and by Klass (both 1961). Gyroscopes of about 2-in. diameter weighing less than $\frac{3}{4}$ lb are now manufactured in large numbers for missiles and rockets with greatly improved performance credited to the gas bearings. Studies by Brehm and by Rothe (also 1961) compare the advantages and limitations of self-acting and pressurized gas bearings. For gas-lubricated pistons see Bush (1961) as well as Kapitza (1934).

Air bearings for dental drills to run 400,000 rpm, and helium-lubricated bearings for gas circulators to cool nuclear reactors are described by Cherubim (1963), together with liquid oxygen turbo-expanders, and motor-blower bearings for missiles, in which the bearings are subject to vibrating loads nine times gravity. Trugman (1965) reports development work on gyro bearings for operation at extremely high speeds, with reaction products of hydrogen peroxide as the working fluid in a 1400 F environment. Sternlicht & Arwas describe gas-bearing turbomachinery.

Pressurized Applications. Mueller built internal grinders running 100,000 rpm (1951). Firth used air-lubricated trunnion bearings in a high-precision electric dynamometer (1955). Fuller gives details of a commercial centrifuge with pressurized bearings (I: 1956). Laub (1960) designed gimbal bearings for inertial guidance systems. Fischer and coauthors applied pressurized bearings to a turbo-expander with 7-lb rotor operating at 55,000 rpm (1961). High-precision measuring devices with pressurized supports were built by Graneek & Kerr (1961). Laub & Norton (1961) applied spherical gas bearings to space-vehicle simulators; see also Wilcock (1965).

Air bearings were used by Hargens & Keiper to support the moving element of a tonometer—an ophthalmic instrument (Sciulli, 1961). Rothe described gimbal bearings for stabilization gyros and accelerometers in space vehicles (1961). Sixsmith, Wilson, & Birmingham report on a helium expansion turbine with gas bearings designed to prevent whirl up to 240,000 rpm (two papers, 1961). Applications to machine tools are described by H. L. Wunsch (1962). See also Levesque (1965); Steranka (1965).

Design of Gas Bearings. Current interest centers in the design approach to gas bearings. See, for example, Pantall on the design of aerodynamic bearings (pages 27–84 in Grassam & Powell, 1964) and Shires on the design of externally pressurized bearings (110–39, *op. cit.*) as well as Ausman, pages 825–53, and Gross, 395–407 (both in HS, 1965).

10. SUMMARY

The technical literature on gas-lubricated bearings has been discussed in chronological order under each of the several headings, treating theory first, experiments later, and applications last. Special consideration has been given to self-acting bearings.

Many references not readily available are abstracted in the Franklin Institute's Bibliography (Sciulli, 1961). Some twenty papers were presented and discussed at the *First International Symposium on Gas Lubricated Bearings* held at Washington, D.C. in October, 1959. The published volume begins with Professor Fuller's "General Review of Gas Bearing Technology" (1961). The state of the art has been summarized by Cole (1959); by Gross in his book (1962) and in two review papers (1962, 1963); and by Ausman in the Houston volume (1965), as well as by Sternlicht & Arwas (1966).

Self-acting gas bearings behave like liquid-film bearings at the lower values of ZN/P, where there is not much compression. At higher values their behavior is markedly influenced by the pressure ratio or "load ratio" P/p_a, where p_a is the ambient pressure. In the high-speed range, where gas bearings are chiefly needed, the coefficient of friction and relative film thickness are practically independent of speed and viscosity.

It has been shown that the performance characteristics of such bearings are fixed by the "compressibility bearing number" Λ—a new parameter, analogous to the Sommerfeld number. A common method of picturing load capacity is to plot P/p_a against Λ. The curves start out from the origin as straight lines, indicating that P is proportional to ZN at low speeds, but they level off at the higher values of Λ.

The viscosity of a gas increases with increasing temperature. For this reason gas-lubricated bearings are of especial value in high-speed and high-temperature applications, as, for example, dental drills running in air-lubricated bearings at 400,000 rpm; helium-lubricated bearings for the blowers in gas-cooled nuclear reactors; and gyro-spin bearings operating at ambient temperatures approaching 1400 F.

REFERENCES

Abbott, W. G., Jr. (1916), *Device for Utilizing Fluid under Pressure for Lubricating Relatively Movable Elements.* U.S. Pat. 1,185,571. (a) 1,337,742 (1920).

Adams, C. R., Dworski, J., & Shoemaker, E. M. (1961), "Externally Pressurized Step Journal Bearings." *JBE, T* **83,** 595–602.

Allen, D. S., Stokes, P. J., & Whitley, S. (1961), "The Performance of Externally Pressurized Bearings Using Simple Orifice Restrictors." *TASLE* **4,** 181–96.

American Institute of Physics (1957), *Handbook,* McGraw-Hill, New York; see vol. 2, 207, vol. 3, 61. (a) 2nd ed., 1961.

American Society of Mechanical Engineers (1963), "Panel: Gas Bearing Applications." *Program, Winter Annual Meeting,* ASME, New York, p. 26.

Andrussow, Leonid (1962), *Progress in International Research on Thermodynamics and Transport Properties.* ASME, New York; see pp. 279–87.

Artobolevski, I. I. & Sheinberg, S. A. (1950), High Speed Slider Bearings with Air Lubrication (Russian), *Vestnik Mashinostroenia* No. 8, 5–12.

Ausman, J. S. (1957), "The Fluid Dynamic Theory of Gas-Lubricated Bearings." *T* **79,** 1218–24.

Ausman, J. S. (1957), "Finite Gas Lubricated Journal Bearing." Paper 22, *CLW,* 39–45; 11, 821.

Ausman, J. S. (1961), "Theory and Design of Self-Acting Gas-Lubricated Journal Bearings Including Misalignment Effects." First *ISGB,* 1959; 161–92. (a) *JBE, T* **82,** 335–41 (1960).

Ausman, J. S. (1961), "An Approximate Analytical Solution for Self-Acting Gas Lubrication of Stepped Sector Thrust Bearings." *TASLE* **4,** 304–13.

Ausman, J. S. (1961), "An Improved Analytical Solution for Self-Acting, Gas Lubricated Journal Bearings of Finite Length." *JBE T* **83,** 188–94.

Ausman, J. S. (1963), "Linearized *ph* Stability Theory for Translatory Half-Speed Whirl of Long, Self-Acting Gas-Lubricated Journal Bearings." *JBE, T* **85,** 611–9.

Ausman, J. S. (1965), "Gas Lubricated Bearings." *HS,* 825–53.

Ausman, J. S. (1965), "On the Behavior of Gas-Lubricated Journal Bearings Subjected to Sinusoidally Time-Varying Loads." *JBE, T* **87,** 589–603.

Barbezat, A. (1912), *Device to Balance Thrust in Turbines.* U.S. Pat. 1,030,153.

Beams, J. W. (1937), "High Rotational Speeds." *JAP* **8,** 797–804.

Bild, C. F. & Vial, P. F. (1953), "A Simple High Speed Air Spinner for Centrifugal Testing of Small Mechanical Devices." *T* **75,** 515–9.

Block, J. R. W. & Braithwaite, J. D. (1962), "Hydrodynamic Gas-Lubricated Bearings." *Sci. Lubn.* **27.** (a) Paper 7, *LWG.*

Boeker, G. F., Fuller, D. D., & Kayan, C. F. (1958), *Gas Lubricated Bearings, A Critical Survey.* WADC Tech. Rep. 58-495, I. Wright Air Development Center; 179 pp.

Brehm, P. (1961), "Gas Bearings, III. Hydrodynamic Type." *Eng. News* **7,** No. 3, 2–5.

Brix, V. H. (1959), "Shaft Stability in Gas Film Bearings." *Eng.* **187,** 178–82.

Brody, S. (1960), "Solution of Reynolds' Equation for a Plain Slider Bearing of Finite Width with an Isothermal Gas Flow." *ASLE* 60 AM 5A–4.

Brubach, H. F. (1947), "Some Laboratory Applications of the Low Friction Properties of the Dry Hypodermic Syringe." *Rev. Sci. Instr.* **18,** 363–6.

Brunner, R. H., Harker, J. M., Haughton, K. E., & Osterlund, A. G. (1959), "A Gas Film Lubrication Study, Part III: Experimental Investigation of Pivoted Slider Bearings." *IBM JRD* **3,** 260–74.

Burgdorfer, Albert (1959), "The Influence of Molecular Mean Free Path on the Performance of Hydrodynamic Gas Lubricated Bearings." *JBE, T* **81,** 94–100.

Bush, Vannevar (1961), *Gas-Lubricated Free Piston . . . .* U.S. Pat. 2,983,098. (a) *Hydraulic Pump.* U.S. Pat. 3,145,660 (1964).

Castelli, V. & Elrod, H. G. (1965), "Solution of the Stability Problem for 360 Deg Self-Acting, Gas-Lubricated Bearings." *JBE, T* **87,** 199–212.

Castelli, V., Stevenson, C. H., & Gunter, E. J., Jr. (1964), "Steady-State Characteristics of Gas-Lubricated, Self-Acting, Partial-Arc Journal Bearings of Finite Width." *TASLE* **7,** 153–67.

Charron, M. E. (1910), "Rôle lubrifiant de l'air dans le frottement des solides. Frottement dans le vide." *CR* **150,** 906–8.

Cheng, H. S. & Pan, C. H. T. (1966), "Stability Analysis of Gas-Lubricated, Self-Acting, Plain, Cylindrical Journal Bearings of Finite Length, Using Galerkin's Method." *JBE, T* **88** (in press).

Cheng, H. S. & Trumpler, P. R. (1963), "Stability of the High Speed Journal Bearing Under Steady Load. Part 2, The Compressible Film." *JBE, T* **85,** 274–80.

Cherubim, Justin (1963), *Solutions of Problems Associated with the Application of Gas-Lubricated Bearings. ASME* (June); 8 pp. +.

Cole, J. A. (1959), "Gas Lubrication." *Research* **12,** 348–55.

Cole, J. A. & Kerr, J. (1957), "Observations on the Performance of Air Lubricated Bearings." Paper 95, *CLW,* 164–70; **11,** 743, 820, 847; Plate 22.

Comings, E. W., Mayland, B. F., & Egly, R. S. (1944), *The Viscosity of Gases at High Pressure. EES Bull.* 354, University of Illinois, Urbana; 68 pp.

Comolet, R. (1957), *Écoulement d'un fluide entre deux plans parallèles. Contribution à l'étude des butées à air.* Publics. sci. et tech. du Ministère de l'Air, No. 334; 68 pp. (a) *CR* **235,** 1190–3 (1952).

Comolet, R. (1961), "Radial Flow of a Compressible Fluid between Parallel Plates: Theoretical Study and Experimental Research on the Thrust Bearing." *First ISGB,* 1959; 242–50.

Constantinescu, V. N. (1955), On the Solution of the Equations of Gas Lubrication for a Limiting Case (Rumanian). *Com. Acad. RPR, Mecanica Applicata* **5,** No. 9, 1317–21.

Constantinescu, V. N. (1955), Gas Lubricated Journal Bearings of Infinite Length (Rumanian). *Studii si Cercetari de Mecanica Applicata, Acad. RPR* **6,** 377–400; **7,** 81–105 (1956).

Constantinescu, V. N. (1956), Pressure Distribution in Gas Bearings of

Finite Length (Rumanian). *Rev. Mécanique Appl., Acad. RPR* **1,** No. 1, 144–55. (a) Com. Acad. RPR, Stiinte *Technice* **6,** No. 5, 655–60.

Constantinescu, V. N. (1958), Two Dimensional Turbulent Lubrication (Rumanian). *Studii si Cercetari de Mecanica Aplicata, Acad. RPR* **9,** No. 1, 1, 139–62; No. 2, 369–76. (a) *JBE, T* **86,** 475–82 (1964).

Constantinescu, V. N. (1959), Dynamic Stability of Gas-Lubricated Bearings. *Rev. Mécanique Appl.* **4,** No. 4, 627–42. (a) *Op. cit.* **6,** No. 3, 317–30 (1961). (b) *JBE, T* **87,** 579–88 (1965).

Constantinescu, V. N. (1964), *Gas Lubrication* (Rumanian). Editura Academiei RPR, Bucarest.

Constantinescu, V. N. (1964), "On Some Secondary Effects in Self-Acting Gas-Lubricated Bearings." *TASLE* **7,** 257–68.

Constantinescu, V. N. (1964), *Theory of Hydrodynamic Lubrication in the Turbulent Regime* (Rumanian). Acad. RPR.

Cooper, S. (1961), "An Assessment of the Value of Theory in Predicting Gas Bearing Performance," *JBE, T* **83,** 195–200.

Corey, T. L., Rowand, H. H., Jr., Kipp, E. M., & Tyler, C. M., Jr. (1956), "Behavior of Air in the Hydrostatic Lubrication of Loaded Spherical Bearings." *T* **78,** 893–8.

Drescher, H. (1953), "Gleitlager mit Luftschmierung." *ZVDI* **95,** 1182–90.

Drescher, H. (1961), "Special Features of Self-Acting Air Bearings and Their Effect on Practical Application." *First ISGB,* 1959; 319–45.

Eckert, E. R. G. & Drake, R. M., Jr. (1959), *Heat and Mass Transfer.* McGraw-Hill, New York; see p. 504.

Elrod, H. G., Jr. & Burgdorfer, A. (1961), "Refinements of the Theory of the Infinitely-Long, Self-Acting, Gas-Lubricated Journal Bearing." *First ISGB,* 1959; 93–118.

Farrand, W. A. (1957), "An Air-Floating Disk Magnetic Memory Unit." *Datamation* **3,** 38–41 (Nov.–Dec.)

Ferranti, F. J., de (1909), *Air Bearing for High Speeds.* U.S. Pat. 930,851.

Firth, D. (1955), "Electric Dynamometer of High Precision—Air Lubricated Trunnion Bearings Employed." *Eng.* **179,** 628–30.

Fischer, G. K., Cherubim, J. L., & Decker, O. (1961), "Some Static and Dynamic Characteristics of High-Speed Shaft Systems Operating with Gas-Lubricated Bearings." *First ISGB,* 1959; 383–417.

Ford, G. W. K., Harris, D. M., & Pantall, D. (1957), "Principles and Applications of Hydrodynamic-Type Gas Bearings." *Proc. IME* **171,** 93–128.

Fox, G. R. & Sneck, H. J. (1961), "Orifice Flows in Externally Pressurized Bearings." *First ISGB,* 1959; 482–96.

Fuller, D. D. (1953), "Low Friction Properties of Air-Lubricated Bearings." *Annals, New York Acad. Sci.* (II) **15,** 93–9. (a) *LE* **9,** 298–301.

Fuller, D. D. (1961), "General Review of Gas Bearing Technology." *First ISGB,* 1959; 1–29.

Fuller, D. D., editor (1961), *First International Symposium on Gas-Lubricated Bearings,* Washington, D.C., 1959; U.S. Gov. Printing Office; 617 pp.

Golubev, I. F. (1959), *Viscosity of Gases and Gaseous Mixtures* (Russian), Fizmatgiz, Moscow; see pp. 98, 142–3.

Gordon, K. M. (1960), "Gas Bearings, Part II; Aid in High Speed Torque Determination." *MPB Eng. News* **6,** No. 5.

Graneek, S. K. & Kerr, J. (1961), "Air Bearings—Research and Applications at National Engineering Laboratory, Scotland." *First ISGB,* 1959; 71–91.

Grassam, N. S. & Powell, J. W. (1964), *Gas Lubricated Bearings.* Butterworth, Inc., Washington, D. C.; 326 pp.

Grinnell, S. K. (1956), "Flow of a Compressible Fluid in a Thin Passage." *T* **78,** 765–71.

Grinnell, S. K. & Richardson, H. H. (1957), "Design Study of a Hydrostatic Gas Bearing with Inherent Orifice Compensation." *T* **79,** 11–22.

Gross, W. A. (1958), *Film Lubrication III. Basic Hydrodynamic Relations.* IBM Res. Lab., San Jose, Calif., Tech. Rep. RJ-R12-117-3; 27 pp. (a) *IV. Compressible Lubrication of Infinitely Long Slider and Journal Bearings.* IBM Res. Paper RJ-RR-117-4; 106 pp.

Gross, W. A. (1959), "A Gas Film Lubrication Study, Part I. Some Theoretical Analyses of Slider Bearings." *IBM JRD* **3,** 237–55.

Gross, W. A. (1960), "Steady Performance Characteristics of Gas-Lubricated Journal Bearings with Slenderness Ratio $L/D = \pi$." *J. Aerospace Sci.* **27,** 869–70.

Gross, W. A. (1961), "Numerical Analysis of Gas-Lubricating Films." *First ISGB,* 1959; 193–223.

Gross, W. A. (1962), "Investigation of Whirl in Externally Pressurized Air-Lubricated Journal Bearings." *JBE, T* **84,** 132–8.

Gross, W. A. (1962), "Gas-Lubricated Bearings." *AMR* **15**(10), 765–9.

Gross, W. A. (1962), *Gas Film Lubrication,* Wiley, New York; 413 pp.

Gross, W. A. (1963), "Gas Bearings: A Survey." *Wear* **6,** 423–43.

Gross, W. A. (1965), "Lubrication of Externally Pressurized Bearings." *HS,* 307–421.

Gross, W. A. & Zachmanoglu, E. C. (1961), "Perturbation Solutions for Gas-Lubricating Films." *JBE, T* **83,** 139–44.

Gunter, E. J., Castelli, C., & Fuller, D. D. (1963), "Theoretical and Experimental Investigation of Gas-Lubricated, Pivoted Pad Journal Bearings." *TASLE* **6,** 324–36.

Gunter, E. J., Jr. & Fuller, D. D. (1963), *Recent Progress on the Development of Gas-Lubricated Bearings for High-Speed Rotating Machinery.* USAF Fluids and Lubricants Conf., San Antonio; 40 pp.

Hagen, H. W. (1951), *Bau und Berechnung luftgelagerte Wellen.* Dissertation, TH Aachen.

Henriot, E. & Hugenard, E. (1925), "Sur la réalisation de très grandes vitesses de rotation." *CR* **180,** 1389–92. (a) *J. phys. et Rad.* (6) **8,** 433–43 (1927).

Hersey, M. D. (1923), "Variation of Fluid Properties in Aerodynamics." *Rep. Internat. Air Cong.,* London; pp. 414–21.

Hersey, M. D. (1938), "Physics of Lubrication, III. Note on the Theory of Air-Lubricated Journal Bearings." *Phys. Rev.* **53,** 926.

Hilsenrath, J. et al. (1955), *Tables of Thermal Properties of Gases.* NBS C 564; 488 pp.

Hughes, W. F. & Osterle, J. F. (1957), "On the Adiabatic Couette Flow of a Compressible Fluid." *T* **79,** 1313–6. (a) *ZAMP* **8,** 89–96.

Hutton, U. O. (1959), "A Tilting Air-Lubricated Piston Gage for Pressures below One-Half Inch of Mercury." *NBS J Res* **63C,** No. 1, 47–57.

Iwasaki, H. (1951), "Measurement of Viscosities of Gases at High Pressure, I. Viscosity of Air at 50°, 100°, and 150°C." *Sci. Reports, Tohoku University,* Japan, (*A*)**3,** 247–57.

Kao, Hsiao-Cho (1963), "A Theory of Self-Acting, Gas-Lubricated Bearings with Heat Transfer through Surfaces." *JBE, T* **85,** 324–8.

Kapitza, P. (1934), "The Liquifaction of Helium by an Adiabatic Method." *PRS* **147,** 189–211. (a) *Nature* **133,** 708–9.

Katto, Y. & Soda, N. (1953), "Theory of Lubrication by Compressible Fluid with Special Reference to Air Bearings." *Proc. Second Japan Nat. Cong. AM,* 1952; 267–70.

Katto, Y. & Soda, N. (1962), "Theoretical Contributions to the Study of Gas-Lubricated Journal Bearings." *JBE T* **84,** 123–31.

Kaye, G. W. C. & Laby, T. R. (1959), *Physical and Chemical Constants.* Longmans, Green, London; 12th ed., see pp. 17, 38, 52, 57–8.

Kerr, J. (1962), "Air-Lubricated Journal Bearings at High Compressibility Numbers." *Proc. IME* **176,** 582–3.

Kestin, J. & Leidenfrost, W. (1959), "An Absolute Determination of the Viscosity of Eleven Gases over a Range of Pressures." *Physica* **25,** 1033–62.

Kestin, J. & Richardson, P. D. (1963), "The Viscosity of Superheated Steam up to 275 Deg C. A Refined Determination." *JHT T* **85,** 295–302.

Keyes, F. G. & Vines, R. G. (1965), "The Thermal Conductivity of Nitrogen and Argon." *JHT, T* **87,** 177–83.

Kingsbury, Albert (1896), "Experiments on the Friction of Screws." *T* **17,** 96–116 (see pp. 115–6).

Kingsbury, Albert (1897), "Experiments with an Air-Lubricated Journal." *JAS NE* **9,** 267–92.

Klein, F. & Sommerfeld, A. (1903), *Über die Theorie des Kreisels,* Teubner, Leipzig; see vol. 2, p. 546.

Kochi, K. C. (1959), "Characteristics of a Self-Lubricated Stepped Thrust Pad of Infinite Width with Compressible Lubricant." *JBE, T* **81,** 135–46.

Kowaski, Y. (1947), On the Air Lubricated Bearings (Japanese). *J. Soc. Precision Mach'y,* Japan, **13,** No. 5–7, 39–44; No. 152–4 (May–July).

Krzywblocki, M. E. (1957), "On the Reiner-Taylor-Saffman Dilemma." *J. Aerospace Sci.* **24,** 915–6.

Larson, R. H. & Richardson, H. H. (1962), "A Preliminary Study of Whirl Instability for Pressurized Gas Bearings." *JBE, T* **84,** 511–20.

Laub, J. H. (1960), "Evaluation of Externally Pressurized Gas Pivot Bearings." *JBE, T* **82,** 276–86. (a) *First ISGB,* 1959; 435–81.

Laub, J. H. (1960), "Elastic Orifices for Gas Bearings." *JBE, T* **82,** 980–2.

Laub, J. H. (1961), "Externally Pressurized Journal Gas Bearings." *TASLE* **4,** 156–71.

Laub, J. H. & Norton, R. H., Jr. (1961), "Externally Pressurized Spherical Gas Bearings." *TASLE* **4,** 172–80.

Lawrence, E. O. & Beams, J. W. (1928), "An Improved Arrangement for Obtaining High Speeds of Rotation." *Phys. Rev.* **31,** 1112. (a) Beams, J. W. et al., *Science* **78,** 328–40 (1933).

Lemmon, J. R. (1962), "Analytical and Experimental Study of Externally Pressurized Air Lubricated Journal Bearings." *JBE, T* **84,** 159–65.

Levene, M. L. (1965), "Air Lubrication, A Development Tool." *ME* **87,** 53–5 (Dec.)

Levesque, G. N. (1965), "The Error Correcting Action of Hydrostatic Bearings." ASME 65-LUBS-12; 5 pp. (a) *ME* **87,** 58 (Aug.)

Licht, Lazar (1961), "Air Hammer Instability in Pressurized-Journal Gas Bearings." *JBE, T* **83,** 235–43.

Licht, L. & Elrod, H. (1960), "A Study of the Stability of Externally Pressurized Gas Bearings." *JAM* **27,** *T* **82,** 250–8.

Licht, L. & Fuller, D. D. (1954), "A Preliminary Investigation of an Air-Lubricated Hydrostatic Thrust Bearing." *ASME* 54-LUB-18. (a) *ME* **77,** 176 (1955).

Licht, L., Fuller, D. D., & Sternlicht, B. (1958), "Self-Excited Vibrations of an Air-Lubricated Thrust Bearing." *T* **80,** 411–14.

Malinowski, S. B. & Pan, C. H. T. (1965), "The Static and Dynamic Characteristics of the Spiral Grooved Thrust Bearing." *JBE, T* **87,** 547–8.

Marks, L. S. (1958), *Mechanical Engineers' Handbook.* McGraw-Hill, New York, 8th ed., pp. 4–67, 95.

McCann, R. A. (1963), "Stability of Unloaded Gas-Lubricated Bearings." *JBE, T* **85,** 513–18.

McKinley, D. K. (1953), "Investigation of a Kingsbury Thrust Bearing Using Air as a Lubricant." *JSAE* **61,** 108–9 (Nov.).

Michael, W. A. (1959), "A Gas Film Lubrication Study, Part II. Numerical Solution of the Reynolds Equation for Finite Slider Bearings." *IBM J. Res. & Dev.* **3,** 256–9.

Mow, C. C. & Saibel, E. (1959). "The Gas Lubricated Finite Slider Bearing." *Proc. Sixth Midwest Conf. Fluid Mech.,* 406–16.

Mueller, P. M. (1951), "Air-Lubricated Bearings." *Product Eng.* **22,** 112–5; **23,** 160–3 (1952).

Nahavandi, A. & Osterle, F. (1961), "A Novel Form of Self-Acting Gas Lubricated Bearing." *TASLE* **4,** 124–30.

Nemeth, Z. N. & Anderson, W. J. (1961), "Experiments with Rotating, 10-Inch Diameter Externally Pressurized Air Thrust Bearings." *First ISGB,* 1959; 361–82.

Osterle, J. F. & Hughes, W. J. (1958), "High Speed Effects in Pneumodynamic Journal Bearing Lubrication." *Appl. Sci. Res.* **7,** 89–99.

Pan, C. H. T. (1962), "On the Time Dependent Effects of Self-Acting Gas

Dynamic Journal Bearings." *ASME* 62-LUB-10; 12 pp. (a) *ME* **84,** 95 (Nov.).

Pan, C. H. T. & Sternlicht, B. (1962), "On the Translatory Whirl Motion of a Vertical Rotor in Plain Cylindrical Gas-Dynamic Journal Bearings." *JBE, T* **84,** 152–8.

Pigott, J. D. & Macks, E. F. (1954), "Air Bearing Studies at Normal and Elevated Temperatures." *LE* **10,** 29–33.

Poritsky, H. (1962), "A Note on the Work Done on the Journal by the Forces Exerted by a Compressible Lubricant in a Journal Bearing." *JBE, T* **84,** 486–90.

Powell, J. W. (1963), "Gas Behaviour and Load Capacity of Hydrodynamic Gas Journal Bearings." Paper 11, *LWG,* 110–22.

Powell, J. W., Moye, M. H., & Dwight, P. R. (1963), "Fundamental Theory and Experiments on Hydrostatic Air Bearings." Paper 10, *LWG,* 94–109.

Raimondi, A. A. (1961), "A Numerical Solution for the Gas Lubricated Full Journal Bearing of Finite Length." *TASLE* **4,** 131–55.

Rasmussen, R. E. H. (1937), "Über die Strömung von Gasen in engen Kanälen." *Ann. Physik* **29,** 665–97.

Reiner, Markus (1957), "Research on the Physics of Air Viscosity." *PRS* **240,** 173–88; **247,** 152–67 (1958).

Reiner, Markus (1961), "The Physics of Air-Viscosity as Related to Gas-Bearing Design." *First ISGB,* 1959; 307–18.

Reiner, Markus (1965), "Second Order Stresses in the Flow of Gases." Paper No. 14, *Fourth Internat. Cong. on Rheology,* Providence, 1963; *Proc.,* Wiley, New York; **1,** 267–79.

Rentzepis, G. M. & Sternlicht, B. (1962), "On the Stability of Rotors in Cylindrical Journal Bearings." *JBE, T* **84,** 521–32.

Reynolds, D. B. & Gross, W. A. (1962), "Experimental Investigation of Whirl in Self-Acting Air-Lubricated Journal Bearings." *TASLE* **5,** 392–403.

Richardson, H. H. (1958), "Static and Dynamic Characteristic of Compensated Gas Bearings." *T* **80,** 1503–16.

Rieger, N. F. (1958), "An Experimental Investigation on the Pressurized Air-Lubricated Journal Bearings in the High-Speed Range." *LWG.* (a) *Sci. Lubn.* **11,** 10 (1959).

Rood, L. H. & Erickson, G. J. (1965), "Dynamic Behavior of Self-Acting Gas-Lubricated Thrust Bearings." *LE* **21,** 144–50.

Rothe, H. C. (1961), "Air Bearings for Guidance Components of Ballistic Missiles and Their Production Aspects." *First ISGB,* 1959; 346–60. (a) *ME* **83** (June), 45–8.

Roudebush, W. H. (1957), *An Analysis of the Effect of Several Parameters on the Stability of an Air-Lubricated Hydrostatic Thrust Bearing.* NACA TN 4095; 34 pp.

Salbu, E. O. J. (1964), "Compressible Squeeze Films and Squeeze Bearings." *JBE, T* **86,** 355–66, 638.

Savet, P. H. (editor) (1961), Gyroscopes: Theory and Design, with Application to Instruments, Guidance, and Control. McGraw-Hill; 402 pp. See Chap. 10 by G. H. Neugebauer.

Sawtelle, W. H. (1909), "Some Points Regarding the Traverse Spindle Grinders." *Am. Machinist* **32,** pt. 2 (Aug 5), 248–9.

Schutten, J., Baron, H., Van Der Hauw, T., & Van Deenen, P. J. (1958), 'Dynamic Measurement of Film Thickness in Gas Bearings." *Appl. Sci. Res. (A)* **7,** 429–36.

Sciulli, E. B. (1961), "A Bibliography on Gas-Lubricated Bearings." *First ISGB,* 1959; Appendix B, 86 pp.; revised from Franklin Institute Report of 1957. (a) Revision by A. Peters, ed. by D. D. Fuller, F. I. Interim Rep. I–A 2049–16; 180 pp. (1961).

Scott, R. D. (1962), "On the Hydrodynamically Supported Magnetic Recording Head." *ASME* 62-LUB-11; 8 pp. (a) *ME* **84,** 72–3 (Oct.).

Sheinberg, S. A. (1947), On the Question of Gaseous Lubrication of Rotating Shafts (Russian). *Trudi Semimara po Teorii Mashin i Mekhanizmov* **2,** pt. 8, 175–206.

Sheinberg, S. A. (1950), Experimental Investigation of Aerodynamic Bearings (Russian). *TIM* **6,** 182–299.

Sheinberg, S. A. (1953), Gas Lubrication of Sliding Bearings—Theory and Calculation (Russian). *TIM* **8,** 107–204.

Sheinberg, S. A. & Kharitonov, A. M. (1958), Aerodynamic Bearings for High Speed Motors and Turbines (Russian). *Vestnik Mashinostroenie, SSSR* **38,** 14–7 (Sept.). (a) *Eng.* **187,** 677–8 (1959).

Sixsmith, H. (1961), "The Theory and Design of a Gas-Lubricated Bearing of High Stability." *First ISGB,* 1959; 418–34. (a) *NBS Tech. News Bull.* **46,** 2–3 (1962). (b) With Wilson & Birmingham, *JFI* **267,** 60–1 (1962).

Shires, G. L. & Pantall, D. (1963), "Aerostatic Jacking of a Vented Aerodynamic Journal Bearing." Paper 9, *LWG,* 84–93.

Slater, J. M. (1957), "Gyroscopes for Inertial Navigators." *ME* **79,** 832–5; 857. See Fig. 3. (a) *Mil. Syst. Design* **3,** 138–9 (1959).

Sneck, H. J. (1965), "The Short Gas Lubricated Journal Bearing." *JBE, T* **87,** 1087–9. (a) **85,** 474–5 (1963).

Sneck, H. J. (1965), "An Approximate Solution for the Infinitely Long, Gas Lubricated Slider Bearing." *JBE, T* **87,** 1085–6.

Snell, L. N. (1959), "Journal Bearings with Gas Lubrication." *Eng.* **188,** 285–6.

Stein, R. A. (1961), "On the Theory of Parallel Surface Thrust Bearings of Finite Width." *ASLE* 61 AM4-A2; 25 pp.

Steranka, Paul (1965), "Life in a Miniature Gas Bearing." *Technology Rev.* **67,** 33–5 (May).

Sternlicht, B. (1961), "Gas-Lubricated Cylindrical Journal Bearings of the Finite Length. Part I. Static Loading." *JAM* **28,** *T* **83,** 535–43.

Sternlicht, B. & Arwas, E. B. (1966), "Modern Gas Bearing Turbomachinery. Part 1. The State-of-the-Art." *ME* **88,** 24–9 (Jan.). "Part 2. Research and Application," 42–8 (Feb.).

Sternlicht, B. & Elwell, R. C. (1958), "Theoretical and Experimental Analysis of Hydrodynamic Gas-Lubricated Journal Bearings." *T* **80,** 865–78; **81,** 272–3 (1959).

Sternlicht, B., Poritsky, H., & Arwas, E. (1961), "Dynamic Stability Aspects of Cylindrical Journal Bearings Using Compressible and Incompressible Fluids." *First ISGB,* 1959; 119–60.

Sternlicht, B. & Winn, L. W. (1963), "On the Load Capacity and Stability of Rotors in Self-Acting Gas Lubricated Plain Cylindrical Journal Bearings." *JBE, T* **85,** 503–12. (a) **86,** 313–20 (1964).

Stewartson, K. (1953), "On the Flow Between Two Rotating Disks." *Proc. Camb. Phil. Soc.* **49,** 333–41.

Stone, William (1921), "A Proposed Method for Solving Some Problems in Lubrication." *Commonwealth Engineer* **9,** 114–22; 139–49.

Stull, J. L., *The Stull-Ealing Linear Air Track.* The Ealing Corporation, Cambridge, Mass.; 20 pp.

Sugimoto, Y. (1951), The Theory of Air Bearings (Japanese). *TJSME* **17,** No. 63, 12–5.

Taylor, G. I. & Saffman, P. G. (1957), "Effects of Compressibility at Low Rynolds Number." *J. Aerospace Sci.* **24,** 553–62.

Tipei, N. (1954), The Equations of Gas Lubrication (Rumanian). *Communicarile Acad. RPR* **4,** Nos. 11–12, 699–704.

Tipei, N. & Constantinescu, V. N. (1961), "On High-Speed Self-Acting Gas Bearings." *First ISGB,* 1959; 225–41.

Toba, K. & Saibel, E. (1961), "The Finite Sector Thrust Gas Lubricated Step Bearing." *TASLE* **4,** 293–3.

Trugman, L. A. (1965), "Some Tests on Hydrodynamic Gas Bearings at Normal and Elevated Temperatures." *LE* **21,** 138–43.

University of Southampton, The (1965), "Gas Bearing Symposium." *Sci. Lubn.* **17,** 27–30.

Vitelozzi, W. J. (1950), *Report of an Investigation on Water and Steam Lubricated Bearings.* Rep. 3229-C, USN EES, Annapolis, Md.

Vohr, J. H. & Chow, C. Y. (1965), "Characteristics of Herringbone-Grooved, Gas-Lubricated Journal Bearings." *JBE, T* **87,** 568–78.

Wahlgren, A. G. F. (1938), *Bearing.* U.S. Pat. 2,113,335.

Wernick, R. J. & Pan, C. H. T. (1964), "Static and Dynamic Characteristics of Self-Acting, Partial-Arc, Gas Journal Bearings." *JBE, T* **86,** 405–13.

Westinghouse, George (1904), *Vertical Fluid Pressure Turbine.* U.S. Pat. 745,400.

Whitley, S. (1961), "Review of Research on Gas Bearings in the United Kingdom Atomic Energy Authority." *First ISGB,* 1959; 30–70.

Whitley, S. & Betts, C. (1959), "Study of Gas Lubricated, Hydrodynamic Full Journal Bearings." *Brit. JAP* **10,** 455–63.

Whitley, S., Bowhill, A. J., & McEwan, Miss P. (1962), "Half-Speed Whirl and Load Capacity of Hydrodynamic Gas Journal Bearings." *Proc. IME* **176,** 554–65. (a) Discussion by J. Kerr, 582–3.

Wilcock, D. F. (1965), "Design and Performance of Gas-Pressurized Spherical, Space-Simulator Bearings." *JBE, T* **87,** 604–12.

Wildmann, M. (1956), "Experiments on Gas Lubricated Journal Bearings." *ASME* 56-LUB-8; 9 pp. (a) Sciulli (1961), p. 599.

Wildmann, M. (1960), "Consideration of the Starting of Gas Lubricated Bearings." *ASME* 60-LUB-11; 5 pp.

Wildmann, M., Glaser, J., Gross, W. A., Moors, D. E., Rood, L., & Cooper, S. (1965), "Gas-Lubricated Stepped Thrust Bearing—A Comprehensive Study." *JBE, T* **87,** 213–29.

Wunsch, H. L. (1962), "The Properties and Applications of Externally Pressurized Air Bearings." *Sci. Lubn.* **14,** 14–23.

Yen, K. T. (1960), "On the Compressibility Effects of the Lubricant for Two-Dimensional Slider Bearings." *JAM* **26,** *T* **81,** 609–12.

chapter X Rolling Contact

Oil-film thrust and journal bearings are seldom loaded above 500 psi of projected area when operating under steady conditions. Gas-film bearings are loaded still more lightly. Diesel bearings can take momentary loads up to 5000 psi; but in these applications the minimum film thickness rarely drops below 0.1 mil (100 micro-in.). More concentrated loading occurs in ball and roller bearings and in gear teeth. The lubricating films are so much thinner here that their load capacity and friction depend on surface finish, elastic deformation, and pressure-viscosity; and occasionally on boundary lubrication.

1. THEORY OF ROLLING CONTACT

A review of the Hertzian stresses in dry contact offers a good background for lubrication studies Hertz (1881).

Dry Contact. Hertz showed that when two parallel cylinders of the same material are in contact without friction, the half-width of the contact area is given by

$$b = 1.52(W_1R/E)^{1/2}. \tag{1}$$

Here W_1 is the load per unit of axial length, R the equivalent radius of contact, and E Young's modulus; while Poisson's ratio ν has been taken equal to $\frac{3}{10}$. The equivalent radius is defined by setting $1/R$ equal to the sum of $1/R_1$ and $1/R_2$, where R_1 and R_2 are the radii of the two cylinders. If the cylinder of radius R_2 is in contact with a plane surface, R_1 is infinite; if it contacts the inside of a hollow cylinder, as in the case of a journal and bearing, R_1 is negative.

The pressure distribution over the contact width is elliptical with a maximum compressive stress, for $\nu = \frac{3}{10}$,

$$p_m = 0.418(W_1E/R)^{1/2}. \tag{2}$$

Belayef found that the maximum shear stress occurs on a plane at 45 deg and at a depth of $0.78b$, where it is equal to about 0.30 of p_m.

Consider two steel cylinders 3 in. in diameter under a load of 1000 lb per inch length. From (1) with $E = 3 \times 10^7$ psi and $R = 0.75$ in., $b = 10.75$ mils; and the whole width is 21.5 mils. From (2) the maximum pressure is 83,600 psi. The mean pressure, $W/2b$, is 46,500 psi.

Hertz gave for the radius of the contact circle between two spheres

$$a = 1.11(WR/E)^{1/3}, \tag{3}$$

and for the maximum pressure

$$p_m = 0.388(WE^2/R^2)^{1/3}, \tag{4}$$

where W is the total load with ν again $\frac{3}{10}$. If the two bodies 1 and 2 are of different materials, $1/E$ in Eqs. (1) through (4) may be taken equal to 1.10 times the mean value of $(1 - \nu_1^2)/E_1$ and $(1 - \nu_2^2)/E_2$.

Tokuhira evaluated the stresses due to barrel-shaped rollers (1948). Radzimovsky (1953) examined stress fluctuations in rolling contact as a basis for fatigue investigations. The effects of friction and tangential force were considered by Palmgren, Carter, Karas (1921, 1926, 1945); Palmgren, Beeching & Nicholls (1945, 1948); Poritsky, Johnson, de Pater, and Goodman (1950, 1955–1962, 1962, 1962).

Theory of Dry Rolling Friction. The nature of rolling friction was discussed in *The Ball Bearing Journal* (1926). Ishlinski, Konvisarov, and others attributed rolling friction to imperfect elasticity (1938, 1952). Reynolds' theory of differential slip was extended by Glagolev, Föppl, and Shchedrov (1945, 1947, 1947). Bikerman calculated the static rolling friction of balls on rough surfaces (1949). Tabor explained rolling friction by elastic hysteresis (1952–1962); Sasaki & Orino by impact of rough surfaces (1962). The theory of hard rollers on a visco-elastic surface by Flom, May, and co-authors predicts a maximum point on the speed curve (1959, 1959, 1962). Plastic rolling was soon analyzed by Merwin & Johnson (1963).

2. RIGID SURFACE LUBRICATION

The film thickness h_0 between two rolling cylinders, or between a cylinder and a plane surface, was calculated by Gümbel (1916), and more precisely by *Engineering* in an editorial often quoted (1916), leading to the expression

$$h_0/R = 2.45\mu U/W_1 . \tag{5}$$

Here μ is the film viscosity, U the sum of the surface velocities U_1 and U_2, and W_1 the load per unit of axial length. As before, R denotes the equivalent radius. The film pressure has a maximum value

$$p_m = 0.28(W_1/\mu U)^{1/2} W_1/R , \tag{6}$$

which occurs slightly upstream from the point of nearest approach, such that the pressure is $p_m/2$ at that point. See Fig. 1 herewith, from Fig. 5, *Engineering* (a) 1920.

Further calculations were made by Büche (1934), Heidebroek (1935), and Peppler (1938); also by Banks & Mill (1954). Although originally derived for pure rolling, (5) and (6) were shown by Peppler to be valid when sliding also occurs. The velocities are considered positive when both surfaces move in the same tangential direction, usually

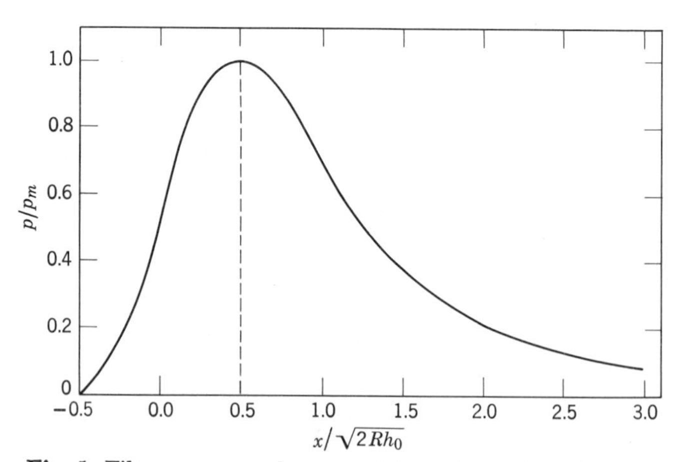

Fig. 1. Film pressure p between rigid surfaces, from *Engineering* (1920). Here p is at distance x from point of minimum film thickness h_0; p_m is its maximum value, and R the equivalent radius of curvature.

toward the right. If moving with equal speeds in opposite directions, $U_1 + U_2 = 0$ so that (5) and (6) make h_0 and $p_m = 0$.

As an example, consider two cylinders 3 in. in diameter, rotating at 600 rpm, one clockwise and the other counterclockwise, such that $R = \frac{3}{4}$ in. and $U = 189$ in./sec. Suppose the viscosity to be that of an SAE 40 oil at 130 F, for which $\mu = 10$ n or micro-reyns. Under a load of 1000 lb/in., Eq. (5) makes $h_0 = 3.47$ micro-inches and the surfaces are on the verge of metallic contact. By Eq. (6), $p_m = 272{,}000$ psi, over three times the maximum Hertz pressure for dry cylinders.

More precise solutions have been worked out by Floberg with the aid of digital computer (1959, 1961). Floberg takes into account the effects of cavitation in the divergent area, whereby the film is broken into strips or streamlets. The number of strips per unit length increases with speed at constant h_0. Further, he considers the effect of a restricted oil supply. Apparently a drop of 3.5 per cent below the maximum rate of oil flow reduces the load capacity to half, with a corresponding increase of friction.

Spheres in rolling contact ride on much thinner films than cylinders of comparable size under equal load. Archard & Kirk (1961) concluded from studies by Howlett (1946) and Kapitza (1955) that for rigid spheres and uniform viscosity

$$h_0/R = 28.4(\mu UR/W)^2 \,. \tag{7}$$

Here W is the load, U the sum velocity, and R the equivalent radius as before. The maximum pressure is given by

$$p_m = 3.6 \times 10^{-3} \frac{W^3}{\mu^2 U^2 R^4} \cdot \tag{8}$$

The authors note that two crossed cylinders of equal radii with axes at right angles are equivalent to a sphere of the same radius contacting a flat surface; the vector sum velocity U then being equal to $\sqrt{2}$ times the sliding velocity of each cylinder. Numerical data show that, in general, only boundary lubrication can be expected from the contact of *rigid* spheres or crossed cylinders.

Friction Calculations. The power loss per unit length by Peppler's solution may be written

$$H_1 = 1.47 U W_1 (\mu U / W_1)^{1/2} (1 + 1.24 s^2) \,, \tag{9}$$

where the slip ratio s is defined as the difference $U_2 - U_1$ divided by the sum velocity U. Now H_1 is the sum of the losses at the two surfaces, such that

$$H_1 = W_1(f_1 U_1 + f_2 U_2) \,, \tag{10}$$

where f_1 and f_2 are the coefficients of friction on the respective surfaces. When the lower surface is stationary, $s = 1$; when there is no slip, a condition of pure rolling, $s = 0$. In either case the coefficient f on a moving surface is H_1/W_1U. Substituting from (9) gives

$$f = k(\mu U/W_1)^{1/2}, \tag{11}$$

where $k = 1.47$ for pure rolling and 3.29 when one surface is stationary. In the case of combined rolling and sliding Peppler showed that

$$f_{1,2} = 1.47(\mu U/W_1)^{1/2}(1 \pm 1.24s) ; \tag{12}$$

the plus sign going with f_1 and the minus with f_2. Equation (12) is derived from the hydrodynamic pressure distribution taking into account that the shear stresses are different on the two moving surfaces. Substituting from (12) into (10) leads to Eq. (9), while (12) reduces to (11) when $s = 0$ or 1. It is interesting that the friction losses are independent of the radii if the surface speeds are known, and provided there is an ample supply of lubricant. Floberg studied the effects of a restricted oil supply on friction as well as on film thickness.

Two problems may be considered in the case of pure rolling: (1) when two equal cylinders are rolling upon each other about fixed axes, and (2) when one cylinder rolls freely along a stationary plane surface. The first problem is solved by writing $U_1 = U_2 = U/2$, so that $s = 0$ and $k = 1.47$. In the example of a 3-in. diameter cylinder rotating at 600 rpm, $U = 189$ in./sec. If again $\mu = 10$ n and the loading is 1000 lb/in., $f = 0.0020$. The second is solved by imparting a finite velocity U_1 to the whole system. This does not cause any internal change but makes the second problem the same as the first, with an identical solution.

Two problems can also be distinguished in the case of pure sliding, (1) when one cylinder is rotating about a fixed axis and against a stationary plane or cylindrical surface, and (2) when the cylinder is just sliding along a stationary surface without rotation. We can visualize these problems by considering the lower surface to be the stationary one, as usual, with $U_1 = 0$; so that $U = U_2$, $s = 1$, and $k = 3.29$. Both problems have the same solution, in our example $f = 0.0032$. Note the greater resistance to sliding as compared with rolling.

Effects of Variable Viscosity. Karlson recognized the problem of rolling contact with pressure-dependent viscosity (1926). His integration was completed later by a graphical method (Hersey & Lowdenslager, 1950). In the meantime Gatcombe offered an approximate solution that stirred up new interest in gear research (1945). The results were plotted in dimensionless coordinates by Hersey & Hopkins (IV:

1945). Blok derived a more exact solution for load capacity in discussing the 1950 reference. He showed that it could not be increased more than two or three fold due to pressure-viscosity as long as the surfaces are rigid. The question was further discussed by McEwen and by Blok using the generalized Karlson equation (1952). When theoretical film pressures approach infinity, the peak of the curve is so much narrower that the load capacity remains finite. Applications were made by Poritsky (1954) and by Kapitza (1955) to the rolling contact of cylinders, and by Korovchinskii to the sliding of spheres (1958); but no gain was found in the load capacity of rigid surfaces beyond the limits indicated by Blok.

Since the pressure coefficient of viscosity diminishes with rising temperature, the beneficial effects of pressure-viscosity are partly offset by any heat developed in the film. Heat generated by shearing is present in every lubricating film, and the effect of adiabatic compression may be appreciable in rolling contact. This contribution is positive from film entrance to the point of maximum pressure, but negative the rest of the way. Thus it tends to equalize film temperatures that would otherwise rise more steeply toward the exit. The fractional increase in absolute temperature per unit rise of pressure under isentropic conditions is initially equal to α/q, where α is the thermal expansivity of the oil and q its heat capacity per unit volume. Starting from 100 F, an initial increase of the order of 1.6 deg F per 1000 psi would be expected in a mineral oil pressures applied instantaneously. The adiabatic temperature rise is partly neutralized by the Joule-Thomson effect (Chapter IV).

Temperature rise due to shearing may be estimated when sufficient data are given. Finston (1951) calculated temperature distribution from an energy equation. In a certain application he showed that the maximum temperature would be found at one-tenth of the distance from one roller surface to the other. Floberg (1961) determined the temperature rise at exit by means of a computer, taking account of cavitation and restricted oil supply. A rough upper limit is fixed at p_m/q by neglecting heat conduction, since the work done per unit volume in expelling an incompressible fluid is equal to the pressure drop.

Non-Newtonian Effects. Gaskell (1950) treats pure rolling symbolically, but with more detail in the case of a Bingham plastic, where a solid core is formed in the interior of the film. Kotova and Deriagin (1957) gave the theory for a cylinder rolling over an ideal or Tresca plastic, a material characterized only by the limiting shear stress, beyond which there is no resistance to deformation. Kotova

(1957) extended the solution to a Bingham plastic. Her results are reducible to those for an ideal plastic and for a Newtonian lubricant.

The Bingham plastic was characterized by the limiting shear stress and the "plastic viscosity" or mobility reciprocal. A solid core is formed as noted by Gaskell. It extends completely across the clearance at the point of maximum pressure and at the tail end of the film. The solution for load capacity per unit length is a linear function of the sum velocity.

The foregoing assumes rigid rollers, lubricant at a uniform temperature, and supply rate ample; but the case of a restricted supply is solved for any given entrance length. Kotova's paper includes a formula for power loss. At high speeds, the loss cannot be much greater than for Newtonian lubricants of comparable viscosity.

Milne (VI: 1957) studied lubrication with a Maxwell material. He assumed a nearly uniform film thickness so that the velocity profile could be taken as a straight line. Dimensionless load capacity and friction values are plotted against the ratio of transit time to relaxation time for two conditions: (1) the lubricant fully relaxed at entry, and (2) unstressed at entry. Load capacity and friction are greater for the unstressed entry. Both curves rise with increasing transit time and approach the Newtonian value as a limit. These results tend to show that the high load capacity of a lubricant under rolling conditions cannot be attributed to shear elasticity. Burton (1960) compared film pressure calculations for Maxwell and Newtonian fluids. He concluded that shear elasticity can reduce load capacity and friction. And he noted that the Maxwell model serves only as a first approximation to the more complex visco-elastic lubricants encountered in practice. The relaxation time will be longer, and the influence of shear elasticity greater for polymers than for small molecules. It will be most important for flow over asperities under a heavy load because the transit times are then relatively short compared to relaxation times. The latter may be as high as 10^{-5} sec in polymer solutions.

A graphical integration for the Maxwell fluid was given by Crouch & Cameron (1960)—it had been thought by Cameron that shear elasticity might be important in gear lubrication (1954). The authors assume linear velocity distribution, a parabolic pressure curve, and pressure-dependent viscosity. They find Maxwell friction dropping to two-thirds of the Newtonian friction for typical parameter values. Lack of correlation with Borsoff's scuffing tests (XI: 1951) was seen to indicate that relaxation properties may not be significant. Sasaki, Mori, & Okino worked out a theory of grease lubrication for two parallel cylinders in rolling contact (1960), treating the grease as a

Bingham plastic. The shape of the solid core is different under three conditions: (1) one cylinder rotating, the other stationary, (2) one rotating at half the speed of the other, and (3) pure rolling. Starting friction is found proportional to the yield shear stress. With increasing speed the plastic viscosity becomes more significant than the yield stress until, at high velocities, the theory of grease lubrication approaches the Newtonian form. Bell discussed the Ree-Eyring fluid (IV: 1962).

3. ELASTOHYDRODYNAMIC LUBRICATION

Concentrated loading often leads to film pressures that are high enough to deform the opposite surfaces appreciably. The elastohydrodynamic problem then requires the simultaneous solution of two equations, one giving the hydrodynamic pressures corresponding to an unknown clearance shape, the other giving the solid contours in terms of the unknown pressure distribution. Peppler (1938) and Uggla (1939) foresaw that under the heaviest loads, film pressures would approach the dry surface Hertzian distribution, and the film would tend toward a uniform thickness. This would account for the fact that hydrodynamic load capacities are greater than shown by rigid surface theory.

Meldahl (1941) and Dörr (1954) solved the elastohydrodynamic problem for two rolling cylinders assuming a uniform viscosity. The solution was extended by Grubin & Vinogradova (1949) and by Petrusevich (1951) to include pressure-viscosity. Grubin's theory assumes that the surfaces are flat and parallel in the Hertz area with a film thickness h_0; but that the film thickness h in the entering section is equal to h_0 plus the Hertz separation. This concept led to the isothermal formula

$$\frac{h_0}{R} = 1.08 \left(\frac{E'R}{W_1}\right)^{1/11} \left(\frac{b_1 U}{R}\right)^{8/11}, \tag{13}$$

where E' is a reduced Young's modulus such that $1/E'$ is the mean value of $(1 - \nu^2)/E$ for the two bodies, ν being Poisson's ratio. The symbols R, U, and W_1 have their usual meanings while μ_1 and b_1 denote the film viscosity and its pressure coefficient, both evaluated at 1 atm. The equation is a milestone in rolling contact theory, although it cannot be completely general because it makes h_0 zero when b_1 is zero; that is, for all lubricants of uniform viscosity. Petrusevich calculated friction and temperature rise as well as film thickness. He found a second pressure maximum having a very sharp peak close to the outlet of the film. Skeptically received, the result has since been confirmed mathematically and experimentally. See now the discussions by Blok and by Misharin (1958), pages 464–6 and 500–1 of the reference. Petrusevich's equation for the coefficient of friction was amplified by Blok and made dimensionally homogeneous. Petrusevich further predicted a second minimum film thickness owing to local deformations near the outlet. This, too, has been confirmed.

Poritsky formulated the elastohydrodynamic problem rigorously and explored several methods for approximate solution (1952). He concludes with a striking example to illustrate the separate and combined effects of elastic deformation and pressure-viscosity. For the gear design chosen, and assuming a required film thickness of 8 microinches, the calculated load capacity for rigid teeth with uniform oil viscosity is 28 lb/in. of face width. This will be raised to 40 or less by adding pressure-viscosity alone, and to 48 allowing for deformation alone; but to 3000 lb/in. by including both pressure-viscosity and elastic deformation. Blok had already shown that the two-body elastohydrodynamic problem could be simplified by substituting for the system a single deformable body contacting a rigid plane surface. The deformable body would be assigned an appropriate "conformity radius" equal to the original equivalent radius R, and an effective elastic modulus (1951). Details are explained in his discussion of Poritsky's paper. Weber & Saalfeld's calculations (1954) showed 50 per cent gain from pressure-viscosity. The general theory was further developed by Kodnir (1960), Korovchinskii (1960), and Saibel (VII: 1960).

Dowson and co-authors came up with a numerical solution that clearly brings out the double maximum as a characteristic of the film pressure at high speeds (1959, 1960). In later publications (1961, 1962) they arrived at a formula for the relative film thickness, equivalent to

$$\frac{h_0}{R} = 0.99\left(\frac{E'R}{W_1}\right)^{0.13}\left(\frac{\mu_1 U}{E'R}\right)^{0.7}(b_1 E')^{0.6}. \tag{14}$$

The similarity of this equation to Grubin's will be apparent. Note the small exponent on W_1 and the larger one on U. It was found to be in fair accord with experiment over the range tested.

Archard & Kirk (1961) observed that hydrodynamic lubrication could exist at spherical, or nominal "point" contacts over a wide range of conditions. Calculations similar to Grubin's indicated a minimum film thickness proportional to the 0.74 power of $\mu_1 b_1 U$, the 0.41 of R, and the 0.074 power of E/W_1. A solution for rolling cylinders by

Archard, Gair & Hirst (1961) confirmed the double maximum for small radii at the higher speeds. Numerical examples are worked out for steel cylinders. Upon substituting the data into Eqs. (13) and (14), it would appear that their calculated film thickness comes about $\frac{1}{3}$ less than by Grubin's formula, and $\frac{1}{6}$ greater than by Dowson's.

Sternlicht and co-authors (1961) solved the Reynolds, energy, and elasticity equations simultaneously by iteration. They treated viscosity as a function of temperature and pressure, neglecting heat conduction. An example is given of steel disks in rolling contact with h_0 = 4 micro-inches. What load capacity may be expected? Assuming a low VI oil, it was found that

1. For rigid surfaces and uniform viscosity, W_1 = 140 lb/in.
2. With elastic deformation and allowing for pressure-viscosity at constant temperature, W_1 goes up to 12,000 lb/in.
3. Allowing for an adiabatic temperature rise, W_1 drops to about 9800 lb/in.

As a second application, two oils of different viscosity index were compared at the same load and inlet temperature. The low VI oil gives the greater film thickness and the lower temperature. See also Christensen (1962).

Stephenson & Osterle (1962) describe a computer solution including compressibility of the lubricant at constant temperature. The effect is to increase the load capacity slightly, and to reduce the sharp pressure peak near outlet of film. Reading from curve at a film thickness of 0.1 mil gives a load capacity of about 14,000 lb/in. The film thickness for this load by Eq. (14) would be 0.06 mil, less than

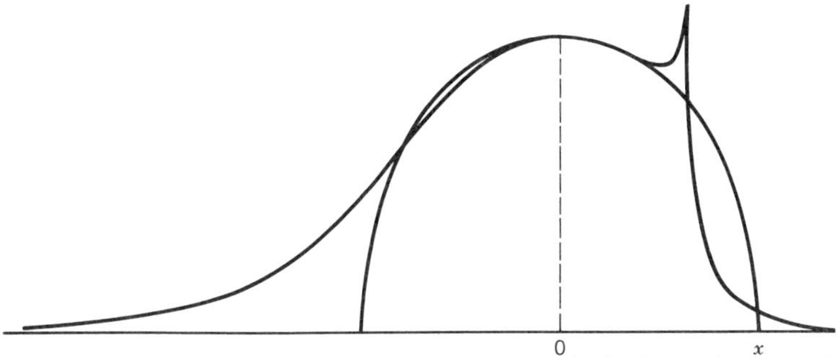

Fig. 2. Film pressure $p(x)$ in a typical elastohydrodynamic contact, from Dowson & Whitaker, *TASLE* **8**, No. 3 (1965)—Hertzian ellipse for comparison.

$\frac{2}{3}$ of their value. See also Johnson and Korovchinskii (both 1963); Archard & Cowling, Cheng & Sternlicht, Dowson & Whitaker, and Smith (all 1965). See Fig. 2.

4. ROLLER EXPERIMENTS WITH FLUID FRICTION

Many tests have been conducted on cylindrical rollers, and other apparatus, to check the foregoing theories and simulate the conditions in gears and rolling bearings.

More than forty investigations made use of disk machines having cylindrical surfaces in contact with or without slip, and operating under steady conditions. Nonuniform or fluctuating conditions were simulated in others, as for example by contact of conical rollers; by eccentric gear drives; or impulsive loading. At least six utilized rolling balls in mutual contact. Some used crowned disks, rollers on skewed axes, or balls loaded between flat plates. Many different materials have been tested, both dry and with a lubricant. The parameters varied include load, rolling speed, sliding speed; temperature, viscosity, and composition of the lubricant; hardness and finish of the solid surfaces. "Rolling speed" usually means the sum, U, of the two surface speeds. "Sliding speed" refers to their difference, $U_2 - U_1$.

Hydrodynamic performance may be reported in two groups: (1) film pressure and film thickness, and (2) frictional resistance.

Film Pressure and Film Thickness. Peppler's measurement (1938) of film pressure confirmed the distribution calculated for rigid cylinders under moderate loading. The points of maximum and zero gage pressure are equally distant from the point of nearest approach. Bergen & Scott (1951) confirmed this for viscous liquids, using 10-in. diameter rolls. They measured film thickness by flattening of solder wires. In rolling plastic materials, however, they found the distance greater to the end of the film. Dynamic squeeze-film action was simulated in Gatcombe's machine by disks having a fluctuating center distance (1951). Banks & Mill (1954) measured 2 atm of negative pressure at the beginning of cavitation. And they photographed the filaments in the cavitated region between transparent rollers. Magnified photos by Miller & Myers (1958) show cavitation with exposures of 10^{-6} sec.

Crook determined film thickness electrically in his disk machines (1957, 1958, 1962). At first he used a bronze disk of 8-in. diameter rolling against a 4-in. steel disk. After running-in, the resistance rose quickly and leveled off at 50,000 ohms, indicating a hydrodynamic film. Next he tried a pair of steel disks 3 in. in diameter, with respective surface speeds about 30 and 20 ft/sec in the same direction. Capacitance measurement revealed a mean film thickness of 40 micro-inches under a loading of 1120 lb/in. With Hertzian deformation the film would be of uniform thickness, 16 mils long in the direction of motion. At light loads h_0 was found proportional to the speed/load ratio as expected from classical theory, but only half as thick as calculated. It was seen that elastic deformation becomes important at loads above 100 lb/in. Graphs then showed film thickness to be a function of μU alone, independent of load and slide/roll ratio over a wide range, and varying approximately as the square root. Crook confirmed the Petrusevich hump or restriction near trailing end. He measured the temperature and velocity profile of the film between rollers, found a considerable rise in temperature with sliding speed, and showed that the important viscosity is that at the surface temperature of the metal.

Additives were found by El Sizi & Shawki (1960) that increased the conductivity of mineral oils, helping to make film measurement by resistance more reliable. The lubricant contained 4 per cent of sodium sulphonate. Test results are plotted over a wide range, each curve for a constant slide/roll ratio. The disks were of 6-in. diameter by about 2-in. face. Measured film thicknesses are nearly proportional to the square root of $\mu U/W_1$, and less than calculated for rigid cylinders. The two companion papers contain extensive test data, followed by discussion.

Archard & Kirk (1961) observed hydrodynamic lubrication at "point" contacts. A crossed-cylinder machine was used with diameters up to 3 in. at speeds from 2.5 to more than 3000 rpm. Cylinders were of hardened steel, finished to a C.L.A. (center-line average) better than 2 micro-inches. Plain turbine oil was supplied by jet at temperatures from 15 to 50 C. The pressure coefficient of viscosity at these temperatures varied from about 3.0×10^{-9} to 2.3×10^{-9} cm^2/dyne. Resistance fell from 10,000 ohms at 300 kg to approximately 1000 ohms at 500-kg load. Capacitance measurements, treated statistically, showed film thickness values nearly proportional to the 0.56 power of $\mu_1 b_1 U$ and the 0.62 power of R. Differences between these exponents and the calculated values were attributed to imperfections in the theory. "It appears," the authors conclude, "that hydrodynamic lubrication occurs much more frequently, and boundary lubrication less frequently than is usually supposed." Later experiments by the same authors (1963) using crossed cylinders of brass, glass, and transparent "perspex" showed that the lubrication was always hydrodynamic.

Floberg confirmed his theory of cavitation by experiments on a cylinder of 80-mm diameter with $L/D = 1$, rotating close to a glass plate (1961). It was seen that the pressure film carries well past the point of nearest approach. The number of oil strips in the cavitation zone varied from 11 at the largest clearance with a low speed to 150 or more at the smallest clearance and a higher speed, as shown photographically. Sasaki, Okamura, & Isogai (1961) experimented with 60-mm diameter hard steel disks, superfinished to a maximum roughness of 0.1 micron (4 micro-inches). Speeds were varied from 100 to nearly 3600 rpm, and other parameters over a wide range. Rolling tests showed that film thicknesses based on electrical resistance increase from 26 to 52 micro-inches as $\mu U/W_1$ varies from 8×10^7 to 100×10^7. Translating the results into gear tooth action, a chart was drawn up for shaft speed against tooth load showing the transition curves from boundary through semifluid to fluid film lubrication. For example, in a certain gear set, using a lubricant of 0.005 kg sec/m^2 or 7.1 n viscosity, with 2 to 1 speed reduction at a tooth load of 100 kg/cm, boundary lubrication would be expected at pinion speeds below 100 rpm; full fluid lubrication above 20,000 rpm.

Sibley and co-authors experimented with crowned rollers using X-rays to determine film thickness (1961, 1962). Disks were nearly 3 in. in diameter, each with a crown radius of 36 in. and face width about $\frac{1}{2}$ in. Of three lubricants in the first series, the viscosity ranged from 11 to 26 cp at inlet temperature. Rolling speeds and loads were such that μ_1/W_1 ranged from 3×10^{-6} to 20×10^{-6}. Minimum film thickness ranged from 3 to 50 micro-inches, respectively 0.3 and 0.7 of the classical values for a rigid cylinder. Thus the benefits of pressure-viscosity and elastic deformation were partly neutralized by crowning. A possible non-Newtonian effect may have been another factor. To compare film thickness readings with Grubin's theory, h_0/R was plotted against a "contact lubrication" number $\mu_1 b_1 U/R$ at constant values of the load number W_1/ER. The points for all three lubricants—mineral oil, diester, and silicone—are represented by a single graph at each load number. The graphs run parallel to those for Grubin's theory except at the lower contact lubrication numbers; but Grubin's values are greater.

Observations by Dowson & Longfield on pressures and temperatures between surfaces in sliding contact (VII: 1963) indicate film thicknesses in accord with those calculated by Dowson & Higginson for the equivalent case of pure rolling (1959). See also Crook (1963); Niemann & Gartner (1965). Lubricant films on rough steel balls in rolling contact were investigated by Tallian, Sibley, and co-authors (1964,

1965). Average film thicknesses, measured electrostatistically by an adaptation of Furey's method, were found to be of the order expected from elastohydrodynamic theory, or somewhat higher. See also Archard (1964, 1965), Kannel et al. (1965), and O'Donoghue & Cameron (1965). The last mentioned work includes film thickness measurement by the voltage-discharge method and metal temperatures by an embedded thermocouple. Results agree well with the theory by Dowson, Higginson, & Whitaker (1962).

Frictional Resistance. *(1) Parallel Cylinders.* Measurements on a David Brown machine with bronze disk of 8-in. diameter, loaded against a 4-in. disk, were reported by Merritt, Hughes, and Watson (1935, 1952, and 1952). Both disks rotate in the same angular direction, with mineral oil lubrication.

In Merritt's tests the rolling velocity of the larger disk is $\frac{2}{3}$ of its sliding velocity, making a slide/roll ratio of 1.5. The coefficient of friction drops from 0.050 to 0.033 with a four-fold increase of viscosity, under a load of 1480 lb/in. at a sliding speed of 1000 ft/min. The oil used has a viscosity of 58 cp at 140 F. Hughes found $f = 0.034$ at 70 C, and 0.065 at 150 C. This was brought down to 0.043 with an additive, but the additive proved ineffective at lower temperatures. Watson found $f = 0.025$ at 93 cp with a slide/roll of 1.6, again fair agreement. All tests showed a drop in friction with increased sliding speed and viscosity, indicating partial hydrodynamic or "mixed" lubrication, probably.

It was found by Kuzmin (1954) that in heavily loaded contacts the friction coefficient can pass through a maximum and drop off at low sliding speeds. This observation was explained by Crook (1961) and confirmed by O'Donoghue & Cameron (1965).

Hard steel rollers were tested in the disk machine by Crook & Shotter (1957) under a load of 3100 lb/in. Watson (1957) tested steel disks under the same load over sliding speeds from less than 100 to more than 5000 ft/min, at a slide/roll of minus 2.14. His oil contained an EP additive. The coefficient on the 8-in. disk dropped from 0.075 to 0.009 over the speed range indicated, in fair accord with earlier reports.

In contrast, Dunk & Hall (1958) observed hydrodynamic friction at light loads. Offset and angle of the resultant force relative to the line of centers were measured, rather than the torque itself. Apparently, for pure rolling, f varied as the cube root, and for pure sliding as the three-fifths power of $\mu_1 U/W_1$.

The friction of steel disks was determined by Misharin (1958) with mineral oils from 1 to 175 cp viscosity. The coefficient dropped from

0.09 to 0.02 with increased rolling speed and viscosity, whereas change of loading had little effect. Mixed friction was indicated with f inversely proportional to the fourth root of $\mu_1 U_s U$, where U_s denotes the sliding and U the sum velocity. Blok showed how the results might be generalized by dimensional analysis (pages 464–6 of Misharin; and Blok, 1962).

The effects of surface finish were investigated by Shotter (1958). He found lowest friction with polished disks and oil supplied on entering side. Watson (1958) tested worm gear lubricants with bronze disks on steel. Mineral oil gave $f = 0.023$ to 0.024 on the bronze disk for various compositions but it rose to 0.050 when the usual steel disk was replaced by one of sulfur-impregnated steel. A light mineral oil gave higher coefficients than the more viscous grades.

Friction measurements were included by El Sizi & Shawki in their investigation (1960). Friction torque changed sign with reversal in the direction of slip. It apparently passed through zero for pure rolling—a finding questioned in the discussion. The coefficient of friction reached a sharp maximum of 0.047 for pure sliding. It dropped from 0.024 to 0.018 at a constant slide-roll of 0.6 as $\mu_1 U/W_1$ advanced from 3.3×10^{-6} to 100×10^{-6}. Genkin, Kuz'min, & Misharin (1960) experimented with rollers of various steels. Oil was fed to the contact at 25 liters/min to control the bulk temperature. Surfaces were run in until friction became constant. Friction was less for polished surfaces than for those initially rougher but well run in. Roller material had little effect, whereas f decreased slightly with increasing load, and markedly with slip speed and viscosity. Observations were said to accord with Petrusevich's theory. Misharin & Sivyakova (1960) simulated worm gear friction by rollers of bronze against steel. Surfaces were lapped without running-in. Seeking the least favorable condition for lubrication, they operated with equal and opposite peripheral speeds. The coefficient remained nearly constant at 0.06 to 0.10 for different lubricants, but rose to 0.14 for incipient seizure.

Benedict & Kelley (1961) measured the coefficients of friction between steel rollers of 2.4- and 3.6-in. diameter. Mineral oils were used. Sliding speeds ranged from $U_s = 18$ to 520 in./sec; sum velocities from $U = 102$ to 890 in./sec; inlet viscosities from $\mu_1 = 3$ to 500 cp; and loads from $W_1 = 1250$ to 5000 lb/in. The data from a large number of tests were plotted as a straight line with f against log x, where x denotes the parameter $\mu_1 U_s U^2/W_1$, or the product of $\mu_1 U/W_1$ and $U_s U$. The trend of the observations could be represented, with moderate scatter, by

$$f = 0.108 - 0.127 \log x \tag{15}$$

over the range of log x from 2 to 8. Since the parameter chosen is not dimensionless, the above units must be retained and viscosities expressed in centipoises. Application should be restricted to tooth pairs having the same ratio of radii as the rollers tested. The graph slopes downward, suggesting mixed lubrication.

Coefficients of friction between lubricated cylinders have also been determined by Sasaki and co-authors (1961) and by Gröbner (1962), including true rolling without slip. For comment on slip-free rolling, see Blok (1962). Tractive capacity and efficiency of friction drives were investigated by Hewko and co-authors (1962). The effects of temperature were studied by Leach & Kelley (1965). Finally O'Donoghue & Cameron (1965) measured disk friction with an Amsler machine. Combining past and present data, the mean coefficient was seen inversely proportional to the one-eighth power of the viscosity, the one-sixth power of the sum velocity, and the square root of the equivalent radius. It was directly proportional to a linear function of roughness. One hopes that the formula can be generalized by dimensional analysis. This completes our review of friction experiments on parallel cylinders.

(2) Other Roller Types. Walker (1947) measured the torque required for driving axially loaded cones to simulate gear teeth. A disk rotated in contact with a plane was used by Cameron (1952), who varied the slide/roll ratio. The coefficient fell to a minimum of 0.026 at 1750 rpm. The tests were continued by Newman (1960), who compared different lubricants and found fluid films at 3000 rpm. A sudden jump in torque was taken as evidence of film failure. A machine with steel cylinders on skewed axes, one running at 50, the other at 300 rpm, was used by E. A. Smith (1952). Colloidal graphite was reported to give less friction than mineral oils, with progressively lower friction at increasing loads.

The total coefficient f_t was measured by Reichenbach (1959) on steel balls of $1\frac{1}{4}$ in. diameter, rolling and spinning in V-grooves. He compared conditions from 0 to 500 lb-load per ball. Mean values of $10^2 f_t$ were respectively 2.4, 1.8, 1.6, and 1.3 for dry rolling, mineral oil, EP gear oil, and a synthetic sperm oil over the stated load range. Values dropped to about half for loads approaching zero, with spin friction greater than pure rolling friction. To determine spin friction, he first measured the rolling friction by loading each ball between a pair of flat steel plates, then subtracted the resistance due to rolling from the total observed. He confirmed the finding by Drutowsky (1959) and others that the use of a lubricant has little effect on pure rolling at low speeds. To show that in many types of rolling the major resistance is due to spinning friction, he recalls a ball thrust bearing of 240 mm diameter for an aircraft turbine operating at 9000 rpm under a load of 2000 lb. Heats generated by pure rolling and spinning were calculated to be 7.5 and 82 Btu/min, respectively. Actual test by Pratt & Whitney Aircraft gave a total of 90 Btu/min.

F. W. Smith (1959, 1960), worked with a novel pair of steel rollers on slightly skewed axes. The upper roller is part of a sphere while the lower one is a cylinder of same diameter. Skewing introduces a force on the upper roller in the direction of its axis. The ratio of this force to the load may be called the coefficient of slip friction, f_s. At a moderate speed and load, the slip friction for castor oil is increased threefold by an increase of C. L. A. roughness from 6 to 50 micro-inches. The fact that roughness makes no difference at the higher rolling speeds reflects increased film thickness. An ester base lubricant performed like the castor oil, allowing for lower viscosity; but the f_s curves for a silicone and for a petroleum oil (1962) are more complex. Smith is inclined toward a non-Newtonian explanation.

It was found by Rouverol & Tanner (1960) that the friction coefficient depends on lubricant, slip velocity, and ball diameter but is practically independent of load and rolling speed. Steel balls from $\frac{1}{4}$ to $\frac{1}{2}$ in. diameter, D, were loaded between the flat surfaces of hardened steel disks. Rolling speeds ranged from 5 to 100 ft/min; slip speeds from 0.1 to nearly 3.5 per cent thereof; and loads from $W/D^2 = 20$ to 330 psi. Eleven oils, mineral and synthetic, were compared with and without EP additives. Viscosities ranged from 2 to 3300 cp at 25 C. Friction coefficients doubled for a medium oil when the slip increased from 0.1 to 3.0 per cent. Increasing the rate of flow of the lubricant reduced the friction coefficients. Only a slight reduction was effected by additives. Sasaki & Okino observed that the rolling resistance of cylinders was from five to twenty times greater using a lubricant than when dry (1962). See also Sibley and Orcutt (1961, 1962).

The X-ray tests on crowned rollers were continued by Bell, Kannel, & Allen (1964). They introduced traction measurements and slip in order to determine the effective viscosity, knowing film thickness. Rollers were driven independently with a known difference in surface speeds. Traction in pounds can then be plotted against shear rate in reciprocal seconds at constant rolling speeds. One such chart is for a maximum Hertz pressure of 100,000 psi at 175 F with curves for rolling speeds of 4500, 6800, and 9100 ft/min. The middle curve shows an increase of traction from 0 to 6.0 lb as the shear rate goes from 0 to 1.5×10^5 sec^{-1}. All curves are linear at first, then bend over as if the film were non-Newtonian or subject to a pronounced temperature effect. An increase of rolling speed from 4500 to 9100 ft/min lowers the traction from 7.0 to 4.2 lb at 10^5 sec^{-1}. Calculation showed non-Newtonian properties to be a more likely explanation than frictional heating for the nonlinearity, but no explanation has been offered for the effect of rolling speed. All tests were run on polyphenyl ether, for which high-pressure data were lacking.

See also the reports of experiments by Orcutt (1965) and by Tallian and co-authors (1965).

5. ROLLER EXPERIMENTS WITH SURFACE DAMAGE

Pitting and Fatigue. Stewart Way discovered from roller tests that pits are often formed by intersecting fatigue cracks (1935); and that these cracks usually start below the surface, in the region of maximum shear stress. It seems that pitting does not occur with unlubricated rollers unless they are rough. Way suggested that the lubricant is forced into the cracks, causing metal to break out at the surface. Further studies were carried out on cylindrical roller machines by Nishihara (1937), Meldahl, and others.

Meldahl (1938) investigated pitting with a machine driven by eccentric gears. He discovered that pitting occurred only on that part of the roller representing the addendum arc and the pitch point in gear teeth. The same arrangement was used by Van Zandt & Kelley (1949). Knowlton & Snyder (1940) made the roller radii equal to those of gear and pinion teeth, respectively, and found that rollers would carry much higher loads than gear teeth. Niemann (1943) tested both ferrous and non-ferrous metals. The constant-load roller machine described by Earle Buckingham (1944) had been used in ASME research, and on thesis work at M.I.T. since 1931. Comparative tests were made on the "surface endurance" of many different materials from plastics to alloy steels, using the same mineral oil. The results were usefully applied in gear design with appropriate correction for curvature of gear teeth based on Hertzian theory. Later reports on cylindrical roller tests were published by Buckingham and Talbourdet (1950, 1954), M. R. Gross (1951), Evans and Tourret (1952), and Chesters (1958). Shotter (1958) found that polished surfaces can easily be separated by an oil film and have a greatly improved pitting resistance. See Scott (1957); Grunberg & Scott (1958).

In the meantime Walker had resorted to conical rollers (1947). Barwell, Milne, and Scott studied pitting and fatigue with the rolling four-ball machine (1956, 1957, 1957). Butler & Carter (1957) determined stress-life relations by the NACA "fatigue spin rig." The cone-

and-three-ball system was used by Milne & Nally (1957). The three-roller principle was introduced by Ryder & Barnes (1956), who loaded a ball between two disks. They were investigating ball life for jet-engine thrust bearings. The same principle was applied by Bauman and by Pinegin with a small test roller between two large drums (1958, 1958). Carter and co-authors (1960) used a five-ball machine to simulate ball-bearing assemblies more closely. Martin & Cameron (1961) investigated the pitting effect of lubricants with an opposed double-roller machine (1961). Endurance limit varied with the fourth root of viscosity above say 20 cp, both for naphthenic and paraffinic types, but the slope of the fatigue line depends on the oil type.

The effect of roughness on plastic-rolling fatigue was studied by Akaoka using a needle specimen loaded between three large disks (1962). Dawson found that an essential parameter in pitting of steel on steel in a disk machine is the ratio of total surface roughness to the calculated oil film thickness (1962). Toroidal contact rollers, essentially crowned disks, were chosen by Greenert for fatigue tests on bearing materials (1962). The rolling four-ball fatigue machine was used by Rounds in determining the effects of lubricants and surface coatings (1962), as well as by Scott & Blackwell (1963) in their comparisons of ball materials and lubricants. Zaretsky and co-authors had investigated the effects of nine lubricants, and other variables, using both the "fatigue spin rig" and the rolling five-ball tests (1961, 1962).

Correlation of theory and experiment was investigated, in the light of recent data, by K. L. Johnson (1963). He saw the possibility of raising the endurance limit by improved surface finish, whereas the use of a more viscous oil leads to an unfavorable stress distribution.

In general, experimenters found they could obtain pits without scuffing by means of pure rolling at moderate loads for a sufficiently long time; and that they could obtain scuffing or scoring by the introduction of a sliding component, increasing the load until visible surface failure occurred.

Roller Scuffing. The terms "scuffing," "scoring," and "galling" refer to a sudden form of wear initiated or aggravated by seizure and melting. Although "scuffing" is a term in common use, "scoring" is preferred by the American Gear Manufacturers Association (1951). This form of surface damage is frequently seen near the tips of gear teeth and on piston rings. To the uninitiated, scuffing might be mistaken for abrasive wear, since it looks like an increasing area of nearly parallel grooves and scratches of an erratic kind. Large numbers of roller tests, usually with the David Brown or similar disk machines, have been made in comparing gear materials and lubricants, and in attempting to discover how to predict scuff loads.

From such tests Hofer concluded that the scuff load, or critical load W at which scuffing begins, is inversely proportional to some power of the sliding speed V, such that WV^n is a constant (1931). Scuffing became a cause for serious concern when hypoid gears were introduced for automotive use. A much larger component of sliding occurs in hypoid than in spur gears. The "extreme pressure" or EP lubricants contain ingredients reacting with the steel surfaces to form protective layers that resist welding. Experiments at the National Bureau of Standards by S. A. McKee and co-authors led to a roller-type disk machine for load capacity tests (1933). It was simplified, standardized, and adopted by the SAE.

A letter from Albert Kingsbury in 1928 suggested that the oil film between disks might be broken down by using equal and opposite surface speeds. McKee's machine was designed with a reversing gear for that purpose but no data were published. The same device was independently tried with partial success by H. Blok (1949). McKee's discussion of Blok's paper throws new light on scuffing conditions. McKee varied the rate of loading, and compared different types of lubricants. Under most conditions the scuff load could be greatly decreased by a faster rate of loading. It was seen that the relative merit of different lubricants depends on the operating conditions, especially on rolling speed and temperature. In the meantime, scuff experiments with test machines were reported by Merritt (1935, 1937), Meldahl (1939), Manhajm & Mills (1945), and Walker (1947). Disk machines proved especially useful in evaluating materials and lubricants for worm gears.

Lane (1951) studied scuff temperatures with a two-ball machine, obtaining data in good accord with Blok's "flash temperature hypothesis" (Chapter XI). The same method was applied with Hughes (XI: 1952). Cameron (1952) confirmed Hofer's law using the disk and plane with combined rolling and sliding. Newman (1960) found that the torque-jumps due to scuffing in this apparatus occurred at lower loads for the higher slide/roll ratios, and at lower loads for mineral than for EP lubricants.

Scuff load data were reported at the Lubrication and Wear Conference by Crook & Shotter (1957), Elliott & Edwards (1957), Milne, Scott, & Macdonald (1957), and Watson (1957). A vast amount of data was presented, all from disk machines except the data by Milne and co-authors, who employed a crossed-cylinder machine. Watson refers to the difficulty of correlating bench tests with service experience (1958).

The question of how to predict scuff loads remained incompletely answered by all these informative investigations. Wear and scuff loads on turboprop lubricants with the McKee and SAE machines are reported by White (1958). Friction and seizure experiments on steel rollers are described by Genkin and co-authors (1960), who find that Blok's critical temperature may be exceeded without seizure when high-pressure hydrodynamic conditions prevail. Jackson and co-authors (1960) compared a series of steels with one silicone and one petroleum lubricant in a roller machine at inlet temperatures up to 700 F. Scuff loads were reported both in terms of the maximum Hertz stress (52,000 to 167,000 psi) and the calculated flash temperature (286 to 898 F). Rollers were of 2.5-in. diameter, one operating at 175 rpm, the other enough faster to establish a rubbing speed of 600 ft/min in most of the tests. In general, scuff loads and flash temperatures increased with the hardness of the steel and stood in nearly the same order with both types of lubricant, but exceptions are noted. Seven combinations of roller material were tested by Misharin & Sivykova (1960) to simulate worm gear contact. Equal and opposite surface speeds were used. The best combination seemed to be sulphide-treated steel on the same, which survived a load of 200 kg/cm without seizure.

Recently the Admiralty Oil Laboratory completed an investigation using disk machines designed with extraordinary care. Disks were of 2- and 4-in. diameter, held in lathe headstocks for positive alignment, and operated over a wide range of variables. The results reported by de Gruchy & Harrison (1963) seemed to show that

1. Scuffing can be indicated by a quick-response temperature recorder, but measurements of instantaneous friction might be better.
2. Partial scuffs with mineral oil "heal" readily, leading to a sudden "final scuff" as the performance criterion.
3. Plastic deformations alter tooth forms in gear-type machines, indicating one advantage of the disk type.
4. Scattered data, attributed to surface condition or hardness variations, requires taking the mean of four tests.
5. Increased oil flow improves load capacity.
6. The lower the slide/roll ratio the higher the scuff load.
7. The A. O. L. machine differentiates between the various lubricants and disk materials, EP oils doing better at high speeds.
8. Scuff load curves drop to a minimum around 1000 ft/min sliding

speed; such a transition also being observed with the I. A. E. gear-type machine.

9. Elastohydrodynamic calculations can be applied to the present data, indicating film thickness and viscosity relations.

10. The trend toward agreement between A. O. L. and I. A. E. machine data offers hope for correlation with the performance of full-scale reduction gearing.

Admiralty authors have clearly left no stone unturned in executing the kind of program now widely approved.

Further experiments on the oil film between rollers were reported by Leach & Kelley (1965) and discussed by other investigators. Contact temperature was seen to be the key to lubricant load capacity.

Roller Wear. Ordinary gradual wear may be shown by roller tests without pitting or scuffing, as found by French using the Amsler machine (1928, 1929). Wear measurements on lubricated surfaces have been reported by many of the pitting and scuffing investigators using later forms of disk machines; for example White, McKee, and associates (1949, 1958); Blok (1949), Evans & Tourret (1952), Onaran (1960), and Crook (1957, 1959). Fretting wear was studied by Calhoun & Murphy (1963) with grease lubrication; the test balls loaded between flats, and oscillated slightly. See also Tallian and coauthors (1964).

6. THEORY OF ROLLING BEARINGS

Ball and roller bearings, or "rolling bearings" for short, have an advantage over plain bearings in their low starting friction. They require less lubricant and less attention. They have been commercially standardized, which simplifies design, application, and replacement. Disadvantages are the space required, since ball and roller diameters are so much greater than the thickness of an oil film, and above all, their limited life.

Mechanical Characteristics. Both ball and roller types are designed for either radial or thrust loading. The "angular contact" ball bearing can take combinations of radial and thrust load. The cage or separator keeps the rolling elements from rubbing against each other. It can be made of low friction, long wearing materials since it carries so little load. Needle bearings are made from small-diameter elongated rollers, usually assembled without separators. In any type the rolling elements roll between an outer and inner race. The outer race is attached to the housing, the inner race to the shaft. Grooved races are required for ball bearings. Roller bearings use cylindrical races for straight rollers, conical for tapered rollers, and curved races for "spherical" rollers. Roller bearings are better suited for the heavier loads because of their nominal "line contact"; ball bearings for the lighter loads and higher speeds, because of their "point contact." The various types and commercial sizes are described in manufacturers' catalogs and in the treatises by Allan (1945), Palmgren (1945), Wilcock & Booser (I: 1957), and others.

The greatest load on any one ball, in a radial bearing of n balls, is commonly taken to be $5W/n$, where W is the total load. This relation is known as Stribeck's formula, from the pioneer investigator of ball bearings (1901). Stribeck calculated 4.37 for the constant, but recommended 5, to allow for practical conditions. He may have been the first to confirm the Hertzian theory of contact stresses and deformations. Other early investigators were Goodman (1912) and Heathcote (1921). Since rolling bearings have to be constructed and assembled with great accuracy, it is important to anticipate their deformations under load. According to Palmgren, the displacement of the bearing center in a roller bearing, in millimeters, will be 0.0006 times $W^{0.9}L^{0.8}$, where W is the radial load in kilograms and L the length of the rollers in millimeters. The deflection of a ball thrust bearing under an axial load W is 0.0024 times the cube root of W^2/d, where d is the ball diameter in millimeters. A rolling element of mass m exerts a centrifugal force $mR\omega^2$ on the outer race; where R is the *pitch radius* of the bearing, and ω the angular velocity of the whole set of rolling elements, whether balls or rollers. By R we understand the radial distance from bearing axis to the centers of the rolling elements; and by ω, the angular velocity of the center of each element around that axis. The mass m can be taken equal to W/g, where g is the gravitational acceleration. Gyroscopic forces act when the axis of rotation is changing, as in aircraft maneuvers.

A general theory of the mechanics of rolling bearings, apart from lubrication, has been developed by A. B. Jones (1960). For the actual motion of balls and rollers, determined experimentally, see Iida (1959), Hirano (1961), and co-authors, and C. F. Smith (1962). For critical speeds of a shaft supported in ball bearings reference is made to Yamamoto (1959). Some understanding of mechanical characteristics is essential to an appreciation of lubrication in rolling bearings, and may conveniently be obtained from Palmgren (1945), Shaw & Macks (V: 1949), or from Chapter 6 of Bisson & Anderson (I: 1964).

Action of the Lubricant. Until Schering & Vieweg discovered that the oil film was acting as an insulator (II: 1926), it had been thought that the rolling elements made direct contact with the bearing race. Osterle (1959) worked out hydrodynamic solutions for the rigid roller bearing with uniform viscosity. Accinelli compared grease and oil lubrication (1958).

Osterle's solution for shaft eccentricity in the clearance-type roller bearing shows that the eccentricity ratio ϵ is a function of ZN/P. It can be seen that for low eccentricities $\epsilon = a/x$, where x denotes 10^5 ZN/P. For high eccentricities close to unity, $1 - \epsilon = bx^2$; but in general, values are to be taken from graphs computed in the paper. An example is given for a bearing with fifteen rollers each of $\frac{3}{4}$-in. diameter by 1 in. long. The inner race is 3.000 and the outer race 4.502 in. in diameter, leaving a play of 2 mils and a radial clearance of 1 mil. The constants a and b are 3.2 and 0.16, respectively. When operating at 1000 rpm with an oil of 10 micro-reyns viscosity (60 cp) under a load of 41 lb, $P = 13.7$ psi on the inner race, and $10^5\ ZN/P = 1.22$. The shaft eccentricity will be 0.90, giving a minimum film thickness of 40 micro-inches. Büche's solution was limited to the friction calculation. In the meantime Hackewitz (1958) had shown how to apply Kapitza's theory to the roller bearing assembly, taking pressure-viscosity into account. Solutions comparable to that of Osterle were published by Sasaki and co-authors (1959, 1962). The first was applied to tapered roller bearings while the second considered non-Newtonian lubricants.

Dowson & Higginson (1961) extended their roller calculations to the assembled roller bearing with special reference to deformation and elastohydrodynamic lubrication (1963). Both clearance and preloaded bearings are treated. It was concluded that the rollers can roll on the races while transmitting forces with negligible slip. And since the pressure-viscosity effect is less than represented by the usual exponential formula, the performance of a roller bearing must be intermediate between the solutions obtained for rigid and elastic components. The elastohydrodynamic solution for minimum film thickness h at the inner race may be written

$$\left(\frac{h}{R}\right)_1 = \frac{1.6G^{0.6}\bar{U}^{0.7}}{\bar{W}^{0.13}}, \qquad (16)$$

where G denotes bE', $\bar{U} = \mu_1 U_1/E'R$, and $\bar{W}$ stands for $W/E'R$. Here b is the pressure-viscosity exponent and U_1 the sum velocity of inner race and roller surface; $1/R$ is the sum of $1/R_1$ and $1/r$, in which R_1 is the radius of the inner race and r that of a roller. If slip is neglected, U_1 can be taken as twice the velocity of the inner race. The remaining quantities are defined as in Eq. (13), except that W is the load per unit of axial length on the most heavily loaded roller. The authors prefer to take this equal to four times the bearing load per unit length divided by the number of rollers, instead of five times as recommended by Stribeck. Equation (16) will be recognized as an adaptation of Eq. (14). A similar ratio applies to the outer race.

In the particular example of a bearing with ten rollers each $\frac{1}{5}$ the diameter of the inner race, which is $2\frac{5}{8}$ in., "perfect geometry" was assumed—neither clearance nor interference. Operating at 500 radians/sec (about 4800 rpm) under a load of 1500 lb/in., the viscosity of μ_1 was taken as 10 cp (1.45 micro-reyns) with $G = 5000$. This value of G was considered representative for mineral oil in a steel bearing. It implies a pressure coefficient of 1.52×10^{-4} $(\text{psi})^{-1}$, or 21.5 per cent per 100 kg/sq cm. The solution by Fig. 19.6 of the reference makes $h_1 = 12.5$ micro-inches.

Confirmatory Observations. Film thicknesses measured by Horsch (1963) in gyro ball bearings ranged from 5 to 20 micro-inches, depending on the amount of lubricant available. Minimum oil requirements for ball bearings had been determined by Booser & Wilcock (1953). Schuller & Anderson observed minimum safe oil flow rates up to 500 F (1961). See also Macks & Nemeth (1950).

Greases are commonly used for lubricating small bearings. White and co-authors (1955) found that oil-soaked felt pads released oil for lubrication under stated conditions very much as a grease does. Milne and associates (1957) observed the movement of grease in rolling bearings, Kotova (1960) developed the mathematical theory, and Booser (1957) reported on practical developments. Unconventional lubricants, including liquid metals and rocket propellants, were investigated by Markert (1957), Butner (1962), and co-authors. See also the study by F. G. Rounds (1963).

Friction Theory. Dry friction calculations were offered by Poritsky and co-authors (1947) as well as by Jones (1959); see also Bisson & Anderson's Chapter 11 (1964). Büche worked out the first hydrodynamic theory of roller bearing friction, limited to light loads (1934). He reported fair agreement with Schneider's experiments (VII: 1930). Osterle's theory, cited above (1959), includes friction torque calculations. In the example of a clearance-type roller bearing with fifteen rollers of $\frac{3}{4}$-in. diameter, the coefficient of friction f_0 at light loads, approaching concentric operation, would be about 4000 ZN/P as compared to 6000 ZN/P by Büche's theory. Osterle notes that Büche neglected slip, while Dowson & Higginson think Osterle overlooked friction in the cavitation zone. These two approximations seem to be in the direction needed for explaining the difference. A comparison with Petroff's equation for plain journal bearings shows that a Petroff bearing would give six times the friction torque of a roller bearing having the same play, or an equal torque if the roller bearing had one-sixth as much clearance. The friction of a hydrodynamic roller bearing with perfect geometry and rigid surfaces (no clearance) approaches infinity. Only elastohydrodynamic theory can explain the operation of "perfect" and preloaded rolling bearings.

Other aspects of friction theory, including cavitation are reported by Palmgren & Snare (1957). Note the discussions by Blok and by Johnson indicating how some of the results might be generalized. Later, Palmgren (3rd ed.) showed that the friction moment could be represented by the sum of two terms, M_0 for the fluid resistance under no load, M_1 for the friction component due to load. The fluid component is proportional to the cube of the pitch diameter, the two-thirds power of viscosity and shaft velocity, and one-third power of the cavitation pressure-defect p; meaning the excess of ambient pressure over the vapor pressure of the lubricant. This factor p can ordinarily be taken equal to 1 atm. Evidently the expression represents the frontal resistance without churning loss. The loading component is proportional to the friction coefficient, pitch circle diameter, and load on the bearing. Constants of proportionality depend on the geometrical factors. See also Sasaki and co-authors (1962).

7. FRICTION TESTS ON ROLLING BEARINGS

Ask a bearing engineer what is really the coefficient of friction for a ball or roller bearing under ordinary conditions, and he may say "from slightly under 0.001 to something over 0.003." Then he starts to give you a list of "ifs" and "buts." Asked how the coefficient varies with load, speed, and temperature, he replies that it is not much affected by load; increases slightly with speed, after passing through a minimum; and decreases with moderate rise of temperature. These facts apply to normal or favorable conditions, as shown by numerous investigations.

Early Tests. Among the first experiments on ball bearings of modern type were those of Hanocq (1927), Maurer & Kelso (1931), Muzzoli (1934), and Delfosse (1936). Hanocq tested ball bearings of 40-mm bore under loads from 35 to 500 kg. The coefficient of friction proved to be linear with speed, varying from 0.0010 at 50 rpm to 0.0025 at 450 under the heaviest load, but it was very much higher under light loads. Maurer found the friction of ball bearings lower than that of plain journal or flexible roller bearings, even though the ball bearings were lubricated with grease, and the rollers with oil.

Muzzoli experimented on both ball and roller bearings, varying the load, speed, lubricant, duration of tests, bearing and cage materials, and details of design. He put the results for ball bearings in the form of a two-term empirical equation for the friction torque. The first term is independent of the load, and approximately proportional to the cube root of the speed; depending also on the ball diameter and number of balls. The second term varies with the five-thirds power of the load, at low speeds. Thus an oil-lubricated bearing of 45-mm bore, groove radius 1.06 times ball radius, gave 0.95 lb-in. of torque at 1000-lb load; 4.2 lb-in. at 3000 lb. The whole expression, quoted by Wilcock & Booser (I: 1957), is proportional to a factor representing the consistency of the lubricant; and to another depending on conformity of ball and groove. A similar expression for the coefficient of friction as a function of load shows a minimum point. Delfosse tested ball bearings like those of Hanocq, using a dynamometer, instead of the deceleration method, to eliminate air resistance. He found four different values of the resisting moment depending on whether friction was measured on the outer or inner race, and with the loaded balls in or out of the oil. Hanocq's linear laws were extended to 2500 rpm under loads from 20 to 320 kg per bearing.

The investigations by Styri are frequently cited (1940). His friction moment M for radial-load ball bearings, in lb-in., can be represented by $aW + bW^2$, where W is the load in pounds. Thus for single-row, deep-groove bearings from 45 to 70-mm bore, at 1800 rpm, $a = 10^{-3}$ and $b = 3 \times 10^{-7}$. For self-aligning bearings of the same size and speed, $a = 0.6 \times 10^{-3}$, $b = 1.3 \times 10^{-7}$. Friction torque increased only slightly with speed and size. Now since $f = 2M/D$, where D is the shaft diameter in inches, the corresponding coefficients for a 2-in. shaft or 50-mm bore under 1000-lb load would be around 0.0014 for the deep-groove and 0.008 for the self-aligning type. The formula does not apply to very light loads, since for any finite torque at zero load the coefficient must approach infinity. The coefficients for angular contact and pure thrust ball bearings would be greater than for the deep-groove radial bearing. Styri advocated the use of *torque* rather than *coefficient* in reporting friction measurements on rolling bearings, both to avoid infinite values at zero load, and because friction coefficients depend on a radius arbitrarily chosen. His results varied but little with speed or viscosity, provided not more than one drop of oil per hour was supplied.

The aircraft gas turbine imposed new requirements. Boyd & Eklund (1951) tested super-precision ball bearings of 40- and 60-mm bore, under moderate loads, up to high speeds with oil-air mist. In the smaller size, torque readings indicate an increase of f from 0.0051 to 0.0086 as speed goes up from 5000 to 36,000 rpm at a constant load of 100 lb. In the larger size, f drops from 0.0036 to 0.0017 at a constant speed of 12,000 rpm as load increases from 500 to 2000 lb. The torque and the coefficient decrease with increasing speed for oil-air mist, but go up, as would be expected, when solid oil is injected. The original report should be consulted for many other combinations. Comparable experiments were reported by Fogg & Webber (1953), mostly on bearings of 2-in. bore lubricated by oil mist. Friction and temperature rise were measured at speeds from 5000 to 30,000 rpm. Thus at 10,550 rpm the coefficients dropped from 0.0028 to 0.0011 when loads were stepped up from 250 to 1500 lb. And the same coefficients were found (up to 750 lb) when the speed was raised to 17,750 rpm. Graphical interpolation brings Fogg & Webber's friction below two-thirds that of Boyd & Eklund. The British authors concluded that boundary properties had more effect than differences in viscosity. Their investigation was continued to determine the influence of design factors.

High Speed Bearings. An advanced design of test machine, with driving motor and test spindle supported on air bearings, was described by Graneek & Wunsch (1954). It was used in measuring the friction torque on precision bearings of 5-mm bore at high speeds. Oil fog was supplied for five minutes; the test readings were taken after running an hour. Increasing the thrust load from 2 to 8 lb raised the friction torque 20 per cent; changing the radial load had little effect. A speed increase from 2500 to 50,000 rpm under a constant thrust load of 2 lb, with no radial load, raised the friction torque threefold, with an increase in f from 0.0045 to 0.0135.

Barwell & Hughes (1955) continued experimenting with high-speed ball bearings. Light and extra-light deep-groove bearings, 5-in. bore, were tested under radial and combined loading with jet oil supply. When the coefficient f was plotted against a dimensionless parameter x defined by $ZUnd/W$, straight lines were obtained. Here U is the pitch circle velocity, n the number of balls, and d their diameter; Z being the viscosity, and W the resultant load on a bearing. Thus f can be represented by $C_1 + C_2x$. For the light bearings, $C_1 = 1.09 \times 10^{-3}$ and $C_2 = 130$. It can be seen that $x = k(ZN/P)$, where k denotes $\pi nD/L$, if D is the pitch diameter and L the axial width of the bearing. For the light bearings, k is 192, for the extra light 358. Data are tabulated from $W = 600$ to 2400 lb, or more. The corresponding x values for the light type range from 10×10^{-6} to over 60×10^{-6}. Friction torques are less for the extra-light bearing.

Small ball bearings lubricated with a light petroleum oil were tested up to 100,000 rpm by Moore & Jones (1955). Friction torque was found to depend more on speed and size than upon oil flow rate. The use of oil mist was not successful, but bearings could be cooled by high flow rates with little increase in friction power loss. Gyroscope drift caused by friction in gimbal bearings was investigated by Lane and co-authors (1959). This effect was greatly reduced by rotating the outer ring at intervals. In experimenting on ball bearings of 3- to 5-mm bore, Lawrence plotted friction torque against the weight of grease contained (1961). Three greases and an instrument oil were compared at a low speed, and again at 15,000 rpm. Shield and cage designs were varied. Some twenty-four charts were required to show all the combinations tried. A three-fold increase in torque was found over the speed range indicated, with a twofold increase from 0 to $\frac{1}{20}$ gram weight of grease, at any one speed and load. Shuller & Anderson (1961) tested 75-mm bore deep-groove ball bearings with a synthetic diester, using oil mist. They found friction torques of about 4 lb-in. at 12,000 rpm under 3000-lb thrust load, when the oil flow was reduced to 0.003 lb/min.

A comparison of numerous fluids using a thrust ball bearing machine was reported by Rounds (1963). Coefficients of friction varied from a static value of 0.145 for glycerol to kinetic values as low as 0.004 for ethylene glycol at a ball velocity of 500 ft/min. Tests were run at 200 F under a Hertz load of 400,000 psi.

Roller Bearings. Among the earlier experiments we recall those of Schneider (VII: 1930) and of Maurer & Kelso (1931), cited previously. The latter employed flexible roller bearings of 0.78 in. diameter by $9\frac{1}{4}$ in. long, lubricated with red engine oil at 90 F. The coefficient of friction rose with speed but fell with increasing load and temperature. An investigation of needle bearings was reported by Ferretti, who worked with a bearing of 35-mm bore containing twenty needles 2-mm diameter, 15-mm long (1932). Tested at 1250 rpm under loads from 7 to 100 kg/sq cm of projected area, these bearings gave coefficients decreasing from 0.036 to 0.008. Curves for f against speed showed the characteristic minimum. Higher values were found for static friction. The tests were accompanied by observations on the motion of the needles. Railway roller bearings of 120-mm bore, grease lubricated, containing sixty-four rollers of 30-mm diameter, 36 mm long were tested by Hanocq (1937). A total play of 0.11 mm gave better performance than closer fits. A mean coefficient just below 0.0019 was found at 740 rpm under loads between 7000 and 8000 kg.

Roller bearing tests were included in the experiments by Styri and by Boyd & Eklund cited previously. In the first-named investigation friction measurements were made on cylindrical roller bearings with scant lubrication. The torque rose from 3.0 to 6.7 lb-in. under radial loads from 2000 to 8000 lb at 1800 rpm. Spherical roller bearings gave friction torques from two to four times as great. The coefficient of friction decreased with increasing load. No difference in friction was noticed between standard and super-finish; a fact attributed to the film of oil. Roller bearings were tested by Boyd and co-author under radial load at high temperatures. Size 306, with straight outer race, showed a no-load loss of 0.05 hp at 36,000 rpm. This type carries eleven rollers of 0.413-in. diameter, mean internal clearance 1.4 mils. An oil-air mist was used at a jet temperature of 200 F and flow rate $\frac{1}{2}$ pint/hr. Size 215, was tested at 12,000 rpm under loads up to 300 lb. Here the internal clearance was 3 mils and oil flow 1.4 pint/hr at the same jet temperature. The effect of the load was to raise the outer-race temperature from 360 to 410 F. Finally a self-aligning bearing of the same size was tested. A load increase from 1000 to 7000 lb raised the outer-race temperature from 276 to 328 F and the friction torque from 9.5 to 13.8 lb-in. Other data are shown by numerous charts. As a result of these tests, and related work, it was seen that aircraft turbine bearings, once a critical component, had now become as reliable as other engine parts.

The investigation by Fogg & Webber (1953) included roller bearings of 2-in. bore at speeds up to 20,000 rpm and loads to 2000 lb. Consideration was given to the effects of varying end-clearance and other design factors. Bearings were lubricated with an oil mist at $\frac{1}{4}$ pint/hr. Early failures were observed under some conditions. In a typical curve for f against load the coefficient drops from 0.0036 to 0.0012 with an increase of load from 500 to 2500 lb. This is for a bearing with eighteen rollers of length and diameter $\frac{5}{16}$ in. Cageless bearings of the same size, using twenty-six rollers, operated satisfactorily at high speeds.

The mean friction torque M on *spherical roller bearings* was determined from 200 to 2000 rpm by Palmgren & Snare using oil bath (1957). Speed had little effect. Results for radial and thrust types were plotted on logarithmic scales with M/WD as ordinate against M_0/WD, where M_0 is the idling or no-load torque, W the radial or thrust load, and D the pitch diameter. The two graphs are nearly identical at mid-range. As M_0 approaches zero, each curve approaches a constant horizontal asymptote representing the load component $M = M_1$. As M_0 approaches infinity, the curves approach a 45-deg asymptote representing the no-load component $M = M_0$. The total resistance M could be approximated by the sum of M_0 and M_1.

Temperature Rise. The temperature of the outer race was determined in all the experiments on ball and roller bearings by Boyd & Eklund, especially as affected by speed, load, and jet temperature. Macks and co-authors (1951, 1952) were able to generalize their test rig and turbo-jet engine data so as to predict temperature rise from operating conditions. As ordinate in a plot of laboratory observations, they took the difference between outer and inner race temperatures, deg F, divided by DN to the 1.2 power (D is the bore diameter in millimeters and N the speed in rpm). As abscissa they chose the product of $Z^{0.25}$ and $d^{0.18}$ divided by $M^{0.36}$, where Z is the oil inlet viscosity in reyns, d the jet diameter in inches, and M the mass flow of oil in pounds per minute. Straight lines were obtained, depending somewhat on surface speed. Engine data gave temperature differences about 40 per cent higher. Test bearings were of 75-mm bore with one-piece, inner-race-riding bronze cages. Getzlaff (1952) plotted curves showing outer-race temperature against oil quantity for a 15-mm bore deep-groove ball bearing of radial clearance 20 microns, under radial and axial loads. The temperature rise is only 3 deg C at 21,000 but reaches 103 deg C at 70,000 rpm. This bearing was provided with a brass cage guided by inner ring, and with indirect air cooling. Fogg and co-author (1953) included temperature observations in their experiments on 2-in. bore light and extra-light ball bearings. At 14,000 rpm under a radial load of 750 lb, the light bearing rose 80 deg C above ambient, while the extra-light rose 120 deg C.

Formulas and constants for calculating thermal equilibrium temperatures from design data and operating conditions may be found in Palmgren's book. Application to a practical example is found in the study of a 10-in. bore spherical roller thrust bearing by T. A. Harris (1964). Friction calculation is based on Palmgren's formulas; heat transfer by Dusinberre's method, analogous to Kirchhoff's solution for electrical networks. Outer ring temperatures found by test were plotted against the product of thrust load and speed. Test loads ranged from 46 to 172 lb, speeds from 150 to 300 rpm. The outer-ring temperature rise varied from 110 to 159 deg F above an ambient of 80 F. The predicted temperatures averaged within 10 deg of those observed.

High-Temperature Operation. Some of the friction tests led to relatively high temperatures at the higher speeds. Still higher bearing temperatures result from operation at high ambients, as in aircraft and other applications. New liquid and solid lubricants, ceramic bearings, and inert atmospheres have been tried with fair success. See, for example, Nemeth and co-authors (1952, 1955), as well as Anderson's Chapter 11 (1964). Favorable operation of steel ball bearings to 1000 F was obtained with graphite or molybdenum disulfide suspended in air or nitrogen. Sorem and co-author (1956) found good results to 600 F by blanketing the bearings with an air-fuel mixture in the absence of liquid lubricants, after prelubricating with a volatile synthetic. Krause and co-authors (1960) found that phthalocyanines offered thermal stability to 1300 F. Eickhoff & White (1961) compared the effects of nitrogen and carbon dioxide, at elevated temperatures, on the life of unlubricated ball bearings under thrust loading. Cageless bearings were found superior in nitrogen, caged bearings in carbon dioxide. Sliney operated bearings to 1250 F with solid lubricants (1961). A ceramic ball bearing developed by Taylor and associates (1963) was successfully tested to 1500 F.

8. LIFE EXPECTANCY

It is hard to believe that ten out of every hundred rolling bearings that are operated under a given load are expected to fail before attaining their rated life. The "rated" or estimated life of such a bearing is, in fact, defined as the number of million revolutions that can be survived by 90 per cent of a large collection. From the statistical study of such tests a curve has been plotted, for each type and load, showing the actual life of the bearing as ordinate against the percentage of the whole lot reaching that life (Fig. 3). A point can be marked on the curve representing median life—the life, in millions of revolutions, reached by 50 per cent of the collection. This median life comes to about five times the rated life. The curve has an ordinate equal to the rated life at an abscissa of 90 per cent. It falls to zero at 100 per cent. It rarely climbs above twenty times the rated life at zero per cent. Palmgren's Eq. (31.01) indicates the form taken by all such curves when the abscissa is expressed by a proper fraction, x, and the ordinate by the ratio, y, of actual to median life. The equation may be put in the form y^e equals 21.8 log $(1/x)$, where e is an empirical constant called the "dispersion exponent." It is customary to take $e = \frac{10}{9}$ for ball and $\frac{9}{8}$ for roller bearings, although it varies slightly for different steels.

A lighter load results in longer life. From large numbers of tests it has been found that the life of a bearing, in millions of revolutions, is inversely proportional to the cube of the load, or $L_N = (C/W)^3$. The constant C has been investigated as a function of the properties of the material, degree of osculation between race and rolling elements, and other factors. Studies have also been made to determine the "equivalent load," or constant load giving the same life as a fluctuating load. See Palmgren and Lundberg (1945, 1949), Shaw & Macks (1949), Jones (1952), and Hustead (1963).

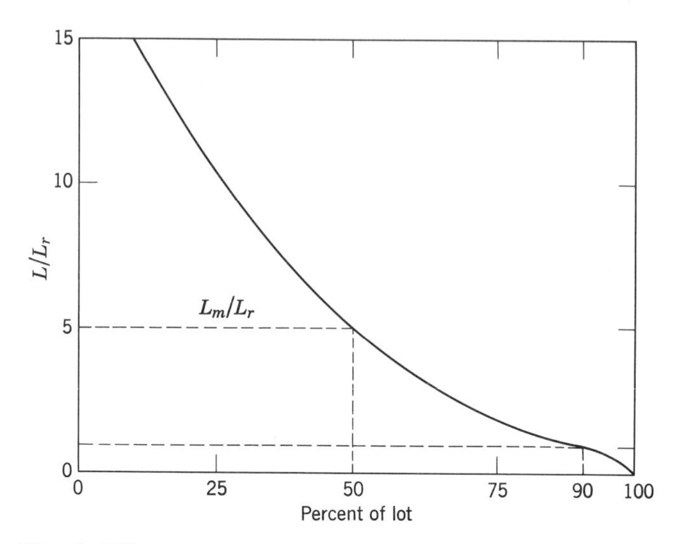

Fig. 3. Life expectancy L for rolling bearing under a given load. See A. Palmgren (SKF, 1959)—L_m, median life; L_r, rated life.

Herculean efforts have long been made to improve the fatigue life of bearing materials. Such research is continuing with special reference to the basic cause of fatigue and the influence of the lubricant. See, for example, Jones & Wilcock (1950), Otterbein (1958), and Schreiber (1960), as well as Anderson's Chapter 12 (1964). Minimum oil requirements for ball bearings had been determined by Booser & Wilcock (1953), leading to a formula for life expectancy in terms of the available quantity of lubricant.

The effects of high temperature have been studied by Moore & Lewis (1957), Glaeser (1960), and Gustafsson (1964); of low temperature by Martin (1959), Wilson (1961), and co-authors. Others have experimented with inert atmospheres (Fricker, 1963; Irving & Scarlett, 1963); and with the effects of electric currents on bearing life (Simpson & Crump, 1963).

9. SPACE APPLICATIONS

Instrument bearings were tested with grease and liquid lubricants at 4000 rpm, at a pressure of 10^{-5} torr,* by Freundlich & Hannan to

* Named for Torricelli, the *torr* is a unit of pressure equal to 1 mm of mercury.

indicate life expectancy (1961). Ball bearings with self-lubricating retainer materials were operated by Bowen (1962) at pressures as low as 7×10^{-9} torr, with temperatures ranging from +300 to —300 F, under Hertz loading close to 200,000 psi. Ball bearings of 30-mm bore were tested at 3450 rpm, 200 F outer-race temperature, down to pressures of 2×10^{-5} torr by Coit & Sorem (1962), who compared a wide variety of lubricants, both petroleum and synthetic. Volatility limits were noted, together with operating time to failure. Larson & Piken (1962) outlined the operating conditions and performance requirements of bearings in typical aerospace projects. They compared a variety of solid lubricants in an R-16 size ball bearing of stainless steel at 1000 rpm, 600 F, under a Hertz loading of 370,000 psi. Friction torque and radial play were measured at the beginning and end of each test. Promising results were obtained with a molybdenum disulfide lubricant.

Long-term operation of self-lubricated ball bearings was investigated by Boes (1963) down to pressures of 2×10^{-8} torr. Special cage materials were compared at temperatures up to 230 C. See also Brown, Burton, & Ku (VIII: 1963). Lewis and co-authors (1963) tested numerous lubricants in instrument-size ball bearings down to low pressures. Special study was given to evaporation rates and the effect of radiation. Gold plating gave less friction than silver at all temperatures up to 1000 F. Lipp & Klemgard (1963) reviewed the techniques for investigating friction and wear in aerospace ball and roller bearings. Instrument-sized ball bearings were tested in pressures down to 10^{-8} torr by Young and co-authors (1963). Lubricant compositions and retainer materials were highly diversified, and exceptionally long lives obtained from favorable combinations. The methods and results make this report one of outstanding interest.

Many investigations not cited here are described in Bisson's Chapter 9 (I: 1964). It may be interesting to look back, a decade hence, to compare the fragmentary and little-coordinated knowledge of today with what may hopefully be available then.

10. SUMMARY

Rolling contact has been reviewed from three standpoints, (1) theory, (2) roller experiments, and (3) rolling bearings. The reference list has been divided accordingly.

The theoretical discussion begins with the Hertzian formulas, and a few hints as to the explanation of rolling friction. Rigid surface lubrication covers the theory for uniform and variable viscosity, including the action of non-Newtonian lubricants. Some contributions to elastohydrodynamic theory are outlined, indicating the great increase of load capacity to be expected as a result of elastic deformation.

Roller experiments are described in two groups, first those accompanied mainly by fluid friction; then the experiments in which surface damage occurs, as for example in disk machine tests for comparing lubricants. Good confirmation of theoretical calculations has been seen in the first group. The results obtained in the second group throw valuable light on the phenomena involved, although it is still difficult to predict scuff loads and wear rates.

The several types of ball and roller bearings are described briefly with a review of calculation methods and test results on film thickness, friction torque, and temperature rise. Minimum film thicknesses up to 20 micro-inches have been measured in gyro ball bearings. Oil requirements have been determined for ball bearings in relation to life expectancy. Friction coefficients below 0.0010 have been found in deep-groove ball bearings of 77-mm bore with a synthetic lubricant applied in the form of an "oil mist," when running at 12,000 rpm under 3000 lb of thrust load. Since they require little or no lubrication for operation over a limited period, it is thought that rolling bearings can serve usefully in space applications.

REFERENCES

Part 1. Theory of Rolling Contact

Archard, J. F. & Cowling, E. W. (1965), "Elastohydrodynamic Lubrication at Point Contacts." Paper No. 3, Symposium on Elastohydrodynamic Lubrication, Leeds (Sept.), 17–26.

Archard, J. F., Gair, F. C., & Hirst, W. (1961), "The Elasto-Hydrodynamic Lubrication of Rollers." *PRS* **262,** 51–72.

Archard, J. F. & Kirk, M. T. (1961), "Lubrication at Point Contacts." *PRS* **261,** 532–50.

Ball Bearing Journal, The (1926), "The Nature of Rolling Resistance." No. 1, pp. 2–6. (a) Pages 30–5 in Palmgren (1954)—See Refs. Part 3.

Banks, W. H. & Mill, G. C. (1954), "Some Observations on the Behaviour of Liquids Between Rotating Rollers." *PRS* **223,** 414–9.

Beeching, R. & Nicholls, W. (1948), "A Theoretical Discussion of Pitting Failures in Gears." *Proc. IME* **158,** 317–26.

Bikerman, J. J. (1949), "Effect of Surface Roughness on Rolling Friction." *JAP* **20,** 971–5.

Blok, Harmen (1951), "Fundamental Mechanical Aspects of Thin Film Lubrication." *Annals of N. Y. Acad. Sci.* **53**, Art. 4, 779–804.

Büche, W. (1934), "Eine hydrodynamische Theorie der Flussigkeitsreibung in Rollenlagern." *Forschung* **5**, 237–44.

Burton, R. A. (1960), "An Analytical Investigation of Visco-Elastic Effects in the Lubrication of Rolling Contact." *TASLE* **3**, 1–10.

Cameron, Alastair (1954), "Surface Failure in Gears." *JIP* **40,** 191–202.

Carter, F. W. (1926), "On the Action of a Locomotive Driving Wheel." *PRS* **229,** 151–7.

Cheng, H. S. (1965), "A Refined Solution to the Thermal Elastohydrodynamic Lubrication of Rolling and Sliding Cylinders." *TASLE* **8,** 397–410.

Cheng, H. S. & Sternlicht, B. (1965), "A Numerical Solution for the Pressure, Temperature and Film Thickness between Two Infinitely Long, Lubricated Rolling and Sliding Cylinders, under Heavy Loads." *JBE, T* **87,** 695–707.

Christensen, H. (1962), "Oil Film in Closing Gap." *PRS* **266,** 312–28.

Crook, A. W. (1961), "A Theoretical Discussion of Friction." *Phil. Trans. Roy. Soc. (London) A* **254,** 223–58.

Crouch, R. F. & Cameron, A. (1960), "Graphical Integration of the Maxwell Fluid Equation and Its Application." *JIP* **46,** 119–25.

Dörr, Johannes (1954), "Schmiermitteldruck und Randverformungen des Rollenlagers." *Ing. Archiv.* **22,** 171–93.

Dowson, D. (1965), "Thin Film Lubrication." *HS,* 215–81.

Dowson, D. & Higginson, G. H. R. (1959), "A Numerical Solution to the Elastohydrodynamic Problem." *J. ME Sci.* **1,** 6–15.

Dowson, D. & Higginson, G. H. R. (1960), "The Effect of Material Properties on the Lubrication of Elastic Rollers." *J. ME Sci.* **2,** 188–94.

Dowson, D. & Higginson, G. H. R. (1961), "New Roller Bearing Lubrication Formula." *Eng.* **192,** 158–9.

Dowson, D., Higginson, G. H. R. & Whitaker, A. (1962), "Elastohydrodynamic Lubrication: A Survey of Isothermal Solutions." *J. ME Sci.* **4,** 121–6.

Dowson, D. & Whitaker, A. V. (1965), "The Isothermal Lubrication of Cylinders." *TASLE* **8,** No. 3, 224–34.

Engineering (1916), "The Lubrication of Gear Teeth," **102,** 119–21. (a) **109,** 599–601 (1920).

Finston, M. (1951), "Thermal Effects in Calendering Viscous Flow." *JAM* **18,** *T* **73,** 12–18.

Floberg, Leif (1959), *Lubrication of a Rotating Cylinder on a Plane Surface, Considering Cavitation.* No. 216, *TCUT;* 40 pp.

Floberg, Leif (1961), *Lubrication of Two Cylindrical Surfaces, Considering Cavitation.* No. 234, *TCUT;* 36 pp.

Flom, D. G. & Bueche, A. M. (1959), "Theory of Rolling Friction for Spheres." *JAP* **30,** 1725–30. (a) Bueche & Flom, *Wear* **2,** 168–82.

Flom, D. G. (1962), "Dynamic Mechanical Losses in Rolling Contacts." Pages 97–112 in *RCP*. (a) *JAP* **31,** 306–14 (1960).

Gaskell, R. E. (1950), "The Calendering of Plastic Materials." *JAM* **17,** *T* **72,** 334–6.

Gatcombe, E. K. (1945), "Lubrication Characteristics of Involute Spur Gears—A Theoretical Investigation." *T* **67,** 177–88.

Glagolev, N. I. (1945), Resistance of Cylindrical Bodies in Rolling (Russian). *PMM* **9** (4) 318–33.

Goodman, L. E. (1962), "Contact Stress Analysis of Normally Loaded Rough Spheres." *JAM* **29,** *T* **84,** 515–22.

Grubin, A. N. & Vinogradova, I. E. (1949), Fundamentals of the Hydrodynamic Theory of Lubrication of Heavily Loaded Cylindrical Surfaces (Russian). In *Investigation of the Contact of Machine Elements* (Russian), ed. by Kh. F. Ketova, TsNIITMASH Book No. 30, Moscow. (a) DSIR Transl. No. 337.

Gümbel, L. (1916), "Über geschmierte Arbeitsräder." *Z. gesamte Turbinenwesen* **13,** 205–9, 220–3, 225–8, 239–41, 245–8, 258–62, 268–72.

Heidebroek, E. (1935), "Zur Theorie der Flussigkeitsreibung zwischen Gleit- und Walzflächen." *Forschung* **6,** 161–8.

Hersey, M. D. & Lowdenslager, D. B. (1950), "Film Thickness between Gear Teeth." *T* **72,** 1035–42.

Hertz, Heinrich (1881), "Über die Berührung fester elastischen Körper." *J. reine u. angew. Math.* **92,** 156–71.

Howlett, J. (1946), "Film Lubrication between Spherical Surfaces, with an Application to the Theory of Four-Ball Testing." *JAP* **17,** 137–49.

Institution of Mechanical Engineers (1965), *Symposium on Elastohydrodynamic Lubrication. LWG.* (a) *Sci. Lubr.* **17,** No. 7, 14–5.

Ishlinski, A. I. (1938), Rolling Friction (Russian). *PMM* **2,** 245–60 (with French resume).

Johnson, K. L. (1955), "Surface Interaction between Elastically Loaded Bodies under Tangential Forces." *PRS* **230,** 531–48.

Johnson, K. L. (1957), "Recent Developments in the Theory of Elastic Contact Stresses: Their Significance in the Study of Surface Breakdown." *CLW*, 620–7; 240, 594, 596, 840, 842; Plate 10.

Johnson, K. L. (1958), "The Effect of Spin upon the Rolling Motion of an Elastic Sphere on a Plane." *JAM* **25** *T* **80,** 332–8. (a) "The Effect of a Tangential Contact Force . . . ," 339–46.

Johnson, K. L. (1959), "The Influence of Elastic Deformation upon the Motion of a Ball Rolling between Two Surfaces." *Proc. IME* **173,** 795–810.

Johnson, K. L. (1962), "Tangential Tractions and Micro-slip in Rolling Contact." Pages 6–28 in *RCP*.

Kapitza, P. L. (1955), Hydrodynamic Theory of Friction in Rolling (Russian). *Zh. Teh. Fizik* **25,** 747–62.

Karas, F. (1941), "Die äussere Reibung beim Walzendruck." *Forschung* **12,** 237–43; 266–74.

Karlson, K. G. (1926), Contact Problem (Swedish), *TTM* **56,** 1–6.

Kodnir, D. S. (1960), A Method for Solving the Contact Hydrodynamic Problem (Russian). *Third All-Union Conf. on Friction and Wear,* Moscow, **3,** 58–66.

Konvisarov, D. (1952), On the Theory of Rolling (Russian). *Izvest. SSSR AN* **83,** No. 3. (a) with Pokrovskaya (1955).

Korovchinskii, M. V. (1958), Possible Limiting Conditions of Hydrodynamic Friction in a Four-Ball Testing Machine (Russian). *TIM* **12,** 242–85. (a) *FWM* **12,** 233–73 (1960).

Korovchinskii, M. V. (1960), Some Problems in the Hydrodynamic Theory of Lubrication, with Deformation of the Bodies Bounding the Lubricant Film (Russian). *Third All-Union Conf. on Friction and Wear* **3,** Moscow, 78–84.

Korovchinskii, M. V. (1963), On Some Problems in Elasto-Rheology, with Application to the Theory of Friction (Russian). *TIM* **17,** 121–64. (a) *FWM* **17,** 114–55 (1964).

Kotova, L. I. & Deriagin, B. V. (1957), Theory of a Cylinder Rolling on a Surface Covered with a Layer of Plastic Lubricant (Russian). *Zh. Tek. Fizik.* **27,** 1261–71. (a) *Soviet Phys.—Tech. Phys.* **2,** 1154–64. (b) *Kotova,* 1424–41.

May, W. D., Morris, E. L., & Atack, D. (1959), "Rolling Friction of a Hard Cylinder Over a Viscoelastic Material." *JAP* **30,** 1713–24.

McEwen, Ewen (1952), "The Effect of Variation of Viscosity with Pressure on the Load Capacity of the Oil Film Between Gear Teeth." *JIP* **38,** 646–50, 668–90, 693–6.

Meldahl, A. (1941), "Contribution to the Theory of the Lubrication of Gears and of the Stressing of the Lubricated Flanks of Gear Teeth." *Brown-Boveri Rev.* **28,** 374–82.

Merwin, J. E. & Johnson, K. L. (1963), "An Analysis of Plastic Deformation in Rolling Contact." *Proc. IME* **177,** 676–90.

Misharin, J. A. (1958), "Influence of the Friction Conditions on the Magnitude of the Friction Coefficient in the Case of Rolling with Sliding." Paper 40, *Proc. ICG,* 159–64; 118, 464, 467, 500–1.

Palmgren, Arvid (1921), Investigations of Rolling Under Tangential Force (Swedish). *TTM* **15,** 129–32. (a) *Ball Bearing Jour.,* SKF, Phila., No. 3, 67–76 (1928).

Pater, A. D. de (1962), "On the Reciprocal Pressure between Two Elastic Bodies." Pages 29–75 in *RCP.*

Peppler, W. (1938), *Druckübertragung an geschmierter zylindrischen Gleit- und Walzflächen.* Forschungsheft **391;** 24 pp.

Petrusevich, A. I. (1951), Basic Conclusions from the Contact-Hydrodynamic Theory of Lubrication (Russian; abridged from thesis). *Izv. AN SSSR Odteleniia Teknishcheskikh Nauk* **2,** 209–23.

Poritsky, Hillel (1950), "Stresses and Deflections of Cylindrical Bodies in Contact, with Application to Contact of Gears and of Locomotive Wheels." *JAM* **17,** *T* **72** 191–201, 465–6.

Poritsky, Hillel (1954), "Lubrication of Gear Teeth, Including the Effect of

Elastic Displacement." *First Nat. Symposium on Friction and Lubrication in Engineering,* ASLE, Chicago (1952); *Proc.*, 98–128.

Radzimovsky, E. I. (1953), *Stress Distribution and Strength Condition of Two Rolling Cylinders Pressed Together.* University of Illinois, EES Bull. 408; 40 pp.

Sasaki, T., Mori, H., & Okino, N. (1960), "Theory of Grease Lubrication of a Cylindrical Roller Bearing." *Bull. JSME* **3,** 212–9.

Sasaki, T., Mori, H., & Okino, N. (1962), "Fluid Lubrication Theory of the Roller Bearing, I . . . Two Rotating Cylinders in Contact." *JBE T* **84,** 166–74; "II . . . Applied to Roller Bearing." *Op. cit.,* 175–80.

Shchedrov, W. S. (1947), Rolling Friction of an Elastic Rough Cylinder on the Real Plane (Russian). *Proc. Second Conf. on Friction and Wear in Machines 1.* Moscow. (a) L. Föppl, Die Strenge Lösung . . . Munich.

Smith, F. W. (1965), "Rolling Contact Lubrication—The Application of Elasto-hydrodynamic Theory." *JBE, T* **87,** 170–6.

Stephenson, R. R. & Osterle, J. F. (1962), "A Direct Solution of the Elastohydrodynamic Lubrication Problem." *TASLE* **5,** 365–74.

Sternlicht, B., Lewis, P., & Flynn, P. (1961), "Theory of Lubrication and Failure of Rolling Contacts." *JBE T* **83,** 213–26.

Tabor, David (1952), "The Mechanism of Rolling Friction." *Phil. Mag.* (7) **43,** 1055–9; **45,** 1081–4 (1954).

Tokuhira, Masatoshi (1948), Theory of Pressures Between Two Barrel-Shaped Rollers in Contact (Japanese). *TJSME* **14,** 12–21.

Uggla, W. R. (1939), Hertzian and Fluid Pressures, *TTM* **69,** 8–11.

Weber, C. & Saalfeld, K. (1954), "Schmierfilm bei Walzen mit Verformung." *ZAMM* **34,** 54–64.

Part 2. Roller Experiments

Akaoka, Jun (1962), "Some Considerations Relating to Plastic Deformation under Rolling Contact." Pages 266–300 in *RCP.* (a) *Bull. JSME* **2,** No. 5; 43–50 (1959).

American Gear Manufacturers Association (1951), *Standard Nomenclature —Gear Tooth Wear and Failure. AGMA* Pub. 110.02 (Dec.).

Anderson, W. J. & Zaretsky, E. V. (1962), "Effect of Several Operating and Processing Variables on Rolling Fatigue," Pages 317–45 in *RCP.*

Archard, J. F. (1965), "Experimental Studies of Elastohydrodynamic Lubrication." Paper R2, Symposium on Elastohydrodynamic Lubrication, Leeds (Sept.), 17–30. (a) Bibliography by Dowson & Archard, 31–5.

Archard, J. F., Hatcher, B. G., & Kirk, M. T. (1964), "Some Experiments upon the Behaviour of Hypoid Oils in Heavily Loaded Contacts." *LWG,* Eastbourne, Paper No. 4, 258. (a) *Sci. Lubn.* **16,** 19 (June).

Archard, J. F. & Kirk, M. T. (1963), "Influence of Elastic Modulus Properties in the Lubrication of Point Contact." Paper 15, *LWG,* 24–31. (a) *JME Sci.* **6,** No. 1, 101–2.

Barwell, F. T. & Scott, D. (1956), "Effect of Lubricant in Pitting Failure in Ball Bearings." *Eng.* **182,** 9–12.

Baugham, R. A. (1960), "Effect of Hardness, Surface Finish and Grain Size on Rolling Contact Fatigue Life . . . ." *JBE, T* **82,** 287–94.

Bell, J. C., Kannel, J. W., & Allen, C. M. (1964), "The Rheological Behavior of the Lubricant in the Contact Zone of a Rolling Contact System." *JBE, T* **86,** 423–35.

Benedict, G. H. & Kelley, B. W. (1961), "Instantaneous Coefficients of Gear Tooth Friction." *TASLE* **4,** 59–70.

Bergen, J. T. & Scott, G. W., Jr. (1951), "Pressure Distribution in the Calendering of Plastic Materials," *JAM* **18,** *T* **73,** 101–6; 317–8.

Blok, Harmen (1949), "Gear Wear as Related to the Viscosity of the Gear Oil." *De Ingenieur* **61,** 39–46. (a) Pages 199–227, 376–7 in *Mechanical Wear,* ed. by J. T. Burwell, Am. Soc. for Metals, Cleveland (1950).

Blok, Harmen (1962), "Hydrodynamic Effects on Friction in Rolling with Slippage." Pages 186–251 in *RCP.*

Buckingham, Earle (1937), "Qualitative Analysis of Wear." Prog. Rep. **15,** *ME* **59,** 576–8.

Buckingham, Earle (1944), "Surface Fatigue of Plastic Materials." Prog. Rep. 16, *T* **66,** 297–310.

Buckingham, E. & Talbourdet, G. J. (1950), "Recent Roll Tests on Endurance Limits of Materials." Pages 289–307, 378 in *Mechanical Wear,* ed. by J. T. Burwell, Am. Soc. for Metals, Cleveland.

Butler, R. H. & Carter, T. L. (1957), *Stress-Life Relation in the Rolling Contact Fatigue Spin Rig. NACA TN* 3930; 23 pp. Cf. (a) Macks, E. F., *LE* **9,** 254–8; (b) *TASLE* **1,** 23–32, 266–72 (1958).

Calhoun, S. F. & Murphy, G. P. (1963), "Effects of Additives Upon Greases." *Sci. Lubn.* **19,** 153–68.

Cameron, A. (1952), "Hydrodynamic Theory in Gear Lubrication." *JIP* **38,** 614–22; 670–1, 673–7, 690, 696.

Carter, T. L., Zaretsky, E. V., & Anderson, W. J. (1960), *Effect of Hardness and other Mechanical Properties on Rolling Contact Fatigue Life of Four High Temperature Bearing Steels, NASA TN* D-270; 51 pp.

Chester, W. T. (1958), "Study of the Surface Fatigue Behavior of Gear Materials with Specimens of Simple Form." Paper 21, *ICG,* 3, 91–8; 55, 380, 386, 389–91, 418, 474–5, 488; Plate 1.

Crook, A. W. (1957), "Simulated Gear Tooth Contact: Some Experiments upon their Lubrication and Sub-Surface Deformation." *Proc. IME* **171,** 187–214.

Crook, A. W. (1958), "The Lubrication of Rollers." *Phil. Trans. Roy. Soc. London* **250** *A,* 387–409. (a) Discussed by J. F. Archer, *ICG,* 402–3; reply, 403–4.

Crook, A. W. (1959), "Some Studies of Wear and Lubrication." *Wear* **2,** 364–93. (a) *Nature* **190,** 1182–3 (1961). (b) *Contemporary Physics* **3,** 257–71 (1962).

Crook, A. W. (1962), "The Lubrication of Rollers. Part II, Film Thickness with Relation to Viscosity and Speed." *Phil. Trans. Roy. Soc. London* **254** *A,* 223–36. (a) "Part III, A Theoretical Discussion of Friction and the Temperatures in the Oil Film." 237–58.

Crook, A. W. (1963), "Developments in Elasto-Hydrodynamic Lubrication." *JIP* **49,** 295–307.

Crook, A. W. & Shotter, B. A. (1957), "Some Scuffing Experiments on a Disk Machine." Paper 6, *CLW,* 205–9; 202, 428, 591, 596–7, 756, 786, 788, 796, 811, 814, 829, 834, 850; Plates 1–3.

Dawson, P. H. (1962), "The Effect of Metallic Contact on the Pitting of Lubricated Rolling Surfaces." *J. ME Sci.* **4,** 16–21.

Drutowski, R. C. (1959), "Energy Losses of Balls Rolling on Plates." *JBE T* **81,** 233–8. (a) Pages 16–35 in *Friction and Wear,* Robert Davies, editor.; Elsevier, Amsterdam.

Dunk, A. C. & Hall, A. S., Jr. (1958), "Resistance to Rolling and Sliding." *T* **80,** 915–20.

El Sizi, S. I., & Shawki, G. S. A. (1960), "Performance Characteristics of Lubricating Oil Film Between Rotating Disks." *JBE, T* **82,** 19–28. (a) "Measurement . . . ." 12–18.

Elliott, J. S. & Edwards, E. D. (1957), "Load Carrying Additives for Steam Turbine Oils." Paper 98, *CLW,* 482–91; 428, 494, 789, 795, 852.

Evans, L. S. & Tourret, R. (1952), "The Wear and Pitting of Bronze Disks Operated under Simulated Worm and Gear Conditions." *JIP* **38,** 652–67, 671–3.

French, H. J., Rosenberg, S. J., Harbaugh, W. L., & Gross, H. C. (1928), "Wear and Mechanical Properties of Railroad Bearing Bronzes at Different Temperatures." *NBS JR* **1,** 343–421.

French, H. J. & Staples, E. M. (1929), "Bearing Bronzes with and without Zinc." *NBS JR* **2,** 1017–38.

Gatcombe, E. K. (1951), "The Non-Steady State Load-Supporting Capacity of Fluid Wedge-Shaped Films." *T* **73,** 1065–75.

Genkin, M. D., Kuz'min, N. F., & Misharin, Yu. A. (1960), A Study of Seizure of Steel Rollers (Russian). *Trans. Third All-Union Conf. on Friction and Wear in Machines, AN SSSR* **1,** 115–22.

Greenert, W. J. (1962), "The Toroid Contact Roller Test as Applied to the Study of Bearing Materials." *JBE, T* **84,** 181–91.

Gröbner, Walter (1962), "Die Reibungzahlen beim technisch reinen Abwälzen geschmierte Rollen." *ZVDI* **104,** 828.

Gross, M. R. (1951), "Laboratory Evaluation of Materials for Marine Propulsion Gears." *Proc. ASTM* **51,** 701–20.

Gruchy, V. J. de & Harrison, P. W. (1963), "Development of an Edge-Type Disk Machine and Preliminary Studies of Various Gear Material-Lubricant Conditions." Paper 14, *LWG,* 3–23.

Grunberg, L. & Scott, D. (1958), "The Acceleration of Pitting Failure by Water in the Lubricant." *JIP* **44,** 406–10.

Hamilton, G. M. (1963), "Plastic Flow in Rollers Loaded Above the Yield Point." *Proc. IME* **177,** 667–75, 686–90.

Hewko, L. O., Rounds, F. G., Jr., & Scott, R. I. (1962), "Tractive Capacity and Efficiency of Rolling Contacts." Pages 157–85 in *RCP.*

Hofer, H. (1931), "Die zulässige Zahnradbeanspruchung und ihre Berechnungsweise im Werkseugmaschinenbau." *Werkstattstechnik* **25,** 128–31.

Hughes, J. R. (1952), "From Test Machine to Gear Box: Problems Associated with the Translation of Laboratory Test Results into Predictions of Field-Service Performance of Gear Lubricants." *JIP* **38,** 712–18.

Jackson, E. G., Muench, C. F., & Scott, E. H. (1960), "Evaluation of Gear Materials Scoring at 700 F." *TASLE* **3,** 69–82.

Johnson, K. L. (1963), "Correlation of Theory and Experiment in Research on Fatigue in Rolling Contact." Paper 14, *Proc. Symposium on Fatigue in Rolling Contact,* IME, 155–9. (a) Cf. Hamilton (1963).

Kannel, J. W., Bell, J. C., & Allen, C. M. (1965), "Methods for Determining Pressure Distribution in Lubricated Rolling Contact." *TASLE* **8,** 250–70.

Kuzmin, N. F. (1954), Coefficient of Friction in a Heavily Loaded Contact (Russian). *Vestn. Mashinostr.* **34,** 8–36.

Lane, T. B. (1951), "Scuffing Temperatures of Boundary Lubricant Films." *Brit. JAP* **2,** Supp. 1, 35–8.

Leach, E. F. & Kelley, B. W. (1965), "Temperature—the Key to Lubricant Capacity." *TASLE* **8,** 271–85.

Manhajn, J. & Mills, H. R. (1945), *Disc Testing Machine for Gear Materials and Lubricants.* IAE; 12 pp.

Martin, J. B. & Cameron, A. (1961), "Effect of Oil on the Pitting of Rollers." *JME Sci.,* **3,** 148–52.

McKee, S. A., Bitner, F. G., & McKee, T. R. (1933), "Apparatus for Determining Load Carrying Capacity of Extreme Pressure Lubricants." *JSAE* **33,** 402–8.

McKee, S. A., Swindells, J. F., White, H. S., & Mountjoy, W. (1949), "Laboratory Wear Tests with Automotive Gear Lubricants." *NBS JR* **42,** 125–30.

Meldahl, A. (1938), "The Brown-Boveri Testing Apparatus for Gear-Wheel Material." *Brown-Boveri Rev.* **38** (Jan.–Feb.). (a) *Eng.* **148,** 63–6 (1939).

Merritt, H. E. (1935), "Worm Gear Performance." *Proc. IME* **129,** 127–94.

Merritt, H. E. (1937), "The Lubrication of Gear Teeth." *GDLL* **2,** 92–103.

Miller, J. C. & Meyers, R. R. (1958), "A Photographic Study of Liquid Flow in a Roll Nip." *Trans. Soc. Rheology* **2,** 77–93.

Milne, A. A. & Nally, M. C. (1957), "Some Studies of Pitting Failure in Rolling Contacts." Paper 54, *CLW,* 459–62; 427, 591, 596–7, 826–7; Plate 1.

Milne, A. A., Scott, D., & MacDonald, D. (1957), "Some Studies of Scuffing with a Crossed-Cylinder Machine." Paper 97, *CLW,* 735–41; 589, 591, 595–6, 804, 814, Plates 1–4.

Misharin, Yu. A. & Sivyakova, A. V. (1960), A Laboratory Study of Anti-Seizure Properties of Certain Materials Used for Worm Gears (Russian). *Trans. Third All-Union Conf. on Friction and Wear in Machines, AN SSR* **1,** 170–3.

Newman, A. D. (1960), "Extreme-Pressure Lubricants for Marine Gears." *Proc. IME* **174,** 241–70.

Niemann, Gustav (1943), "Wälzenfestigkeit und Grubchenbildung von Zahnrad und Walzlagerwerkstoffen." *ZVDI* **87,** 521–3.

Niemann, G. & Gartner, F. (1965), "Distribution of Hydrodynamic Pressure on Counterformel Line Contacts." *TASLE* **8,** 235–49.

Nishihara, T. & Kobayashi, T. (1937), "Pitting of Steel under Lubricated Rolling Contact and Allowable Pressure on Tooth Profiles." *TJSME* **3,** 292–8; S-73.

O'Donoghue, J. P. & Cameron, A. (1965), "Friction and Temperature in Rolling Sliding Contacts." *ASLE* 65 LC-15; 8 pp.

Onaran, Kasif (1960), *Measurement of Wear and Friction of Some Steel and Aluminum Alloys under Rolling Conditions.* Thesis, Tech. Univ. Istanbul; 73 pp.

Orcutt, F. K. (1965), "Experimental Study of Elastohydrodynamic Lubrication." *TASLE* **8,** 381–96.

Pinegin, S. V. (1958), "On Methods of Testing Contact Fatigue of Hard Steel." Paper 41, *ICG,* 4, 110–113; 55, 402, 503.

Poritsky, H., Hewlett, C. W., & Coleman, R., Jr. (1947), "Sliding Friction of Ball Bearings of Pivot Type." *JAM* **14,** *T* **69,** A261–A268.

Reichenbach, G. S. (1959), "Importance of Spinning Friction in Thrust-Carrying Ball Bearings." *JBE, T* **82,** 295–301.

Rounds, F. J., Jr. (1962), "Effects of Lubricants and Surface Coatings on Life as Measured on the Four-Ball Fatigue Test Machine." Pages 346–64 in *RCP.*

Rouverol, W. S. & Tanner, R. I. (1960), "A Brief Examination of Factors Affecting Tractive Friction Coefficients of Spheres Rolling on Flat Plates." *TASLE* **3,** 11–17.

Ryder, E. A. & Barnes, G. C. (1956), "A Rapid Fatigue Test for Rolling Contact Materials." *Bull. ASTM* **217,** 63–4.

Sasaki, T., Okamura, K., & Isogai, R. (1961), "Fundamental Research in Gear Lubrication." *Bull. JSME* **4,** 382–94.

Scott, D. (1957), "Study of the Effect of Lubricant on Pitting Failure of Balls." Paper 58, *CLW,* 453–68; 427, 591, 596–7, 782, 798, 825, 851, Plates 1, 2.

Scott, D. & Blackwell, J. (1963), "Accelerated Test for the Study of Materials under Rolling Contact." Paper 26, *LWG,* 14–21.

Shotter, B. A. (1958), "Experiments with a Disc Machine to Determine the Possible Influence of Surface Finish on Gear Tooth Performance." Paper 15, *ICG,* 120–5; 5, 118, 398, 401–2, 467, 469, 504; Plates 1, 2.

Sibley, L. B. (1962), Discussion on "Hydrodynamic Effects on Friction in Rolling . . . ." Pages 243–8 of Blok *RCP* (1962).

Sibley, L. B. & Orcutt, F. K. (1961), "Elasto-Hydrodynamic Lubrication of Rolling-Contact Surfaces." *TASLE* **4,** 234–49.

Smith, E. A. (1952), "Note on Performance of Graphited Oil." *JIP* **38,** 650–2, 696–7.

Smith, F. W. (1959), "Lubricant Behaviour in Concentrated Contact—The Castor Oil-Steel System." *Wear* **2,** 250–63.

Smith, F. W. (1960), "Lubricant Behavior in Concentrated Contact Systems —Some Rheological Problems." *TASLE* **3,** 18–25.

Smith, F. W. (1962), "The Effect of Temperature in Concentrated Contact Lubrication." *TASLE* **5,** 142–8.

Tabor, David (1955), "The Mechanism of Rolling Friction, II. The Elastic Range." *PRS* **229,** 198–220. (b) Pages 1–5 in RCP (1962).

Talbourdet, G. J. (1954), "A Progress Report on the Surface Endurance Limits of Engineering Materials." *ASME* 54-LUB-14; 13 pp. +. (a) *ME* **77,** 46 (1955).

Tallian, T., Brady, E., McCool, J., & Sibley, L. B. (1965), "Lubricant Film Thickness and Wear in Rolling Point Contact." *ASLE* 65 AM 4 A-4.

Tallian, T. E., Chiu, Y. P., Kamenshini, J. A., Sibley, L. B., Sindlinger, N. E., & Huttenlocher, D. F. (1964), "Lubricant Films in Rolling Contact on Rough Surfaces." *TASLE* **7,** 109–26, 407.

Van Zandt, R. P. & Kelley, B. W. (1949), "Gear Testing Methods for Development of Heavy Duty Gearing." *SAE Trans.* **3,** 354–68.

Walker, Harry (1947), "Testing Machine . . . ." *The Engineer* **183,** 486–8.

Watson, H. J. (1952), "The Testing and Selection of Gear Lubricants." *JIP* **38,** 703, 763–74, 789–802.

Watson, H. J. (1957), "Testing of Marine Main-Propulsion-Gear Lubricants in a Disc Machine." Paper 68, *CLW,* 469–76; 428, 780, 785–6, 788, 854.

Watson, H. J. (1958), "The Choice of Gear Lubricants." Paper 28, *ICG,* 5, 126–34; 118–9, 396, 398–9, 463, 469, 479, 483, 508, 513; Plate 3.

Way, Stewart (1935), "Pitting Due to Rolling Contact." *JAM* **2,** *T* **57,** A 49-A 58.

White, H. S. (1958), "Laboratory Tests with Turboprop Lubricants." *TASLE* **1,** 51–67.

Zaretsky, E. V., Anderson, W. J., & Parker, R. J. (1962), *The Effect of Nine Lubricants on Rolling Contact Fatigue Life.* NASA TN D-1404; 41 pp.

Zaretsky, E. V., Sibley, L. B., & Anderson, W. J. (1962), "The Role of Elastohydrodynamic Lubrication in Rolling Contact Fatigue." *JBE,* **85,** 439–50.

Part 3. Rolling Bearings

Accinelli, J. B. (1958), "Grease Lubrication of Ultra-High Speed Rolling Contact Bearings." *TASLE* **1,** No. 1, 10–16.

Allan, R. K. (1945), *Rolling Bearings.* Pitman, London; 401 pp. (a) 3rd ed. (1965); 395 pp.

Anderson, W. J. (1964), "Rolling Element Bearings." *ABT,* Ch. 6, 139–73. "Fatigue in Rolling Element Bearings." Ch. 12, 371–450. "Extreme Temperature Bearings." Ch. 11, 309–70.

Barwell, F. T. & Hughes, M. J. (1955), "Some Further Tests on High-Speed Ball Bearings," *Proc. IME* **169,** 699–715. (a) *Proc. Seventh ICAM,* London (1948), 257–67.

Bisson, E. E. (1964), "Friction and Bearing Problems in the Vacuum and Radiation Environments of Space." *ABT,* Ch. 9, 259–87.

Boes, D. J. (1963), "Self-Lubricated Bearings . . . ." *LE* **19,** 137–42. (a) *TASLE* **4,** 213–9 (1961).

Booser, E. R. (1957), "Recent Advances in Grease Lubrication of Ball Bearings." Paper 32, *CLW,* 430–7; 427, 781.

Booser, E. R. & Wilcock, D. F. (1953), "Minimum Oil Requirements of Ball Bearings." *LE* **9,** 140–3, 156.

Bowen, P. H. (1962), "Dry Lubricated Bearings for Operation in a Vacuum." *TASLE* **5,** No. 2, 315–26. (a) *ASLE* 64 AM 3A2; 14 pp.

Boyd, J. & Eklund, P. R. (1951), "Some Performance Characteristics of Ball and Roller Bearings for Aviation Gas Turbines." *ASME* 51-A-78; 17 pp.

Büche, W. (1934), "Eine hydrodynamische Theorie der Flüssigkeitsreibung in Rollenlagern." *Forschung* **5,** 237–44.

Butner, M. F. & Rosenberg, J. C. (1962), "Lubrication of Bearings with Rocket Propellants." *LE* **18,** 17–24.

Coit, R. A. & Sorem, S. S. (1962), "Anti-Friction Bearing Lubricant Requirements in High Altitude Environments." *LE* **18,** 438–42.

Delfosse, Marcel (1936), *Sur le couple des roulements à billes.* Pub. sci. et tech. du Ministère de l'Air, Paris, No. 83; 64 pp.

Dowson, D. & Higginson, G. R. (1963), "Theory of Roller-Bearing Lubrication and Deformation." Paper 19, *LWG,* 58–69.

Eickhof, K. G. & White, A. (1961), "The Performance of Ball Bearings in Nitrogen and Carbon Dioxide at Elevated Temperatures." *TASLE* **4,** 39–49.

Ferretti, Pericle (1932), Experiments with Needle Bearings (Portuguese). *Revista Aeronautica* **8,** No. 10; 47–71. (a) *NACA TM* 707 (1933); 18 pp.

Fogg, A. & Webber, J. S. (1953), "Lubrication of Ball and Roller Bearings at High Speed." *JIP* **39,** 743–64. (a) "Influence of Design Factors . . . ." *Proc. IME* **169,** 716–45 (1955).

Freundlich, M. M. & Hannan, C. H. (1961), "Problems of Lubrication in Space." *LE* **17,** 72–7.

Fricker, H. W. (1963), "Bearings and Gears for Operation in Inert Gases." Paper 30, *LWG,* 46–59.

Getzlaff, G. (1952), "Das Verhalten von Wälzlagern bei sehr hohen Drehzahlen." *Konstruktion* **4,** No. 9, 280–8. (a) *JASNE* **65,** 328–38 (1953).

Glaeser, W. A. (1960), "The Performance of Heavily Loaded Oscillating Roller Bearings Operating from 300 F to 600 F." *TASLE* **3,** No. 2, 203–7.

Goodman, John (1912), "Roller and Ball Bearings." *Proc. ICE* **189,** 82–166.

Graneek, M. & Wunsch, H. L. (1954), "Testing Ball Bearings at High Rotational Speeds." *Eng.* **178,** 695–7.

Gustafson, J. H. (1964), "High Temperature Oil Evaluation in Full-Scale Bearing Fatigue Tests." *LE* **20,** 65–8.

Hackewitz, F. W. (1958), "The Influence of Kapitza's Viscosity on the Hydrodynamic Lubrication of a Cylindrical Roller Bearing . . . ." *JAM* **25,** *T* **80,** 620–2. (a) **28,** *T* **83,** 297–9 (1961).

Hanocq, Ch. (1937), "Étude expérimental des paliers de transmission." *RUM* Apr. 1, May 1. (a) *GDLL* **2,** 75–80.

Harris, T. A. (1964), "Prediction of Temperature in a Rolling Contact Bearing Assembly." *LE* **20,** 145–50.

Heathcote, H. L. (1921), "The Ball Bearing: In the Making, under Test, and on Service." *Proc. IAE* **15,** 569–702.

Hirano, F. & Tanone, H. (1961), "Motion of Ball in Ball Bearing." *Wear* **4,** 177–97.

Horsch, J. D. (1963), "Correlation of Gyro Spin-Axis Ball Bearing Performance with the Dynamic Lubricating Film." *TASLE* **6,** 112–24.

Hustead, T. E. (1963), "Consideration of Roller Bearing Load Rating Formula." *TSAE* **71,** 202–8.

Iida, K. & Igarashi, A. (1959), "On the Behavior of Rollers in a Cylindrical Roller Bearing." *Bull. JSME* **2,** No. 8, 538–45.

Irving, R. & Scarlett, N. A. (1963), "Wear Problems with Grease-Lubricated Rolling Bearings in Inert Atmospheres." Paper 29, *LWG,* 39–45.

Jones, A. B. (1952), "Life of High Speed Ball Bearings." *T* **74,** 695–703.

Jones, A. B. (1959), "Ball Motion and Sliding Friction in Ball Bearings." *JBE T* **81,** 1–12.

Jones, A. B. (1960), "A General Theory for Elastically Constrained Ball and Radial Roller Bearings under Arbitrary Load and Speed Conditions." *JBE T* **82,** 309–20.

Jones, F. G. & Wilcock, D. F. (1950), "The Mechanism of Lubrication Failure in High Speed Ball Bearings." *T* **74,** 817–23.

Kotova, L. I. (1960), Theory of the Lubrication of a Cylindrical Roller Bearing with a Visco-Plastic Lubricant (Russian). *Trans. Third All-Union Conf. on Friction and Wear, AN, SSSR* **3,** 84–95.

Krause, H. H., Cosgrove, S. I., & Allan, C. M. (1960), "Phthalocyamines Promise Good 1000 F Lubricants." *Space/Aeronautics* **34,** 161–5.

Lane, A. W., Klemes, M. S., & Zeigler, E. L. (1959), "Achieving Extremely Accurate Non-Floated Gyros." *Aero Space Eng.* **18,** 43–6.

Larson, R. H. & Piken, A. G. (1962), "Bearing and Lubricant Requirements for Some Aerospace Projects . . . ." *TASLE* **5,** No. 1, 1–7.

Lawrence, J. C. (1961), "Study of Torque Characteristics of Grease-Packed R-2 and R-3 Ball Bearings." *LE* **17,** 484–7.

Lewis, P., Murray, S. F., Peterson, M. B., & Esten, H. (1963), "Lubricant Evaluation for Bearing Systems Operating in Spatial Environments." *TASLE* **6,** No. 1, 67–79.

Lipp, L. C. & Klemgard, E. N. (1963), "A Review of Techniques for Investigation of Friction and Wear in Aerospace Ball and Roller Bearings." *LE* **19,** 495–502.

Lundberg, G. & Palmgren, A. (1949), "Dynamic Capacity of Rolling Bearings." *JAM* **16,** *T* **71,** 165–72. (a) *Forschung* **18,** 97–105 (1952).

Macks, E. F. & Nemeth, Z. N. (1950), "Investigation of 77 mm-Bore Cylindrical Roller Bearings at High Speeds . . . ." I. *NACA TN* 2128; 54 pp. II. *TN* 2216; 28 pp. (a) "Influence of Viscosity . . . ." *TN* 2636 (1952); 47 pp.

Macks, E. F. & Nemeth, Z. N. (1951), "Lubrication and Cooling Studies of Cylindrical Roller Bearings at High Speeds." *NACA TN* 2420, 49 pp. (a) *TN* 2636; 47 pp. (1952). (b) *LE* **9,** 263–8 (1953).

Macks, E. F., Nemeth, Z. N., & Anderson, W. J. (1952), "Operating Characteristics of Cylindrical Roller Bearings at High Speeds." *T* **74,** 705–13 (a) *NACA Rep.* 1084; 12 pp.

Market, W., Jr., & Ferguson, K. M. (1957), "Use of Rolling Contact Bearings in Low Viscosity Liquid Metal Lubricants." *LE* **13,** 285–90.

Martin, K. B. & Jacobs, R. B. (1959), "Testing and Operation of Ball Bearings Submerged in Liquified Gases." *TASLE* **2,** No. 1, 101–7.

Maurer, E. R. & Kelso, L. E. A. (1931), *Friction of Some Babbitt, Roller and Ball Bearings.* Bull. University Wisconsin, EES No. 72; 48 pp.

Milne, A. A., Scott, D., & Scott, Mrs. H. M. (1957), "Observations on the Movement and Structure of Grease in Rolling Bearings." Paper 45, *CLW,* 450–3; 426; 781, 842; Plates 1–4.

Moore, C. C. & Jones, F. C. (1955), "Operating Characteristics of High Speed Ball Bearings at High Oil Flow Rates." *ASME* 55 LUB 10; 6 pp. + (a) *ME* **77,** 1004.

Moore, C. C. & Lewis, P. (1957), "Current Development Problems in High Temperature Aircraft Roller Bearings." Paper 34, *CLW,* 438–43; 427, 782, 788, 825; Plates 1–3.

Muzzoli, Manlio (1934), "L'attrito nei cuscinetti a rotolamento." *Ricerche de Ingegneria* **12,** No. 5, 205–37; **16,** No. 6, 165–78 (1937).

Nemeth, Z. N. & Anderson, W. J. (1955), "Temperature Limitations of Petroleum, Synthetic and Other Lubricants in Rolling Contact Bearings." *TASLE* **63,** 556–66. (a) *NACA TN* 3337; 31 pp. (b) *LE* **11,** 267–73. (c) **12,** 267–73 (1956).

Nemeth, Z. N., Macks, E. F., & Anderson, W. J. (1952), "Investigation of 75-Millimeter-Bore Deep-Groove Ball Bearings under Radial Load at High Speeds." I. *NACA TN,* 2841; 30 pp. II. *TN* 3003; 33 pp. (1953).

Osterle, J. F. (1959), "On the Hydrodynamic Lubrication of Roller-Bearings." *Wear* **2,** No. 3, 195–202.

Otterbein, M. E. (1958), "The Effect of Aircraft Gas Turbine Oils on Roller Bearing Fatigue Life." *TASLE* **1,** 33–40.

Palmgren, Arvid (1945), *Ball and Roller Bearing Engineering* (Swedish), trans. by G. Palmgren and B. Ruley, SKF Industries, Inc., Philadelphia. (a) 3rd ed., 1959; 264 pp.

Palmgren, A. & Snare, B. (1957), "Influence of Load and Motion on the Lubrication and Wear of Rolling Bearings." Paper 52, *CLW,* 454–8; 202, 427, 782–7, 838, 880.

Rounds, F. G. (1963), "Influence of Glycol Molecular Configuration on Friction." *TASLE* **6,** 89–101.

Sasaki, T., Okino, M., & Fujita, T. (1959), "Lubrication of High Speed Tapered Roller Bearing by Atomized Oil." *Bull. JSME* **2,** 223–9.

Sasaki, T. & Okino, N. (1962), "Rolling Friction at High Speed." *Bull. JSME* **5,** No. 18, 360–73. (a) *TJSME* **17,** 4–8 (1951).

Schreiber, H. H. & Ulsenheimer, G. (1960), "Zur Frage der Ermüdungserscheinungen bei Wälzlagern." *Wear* **3,** 122–41.

Schuller, F. T. & Anderson, W. J. (1961), "Operating Characteristics of 75-Millimeter Bore Ball Bearings at Minimum Oil Flow Rates over Temperature Range to 500 F." *LE* **17,** 291–8. (a) *NACA* Rep. 1177 (1954).

Shaw, M. C. & Macks, E. F. (1949), "Rolling Contact Bearings." *ALB,* Ch. 10, 389–437.

Simpson, F. F. & Crump, W. J. J. (1963), "Effect of Electric Currents on the Life of Rolling Contact Bearings." Paper 27, *LWG,* 22–7. (a) Paper 85.

Sliney, H. E. (1961), "Bearings Run at 1250 Deg F with Solid Lubricant." *Space/Aeronautics* **35,** 91–2, 94, 96, 98, 100.

Smith, C. F. (1962), "Some Aspects of the Performance of High-Speed Lightly Loaded Cylindrical Roller Bearings." *Proc. IME* **176,** 566–81; 582–601.

Sorem, S. S. & Cattaneo, A. G. (1956), "High-Temperature Bearing Operation in the Absence of Liquid Lubricants." *LE* **12,** 258–60.

Stribeck, R. (1901), "Kugellager für beliebige Belastungen." *ZVDI* **45,** No. 3, 73–9, 118–25. (a) Trans. by H. Hess, *T* **29,** 420–67 (1907).

Styri, Haakon (1940), "Friction Torque in Ball and Roller Bearings." *ME* **62,** 886–90; **63,** 737–42.

Taylor, K. M., Sibley, L. B., & Lawrence, J. C. (1963), "Development of a Ceramic Rolling Contact Bearing for High Temperature Use." *Wear* **6,** No. 3, 226–40.

White, H. S., Swindells, J. F., & Belcher, H. V. (1955), "Oil Soaked Felt-Pad Lubrication of Ball Bearings at High Speed and High Temperature." *LE* **11,** 182–6. (a) *NBS JR 63 C,* No. 1, 19–29 (1959).

Wilson, W. A., Martin, K. B., Brennan, J. A., & Birmingham, B. W. (1961), "Evaluation of Ball Bearing Separator Materials Operating Submerged in Liquid Nitrogen." *TASLE* **4,** No. 1, 50–8.

Yamamoto, Toshio (1959), "On Critical Speeds of a Shaft Supported by Ball Bearings." *JAM* **26,** *T* **81,** 199–204. (a) *ASME* 64-APM-32; 8 pp.

Young, W. C., Clauss, F. J., & Drake, S. P. (1963), "Lubrication of Ball Bearings for Space Application." *TASLE* **6,** No. 3, 178–91.

chapter XI Gear Lubrication

1. Gear geometry. 2. Contact stresses. 3. Film thickness theory. 4. Film thickness experiments. 5. Friction loss, theoretical. 6. Friction loss, experimental. 7. Gear failures. 8. Summary.

Gears are used to transmit motion from one shaft to another, usually with a change in speed. In a watch or clock the gear train effects a great change in speed while transmitting practically no power. At the other extreme, reduction gears are used in a steamship to reduce the speed of a turbine shaft to that required by the propeller. If the speed reduction is in the ratio of five to one, the torque supplied by the turbine will be stepped up fivefold, except for friction losses. Other applications of gearing are to aircraft, automobiles, and machine tools. Good descriptions and illustrations may be seen in the treatise by D. W. Dudley (1954) and in recent textbooks.

1. GEAR GEOMETRY

Spur, helical, and bevel gearing are relatively simple types, having their axes in the same plane. Worm and hypoid gears are more complicated, having non-coplanar axes, as required, for example, in automobile rear axles.

Spur gear teeth are now commonly designed with involute profiles. It will be recalled that an *involute* is the curve traced by the end of a string unwrapped from a circular cylinder. The cylinder is the *base circle* in a cross section of the gear wheel. The tips of the teeth lie on an outer circle. In a pair of meshing gears the tip circles overlap, but the base circles of external gears do not meet. The respective gears are best identified by their *pitch circles*, which are tangent one to the other. The point of tangency falls on the line of centers at an intersection called the *pitch point.* Tooth height above pitch circle is known as the *addendum;* depth from pitch circle to root, as the *dedendum.* A line drawn tangent to the two base circles, passing through the pitch point, is called the *line of action.* It is the locus of the points of contact of any pair of mating teeth, and lies in the direction of the normal force exerted by one tooth on the other. The angle between the line of action and a normal to the line of centers at the pitch point is called the *pressure angle* (Fig. 1).

Involute gear designs are well characterized by stating the pitch diameter D and number of teeth; the addendum, dedendum, pressure angle ϕ, helix angle ψ, and face width or axial dimension L. The *diametral* pitch is often given, meaning the ratio of the number of teeth to the diameter. The *circular pitch* is the distance from one tooth to a corresponding point on the other measured on the pitch circle. In a pair of meshing gears the smaller one is called the *pinion,* the larger the *gear.* The ratio of the number of teeth in the gear to that in the pinion is called the *gear ratio.* If the pinion is driving and the gear ratio is 5, the speed reduction will be 5 to 1 between pinion shaft and gear shaft.

In practice, allowances are made for backlash, clearance between

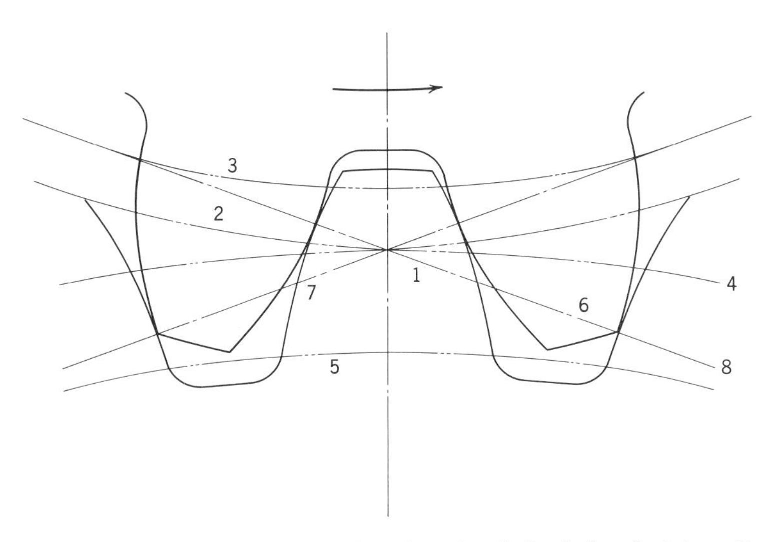

Fig. 1. Involute gears: 1, pitch point; 2, pitch circle of pinion; 3, base circle of pinion; 4, pitch circle of gear; 5, base circle of gear; 6, pinion tooth; 7, gear tooth; 8, line of action.

tip and root, and *tip relief*. The last mentioned is a departure from involute profile near the tips to avoid interference. Similar definitions apply to helical and double helical or *herringbone* gears, the latter preferred in large installations to eliminate thrust. These definitions and the formulas that can be derived from them are discussed at length by Earle Buckingham (1949) and more briefly by Dudley in his book; see also AGMA (1965). For example, in setting up disk experiments to simulate gear tooth contact, a knowledge of tooth radius r at the pitch line is required. A sketch based on the definitions will show that $r = (D/2) \sin \phi$. The construction can be extended to give the radius at other points on the profile.

2. CONTACT STRESSES

Let W denote the normal force per unit of face width exerted by one tooth on another while they are in contact at the pitch point; that is, the force along the line of action. The component W_t, tangent to the pitch circle and needed for computing torque, is $W \cos \phi$. This is called the tangential load. Let it be required to calculate the maximum Hertz compressive stress at the point of contact, or strictly speaking, at the line of contact.

Two geometrically similar gear sets of different sizes will experience the same Hertz stress at corresponding points if they are of the same material and loaded in proportion to the squares of their diameters, so that the load per unit of projected area is the same for both. Let P denote the tangential load per unit area. Any area may be chosen arbitrarily. It is customary to take the product of face width by pinion pitch diameter, which might be called the pitch area of the pinion. A formula can be worked out for the compressive stress in terms of P, E, m, and ϕ, where E is the mean elastic modulus of gear and pinion, m the gear ratio, and ϕ the pressure angle—straight spur gears being understood when not otherwise indicated. It is only necessary to apply the Hertz equation (2) of Chapter X to the contact of gear teeth instead of cylindrical rollers. The formula gives the maximum compressive stress p_m (often denoted by s_c) in terms of the normal load W per unit width, the modulus E, and the relative or equivalent radius of curvature R defined in Chapter X; assuming Poisson's ratio 0.30. Replace W by its equal, $W_t/\cos \phi$, and $1/R$ by the sum of $2/D \sin \phi$ and $2/mD \sin \phi$. The equation then takes the form

$$p_m = \frac{0.592 \sqrt{E}}{\sin \phi \cos \phi} P \left(\frac{m + 1}{m} \right). \tag{1}$$

The K-Factor. Gear people are constantly talking about a mysterious "K-factor." They will tell you how much it has been increased in the past several years because of better gear design, smoother and harder surfaces, smaller manufacturing errors, and improved lubricants. The *K-factor* is a convenient way of describing the contact stress on the gear and pinion teeth. It is defined by the expression $P(m+1)/m$ in Eq. (1). Thus

$$p_m = C\sqrt{K}\,, \tag{2}$$

where C_ϕ is the constant factor of Eq. (1) for any pressure angle ϕ. Note that K has the dimensions of pressure and varies only from $2P$ to P when the pinion drives, while m ranges from unity to infinity. Thus for $m = 2$, $K = 1.5P$. For internal gears, replace $m+1$ by $m-1$.

Helical and Other Gears. Consider an example in which the gear and pinion are of steel, so that $E = 3 \times 10^7$ psi. If the pressure angle is 20 deg, C_ϕ will be 5715. Equation (2) holds for both spur and helical gears. In the case of helical gears C_ϕ is equal to the constant of Eq. (1) multiplied by cos ψ, where ψ is the helix angle, and divided by the square root of m_p, the profile contact ratio. This ratio is the average number of teeth in contact. For a 30-deg helix, m_p is 1.41, making $C_\phi = 4160$. Thus for a given tangential load, the contact stress is less in helical gears. A table for helix angles up to 45 deg appears on page 51 of Dudley (1954), with ϕ from $14\frac{1}{2}$ to 25 deg, from which C_ϕ can easily be computed.

Analogous formulas and tables for non-coplanar gears are given in the references cited and in the *Gear Handbook* (Dudley, 1962). See also the hypoid gear calculations by Powell & Barton (1960) and by Coleman (1964).

One need only read the IME lecture by A. W. Davis (1956) to realize how many nonmathematical factors must be dealt with in the art of gear design—manufacturing errors and misalignment, for example. Practical experiences in the design and operation of gears were discussed at the IME International Conference on Gearing (1958). The need for uniformity in gear design practice is reflected in the design standards issued at intervals by the American Gear Manufacturers Association (Wellauer, 1965).

Dynamic Loads. Thus far steady loads only have been assumed to act on the gear teeth. Actually, the dynamic loads caused by shaft vibrations and especially by errors in tooth spacing can introduce maximum stresses several times as great as those resulting from the mean torque transmitted. Pioneering research on this problem is credited to Earle Buckingham (1931). Theoretical and experimental investigations followed by Reswick (1955), Niemann & Rettig (1957), Harris (1958), Tuplin (1958), and Attia (1959); also, more recently, by Utagawa, Gregory, and co-authors (1962, 1963). Applications to gear design are described by Buckingham in Chapters 20 and 21 of (1949); and by Dudley, pages 46, 298–9, and 327–32 of (1954). The Buckingham equation is well presented by Shigley (1963) in his Chapter 11.

3. FILM THICKNESS THEORY

The importance of the lubricant as one of the component materials in gear design has been emphasized by Almen (1942) and by Blok (1951). Formulas are given by Blok for the amount of lubricant retained in the gear mesh after some has been thrown off by centrifugal force. From a study of gear motions, Michell concluded that "true fluent lubrication" cannot be expected (pages 234–6 in I: 1950).

It has usually been assumed that tooth contact can be simulated by circular cylinders rotating about fixed axes. Such applications have been worked out in detail by Gatcombe (X: 1945), Niemann (1955), and others. Niemann starts with the classical formula for minimum film thickness, Eq. (5) of Chapter X. He then converts roller geometry into gear geometry for each contact position x, or fractional distance along line of action from its point of tangency on base circle of pinion to that on base circle of gear. It follows that for rigid teeth and uniform viscosity we should expect

$$h_0/D = c(ZN/P)\ ; \tag{3}$$

where for any gear ratio m and pressure angle ϕ,

$$c = 3.85x(1 - x)[1 + (m - 1)x]\frac{(m + 1)^2}{m}\sin^2\phi\cos\phi\,. \tag{4}$$

In gears of equal size ($m = 1$) the minimum film thickness will be greatest at pitch line engagement, where $x = \frac{1}{2}$. Under this condition $c = 3.85 \sin^2 \phi \cos \phi$; and the parabolic average for all positions of engagement is $\frac{2}{3}$ thereof, or $2.57 \sin^2 \phi \cos \phi$.

Gatcombe states that he found about 18 per cent greater theoretical load capacity for helical gears than for straight spur gears. Gatcombe & Prowell (1960) showed how to calculate Hertzian contact times in order to compare them with visco-elastic relaxation times in gear oils. It is a moot question whether transient elastic properties are significant in gear lubrication. In the example of a rocket motor operating at 27,000 rpm, the authors calculated the duration of contact to be from 3 to 8 micro-seconds—slightly more than the probable relaxation time in a lubricating oil. Gatcombe challenged theoreticians to extend their solutions to a nonsteady state (1954), including squeeze-film action as indicated by his experimental study (X: 1951). Osterle and Stephenson calculated the small effect resulting from profile differences between the true involute and the arc commonly assumed, but retained the concept of fixed axes (1959, 1961). A solution for the more general case by Adkins & Radzimovsky (1963) is limited to light loads. Gear calculations for heavier loads would require elastohydrodynamic theory.

4. FILM THICKNESS EXPERIMENTS

Hydrodynamic action between gear teeth has been found experimentally. At least nine investigations should be cited here. Electrical resistance methods were used in four, breakdown voltage in the others. Lane & Hughes (1952) found the resistance between meshing teeth of spur gears lowest at the location of highest sliding speed. It was a maximum at the pitch line, where there is no sliding. Beurlein (1955) concluded that mixed lubrication predominates. Campbell & MacDonald (1961) tested hypoid gears in full-sized rear axles under practical operating loads and speeds. They observed hydrodynamic films during a part of the meshing cycle except at the highest loads and lowest speeds. Comparing a mineral with an EP oil, they confirmed the importance of running-in, and credited EP lubricants with preserving a surface finish conducive to hydrodynamic lubrication. These investigators measured electrical resistance.

MacConochie & Shakib (1963), using a voltage method with continuous discharge, tested a twenty-eight-tooth pinion of 7-in. pitch diameter with a thirty-two-tooth gear. They compared a mineral with an EP oil over a wide range of conditions and reported the percentage of time during which boundary lubrication prevailed. This was defined by zero resistance on the oscilloscope. A chart is given showing the per cent of boundary operation against load for three viscosities of mineral oil at a constant pinion speed of 335 rpm. The top curve climbs from zero to 80 per cent at 1600 lb/in. of tooth loading for a viscosity of 27 cp. Lower curves are for higher viscosities. Much lower percentages of boundary operation are reported for the EP lubricant.

Actual film thicknesses were measured in the other investigations,

TABLE 1. Film Thickness between Gear Teeth

Year	Authors	Micro-inches		$10^6 \times$	
		Finish	h_0	h_0/D	ZN/P
	Thickness at pitch line:				
1958	Cameron & MacConochie	20	120	57	1.9
1961	MacConochie & Newman	70	71	26	1.5
1963	Ibrahim & Cameron	17	70	14	2.5
	Mean	54	87	32	2.0
	Overall average thickness:				
1962	Sasaki, Okamura, Konishi, & Mishizawa	35	31	7	11.0
	Thickness where minimum:				
1963	Dareing & Radzimovsky	30	25	5	43.0

as indicated by Table 1. Representative values only are given here. They are listed in the decreasing order of film thickness, h_0, micro-inches. The corresponding h_0/D and ZN/P values are dimensionless, the same in English or metric units. Here D is the pitch diameter, Z the approximate film viscosity at atmospheric pressure, N the pinion speed in revolutions per unit time, and P the tangential load per unit of pitch area as before. Pitch diameters ranged from 2.75 to 5.0 in. Surface finish has been estimated in rms (root mean square) micro-inches. Graphs showing the increase of h_0/D with ZN/P could be plotted from data given by each investigation. Note that the average and minimum film thicknesses are small compared to those observed at the pitch line. The evidence for hydrodynamic action is convincing.

The first three investigations of Table 1, including Ibrahim & Cameron (1963), utilized the electrical discharge method of Brix and Cameron (Cameron, X: 1954). Adkins & Radzimovsky (1965) considered it essential not to disturb the film before measurements are made, so determined the breakdown voltage immediately before discharge. Sasaki and associates measured electrical resistance (1962). They found boundary lubrication for ZN/P values below 5×10^{-6}, and concluded that full fluid lubrication could not be expected with commercially finished surfaces unless ZN/P is greater than 5×10^{-4}.

It is interesting to compare the h_0/D values with those calculated by Niemann's formula. Cameron & MacConochie (1958) found h_0/D closely proportional to ZN/P in one series of tests, but varying about as the 0.15 power in another. The Sasaki reference makes film thickness nearly proportional to the square root of ZN/P at low values. These facts point to the possible influence of pressure-viscosity and elastohydrodynamic action at the heavier loads.

A full review of hydrodynamic and other phenomena in gear tooth contacts is given by Merritt (1962) and followed by a valuable discussion.

5. FRICTION LOSS, THEORETICAL

Early calculations for cycloidal and involute gears assumed a constant coefficient of friction. Blok (1937) quotes a "well-known" formula for efficiency, η, of spur gears from which the loss ratio, $1 - \eta$, would be ρf divided by $r \cos \theta$, where ρ is half the effective line of action, r the pitch radius, θ the pressure angle, and f the coefficient of friction. E. Pistolesi works out a calculation that takes rolling friction into account. It will be found on pages 3–8 of his introduction to Giovannozzi's hydrodynamic theory (1939).

Giovannozzi developed the hydrodynamic theory of friction losses in gears with rigid teeth, finding loss ratios proportional to the square root of ZN/P. In the particular example of two equal gears 10.8 cm in pitch diameter, thirty-four teeth each, pressure angle 15 deg, normal tooth load 40 kg/cm, speed 660 rpm, and lubricated with an oil of 10^{-6} kg-sec/sq cm viscosity, he calculated the loss to be 2.46 per cent.

Buckingham's theory of power loss (1949), credited in part to Professor William H. Clapp of Stanford, treats the coefficient as a constant equal to f_a during the approach motion, and f_r during recession. The respective coefficients are to be read from empirical curves plotted against sliding speed. Power loss H is expressed by its ratio to the input power H_1 or W_tLV, where W_t is the tangential load at the pitch circle per unit of face width L, and V is the linear velocity at the pitch radius. Efficiency is then found by taking 1 minus H/H_1, or 100 less H/H_1 per cent. Buckingham's calculation gives the loss ratio in spur gears, disregarding the loss at no load, as a function of m, f_a and f_r, β_a and β_r; where β_a and β_r are, respectively, the arcs of approach and recess for the driver. In estimating the amount of frictional heat to be dissipated, it is sufficient to assume a mean coefficient

f, and a mean value $\beta/2$ rad for the arcs of approach and recess; whereby

$$\frac{H}{H_1} = \frac{\beta}{4}\left(\frac{m+1}{m}\right)f, \tag{5}$$

in which

$$\beta = [\sqrt{m(m+s)} + \sqrt{1+s} - (m+1)] \sec \phi. \tag{6}$$

Here s has been written for $2\alpha/\sin^2 \phi$, where α is the addendum ratio of the driver, $2a/D$. The addendum height a is assumed the same for pinion and gear.

The coefficient f is to be read from Buckingham's curve at a mean sliding speed

$$V_s = \frac{V}{4}\left(\frac{m+1}{m}\right)\beta \cos \phi. \tag{7}$$

He found that the curve could be fitted by the equation

$$f = 0.05e^{-V_s/8} + 0.002V_s^{1/2}, \tag{8}$$

where V_s is measured in feet per minute. It was successfully applied up to 2500 ft/min, at loads from 0 to 1456 lb on gears of 1.25 in. face having pinion pitch diameters of 6.0 in., with $m = \frac{8}{3}$ and pressure angles from $14\frac{1}{2}$ to 20 deg. The mean coefficient drops to a minimum of 0.012 at 30 ft/min. It was inferred that hydrodynamic lubrication prevailed at the higher speeds, with mixed film conditions at lower speeds.

In an example of the more exact calculation, Buckingham compares two sets of gears each with pinion pitch diameters of 6.0 in., $m = 5$, $a = \frac{1}{4}$ in., $V = 1500$ ft/min; one having $\phi = 14\frac{1}{2}$, the other 20 deg. In the first case he finds $f_a = 0.044$, $f_r = 0.020$, and $H/H_1 = 0.59$ per cent. In the second case he finds $f_a = 0.038$, $f_r = 0.018$, and $H/H_1 = 0.41$ per cent. The set with the greater pressure angle generates less heat. By the short method, Eq. (5), we obtain $f = 0.032$ and $H/H_1 = 0.52$ per cent in the first case; $f = 0.027$ and $H/H_1 = 0.36$ per cent in the second case.

It is interesting to note from Eq. (6) that as the teeth get smaller in a series of designs, α and therefore s approaching zero, the arc of action β approaches $s \tan \phi$. The power loss at a given speed then diminishes until after the coefficient of friction has reached a minimum value. Buckingham's theory makes no reference to the viscosity of the lubricant (possibly SAE 30); yet it can hardly fail to agree with the Lewis machine tests, since the f curve was derived from the measured friction moment. The same method was extended to helical and non-coplanar gears. An equivalent treatment is found in Chapter 23 of Merritt's book (1942).

Niemann's theory, like that of Giovannozzi, leads to a hydrodynamic expression

$$H/H_1 = c' \sqrt{ZN/P}\ , \tag{9}$$

where c', like the constant c of Eq. (3), is a function of the gear geometry, but the same for geometrically similar gear pairs of whatever size. In helical gears, H/H_1 varies inversely with the cosine of the helix angle. The derivation, limited to light loads, is analogous to that of the film thickness equation (3). A rapid succession of equilibrium states having different h_0 values are combined in a continuous manner without regard to squeeze-films. Yet, as pointed out by Niemann (1955), the friction moment corresponding to Eq. (9) agreed with experimental values at the right of the minimum published by Dudley in his Fig. 7–13 (1954). Niemann's treatment was later extended to include mixed film conditions at the left of the minimum (1958).

Two methods of estimating power loss in terms of a known coefficient of friction, or vice versa, are given by Shipley in Chapter 14 of the Gear Handbook (Dudley, 1962). The first and more elaborate method—derived by Shipley (1958)—seems to be equivalent to Buckingham's. It reduces to Eq. (5) when $m = 1$. The second method can be expressed by the formula

$$H/H_1 = f/M\ , \tag{10}$$

where M is the "mechanical advantage of the mesh," to be read from his Fig. 14-3. Table 2, based on Merritt's calculation (1941), is of-

TABLE 2. Data from Merritt for Shipley's Formula
(Pressure angle 20 deg)

Pinion Teeth	Mechanical Advantage, M				
	$m = 1$	2	4	6	8
10	3.0	3.2	3.5	3.8	4.0
25	5.2	6.5	7.3	7.4	8.0
50	9.2	12.0	13.7	14.6	15.1
90	15.8	20.9	23.9	25.5	26.5

fered as a simplified substitute for Shipley's diagram. The required values of the coefficient may be estimated from the curves for f against V, the pitch line velocity. Representative values for hardened gears from Shipley's Fig. 14-2 are given in Table 3 for an oil specified therein with viscosity at inlet temperature approximating 34 cp. The load range in our table is indicated by the K-factor.

Applying Eq. (10) to Buckingham's example of a 6.0-in. diameter pinion with twenty-four teeth and a pressure angle of 20 deg, we recall that $m = 5$ and $f = 0.027$. Interpolating in Table 2 gives $M = 7.4$ from which $H/H_1 = 0.36$ per cent, as before. To check the f value we can assume a mean tangential load of $\frac{1456}{2}$ or 728 lb on a face width of 1.25 in. This makes $P = 97.1$ psi, from which $K = 117$. Enter Table 3 with $V = 1500$ ft/min. Extrapolation gives roughly $f = 0.033$ as compared to 0.027 from Buckingham's curve—which may represent test data on softer gear materials.

Similar relations between f and H were worked out by Sasaki and co-authors (1962). Effects of tooth deflection were studied by Tso & Prowell (1961). Friction losses in gear trains have been calculated by Merritt (1941), Poppinga (1950), Kudryavtsev (1953), Clausen (1954), Shipley (1958), and Radzimovsky (1959) from known coefficients of friction.

Merritt, Walker, Buckingham, Wellauer, and Botstiber have calculated heat transfer and thermal equilibrium in gear boxes (1942, 1944, 1949, 1952, and 1956, respectively). See also pages 19–23 of Niemann & Ohlendorf (1958). Welch & Boron (1961) developed a theory of instability based on thermal expansion in high-power, high-speed reduction gears. It was confirmed experimentally.

TABLE 3. Some Friction Values by Shipley
(Inlet 120 F, oil of 300 SUS/100 F)

Pitch Line V, ft/min	Coefficient, f		
	$K = 200$	500	800[b]
1000	0.039	0.048	0.058
2000	0.033	0.042	0.051
4000	0.026[a]	0.033[a]	0.040
5000	0.028	0.034	0.039[a]

[a] Minimum point. [b] K-factors in psi.

6. FRICTION LOSS, EXPERIMENTAL

Reference has been made to friction measurements by Buckingham, Dudley, and Shipley. Buckingham used the Lewis machine at M.I.T. (Lewis, 1914, 1923). A deceleration method was applied (Buckingham, 1931). This machine is based on the four-square or back-to-back principle. It is a "circulating power" device, analogous to that used by Hopkinson in testing motors and generators; and much more accurate than the older method of subtracting output from input. Two pairs of gears are locked together by a torque coupling, such that one pair loads the other. The only power supplied is that needed to overcome friction. A broad patent on the four-square method, applied for by Albert Kingsbury in 1910, was granted in 1916. This method is now in common use. Recent developments are described by Shipley (1958), pages 77–82.

Spur and Helical Gears. Chronologically, modern test results on spur gears began with the thesis by Green & Doble (1910). They used an early form of the Lewis machine in an investigation under Professor Lanza. Noteworthy tests have since been reported by Rickli (1911), Donald (1922), Ham & Huckert (1925), Kutzbach (1926), and McKee (1929, 1931). See also Allen & Roys (1918) for a novel test method applied to bevel and worm gears; and Emmet for some of the earliest efficiency data on marine gears (1920).

Rickli tested double-helical gears with pinion pitch diameters of 3 and 4 in., $m = 10$ and 5, respectively. Friction losses varied from 2.5 to 1.4 per cent. Donald's tests were on single-reduction units of 22,500 hp for light cruisers. Pinion diameters were 13.4 in. with a 48-in. face, pressure angle 25 deg, and m about 6.3. Full load losses dropped to 0.7 per cent when lubricating with a relatively light oil. Each set comprised a gear meshing with two pinions. No deduction was made for bearing losses. The no-load loss was about three-fourths that of full load under most operating conditions.

The investigation of spur gear efficiency by Ham and Huckert made use of a Lewis machine with pinion diameters of 3 in. and $4\frac{1}{2}$ in., $m = 6$ and 4, respectively, pressure angles $14\frac{1}{2}$ to $22\frac{1}{2}$ deg, and various addendum heights. Friction force on teeth was found by subtracting the no-load value. Power loss rose from 0.75 to 1.25 per cent while loads were increased from 100 to 1500 lb. It seemed to be independent of speed from 80 to 1300 rpm of pinion. A lubricant of 46 SUS at 210 F was used because a lighter grade was thrown off the gears too easily, and heavier grades were too viscous for uniform distribution. Test pinions were of forged steel, gears of cast iron. Addendums with the greater sliding action showed the greater tooth friction. Badly worn teeth gave 30 per cent more friction at full load. The lowest wear rates were found when the efficiency was highest.

An improved form of Rikli's machine was used by Kutzbach in testing gears of 81-mm diameter at 1500 rpm. Friction loss on hardened steel gears averaged 1.02 per cent at 100-kg tooth load, 1.05 at 300 kg, with coefficients usually from 0.03 to 0.04. Bearing friction was determined experimentally. Textile composition and aluminum alloy gears gave lower friction but shorter life.

A comparison of spur gears and worm gears for electric street railway application was conducted by S. A. McKee at the National Bureau of Standards (1929, 1931). Maximum efficiencies fell within the working range of speeds and loads, and were found to be slightly over 98 per cent for the single reduction gears, 96 for double reductions, and 95 for the worm gear sets, before correction for bearing friction and churning losses. The single reductions were lubricated in these tests with a heavy mineral oil containing a small percentage of soap, its viscosity being 690 poises at 104 F and 140 at 130 F. Operating temperatures at the various loads ranged from 92 to 112 F as judged by thermocouples in bath. Efficiency and power loss were plotted against car speed at constant values of the load, or tractive effort at wheel rim. Temperatures were recorded for each of the runs at constant load so that viscosities could be read from a chart. Speeds ranged from about 3 to 20 mph, and loads from 500 to 2000 lb at the rim of a 26-in. diameter wheel, in addition to the no-load run.

To evaluate ZN/P, note that the pinions of the single-reduction unit were of 3.5-in. diameter with a face width of 3.75 in. and had fourteen teeth. The gears had sixty-nine teeth, making $m = 4.93$. Accordingly, the pinion speed N ranged from 3 to over 21 rps, and the tooth load P from 0 to 230 psi, making the K-factor 277 psi at full load. Dimensionless ZN/P values ranged from about 10^{-4} to nearly forty times that figure in dropping from full to one-quarter load. Curves have been plotted in Fig. 2 for the loss ratio H/H_1 against ZN/P at constant load fractions from one-quarter to full. In computing the loss ratio, H is taken from the dynamometer input in inch-pounds per second after correcting for bearing friction; H_1 is equal to $143NP$ in the same units. Bearing losses were estimated to vary from 5 per cent at full load to 9 at no load for a mean pinion speed.

The data fall on two curves, the upper one representing full load and quarter load tests, the lower curve those at intermediate loads. Both curves pass through a minimum near $ZN/P = 5 \times 10^{-4}$, reflecting a mixed film condition gradually changing to hydrodynamic. The

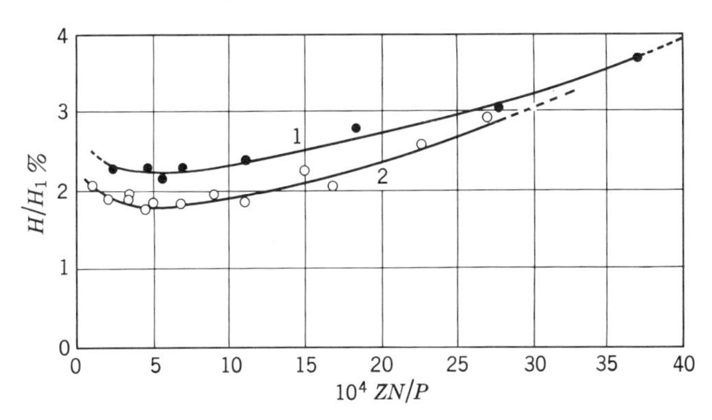

Fig. 2. Total loss in spur gear tests at high ZN/P, corrected for bearing friction (McKee): 1. full and one-quarter load; 2, intermediate loads.

intermediate loads provide optimum loading for maximum efficiency. Presumably at higher ZN/P values, the curves would bend over as indicated in Niemann's Eq. (9) above. No-load losses, which include churning in the oil bath, were found to vary from one-fourth to one-third of the full-load power loss. An approximate correction for such losses may be taken equal to 0.29 of the ordinates of the upper curve. Smoothed values of net loss obtained in this way are recorded in Table 4. If our interpretation is correct, Niemann's equation for rigid teeth leads to a loss ratio considerably higher than shown by McKee's

TABLE 4. Net Friction Loss in Spur Gears
(From McKee data with high-viscosity oil)

$ZN/P \times 10^4$	H/H_1 per cent	
	Full and $\frac{1}{4}$ load	$\frac{2}{5}$ to $\frac{4}{5}$ load
1	1.66	1.36
3	1.59	1.16
5	1.56	1.15
10	1.62	1.22
20	1.91	1.53
30	2.27	1.98
40	2.73	–

data. The difference may be due in part to the beneficial effect of elastic deformation. Each of the three gear sets was also tested with a second lubricant. Results will be found in the later reference (McKee, 1931).

Efficiency tests on two sets of helical gears at the National Physical Laboratory were reported by Hyde, Tomlinson, & Allan (1932). One set comprised cast-iron gears with seventy-six pinion teeth of 4-in. face, pitch diameter about 12.7 in., $m = \frac{153}{76}$. The other set was of hardened alloy steel with ground surfaces, thirty-seven pinion teeth of $\frac{3}{4}$-in. face, pitch diameter 5.82 in., $m = \frac{38}{37}$, and pressure angle 20 deg. It was found in both sets that the loss ratio increased with increasing speed, but decreased with increasing load as expected hydrodynamically; and that friction could be reduced by decreasing the oil supply. Teeth failed by fatigue before any appreciable wear occurred.

The loss ratio in the cast-iron gears increased from 1.6 to 3.8 per cent with a speed increase from 800 to 1600 rpm at 112-lb/in. tooth load; and from 0.56 to about 0.85 per cent with the same speed increase under a load of 709 lb/in. Running-in reduced the friction some 25 per cent. In the steel gears, maximum efficiencies approached 99.4 per cent at loads from 500 to 1500 lb/in., irrespective of speed. Commenting on the NPL experiments, *Engineering* points out that the calculated film thicknesses between the gear teeth (even if rigid) would be of the order of 20 micro-inches, which could explain the absence of wear (Hyde et al., 1932).

Direct measurement of the normal and tangential forces on gear teeth was accomplished by Dietrich (1939), using a piezoelectric arrangement. Oscillograms showing friction reversal at the pitch line led to a realization that toothed gears are inherently designed to create sound vibrations. Dietrich found that inaccurate gears can sometimes be paired so as to neutralize tooth errors. Test gears were each of 120-mm pitch diameter with twenty-four teeth. Both steel gears with ground surfaces and plastic gears were tested. Speeds ranged from 40 to 300 rpm, temperatures from 20 to 80 C. Seven oils were compared, some with additives. The coefficient of friction, f, decreased slightly with increasing speed and viscosity, indicating that the film was probably not fully hydrodynamic. Blok (1951) plotted Dietrich's f for "Gear Oil C" and castor oil at the pitch line against the parameter ZU/W, where U is the "sum" velocity for nearly pure rolling, and W the normal load per unit width. Conversion to ZN/P, with consistent units, leads to the values in Table 5, plotted in Fig. 3. It thus appears that the castor oil and gear oil curves are headed for minimum co-

TABLE 5. Gear Tooth Friction by Dietrich

$ZN/P \times 10^6$	Friction coefficient, f	
	Castor Oil	Gear Oil
0.15	0.089	0.095
0.50	0.071	0.090
5.00	0.041	0.067
15.0	0.032	0.055

efficients of 0.029 and 0.045 at 10^6 $ZN/P = 45$ and 100, respectively. The curves may be expected to coalesce at higher ZN/P's and to approach the hydrodynamic curve asymptotically. Comparable values had been found by Blok in the lower ZN/P range (Fig. 4 of his 1937).

It gradually came to be appreciated that the measurement of gear tooth friction is of value not only for reducing the expense of power loss, but also for controlling surface temperatures and their detrimental effects, and as a means for comparing theory with experiment.

Full-Scale Marine Gearing. At least seven full-scale efficiency tests of marine reduction gears have been reported since those of Donald

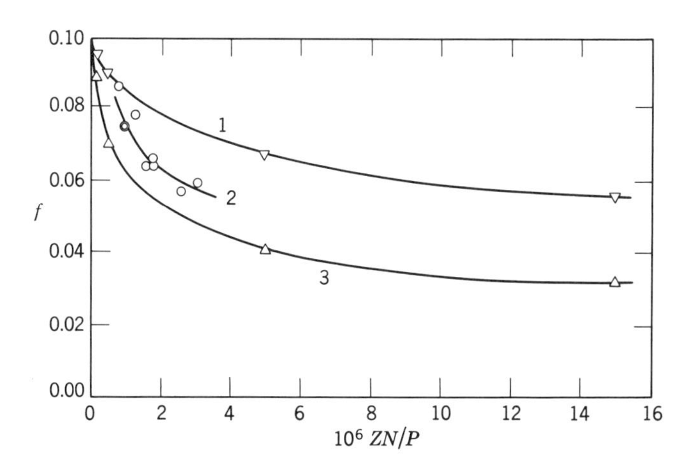

Fig. 3. Friction coefficient from spur gear tests at low ZN/P; 1, Dietrich, "gear oil;" 2, Sasaki et al., mineral oil; 3, Dietrich, castor oil.

(1922). The first four reports were by Douglas (1940), Michel (1941), Davies & Semar (1947), and Waller (1948). Facilities for comparable tests at the U. S. Naval Boiler and Turbine Laboratory were described by Monk (1949), but without reporting the data on friction losses.

Douglas said that the power loss at the teeth of a helical marine gear set is about 0.4 of 1 per cent of the transmitted power for both single and double reductions, the greater part of the loss being in the journal bearings. Michel states that the efficiency curves for destroyer single and double reduction gears show typical full-power efficiencies of 98.3 and 97.0 per cent, respectively; but that these drop to 97 and 91.6 per cent when cruising at one-twentieth power. Yet most of the operation of a surface warship is at cruising speeds. The efficiency of main reduction gears can be raised 8.5 per cent at 9 knots, and 1.9 per cent at full power by running with oil at 110 F inlet and 170 F outlet temperature, instead of the usual 90 F inlet; assuming the use of Navy Symbol 2190 T oil, for which he gives a viscosity-temperature curve. Michel suggests, therefore, heating the oil at lower powers and cooling at the higher powers; or at least shutting off the cooling water at low power. He quotes Douglas on the greater loss in journal bearings and reminds us that a marine single reduction gear has six or eight such bearings. A typical double reduction has twenty-two! The data from Davies and Semar indicate that the spray and mesh losses in modern reduction units are about 30 per cent of the power loss.

An investigation of double reduction gears of the locked-train type for destroyers was reported by Waller. They were tested up to 26,000 hp input, showing an efficiency of 98.2 per cent at the final shaft speed of 400 rpm. Pinion pitch diameters were 9.6 and 11.4 in. in the first, and about 15.1 in the second reduction unit, with m = 2.56, 2.16, and 5.71, respectively. Active face width was about 21.9 in. in the single and 19.5 in the second reduction, K-factors ranging from 89 to 102 psi. Bearing diameters and loading are reported in detail, but apparently without mention of the clearance, viscosity, and temperature data necessary in calculation.

The three remaining reports on marine reduction gears are by Cameron & Newman (1953), the *Marine Engineer and Naval Architect* (1956), and *Engineering* (1959). Cameron and co-author give friction coefficients of 0.047 for hobbed pinions of nickel-steel meshing with hobbed gears of forged steel, as compared to 0.030 for ground case-hardened teeth. They find bearing losses as high as 65 to 85 per cent of the total, but are able to estimate them theoretically. A new design of double reduction dual-tandem gearing is described in the second reference. Tests conducted at the Pametrada Research Station show overall efficiencies from 96.5 to 97.2 per cent at shaft horsepowers from 11,500 to 30,000, with K-factors above 450. These details are impressive but might wishfully have included further records of gear and bearing geometry, and viscosity of the lubricant, needed in calculation. The third reference (1959) gives David Brown tests on a fifty-one tooth pinion of 4.83-in. pitch diameter driving a 305-tooth wheel of 12.5-in. face and 30-deg helix angle. The unit transmits 1000 kw under full load at a pinion speed of 9000 rpm with an efficiency of 98.1 per cent; 96.6 at half load, 93.7 at one-quarter, assuming 15 per cent bearing loss. Oil viscosity met the "Grade 1" specification.

Precision Tests on Spur Gears. In the meantime extensive friction tests on spur gears were being reported by Niemann & Ohlendorf (1958). Pitch diameters were 90 mm with $m = 1$, face width 20 mm, contact ratio 1.5, and the pressure angle usually 22.4 deg. Most of the gears had twenty teeth; some, from 15 to 30. All but one set were of case-hardened alloy steel; that one left soft for comparison. Tooth surfaces were finished by a variety of methods including fine and coarse grinding, phosphating and electropolish. Roughness, specified by the double-mean-depth, ranged from 1.0 to 7.4 microns. Lubricants comprised two mineral oils, A and B, each with and without additive; and a polyether synthetic, F. Viscosities at 50 C were 2.6, 1.15, and 0.50×10^{-6} kg-sec/sq cm, respectively. Corresponding values of the pressure-temperature constant α were 0.261, 0.235, and 0.100 sq cm dC/kg in the formula for relative viscosity with exponent $\alpha p/(T + \beta)$, where p is the gage pressure, kg per sq cm; T the temperature, C; and $\beta = 52$ dC. Operating conditions covered the speed range from 100 to 2000 cm/sec at the pitch circle, and the loading range from 10 to 250 kg/cm normal tooth load. Inlet viscosities were fixed at two levels, 0.4 and 2.5 kg-sec per sq cm $\times 10^{-6}$.

Under these test conditions both friction loss and temperature rise are shown by a large number of charts, each containing a family of curves. Preliminary charts show how the corrected tooth friction is obtained from measurements of total, no-load, and bearing friction. Eight charts are given for the corrected loss ratio, H/H_1 per cent, against speed at six constant loads. These are followed by charts showing thermocouple-in-tooth temperature rise above inlet oil temperature. Three charts show the effect of oil flow on temperature rise. Equations are then derived for mixed film lubrication, utilizing empirical values for the coefficient of boundary friction. Calculations are shown to be in reasonable agreement with test results. A table is given for the characteristic friction loss in every new test series relative to that in the first series, in order to bring out the influence of tooth hardness and finish. It is clearly seen that friction is less with the harder and smoother surfaces.

In each series of tests the minimum point shifts toward lower speeds at the lighter loads. Something might be learned by plotting the loss ratio against ZN/P as in Tables 4 and 5, holding other dimensionless variables constant. A good start was made by Blok (X: 1962) in his review of Ohlendorf's thesis, where he correlates four series of tests in terms of inlet viscosity and other factors. Blok recommends that such tests be extended to larger gears with different reduction ratios. If the elastic modulus could also be varied, we might gain a better understanding.

A test on hardened steel gears of thirty teeth each, 1:1 ratio, 6-in. pitch diameter, 1-in. face, and 20-deg pressure angle was reported by Shipley in 1959. Gears were ground to 11 micro-inches and lubricated with a light mineral oil. Speeds ranged from 500 to 5000 ft/min, loads from zero to 1200 psi K-factor. Mesh losses were corrected for bearing friction and no-load loss. Mean coefficients of tooth friction then varied from about 0.03 to 0.05. Values were computed from the measured power loss by Shipley's formula (1958). These coefficients show the usual minimum point when plotted against speed at constant load.

Spur gears of 120-mm pitch diameter, face 10 mm, $m = 1$, and thirty teeth each were tested by Sasaki and co-authors (1962). Standard involute teeth were used. Pressure angle was 20 deg, effective mating length along line of action 21.6 mm, and contact ratio 1.83. Tooth surfaces were carburized and ground to a finish of 2 to 3 microns maximum height for the principal tests. Two lubricants were compared. Three methods of lubricating were tried—bath, jet, and mist. Gears in bath were immersed to the dedendum; jet rate 800 ml/min; oil mist, 0.5 ml/min.

Gears were supported by preloaded ball bearings to make friction variations negligible. In the principal tests, speeds ranged from 1250 to 2250 rpm with mean normal tooth loads, P_n, from 5400 to 13,000 kg/m at each speed. Taking the viscosity η into account, these operating conditions provide a four-fold range in the authors' parameter $\eta v_r/P_n$. Conversion to ZN/P is effected by noting that $\eta = Z$, $P_n = PD/\cos\phi$, where our P, as before, denotes the tangential load per unit of "pitch area," D is the pitch diameter, and ϕ the pressure angle. Authors' mean rolling speed v_r is 3.72 m/sec when the pinion speed $N = \frac{1730}{60}$ or 28.8 rps; so that $v_r/N = 0.129$ m. If we have understood the authors' units and definitions, $\eta v_r/P_n = 1.01\ ZN/P$. When consistent units are taken, ZN/P will be the same in English or metric systems.

It was calculated that the mean coefficient of tooth friction for the gears under test would be 7.56 times the loss ratio H/H_1. The authors preferred to report their results in terms of the friction coefficient to facilitate comparison with roller experiments. Table 6 shows the principal test results corrected for bearing friction and churning loss. While bath and jet values differed only slightly, the coefficients were more than 50 per cent greater with mist. Bath values are plotted in Fig. 3. There is a drop toward minimum friction at the higher values of ZN/P. Applying the relation connecting loss ratio with friction coefficient, we see that the gear efficiencies vary from 98.9 to 99.3 per cent.

In Table 6, full load represents 130 kg/cm normal tooth loading. In a test series at lighter loads, 4 to 40 kg/cm, the friction moment varied from 0.25 to 2.5 kg/cm, indicating a constant coefficient. These observations were at speeds averaging 2100 rpm. In general, the friction coefficient was seen to be about twice as great in actual gears as it had been in the roller experiments (Sasaki, X: 1961). This was attributed to the greater roughness of the gear surfaces as well as to discontinuous contact, dynamic loading, and reversal of sliding at the pitch line. Two series of roller tests were made to determine the influence of roughness over a wide range; one series at mean values of ZN/P =

TABLE 6. Gear Tooth Friction Data from Sasaki, Okamura, Konishi & Nishizawa
(ZN/P values averaged from bath and jet tests)

Speed, rpm	Load, %	$ZN/P \times 10^6$	f bath	f jet
1250	42	1.8	0.066	0.059
	71	1.0	0.075	0.069
	100	0.8	0.086	0.081
1730	42	2.6	0.057	0.058
	71	1.6	0.064	0.051
	100	1.0	0.075	0.075
2250	42	3.1	0.059	0.068
	71	1.8	0.064	0.067
	100	1.3	0.078	0.073

1.3×10^{-6} the other at 1.0×10^{-4}. In the first series the coefficient f rose from 0.026 at practically zero to 0.048 at $H_{\max} = 10$ microns and leveled off to 0.049 at 20 microns. In the second series, where the lubrication parameter was eighty times more favorable, f started from 0.014, rose to 0.030 at 10 microns, and leveled off to 0.037. Sasaki and co-authors concluded that full fluid lubrication can hardly be expected in small, commercially finished gears since that would require ZN/P values of the order of 5×10^{-4}, as in McKee's tests; more than one-hundred times any in Table 6.

Worm and Hypoid Gears. Friction power loss is usually greater in worm and hypoid than in spur or bevel gears because of the relatively greater sliding motion. William H. Kenerson investigated worm gear friction, using a diaphragm transmission dynamometer of original design (1912). The phosphor-bronze wheel was of about 11-in. pitch diameter with forty-three teeth and a face width of $2\frac{5}{8}$ in. It was driven by a steel worm of 0.57-in. tooth depth. Full power efficiency came to 97.6 per cent using "600 W" cylinder oil, which we assume to be of about 1400-cp viscosity at 77 F, and 140 at 130 F.

Worm gear tests by Allen & Roys (1918) are compared with others in the report by Earle Buckingham (1929). A study of worm gear friction with different lubricants had been conducted by J. H. Hyde at the National Physical Laboratory (1920). The Lanchester machine used by NPL gives the ratio of torque in the worm shaft to that in the wheel shaft. The worm and gear under test had a 34:9 ratio. Thermocouple temperatures were observed at the top of the gear box to determine the temperature of the oil thrown off, as well as in the sump under the worm.

Hyde's principal tests were run at two speeds, $N = 1080$ and 1500 rpm of the worm shaft; and at two loads, $P = 3360$ and 4480 psi on the projected tooth area. Seven oils were compared, four fatty and three mineral. Many physical properties were determined, including pressure-viscosity. The usual procedure was to plot efficiency against operating temperature. This temperature rose gradually, owing to frictional heat, until a sudden drop in efficiency occurred. The run would then be stopped to avoid damaging the gears. It can be seen from Hyde's charts that the minimum power loss H/H_1 averaged 4.1 per cent for the fatty oils, over the ZN/P range from 10 to 60 cp rpm/psi; and 5.0 per cent for the mineral oils from 30 to 110. It was concluded that fatty oils are superior to mineral oils for lubricating worm gears; but it could not be said how much of this benefit, a matter of 18 or 20 per cent, should be attributed to boundary properties and how much, if any, to the lower pressure-viscosity coefficients of fatty oils.

Friction losses in four worm drives were compared in the survey report by Earle Buckingham's committee. Worms ranged from 2.0 to 3.1-in. pitch diameter, gears from 10.6 to 14.3 in. The worms had from one to nine threads, the gear wheels from twenty-nine to forty teeth. Contact lines were carefully determined, and are shown on diagrams to aid in evaluating design details. Operating conditions were: speed of worm shaft, 645 to 1770 rpm; tooth load, 32 to 2510 lb; temperature 74 to 178 F. Three gear sets were lubricated with the conventional "600 W," and also, in two of these, the mineral oil was compounded with 40 per cent of castor oil. The remaining set was lubricated with a heavy gear oil. Friction coefficients varied from 0.027 to 1.08, the high value corresponding to the low 32-lb load. In general, when friction was plotted against speed, the curves showed a minimum point. The speed of minimum friction was greater at the higher loads. This indicates that the relations might be brought out more clearly by plotting coefficients, or loss ratios, against ZN/P. The effect of compounding the lubricant was to reduce the minimum friction to less than three-fourths of its value for a straight mineral oil.

Worm gears with reduction ratio 8.5 to 1, included in the Bureau of Standards experiments (McKee, 1929, 1931) gave 2.7-fold greater loss than spur gears at full load, although operating at a higher temperature with a viscosity of about 20 cp. This was evidently too low a viscosity for maximum efficiency. The lubricant was a mineral oil of 136 cp at 130 F and 26 at 210, containing a small percentage of fatty oil. Worm gear loss was reduced about 11 per cent by changing to castor oil at a lower temperature (1931).

The efficiency formula of W. H. Hines (1932) was discussed by Watson in the light of Merritt's investigations. A chart is given for the coefficient of friction against rubbing speed for constant ratios of the sliding velocity of the wheel to that of the worm, using "600 W" as the lubricant. The coefficients are assumed to be independent of load. They vary from 0.033 to 0.48 at 500 ft/min, and from 0.019 to 0.025 at 2000 ft/min. The coefficients decrease as the sliding ratio increases from 1.2 to 3.4. The rubbing speed in feet per minute is shown to be equal to 0.262 DN divided by cos λ, where λ is the lead angle, D the pitch diameter of the worm in feet, and N its speed in rpm. Efficiency is approximately equal to tan λ divided by tan $(\lambda + f)$, where f is the angle of friction; or nearly enough, the coefficient of friction. Another efficiency formula, and its experimental verification, are reported in the *Pacific Marine Review* (1955): to estimate efficiency, take 100 per cent less half the gear ratio.

In his friction tests on a standard type of steel worm driving a bronze gear at 7-in. center distance, Merritt (X: 1935) found a maximum efficiency of 97.3 per cent at 1130 rpm. With a specially designed worm he reached 97.56 per cent. These tests were made under a constant load transmitting 97.6 hp per 1000 rpm, using a "castor base" lubricant. Averaging the results of a large number of practical worm gear tests, Merritt offers a curve for the coefficient of friction against sliding speed in feet per minute. The curve starts from zero speed with $f = 0.150$; then drops gradually to 0.030 at 500 ft/min, 0.022 at 1000, and 0.014 at 300.

When testing the worm gear drive in a standard automobile rear axle, Schlesman (1935) compared five lubricants. These laboratory tests were continued for the equivalent of 9000 miles on the road, while transmitting 60 hp at 230 F. The lubricants were castor oil, EP, straight mineral, a compounded oil, and a grease. Friction losses during the latter half of the mileage averaged 9.3, 9.6, 10.9, 10.9+, and 11.7 per cent, respectively. Castor oil had about two-thirds the viscosity of the other three oils, which were of nearly equal viscosities. Again castor oil gives the lowest friction. Investigators are well aware of the deterioration of vegetable oils; and advocate them only to reduce friction in usage of short duration, and in scientific studies where a wide range of properties is called for.

Reporting on friction tests at M.I.T., Buckingham found that his worm gear coefficients (1949) conformed to the empirical equation

$$f = 0.20e^{-0.17v} + 0.0013v \,, \tag{11}$$

where v has been written for the square root of V_s, the sliding speed in feet per minute. This equation has nearly the same form as Eq. (8) above. It leads to a minimum friction at 370 ft/min. Observed values fell around 300 ft/min at the light loads and 500 at heavier loads; a trend that might be expected from ZN/P relations. Presumably "600 W" was used here, as in other work. Note that minimum friction requires ten or fifteen times the sliding speed in worm gears as in spur gears.

Three types of worm design—the enveloping cone, plain cylindrical involute, and hollow water-cooled type—were compared for performance at several speeds and loads in Niemann's investigation (1955). Gear ratios were about 10:1, pitch diameter of worm 67 mm, and of worm wheel 289 mm. Friction losses varied from 5.8 to 11.0 per cent at 1100 rpm of the worm shaft; and from 6.4 to 11.3 per cent at 534 rpm. In each run the water-cooled worm gave the least friction, the plain worm the greatest. Losses at the lower speed dropped noticeably upon raising the load 50 per cent. Only the first slight trace of pitting and wear could be detected at the higher load in the water-cooled job, whereas the other two sets were severely scored and worn.

Some investigators have taken the equilibrium temperature rise as a convenient measure of power loss. Tourret (1955) used this method in experimenting on small industrial worm gears, as well as on the larger worm drives in heavy tank trucks. He found that friction increased with the viscosity grade of the lubricant, with tooth load or output torque, and with speed. For a given lubricant, the friction loss was only slightly increased if the speed were doubled while the load was simultaneously doubled. This fact recalls the ZN/P relation, so again it would be of interest to replot the data on that basis. No-load losses were high, and attributed mainly to oil churning.

The sensitivity of worm gears to the slightest geometrical or adjustment error was emphasized by Wakuri & Ueno (1958). They recommend a good "entry gap" between gear flank and worm contact, and provided such by a crown of 0.04 mm. Their tests were conducted on a two-thread steel worm of 18.5-deg lead angle, 30.6-mm pitch, and 44.5-mm face width, hardened and ground, driving a bronze gear of twenty-five teeth at 151-mm center distance. An efficiency of 90 per cent was usually obtained with a specially good entry gap, but only 84 per cent without it. The efficiency of a good set, transmitting 14 hp at 1800 rpm, was seen to rise from 89.1 to 92.4 per cent with an increase in viscosity from 10 to 50 cp. Since efficiency is greater at higher values of the lubrication parameter, the authors infer that boundary lubrication predominates. They conclude that running-in is of especial importance for worm gears.

A lively discussion was initiated by Whittle's report (1961) on the lubrication of automotive worm gears. He plots friction loss against ZN/P at a constant input of 35 hp. To find H/H_1, divide his ordinates by 35. The loss curve for mineral oils falls higher up than that for nonmineral oils, each curve exhibiting the familiar minimum. Graphs tend to coalesce at high ZN/P's. Whittle attributed the differences to pressure-viscosity. When he estimated film pressures and tried to plot against ZN/P using a pressure-corrected viscosity, the resulting curves were said to come closer together. Discussion brought out that boundary properties may also be influential.

Two curves for f against sliding speed are given by Shipley (1962). The lower curve, for a case-hardened steel worm of ground finish driving a phosphor-bronze wheel, is practically the same as Merritt's "average curve." The upper one, for cast-iron worm and wheel, shows 30 per cent higher friction. Both curves are for use with a "good grade of petroleum oil."

7. GEAR FAILURES

Fracture of gear teeth is now under control, thanks to the work of the ASME Research Committee on Strength of Gear Teeth. Fourteen progress reports were published from 1923 to 1929. They are listed by Earle Buckingham (1931, pages 10–11). Attention has shifted to surface endurance including fatigue, scuffing or scoring, and ordinary gradual wear. A good understanding of these problems was initiated by Progress Report No. 15, "Qualitative Analysis of Wear" (Buckingham, X: 1937). A research program on Naval gearing was outlined by Commander R. T. Simpson (1949). See also the papers by Monk, Thomas, & Atkinson (1952) and by Zrodowski (1957), presented before the Society of Naval Architects and Marine Engineers. A general discussion of gear failures has been given by Frederick & Newman (1958).

Fatigue. Pitting of tractor gears was carefully observed in dynamometer tests by Van Zandt & Kelley (1949) to supplement their use of the Meldahl type disk machine. Fatigue failures in automotive and aircraft gearing are described in a chapter by Almen (1950) illustrated by photographs. Gear troubles in British ships are reported by Commander Joughin (1951). Improvement was sought by reduction of manufacturing errors. Discussion indicated the prevalence of pitting in merchant ships, where root fractures resulted from propagation of cracks started by pitting. Further aspects of fatigue failure are described by Cameron (X: 1954). Gear casualties in ships of the United States Navy during the last world war, described by Braley and Berg (1958), are reflected in fifty-eight reports of damaged teeth, sixty-seven of bearing failures, and seventeen of abnormal wear (these include six of pitting and four of flaking). Tooth damage was often caused by mishandling on shipboard; misalignment due to bearing failure; or by incredible kinds of foreign matter passing through the gear mesh.

A new point of view was introduced by Professor Blok in a theoretical comparison of the several types of "barrier" to greater load capacity (1958). Reasoning by dimensions he showed that the power transmissible without breakage, P_b, and that transmissible without pitting, P_p, are both proportional to ND^3, where N is the pinion speed and D a chosen linear dimension, say the pinion pitch diameter. Each proportionality constant is the same for all geometrically similar gear drives if the physical properties are similarly distributed; and provided we disregard the effect of the lubricant on pitting, and that of dynamic loads on breakage. It follows that if the pitting barrier is at a lower level than the breakage barrier in a particular gear drive, it will remain so in similar drives whether larger or smaller. Blok notes that pitting life is greater with higher viscosity.

Test curves wherein the load capacity for pitting was doubled by changing to a high viscosity lubricant are given by Shipley (1958), the K-factor being raised safely from 250 to 500 psi. See also Pohl (1954), Cleare (1958), and Frederick & Newman (1958). The last investigators report that heavy scuffing can lead to pitting by reducing the effective cross section. It will be recalled from Merritt's experience that steel worm and bronze wheel combinations fail by pitting of the bronze before scuffing begins.

Gear Failure by Scuffing. The load at which scuffing begins—the "scuff-load" for short—depends on the design, material, hardness, and finish of the teeth; in addition to manufacturing errors and misalignment. For a particular gear set, interest centers on how the scuff load varies with speed, viscosity grade, and bulk metal temperature, oil supply temperature, and composition of the lubricant. Some indications have been gained from the roller experiments reviewed in Chapter X, but we turn to actual gear tests for more convincing results.

Gear scuffing *par excellence* occurs in the worm and hypoid types, where sliding is prominent. Maximum loads before seizure were recorded by Schlesman in his worm-gear rear axle tests (1935). Comparing five lubricants, he found that castor oil gave the highest and an EP the next highest load capacity, whereas grease stood at the bottom of the list; practically the same order of merit as their friction-reducing ability. Gleason techniques for comparing EP lubricants in full-sized axles are described by Cowell (1958) with test results on four lubricants. Radioactive tracers were used by R. B. Campbell (1963) in a study of hypoid gear lubrication. A substantial sulfur-containing film was gradually built up when the gears were lubricated with a mineral oil containing elemental sulfur.

Scuffing of spur and helical gears occurs frequently in service, but we have little information as to the loads at which it begins. Almen showed examples of scuffed automotive gears which he explained by the "PVT" criterion (1950). Here P is the Hertz compressive stress, V the rubbing velocity, and T the distance along the path of action. Discussion brought out equally good or better correlation with Blok's "square root of P^3V" rule, derived from the flash-temperature hypothesis. Joughin (1951) attributed scuffing in Naval gears to oil-film breakdown caused by overheating of high spots. This was aggravated by surface undulations caused by imperfect temperature control in cutting the gears. Scuffing was said to occur at low power in ships that had passed full load tests without scuffing. Frederick & Newman (1958) found greater susceptibility to scuffing when the mating gears are of similar composition. Full-sized aircraft gear tests up to high temperatures are reported by Shipley (1958). Scuff temperatures of case-carburized gears varied from 360 F at 1000 ft/min pitch line velocity to 490 at 18,000 under a K-factor load of 700 psi, when lubricated with a "Mil. Spec." oil. The temperature for a silicone oil was 400 F from 2000 to 8000 ft/min; but it rose to 630 F at 15,000, when nitrided gears were substituted. Newman (X: 1960) reports that it is not possible to operate main reduction gearing, in a ship, at nominal loads that would be safe in auxiliary gearing—"pinion deflections may double the tooth loading, not to mention dynamic loads."

Data from Gear Test Machines. Quantitative data on scuff loads in gearing come almost entirely from laboratory tests on small-sized gear sets. Among the better known test machines of that class are the Ryder, IAE, and FZG. The Ryder machine and its modifications were first described by E. A. Ryder (1947, 1952); later by Baber, Ku, and co-authors (1960, 1960); and more recently by Beane & Lawler (1964), Droegmueller & Huff (1964). The standard Ryder rig has a wide test gear and narrow test pinion of twenty-eight teeth each, with 3.5-in. center distance, pressure angle 22.5 deg. The "research Ryder" or "WADD high-temperature gear machine" is basically the same, but more versatile. Droegmueller experimented with a hunting-tooth rig: twenty-seven teeth on the wider gear; twenty-eight, as before, on the other. The IAE rig (Institution of Automobile Engineers) has been described by the Institute of Petroleum and by Mansion (both 1952), Hughes (1953, 1958), and others. It evolved from a smaller machine built by H. Blok. The IAE operates with a hunting tooth and 13:12 gear ratio on 3.25-in. centers; pressure angle about 26 deg. The FZG (Forschungsstelle für Zahnräder und Getriebebau) "short test" has been described by Niemann and co-authors (1954, 1961). It operates on 3.6-in. centers with a 3:2 gear ratio, and 20-deg pressure angle. Scuff loads are recorded in the Ryder and IAE rigs when a predetermined area has been visibly scuffed; but in the FZG, only when the measured wear rate takes an upward jump.

Several investigations utilizing a spur gear test machine on 3-in. centers with a 19:17 gear ratio and 20-deg pressure angle have been published by Borsoff and co-authors (1951, 1959). Test gears were case hardened and finished as in aircraft practice. The power transmitted when scoring or scuffing began was plotted against the product of viscosity and speed, using, at first, only straight mineral oils. The curve is convex upward, leveling off at high ZN. Since the power transmitted is proportional to PN, the product of load and speed, it can be seen that the scuff load is nearly proportional to the square root of Z/N, as noted by Ryder in his discussion. The inverse relation to speed implies a maximum scuff load at some moderate speed, since load capacity must approach zero at zero speed. Continuing, the Borsoff group found extra-high scuff loads for EP oils, the points falling higher up on the chart. They found extra-low values for glycerin and silicones, showing that the action is not purely hydrodynamic at the moment of scuffing.

Blok's Hypothesis. Other investigations followed, all tending to support Blok's "flash-temperature hypothesis." This proposition assumes that every mineral oil has a critical failure temperature (1940). Barwell & Milne (1952) had found from their sliding experiments that the temperature rise varies with the square root of the rubbing speed, in accord with Blok's calculation. They measured fluid pressures up to 5600 psi under scuffing conditions, this pressure being transmitted to the gage through a hole of 0.03-in. diameter. Like Borsoff and others, they picture scuffing as a combination of melting, welding, and transfer of metal. Load capacity was improved by a slight roughening across the direction of motion. From experiments on a Thornton high-speed gear rig, Hughes (X: 1952) found scuff loads decreasing with an increase of pinion speed. In this rig the pitch diameters of pinion and gear are 4.8 and 5.2 in. Hutt experimented with the IAE rig (1952). Plotting scuff load W against viscosity Z, bulk oil temperature T, and pinion speed N, he found that for light mineral oils

$$W = (aZ^{1/2} - bT)/N^{2/3}\,, \tag{12}$$

where a and b are constants for the particular gear set. The speed exponent $-\frac{2}{3}$ conforms to Blok's theory. Slightly lower slopes were found for the W, N curves with higher viscosity oils.

Two oils were compared at two speeds by Lane & Hughes in the IAE machine (1952). One oil had about twice the viscosity of the other. The heavier oil showed about one-fifth greater load capacity of the two at 200 rpm, but the same, or less, at 6000. At the higher speed, the mean scuff load of the two oils dropped to three-eighths of its low speed value. Plotting scuff load against oil temperature at 4000 rpm showed a drop to 60 per cent of the initial load when the temperature was raised from 30 to 130 C. Plotting load against viscosity at constant speed showed over three-fold increase in load capacity for a twelve-fold increase in viscosity. These facts combine to support Blok's theory, and to show that $a = 1400$, $b = 40$ in Eq. (12). Here W is the lever load (tangential tooth load in pounds per inch divided by 61); Z the kinematic value in centistokes at 60 C; T the bulk gear temperature, C; and N the pinion speed in rpm. Finally, a collection of data from the present and other investigations was plotted against the values predicted by Blok's theory, showing satisfactory agreement.

IAE and Other Research. Fifteen years of gear research by the I.A.E. and the Motor Industry Research Association are summarized by Mansion (1952). Scuffing is described as a rapid form of local welding, and tearing out of particles from the opposed surfaces, due to breakdown of the boundary lubricating film. The question of oil additives for aircraft reduction gears became important beginning in 1943–1944. It was seen that the I.A.E. rig correlated better than other bench tests with actual gear performance. In this machine, load is stepped up after each 5-min run, with visual inspection between runs. The principal findings were: (1) Scuff load varies with square root of viscosity when comparing different oils at the same temperature. (2) The change in viscosity with temperature for a given oil has much less effect. (3) Castor-oil blends gave from two to three times the load capacity of SAE 50 mineral oil. (4) Some additives can increase load capacity threefold. (5) Low capacity mineral oils show the greatest response. (6) Chemical surface treatments can nearly double the load capacity; but the effect of some wears off after 100 hr. Electrodeposited metallic platings are beneficial. (7) The scuff load varies almost inversely with speed from 1000 to 4000 rpm. (8) Position of oil jet has little effect.

Scuff loads on five lubricants were compared at 6000 rpm by Hughes & Tourret (1953) in the Thornton rig. A low-viscosity oil scuffed at 20 lb on the lever, a medium oil at 50, and a high-viscosity oil at 70 lb. Addition of tricresyl phosphate raised the load capacity of the medium oil from 50 to 90 lb. A proprietary additive carried the load beyond the range of the machine. With each of these lubricants the scuff load was seen to vary inversely as the two-thirds power of the speed from 3000 to 12,000 rpm. These data, too, conform to Blok's theory.

The theory was further confirmed by Kelley (1953), who compared different lubricants in the gear test rig described by Van Zandt & Kelley (X: 1949). Kelley introduced an empirical correction for tooth finish in Blok's formula, later expressed neatly by setting f equal to a constant divided by $1 - s/50$ from $s = 10$ to 35, where s is the surface roughness in rms micro-inches. The constant is usually taken equal to 0.06. Continuing his experiments, Kelley tested an SAE 30 mineral oil with gears of varying geometry, pitch diameter, and surface finish; and with ground steel rollers. The calculated total temperature, bulk plus flash, was plotted against the known bulk temperatures from 125 to 250 F. All points fell near a horizontal line at 500 F, with scatter rarely over ±50 deg F. This result conforms well to the hypothesis that each mineral oil has a characteristic critical temperature, independent of gear design and operating conditions.

An extensive collection of scuff load data from testing nonreactive oils on the I.A.E. machine was presented by Hughes & Waite (1958). When scuff load is plotted against speed for a representative mineral oil, the points follow Blok's calculation to 4000 rpm, but fall higher up at greater speeds. The authors advocate a general equation of the form

$$S_c = k \sqrt{\nu}/c^{a\theta} v_s^{\,x} \,. \tag{13}$$

Here S_c is the scuff load, per unit face width, divided by the relative radius of curvature, ν the kinematic value at say 210 F, and v_s the sliding speed. The constant a will be 0.006 when the bulk temperature θ is in deg C. For speeds below 2300 ft/min, $x = \frac{2}{3}$ and $k = 2.8 \times 10^5$; above 2300 ft/min, $x = 0.4$ and $k = 4.8 \times 10^4$. If ν is then taken in centistokes, the stress criterion S_c will come out in psi. Shipley (1958) found the scuff temperature for aircraft gears constant for any given oil from 500 to 5500 ft/min and above. Comparing five oils from SAE 10 to 90, he found critical temperatures varying from 200 to 500 F with increasing viscosity. His data confirm the Kelley-Blok formulas (1953).

Not only gear tests, but roller experiments like those of Leach & Kelley (X: 1965) tend to support flash temperature theory, leaving the scuff problem open to further investigation.

A gear test machine standardized in Russia has been described by Genkin & Misharin (1960). It is similar to the I.A.E. but with a 44:22 gear ratio; and to make sure of obtaining seizure before pitting or breakage, the tooth height is greater than usual, with increasing sliding. Gears are ground to a "Class 7" finish, roughness height not exceeding 6.3 microns or $\frac{1}{4}$ mil. Pinion speed is set at 11,900 rpm. Running-in is continued for 2 hr to permit healing of limited seizure and insure thermal equilibrium. The results are said to agree well with tests on the same oils in full-sized gear drives. In a second paper (1960) data are given that seem to confirm Blok's theory with three qualifications: (1) since the pinion and gear tooth temperatures measured by thermocouples are different, a "reduced temperature" is defined for use in calculating the temperature flash; (2) the critical temperature is not precisely fixed for a given lubricant but depends on the friction conditions, especially speed, since greater film thickness increases the critical temperature; and (3) strictly, Blok's theory applies only to boundary lubrication.

Scuff loads increase with viscosity and decrease with rising temperature, according to Ku & Baber (1960). Different forms of Ryder machine agreed reasonably well. A minimum point was found in the neighborhood of 10,000 rpm when plotting scuff loads against speed. This confirmed Borsoff (1959) where other investigators found no upturn. The load capacity curve for a silicone-diester on this chart came distinctly higher than the curves for "Mil. Spec." oils.

The FZG machine, like the Russian, has long-addendum teeth. Tests reported by Niemann and co-authors (1961) show the influence of design factors, speed, temperature, viscosity grade, material and roughness of teeth. In the discussion, equations for the actual and permissible surface temperatures were given by Baniak & Fein. Their simultaneous solution leads to a useful curve for load against speed, delineating a safe region bounded by the wear zone at low speeds, and by the scuff region at high speeds.

Recent investigations include the effect of inert atmospheres on high-temperature gear operation. Both the standard and the "research Ryder" or WADD machine were used by Ku (1960), Baber (1960), and associates. Two mineral oils were tested, both straight and with EP additives. The mineral oil load capacities increased nearly 100 per cent when operated in atmospheres of argon or nitrogen; EP loads nearly 50 per cent. The tests were extended to synthetics with inconclusive results. Various gear tests were conducted by Fricker (X: 1963) in helium and in a vacuum, at 150 C. A life of 90 to 300 hr was obtained from 50 to 400 rpm after rubbing molybdenum disulfide on the tooth flanks. It was concluded that the problems of operating a high-temperature, gas-cooled reactor had been successfully overcome. Beane & Lawler (1964) compared eight classes of oils over a wide temperature range. All curves show a minimum load capacity between 400 and 500 F.

Flash Temperature Theory. Exactly what is Blok's theory of failure by scuffing, so frequently mentioned here? It consists of two parts: the critical-temperature *hypothesis*, and the temperature-flash *calculation*. The hypothesis is that failure occurs in a mineral or other nonreactive oil when the rubbing surfaces reach a "critical temperature" characteristic of the particular oil. The calculation, outlined in Chapter III, serves to predict the "temperature flash," or transient temperature rise in the contact area above the bulk temperature of the gear tooth. The theory was first applied to the four-ball test (1940). Thus it is not limited to gear friction; a convincing demonstration was given by Lane (X: 1951) using the two-ball machine.

Blok's calculation of the temperature flash appeared in two papers (III: 1937) and was confirmed by gear tooth experiments published the same year. It was further discussed in later reports (1958, III: 1963) and may now be written, for mating gears of the same material,

$$T_f = 0.62 f W^{3/4} (\sqrt{v_1} - \sqrt{v_2}) (E/R)^{1/4} / c^{1/2} . \tag{14}$$

Here f is the coefficient of friction; W, the normal tooth load per unit width; v_1 and v_2, the larger and smaller surface velocities; E, Young's modulus; R, the conformity or "equivalent" radius; and c, the product of thermal conductivity by heat capacity of the metal per unit volume. Since the equation is dimensionally homogeneous, it will lead to the same result in any consistent system of units. Equation (14) is derived from the Chapter III equation by substituting an expression for the width of the Hertzian contact band. The difficult factor to evaluate is the coefficient f. We have seen that Kelley (1953) took f equal to 0.06 for perfectly smooth surfaces, with higher values for practical surfaces. This seems reasonable, yet Kelley and co-authors find wide variation of friction coefficients with sliding velocity and sum velocity in their roller experiments (Benedict, X: 1961; Leach, X: 1965).

Scuffing is to be expected when the sum of the bulk gear temperature and the temperature flash by Eq. (14) exceeds the critical temperature for a given oil. It has been found that the critical temperature varies approximately as the square root of the viscosity grade, or viscosity at a fixed temperature. This is not considered a hydrodynamic effect, except in some slight degree; but merely as a parallel indication of some property not yet identified, possibly one that affects volatility or thermal instability. Definite predictions have been confined to mineral oils, although nonreactive synthetics appear to behave in a similar manner.

The "barrier" philosophy applied to breakage and pitting was extended by Blok to include contact and scuffing (1958). Consider a series of homologous, or completely similar, gear sets of varying size D and speed N, lubricated with the same oil at the same supply temperature. It is shown that the power transmissible for a given film thickness at incipient contact of asperities will be proportional to the product $N^9 D^{14}$. The reasoning is based in part on Grubin's equation (X: 1949). If, however, the film thickness is required to remain a constant fraction of the size D, the power transmissible will be proportional to ND^3. With the introduction of the "scuffing barrier,"

Blok shows by reference to flash temperature theory that the power transmissible varies with $N^{1/3} D^{5/3}$. Compared with the breakage and pitting relations, the present exponents are smaller. Blok concludes that the contact barrier will be the most restrictive one at low speeds, and the scuffing barrier at high speeds.

Gradual Wear. Ordinary gear wear can occur after long operation under heavy loads at low speed, or when suffering from misalignment, abrasives, or lack of oil. Schlesman (1935) compared four lubricants in a standard automotive worm gear. During 9000 miles of equivalent operation in the laboratory the measured wear came to 16.1 mils for an EP lubricant, 9.5 for castor oil, 3.7 for a mineral oil, and 1.5 for a specially compounded oil. Blok's discussion of the contact barrier may be of interest as applied to gradual wear.

Testing aircraft gears at elevated temperatures, Shipley (1958) found that a silicone oil gave nearly twice the wear rate of a diester, while a "Mil. Spec." oil gave relatively little wear. Continuing his tests with an improved machine, Borsoff (1959) noted that most of the wear occurred at the beginning of a run. The higher the load capacity the smaller the wear. Gears lubricated by unreactive oils were found to wear mainly by scoring, with a maximum wear rate at 5000 rpm. See also Niemann and co-authors (1954, 1961), who measured gear wear in the FZG tests.

Two different rigs were used by H. W. Fricker in his reactor bearing and gear tests cited previously (X: 1963). Gears were operated at 150 C in helium or vacuum environments. In one test the teeth were slightly greased with Nucleol G 121; in the remaining tests the flanks were rubbed with molybdenum disulfide powder. Except where there was misalignment, the wear rate was found practically negligible.

8. SUMMARY

Geometrical definitions and stress calculations are followed by the theory and measurement of film thickness between gear teeth. Publications on these topics are listed in Part 1 of the references.

Theoretical and experimental investigations on friction loss in gearing are reviewed in chronological order. The corresponding references are listed in Part 2. Curves are plotted for the percentage power loss against ZN/P, where P is taken as the tangential load per unit of pitch area; this area being defined as the product of pitch diameter by face width. Friction loss in the gear mesh rarely exceeds 1 per cent for spur and helical gears, but is greater for worm and hypoids.

It appears that the coefficient of friction, f, is nearly proportional to the square root of ZN/P under the fluid film conditions prevailing at higher speeds. The curves pass through a characteristic minimum point at lower speeds, with f dropping to minimum values from say 0.03 to 0.06. The left-hand branch indicates elastohydrodynamic and mixed film lubrication, conditions sensitive to elastic deformation, pressure-viscosity, and surface roughness. Data are presented in graphical and tabular form to show representative test results.

The chapter concludes with a review of gear failures. They may be attributed mainly to fatigue, scuffing, and gradual wear. Scuff loads require further research but are seen to accord well with Blok's "flash-temperature hypothesis." Reports on gear failures are listed in Part 3 of the references. The terminology and descriptions of the American Gear Manufacturers' Association are recommended.

REFERENCES

Part 1. Gear Geometry, Loading, and Film Thickness

Adkins, R. W. & Radzimovsky, E. H. (1965), "Lubrication Phenomena in Spur Gears: Capacity, Film Thickness Variation and Efficiency." *JBE, T* **87,** 655–65. (a) *ASME* 63-WA-85; 8 pp. (1963).

Almen, J. O. (1942), "Dimensional Value of Lubricants in Gear Design." *TSAE* **50,** 373–80.

American Gear Manufacturers Association (1965), *Gear Nomenclature . . .* AGMA 112.04 (June).

Attia, A. Y. (1959), "Dynamic Loading on Spur Gear Teeth." *JEI, T* **81,** 1–9.

Beuerlein, P. (1955), "Moderne Schmierungsfragen bei grossen Getrieben." *Erdöl u. Kohle* **8,** 473–8. (a) *ZVDI* **98,** 318 (1956).

Blok, Harmen (1951), "Gear Lubricant—A Constructional Gear Material." *De Ingenieur* **63,** No. 39; 53–64.

Buckingham, Earle (1931), *Dynamic Loads on Gear Teeth.* ASME Research Publication, New York; 71 pp.

Buckingham, Earle (1949), *Analytical Mechanics of Gears.* McGraw-Hill, New York; 546 pp.

Cameron, A. & MacConochie, I. O. (1958), "The Measurement of Oil Film Thickness on Gear Teeth." *ASME* 58-A-142.

Campbell, R. B. & MacDonald, D. (1961), "Detection of Oil Films between Hypoid Gear Teeth Using Contact Resistance Measurements." *JIP* **47,** 365–74.

Coleman, Wells (1964), "Contact Pressure and Sliding Velocities on Hypoid Gear Teeth." *LE* **20,** 189–94.

Dareing, D. W. & Radzimovsky, E. I. (1963), "Experimental Investigation of the Minimum Oil-Film Thickness in Spur Gears." *JBE, T* **85,** 451–6.

Davis, A. W. (1956), "Marine Reduction Gearing." *Proc. IME* **170,** 477–98. (a) *Eng.* **181,** 107–10.

Dudley, D. W. (1954), *Practical Gear Design.* McGraw-Hill, New York; 335 pp.

Dudley, D. W., editor (1962), *Gear Handbook.* McGraw-Hill, New York; 926 pp.

Gatcombe, E. K. (1954), "On the Need for a Non-Steady State Theory for Lubrication Hydrodynamics." Pages 92–7 in *Fundamentals of Friction and Lubrication in Engineering.* First ASLE National Symposium (held in 1952). ASLE, Chicago.

Gatcombe, E. K. & Prowell, R. W. (1960), "Rocket Motor Gear Tooth Analysis—Hertzian Contact Stresses and Times." *JEI, T* **82,** 223–30. Cf. (a) *LE* **16,** 308–11.

Gregory, R. W., Harris, S. L., & Munro, R. G. (1963), "Dynamic Behavior of Spur Gears." *Chartered ME* **10,** 493–4. (a) *Proc. IME* **178,** pt. 1, 207–26.

Harris, S. L. (1958), "Dynamic Loads on the Teeth of Spur Gears." *Proc. IME* **172,** 87–112.

Ibrahim, M. & Cameron, A. (1963), "Oil Film Thickness and the Mechanism of Scuffing in Gear Tests." Paper 20, *LWG,* 70–80.

IME, International Conference on Gearing (1958). London; 553 pp. + plates. See pp. 116–64, 394–404, 483–509.

Lane, T. B. & Hughes, J. R. (1952), "A Study of the Oil Film Formation in Gears by Electrical Resistance Method." *Brit. JAP* **3,** 315–8.

MacConochie, I. O. & Newman, A. D. (1961), "The Effect of Lubricant Viscosity on the Lubrication of Gear Teeth." *Wear* **4,** 10–21.

MacConochie, I. O. & Shakib, I. D. (1963), "Some Optimum Conditions for the Lubrication of Gear Teeth." *TASLE* **6,** 141–6.

Merritt, H. E. (1962), "Gear Tooth Contact Phenomena." *Proc. IME* **176,** 141–64.

Niemann, Gustav (1955), "Schmierfilmbildung, Verlustleistung und Schadensgrenzen bei Zahrnrädern mit Evolventenverzahnung," *ZVDI* **97,** 305–8.

Niemann, G. & Rettig, H. (1957), "Dynamische Zahnkrafte . . ." *ZVDI* **99,** 89–96; 131–7.

Osterle, J. F. (1959), "Film Geometry Effects on Hydrodynamic Gear Lubrication." *Wear* **2,** 416–22.

Powell, D. C. & Barton, H. H. (1960), "Analytical Study of Surface Loading and Sliding Velocity of Automotive Hypoid Gears." *TASLE* **2,** 173–83.

Reswick, J. B. (1955), "Dynamic Loads on Spur Gear Teeth." *T* **77,** 635–44.

Sasaki, T., Okamura, K., Konishi, T., & Nishizawa, Y. (1962), "Lubrication Performance on Gear Tooth Surface—2nd Report . . ." *Bull. JSME* **5,** No. 19, 561–70.

Shigley, J. E. (1963), *Mechanical Engineering Design.* McGraw-Hill, New York; 631 pp.

Stephenson, R. R. & Osterle, J. F. (1961), "The Load Capacity of Gear Teeth Along the Line of Action." *Wear* **4,** 56–63.

Tuplin, W. A. (1958), "Dynamic Loads on Gear Teeth." Paper 17, *ICG*, 1, 24–30; 16, 377–9, 506, 513, 516. (a) *Gear Load Capacity,* Wiley, New York; 177 pp. (1962).

Utagawa, M. & Harada, T. (1962), "Dynamic Loads on Spur Gear Teeth Having Pitch Errors at High Speed." *Bull. JSME* **5,** 374–81.

Part 2. Friction Loss in Gearing

Allen, C. M. & Roys, F. W. (1918), "Efficiency of Gear Drives." *T* **40,** 101–16.

Botstiber, D. W. & Kingston, L. (1956), "Lubrication of High Capacity Gear Drives." *Product Eng.* **27,** 173–9 (May).

Buckingham, Earle (Chairman) (1929), "Worm Gears—a Study and Review of Existing Data." *ME* **51,** 210–7.

Cameron, A. & Newman, A. D. (1953), "Back-to-Back Testing of Marine Reduction Gears." *Eng.* **175,** 458–9.

Clausen, H. (1954), "The Efficiency of Gear Trains." *Eng.* **177,** 366–7.

Davies, J. A. & Semar, H. W. (1947), "Mechanical Reduction Gears." *TSNAME* **55,** 244–83.

Dietrich, G. (1939), *Reibungskrafte, Laufunruhe und Geräuschbildung an Zahnrädern.* Deutsche Kraftfahrtforschung No. 25, Berlin; 38 pp.

Donald, H. G. (1922), "Efficiency Test of Two 22,500 H.P. Single Reduction Gear Units by the Method of Losses." *J. ASNE* **34,** 499–526.

Douglas, L. M. (1940), "Mechanical Losses for Large Power Transmission." *Trans. N. E. Coast Inst. of Engs. and Shipbuilders* **57,** 69–74 (Dec.).

Emmet, W. L. R. (1920), "Electrical Propulsion of Merchant Ships." *G. E. Review* **23,** 60–6.

Engineering (1959), "Full Load Efficiency Testing of Turbine Gear Units." **187,** 306–7.

Giovannozzi, Renato (1939), *Teoria della lubrificazione negli ingranaggi.* Pubbl. d. R. Scuola d'Ingegneria di Pisa (nona serie) No. 401; 49 pp.

Green, C. E. & Doble, C. F. (1910), *A Comparison of the Running Properties of Three Types of Involute Gearing.* S. B. Thesis, Dept. of Mech. Eng., M.I.T.; 16 pp. (a) *T* **32,** 823–35, 850–1.

Ham, C. W. & Huckert, J. W. (1925), *An Investigation of the Efficiency and Durability of Spur Gears.* Bull. 149, EES, University of Illinois, Urbana; 92 pp.

Himes, W. H. (1932), "Efficiency of Worm Gears." *Mach. Design* **4,** 28–31 (Feb.); 42–6 (Apr.). (a) Watson, H. J., **4,** 31–2, 72 (Oct.). (b) *ME* **55,** 47 (1933).

Hyde, J. H. (1920), "The Variation in Efficiency of a Worm Gear Due to Differences in the Lubricant Employed." App. 2–6, 8–10, 15; pp. 50–79, 85–96, 110–12 in *Rep. of Lubricants and Lubrication Inquiry Com.,* DSIR, London.

Hyde, J. H., Tomlinson, G. A., & Allan, G. W. C. (1932), "An Investigation of the Performance of Gears." *Proc. IAE* **26,** 416–48. (a) *Eng.* **133,** 489–90.

Kenerson, W. H. (1912), "Investigation of Efficiency of Worm Gearing for Automobile Transmission." *T* **34,** 919–67. (a) *T* **31,** 171–9 (1909).

Kingsbury, Albert (1916), *Method and Means for Testing Gearing.* U.S. Patent No. 1,198,637 (Applic. 1910); 3 pp. +

Kudryavtsev, V. N. (1953), Determination of the Efficiency of Planetary Transmission Gears Taking into Account the Loss in the Gears and Bearings (Russian). *AN SSSR, Trudi Sem. Teorii Mash. Mekh.* **13,** 52; 5–23.

Kutzbach, K. (1926), "Reibung und Abnützung von Zahnrädern. . . ." *ZVDI* **70,** 999–1003.

Lewis, Wilfred (1914), "Gear Testing Machine." *T* **36,** 231–7.

Lewis, Wilfred (1923), "Modern Problems in Gear Testing and a Proposed Testing Machine." *Am. Mach.* **59,** 875–81.

Marine Engineer and Naval Architect (1956), "Y.E.A.D-1—A Naval Research Project." Annual Steam Number **79,** No. 961; 375–84. (a) *J. ASNE* **69,** 269–78 (1957).

McKee, S. A. (1929), "Progress on Efficiency Tests of Electric Street-Car Trucks." *Proc. Am. Elect. Ry. Assoc.* **27,** 554–75. (a) **28,** 85–6, 121–35 (1930).

McKee, S. A. & McKee, T. R. (1931), "Power Losses in Electric Street Car Reduction Gearing." *Proc. Am. Elect. Ry. Assoc.* **29,** 625–44.

Merritt, H. E. (1941), "Epicyclic Gear Trans." *The Engineer* **171,** 190–2, 213–5.

Merritt, H. E. (1942), *Gears.* Pitman, London; 420 pp. (a) 3rd ed., 1954; 527 pp., reprinted 1961.

Michel, R. (1941), "Efficiency of Double Reduction Gears as Influenced by Lubricating Oil Temperatures." *J. ASNE* **53,** 756–58.

Monk, Ivan (1949), "Marine-Propulsion Gear Testing at the Naval Boiler and Turbine Laboratory." *T* **71,** 487–99.

Niemann, Gustav (1955), "Grenzleistungen für gekühlte Schneckentriebe." *ZVDI* **97,** 308.

Niemann, G. & Ohlendorf, H. (1958), "Verlustleistung und Erwärmung von Stirnradgetrieben." *Mitt. a.d. Forschungsstelle Zahnräder u. Getriebebau der T. H. München,* 15–23. (a) Ohlendorf, Hermann, Dissertation. (b) *ZVDI* **102,** 216–24 (1960).

Pacific Marine Review (1935), "Efficiency Formula Verified for Worm Gears by Input-Output Test," **32,** 218–9 (July).

Poppinga, R. (1950), "The Efficiency of Planetary Gear Trains." *Glaser's Ann.* **74,** 139–45. (a) *Engs. Digest* **11,** 421–5.

Radzimovsky, E. I. (1959), "Planetary Gear Drives." *Mach. Design* **31,** 144–53 (June 11).

Rickli, H. (1911), "Bestimmung des Wirkungsgrades von Zahnrädern." *ZVDI* **55,** 1435–8.

Schlesman, C. H. (1935), "The Lubrication Requirements of Automotive Worm Gearing." *JSAE* **36,** 147–58.

Shipley, E. E. (1958), "12 Ways to Load-Test Gears." *Product Eng.* **29,** 77–82 (Jan. 6).

Shipley, E. E. (1958), "How to Predict Efficiency of Gear Trains." *Product Eng.* **29,** 44–5 (Aug. 4).

Shipley, E. E. (1959), "Efficiency of Involute Spur Gears." *ASLE* 59 GS-6; 15 pp.

Shipley, E. E. (1962), "Loaded Gears in Action." Chapter 14; 60 pp. in *Gear Handbook,* D. W. Dudley, editor.

Tourret, R. (1955), "Worm-Gear Lubrication. Effect of Oil Viscosity on Power Losses and Wear." *Eng.* **180,** 888–91.

Tso, L. N. & Prowell, R. W. (1961), "A Study of Friction Loss for Spur Gear Teeth." *ASME* 61-WA-85; 9 pp.

Wakuri, A. & Ueno, T. (1958), "On the Lubrication of Worm Gears." *Bull. JSME* **1,** 189–93.

Walker, Harry (1944), "The Thermal Rating of Worm Gear Boxes." *Eng.* **157,** 78–80, 97–100.

Waller, C. R. (1948), "Propulsion Gearings." *T SNAME* **56,** 544–70.

Welch, W. P. & Boron, J. F. (1961), "Thermal Instability in High-Speed Gearing." *JEP, T* **83,** 91–107.

Wellauer, E. J. (1952), "Solving the Thermal Problem for Enclosed Gear Drives." *Mach. Design* **24,** 123–7 (Mar.).

Whittle, J. (1961), "Lubrication of Automotive Worm Gears." *Proc. IME Autom. Div.,* No. 3, 119–46.

Part 3. Gear Failures

Almen, J. O. (1950), "Surface Deterioration of Gear Teeth." Pages 229–88 in *Mechanical Wear,* ed. by J. T. Burwell, Am. Soc. for Metals, Cleveland.

Baber, B. B., Lawler, C. W., Smith, H. R., Beane, G. A., & Ku, P. M. (1960), "Gear Lubrication in Inert Atmospheres." *TASLE* **3,** 142–8.

Barwell, F. T. & Milne, A. A. (1952), "Criteria Governing Scuffing Failures." *JIP* **38,** 624–32, 668–98.

Beane, G. A. & Lawler, C. W. (1964), "Lubricant Evaluation with the WADD High Temperature Gear Machine." *IP Gear Symposium,* I.37-I.53 (Oct.).

Blok, H. (1937), "Measurement of Temperature Flashes on Gear Teeth under Extreme Pressure Conditions." *GDLL* **2,** 14–20.

Blok, H. (1940), "Seizure-Delay Method for Determining the Seizure Protection of EP Lubricants." *JSAE* **44,** 193–200, 220 (May).

Blok, H. (1958), "Lubrication as a Gear Design Factor." Paper 33, *ICG,* 144–58; 117–9, 397, 466–7, 470, 479, 483, 513, 515.

Boner, C. J. (1954), *Gear and Transmission Lubricants.* Reinhold, New York, 493 pp.

Borsoff, V. N. (1959), "On the Mechanism of Gear Lubrication." *T* **81,** 79–93. (a) With F. C. Younger, *LE* **9,** 259–62; 271 (1953). (b) With C. D. Wagner, *LE* **13,** 91–9 (1957). (c) *LE* **18,** 266–70 (1962).

Borsoff, V. N., Accinelli, J. B., & Cattaneo, A. G. (1951), "The Effect of Oil Viscosity on the Power Transmitting Capacity of Spur Gears." *T* **73,** 687–96.

Braley, W. W. & Berg, M. S. (1958), "Design and Service Experience with U.S. Naval Gears." Paper 3, *ICG,* 287–95; 285, 436–40, 442–3, 448–51, 457, 513; Plates 1, 2.

Campbell, R. B. (1963), "Study of Hypoid Gear Lubrication Using Radioactive Tracers." Paper 25, *LWG,* 9–13.

Cleare, G. V. (1958), "Gear Tooth Failures in Highly Stressed Transmission Gears," Paper 1, *ICG,* 57–61; 53–6, 385–6, 472, 476, 490, 522; Plates 1–4.

Cowell, W. T. (1958), "Laboratory Testing of Final Drive and Aircraft Gear Assemblies." *Paper 19, ICG,* 354–61; 340, 454–5, 457–8, 476.

David, V. W., Hughes, J. R., & Reece, D. (1956), "Some Lubrication Problems of Aviation Gas Turbines." *JIP* **42,** 330–6.

Droegmueller, E. A. & Huff, R. G. (1964), "The Effect of Gear Machines on Scuff Limited Loads." *ASLE* 64 AM 1B-2; 5 pp.

Frederick, S. H. & Newman, A. D. (1958), "Gear Failures." Paper 13, *ICG,* 77–81; 53, 55–6, 381, 385–9, 420, 451, 463–4, 475, 490, 492, 522–3; Plates 5–8.

Genkin, M. D. & Misharin, Yu. A. (1960), A Method of Studying Lubricating Properties of Oils in a Gear Test Rig (Russian). *T. Third All-Union Conf. on Friction and Wear in Machines* (1958). AN SSSR **1,** 122–7. (a) TT-1057, Nat. Res. Council, Ottawa, Canada; 10 pp. (1963).

Genkin, M. D. & Misharin, Yu. A. (1960), An Examination of the Hypothesis of Blok on Critical Temperatures in Gearing Seizures (Russian). *Op. cit.,* **2,** 188–98. (a) TT-1059, Nat. Res. Council, Ottawa, Canada; 14 pp. (1963).

Hughes, J. R. & Tourret, R. (1953), "Mechanical Testing of Gear Lubricants." *Eng.* **175,** 200–3; 257–9.

Hughes, J. R. & Waight, F. H. (1958), "The Lubrication of Spur Gears." Paper 25, *ICG,* 135–43; 117–8, 394–5, 397–9, 463, 475, 494, 515; Plates **4, 5.**

Hutt, E. T. (1952), "Lubrication and the Load-Carrying Capacity of Gears." *LE* **8,** 180–2; 201–3.

Institute of Petroleum (1952), "The Development of a Test Method for Gear Lubricants on the IAE Machine." *JIP* **38,** 703–11, 783–92. (a) **45,** 338–41 (1959).

Joughin, J. H. (1951), "Naval Gearing—War Experience and Present Development." *Proc. IME* **164,** 157–76.

Kelley, B. (1953), "A New Look at the Scoring Phenomena of Gears." *TSAE* **61,** 175–88.

Ku, P. M. & Baber, B. B. (1960), "Effects of Lubricants on Gear-Tooth Scuffing." *TASLE* **2,** 184–94.

Lane, T. B. & Hughes, J. R. (1952), "A Practical Application of the Flash-Temperature Hypothesis to Gear Lubrication." Third World Petroleum Cong., 1951; *Proc.,* Sect. VII, 320–5.

Mansion, H. D. (1952), "Some Factors Affecting Gear Scuffing." *JIP* **38,** 633–45; 692–3.

Monk, I., Thomas, L. I., & Atkinson, C. C. (1952), "Recent Developments in Naval Propulsion Gears." *T. SNAME* **60,** 273–313.

Niemann, G. & Rettig, H. (1954), "Der FZG-Zahnradkurztest zur Prüfung von Getriebölen; Beschreibung des Testes, Testbedingungen und Testergebnisse." *Erdöl u. Kohle* **7,** No. 10, 640–2.

Niemann, G., Rettig, H., & Lechner, G. (1961), "Experiences with Scuffing Tests on Gear Oils in Back to Back Rigs." *TASLE* **4,** 71–86. (a) *TSAE* **71,** 169–84, 201.

Pohl, W. (1954), "Quelques aspects de la lubrification des engrénages dans l'industrie." *RUM* (9) **10,** 331–40.

Ryder, E. A. (1947), "A Gear Lubricant Tester—Measures Tooth Strength or Surface Effects." *Bull. ASTM* **148,** 69–73. (a) **184,** 41–3 (1952).

Shipley, E. E. (1958), "Design and Testing Considerations of Lubricants for Gear Applications." *LE* **14,** 148–52; 191. (a) **15,** 98–103, 119, 191 (1959).

Simpson, R. T. (1949), "Lubrication of Naval Gearing." *ASTM Tech. Pub.* No. 92, 3–8.

Van Zandt, R. P. & Kelley, B. W. (1949), "Gear Testing Methods for Development of Heavy Duty Gearing." *JSAE* **57,** 38–40.

Wellauer, E. J. (1965), "Gears—New Rating System Simplifies Design." *ME* **87,** 21–7 (Feb.).

Zrodowski, J. J. (1957), "Progress and Operating Experience with Modern Ship Propulsion Gears." *TSNAME* **65,** 839–82.

Chapter XII Vibrations and Whirl

1. General survey. 2. Experimental work to 1954. 3. Later investigations. 4. Varieties of whirl. 5. Field experience. 6. Analytical studies to 1953. 7. Later calculations. 8. Summary.

This chapter is a guide to the literature on journal-bearing vibrations with more particular consideration of liquid-lubricated bearings. Gas-film bearings are discussed in Chapter IX.

Experimental and analytical investigations are separately described, each in chronological order. No attempt is made here to reconcile divergent results; but from the references given the reader should be able to find his way around, and draw his own conclusions.

1. GENERAL SURVEY

Nonfluid Vibrations. The dynamic action of an unbalanced rotor is like that of a dynamic load rotating with the shaft. The resulting vibration is called "synchronous whirl" since it has the frequency of the shaft rotation. An "inversion speed" may be reached at which the rotor begins to spin about its center of gravity instead of about an axis through its center of figure. Grobel, Church, and others have described modern practice in balancing large rotors (1953, 1961). Rotors are considered balanced unless otherwise stated.

Shaft flexibility in conjunction with the rotor mass imparts a "critical speed" or natural frequency of vibration to the system. Critical speeds were explained by Rankine at an early date (1869). If the shaft stiffness, or force per unit deflection is k, and the entire rotating mass m is concentrated halfway between supports, the lowest natural frequency is given by Dunkerley (1894) as $n_0 = \omega_0/2\pi$, where ω_0 is the square root of k/m. The expressions are more complicated when there are several masses at stated positions along the shaft. Dunkerley's formulas, based upon experiments suggested by Osborne Reynolds, give the resultant frequency in terms of known frequencies with the separate masses taken one at a time.

It can be shown by dimensional reasoning that all solutions for steady-state, undamped vibrations are special cases of the formula

$$n_0 = \text{const } (k/m)^{1/2}. \tag{1}$$

The constant is the same for all systems that have the same "generalized shape"—that is, the same geometrical shape with a similar distribution of the density and elastic constants (Hersey, **1917**).

A prolific literature has grown up around the calculation of critical speeds without reference to lubrication. General methods are reviewed by Den Hartog (1940), Biezeno & Grammel (1953), and others. Skew stiffness, gyroscopic action, torsional coupling, variation with time, and yielding of supports are treated by later authors. Conditions for stability were given by Routh (1877).

Professor Newkirk was a pioneer in the study of vibrations. One of his first discoveries was the effect of frictional slip between wheel hub and shaft. This he called "cramped shaft whirl" (**1924**). A similar effect, though less pronounced, can be caused by "internal friction," or elastic hysteresis in the shaft. That was soon demonstrated and explained by Kimball (1924). These effects are further discussed by Den Hartog (1958) and by Ehrich (1964).

Dry-bearing and boundary friction cause the journal to climb uphill backwards, in a manner contrary to fluid film lubrication (Fig. 2 of Chapter I). This has been known to induce vibrations both by "stick-slip," and by rolling completely around the clearance space (Fujii, 1951; Sines, 1954). For vibrations in a vacuum see Demorest & McKannan (1963).

Fluid Film Whirl. Harrison treated the motion of a rigid shaft without mass (VI: 1919). He showed that the hydrodynamic equations for a Sommerfeld bearing are satisfied by an orbital motion of the journal axis when the bearing is under a constant load. If the eccentricity ratio is close to zero or unity, his equations lead to an orbital frequency approaching half the speed of the shaft. The Harrison effect may be considered a special case of "half-frequency whirl."

Professor Stodola, from his long experience in the testing and design of steam turbines, had come to suspect the existence of a new critical speed, not depending on rotor unbalance or shaft flexibility. To account for it, he introduced the idea of the "oil spring." In his paper of May, 1925 he showed that the oil film supporting a rigid shaft must act like a nonlinear spring having a definite stiffness for small vibrations at each eccentricity. The Stodola effect was confirmed experimentally the following year, and is now taken as basic in analytical work.

Newkirk & Taylor in their paper of August, 1925 reported a new type of vibration depending on shaft flexibility. They called it "resonant whip" or preferably, "resonant whirl." They defined "whirl" by saying that "whirling is viewed as a composite of two vibrations in planes 90 deg apart and with a phase difference of one-quarter period." The newly discovered effect does not occur with a rigid shaft; or even with a flexible shaft *if the oil supply is cut off*. The Newkirk & Taylor effect differs from the Harrison effect and the Stodola effect in its requirement of a flexible shaft.

Fluid film whirl is recognized now as the principal cause of vibration troubles in properly balanced commercial rotating machinery.

2. EXPERIMENTAL WORK TO 1954

The Newkirk & Taylor experiments included several different bearing and rotor combinations, always carefully balanced. Whirl orbits were shown photographically by means of the spherical mirror system described by Newkirk in his paper of 1924. A relatively slight whirl amplitude is observed at the shaft critical speed, where the vibration is synchronous. A sharp jump in whirl amplitude then occurs at about *twice the shaft critical*. This is the newly discovered oil-resonance effect, a vigorous whip that continues to increase in amplitude with increasing shaft speeds under moderate load. The whirl frequency at oil-resonance speed is close to the shaft critical, or about half the running speed, hence the term "half-frequency whirl." It increases only slightly at higher speed. Whipping could be reduced by operating below twice the shaft critical; by increasing the load or shortening the bearing; by misalignment, by decreasing clearance, or by shutting off the oil supply. The bearings ranged in diameter from 1 to 2 in., lengths from $\frac{3}{8}$ to 5 in., and clearances from 2 to 15 mils on the diameter. Critical speeds ranged from 800 to 1700 rpm. Loading was varied up to 85 psi and speeds to 5000 rpm. Whirl amplitudes as great as 85 mils were observed.

Hummel (III: 1926) carried out the experiments with a rigid shaft proposed by Stodola. Whirl photographs were secured by an adaptation of Newkirk's method. Some thirty-five half-bearings were tested, with varying cap clearances and side reliefs. Diameters ranged from

25 to 40 mm; L/D from 1/3 to 5/4. Clearance ratios were from 1/22 to 1/40—notably higher than in commercial practice. Loads ranged from 0.2 to 2.8 kg/sq cm. Speeds were carried above 4000 rpm. Numerous photographs show the paths of the journal center. The characteristic shape is a narrow horizontal crescent, rather than a full-fashioned orbit. Whirl amplitudes are plotted against shaft speed, showing peaks near 2000 and slightly under 4000 rpm in most tests—the more violent ones at the higher speed. Let ω denote the whirl frequency in radians per second. Hummel plotted $\omega^2(c/g)$ against a dimensionless variable Γ, where c is the radial clearance, g the acceleration of gravity, and

$$\Gamma \equiv \frac{c}{g}\left(\frac{P}{Z}\right)^2\left(\frac{c}{r}\right) \tag{2}$$

in our notation. Hummel's number Γ may also be expressed by c/g times $(N/S)^2$, where N is the running speed and S the Sommerfeld variable. When the observed values are plotted on Hummel's chart, a curve is obtained with two branches, in agreement with Stodola's theory. It was found that whirling could be reduced by certain close-fitting caps, and a generous side relief.

Other Early Tests. Newkirk tested a short, stiff shaft having its first critical far above the limits of operation (1931). Whirling occurred at low speeds with a frequency slightly below half running speed—"half-frequency whirl" again. Test bearings were of 2-in. diameter, $L/D = \frac{9}{16}$. A light load, 4 psi, was applied at speeds up to 30,000 rpm. Further tests on the same shaft are described by Newkirk & Grobel (1934). A "nonwhirling bearing" was developed by means of a wide groove and dam in the cap. This reduced the tendency to whirl by maintaining a high eccentricity. See also Newkirk (1937). Stone reported whirl in his sleeve bearing tests (1935). Rumpf noted quieter running with the lemon-shaped bearings than with those of conventional design (III: 1938). Here again pressure is generated at the crown of the bearing.

Whirl Testing Continued. Hagg's studies first appeared in 1946 and 1948. His calculations for the rigid shaft in a partial bearing agreed with experiment, and his tests indicated that tilting-pad journal bearings were free from whirl. Manabe experimented on transient conditions (1949).

Simons (VIII: 1950) obtained whirl of constant amplitude, with a stiff shaft in an unloaded bearing, at a frequency slightly under half the shaft speed. He found the most rapid damping of whirl at high loads, speeds, and inlet oil pressures, but with low-viscosity lubricants.

Dayton & Simons (1951) saw that the whirling of a stiff shaft can occur at all rotational speeds with a frequency equal to half the speed of the shaft—another confirmation of half-frequency whirl. They noted that whirl can be reduced by unbalance—adding a load rotating at the shaft speed—and they confirmed Hagg's finding that the tilting-pad bearing apparently does not excite whirl.

Linn & Prohl found that the critical speed of a shaft is lowered by flexible supports (1951). Critical speeds were below calculated values, but, as noted by Professor Newkirk, this might have been due, in part, to the yielding of the oil film. Hagg & Warner (1953) published steam-turbine tests up to 12,000 rpm. Whirl frequencies were observed equal to the resonant frequency of the system. The calculated and observed whirl thresholds were in reasonable agreement at different L/D's and rotor stiffnesses.

Pinkus (1953) reported briefly on two rotors having shaft criticals of 4000 and 6100 rpm. Half-frequency whirl appeared at speeds below twice critical, but went over into resonant whirl as the speed passed through twice critical. Also, a disturbance that develops when the oil is heated up can be inhibited by a more viscous oil. In both respects, Professor Newkirk points out (1956), the stiff rotors tested by Pinkus behaved differently from the more flexible rotors in his own experiments.

Newkirk's program was resumed by Newkirk & Lewis (1954) and continued by Lewis & Fulton (1956). Many combinations of operating and design variables were tested. The range of stable operation could be extended to higher speeds by going to shorter bearings of larger clearance, operating under greater loads at a lower viscosity. Disturbances rarely developed at shaft speeds below twice critical. Whirl frequencies were always equal to the natural frequencies—which, however, increased slightly with increasing speeds. Barwell, reviewing some of these data in his book, plots out the ratio of whirl frequency to the shaft speed at which whirling begins, n/N in our notation. For abscissa he takes a "duty parameter" equal to πS times $(L/D)^2$, an expression proportional to Ocvirk's number. The n/N ratio varies from about $\frac{1}{4}$ at low values of the parameter up to nearly $\frac{1}{2}$ at the higher values, where it levels off.

Sines' Experiments. Sines (1954) experimented with a small turbine rotor on a stiff shaft running in porous bushings at speeds up to 90,000 rpm. Synchronous whirl, only, could be found. It was attributed in different tests to unbalance, to the Stodola effect, and to boundary friction; there were no signs of resonant or of half-frequency whirl. A "high-oiliness" oil, such as mineral oil plus oleic acid, was found necessary in order to insure long life with smooth operation. Sines' technique revealed different whirl speeds in different parts of the orbit.

3. LATER INVESTIGATIONS

Shawki (1955) observed both transient and steady whirling of a rigid journal under no load. Whirl frequencies were from 0.491 to 0.499 of the shaft speed, confirming Simons' experiments and the concept of half-frequency whirl. No film failure occurred, the eccentricity ratio not exceeding 0.81. Shawki continued testing under unidirectional load (VIII: 1956). Subharmonic vibrations were induced by load speeds equal to submultiples of half running speed.

Boeker & Sternlicht (1956) experimented with rigid vertical shafts undergoing "translational" whirl. Tested with a plain bearing and air lubrication, the shaft whirled at all speeds down to the lowest; but with water only at speeds above 130 rpm. Whirl frequency in air varied from 0.41 to 0.50 of the shaft speed. A journal designed with Rayleigh surfaces was tested in water, showing a marked contrast with the plain journal. The new type journal whirled only at speeds above 2700 rpm. Results on both types agreed with Poritsky's criterion (1953). Further confirmation was obtained in turbine tests reported by Sternlicht (1955). Professor Fuller on page 187 of his 1956 book described tests showing stability of pivoted-shoe bearings with low-viscosity lubricants.

Hagg & Sankey (1956) report experiments on two types of bearings operating with unbalanced rotor on rigid shaft. One was a partial of 150 deg, the other a four-shoe pivoted type. Both were of 3-in. diameter with L/D near unity. Dynamic properties of the oil film are represented by plotting Kc/W and $Bc\omega/W$ against the Sommerfeld variable. Here K is the oil-film stiffness and B the damping constant for a radial clearance c, load W, and shaft speed ω. Tests were also included on a 16 by 16 in. bearing. Data agreed closely with the calculated whirl frequency, showing an amplitude slightly below calculated value.

In a final report, Pinkus (1956) tested six types of bearings with a flexible shaft. He rates them in decreasing order of stability: (1) three-lobe designs, (2) tilting pad, (3) pressure types, (4) elliptical, (5) three axial grooves, and (6) the plain sleeve bearing. Under "pressure types" he includes both self-acting and externally pressurized bearings. He found that resonant whip persists over a wide range of speeds; the range being less at the higher loads and oil temperatures.

Cameron & Solomon (1957) tested a rigid journal of 6-in. diameter with $L/D = 1$ and clearance ratios as high as 1/12. The oil-film criticals came below 3000 rpm. The ratio of vibration frequency to shaft speed was always about 0.49, closely corresponding to half-frequency whirl. The existence of steady whirl at eccentricity ratios far below 0.7 confirmed Cameron's extension of Hummel's theory. Film pressures and journal eccentricities were measured. Resonant vibrations occurred over a wide frequency band, calculated frequencies corresponding to the maximum disturbance.

Cole experimented with stiff shafts in a transparent full bearing (1957). The test bearing was of 1.0-in. diameter with $L/D = 1$. Clearance ratios were varied from 1 to 4 mils/in. Speeds ranged from 200 to 5000 rpm. with loads up to 100 psi. Whirling occurred only with light loads corresponding to Δ values below 0.2, where Δ is a load parameter equal to $1/\pi S$. Whirl-free or stable operation occurred only with cavitated films; and usually with eccentricity ratios of the order of 1/10. The ratio of whirl frequency to shaft speed varied only from 0.480 to 0.506. Cole's very interesting chart of stability zones reveals wide differences between the findings of various investigators.

In a first report on whirl research, E. H. Hull (1958) applied rotating loads having different ratios of load speeds to shaft speed, and so obtained a series of oil-film resonances with a rigid shaft. The whirl path takes the form of an epitrochoid (an orbit with internal loops) when the frequency ratio is positive, load vector and journal rotating in the same direction. It takes the form of an hypotrochoid (external loops) when the ratio is negative. The figures shown are reminiscent of similar effects noted by Pinkus and by Shawki (both 1956). The number of loops can be predicted by an empirical formula. The particular case of a load rotating at half the shaft speed corresponds to normal half-frequency whirl in a circular orbit. In a second report, Hull (1960) demonstrates the use of a rotating sleeve between journal and bearing to raise the whirl frequency in a vertical or unloaded bearing. In a third report Hull (1961) described experiments on shaft whirling as influenced by stiffness asymmetry. Tondl, in the meantime, published a series of whirl tests on bearings of both circular and elliptical sections (1960), leading to the design of a bearing with elastic elements. See also Morrison (1962) and Parzewski & Cameron (1962).

An unloaded vertical journal of 3-in. diameter was tested by Bowman and co-authors in a study of factors affecting whirl instability (1964). In the laminar flow region stable conditions were found up to forty times the natural frequency, while the onset speed of whirl increased with increasing viscosity. In the Taylor vortex region stable conditions were found at shaft speeds up to twenty times the natural frequency. The onset speed increased with viscosity and with reduction of clearance ratio. It may be considered to occur at twice the "effective" natural frequency determined by oil-film stiffness in conjunction with shaft stiffness. The relative or "specific" onset speed n_c correlates inversely with a dimensionless parameter taken as the product of mN_n/ZL by the square of the clearance ratio. Here n_c is the ratio of shaft onset speed to the natural frequency N_n, and m is the rotor mass, Z the film viscosity, and L the length of the bearing. See Cheng & Trumpler on high speed bearings (1963); Cooper on control of vibrations (1964).

4. VARIETIES OF WHIRL

According to a classification by Professor Newkirk (1956, 1957) there are basically three types of whirl, quite apart from unbalance and nonfluid vibrations:

1. A "half-frequency whirl" that may occur with rigid shaft at any running speed, having a frequency just under half the speed of the shaft.

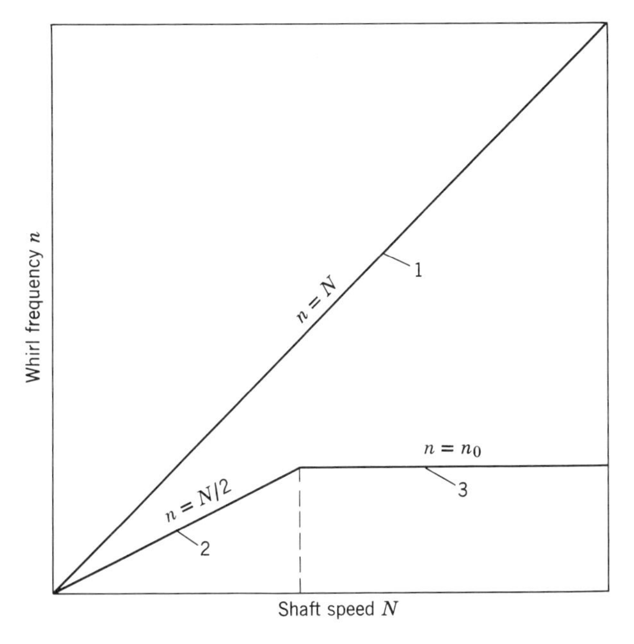

Fig. 1. Three types of shaft vibration. From Wilcock & Booser (I: 1957), copyright © 1957, McGraw-Hill, Inc.; used by permission. (1, Synchronous whirl due to unbalance; 2, half-frequency whirl caused by oil film; 3, resonant whip of shaft having a natural frequency n_0).

2. A "resonant whirl" (the Newkirk & Taylor effect), occurring at running speeds not less than twice the natural frequency of the rotor due to shaft flexibility, the whirl frequency being equal to the natural frequency.

3. "Oil-film criticals" (the Stodola effect), depending on the spring constant of the film.

Other disturbances that are difficult to identify may be combinations of the foregoing or involve nonfluid vibrations. Analytical studies help to indicate which effect will predominate in a given problem. Similar effects met in a gas-lubricated bearing are noted in Chapter IX.

The three main types of shaft vibration are neatly visualized by Wilcock & Booser (I: 1957) in a diagram of whirl frequency against shaft speed, Fig. 1 herewith (Fig. 8-6 of the reference).

5. FIELD EXPERIENCE

Not every aspect of whirl is revealed by laboratory-type experiments or by analytical study. Field experience is equally important. The following general comments were kindly contributed by Professor Newkirk:

> The driving forces that produce these whirls are relatively small and they may be offset by damping forces arising in the structure. Erratic variations in occurrence of the whirling are caused by changes in alignment of multi-bearing units that change the loads on individual bearings.
>
> Whirling of the journal is often serious when the whirl frequency is equal to the natural frequency of the rotating shaft so that a resonant condition develops. A similar situation develops if some other part of the machine or structure responds. Violent whirling may also occur when the speed exceeds twice critical. The urge to whirl in the bearing supports a shaft whirl of relatively large amplitude, and of a frequency equal to the natural frequency of the whirling shaft. At these higher speeds the shaft whirl becomes dominant, determining the frequency and relegating the journal whirl to a supporting role.
>
> Operation of machines above twice "critical speed" is not uncommon but special expedients or design features are necessary. When operating above twice critical speed a shock may start severe whirling of a rotor that would be stable otherwise.

How to Minimize Whirl. To the practical mind, the immediate question is how to avoid or minimize whirl. This can be done by running at a low speed. More sophisticated remedies are (1) use of heavier load per unit area, (2) stiffening the shaft and the bearing supports, (3) reducing the rotor mass, (4) varying the clearance, (5) varying the oil viscosity by raising it in synchronous and half-frequency whirl, (6) controlled misalignment, and (7) changing the design or type of bearing.

The tilting-pad type, and bearings designed to develop fluid pressure on the unloaded side, like that of Newkirk & Grobel, are widely favored. Other designs are the Rayleigh step and a "displaced" bearing in which the two halves are relatively displaced at right angles to the load (Howarth, 1937; Schnittger, 1959; Wilcock, 1961). An intermediate sleeve, floating or driven, has been found effective (Hull, 1961), as well as a "rippled" bearing surface—page 804 of Shawki, 1957.

Methods of minimizing whirl are reviewed by Wilcock & Booser (I: 1957). One guiding principle is to bring the resultant fluid pressure closer to the line of centers; another is to increase the damping.

6. ANALYTICAL STUDIES TO 1953

From Harrison to Kapitza. Harrison was the first to apply hydrodynamic theory to the motion of the journal axis (VI: 1919). This he did by extending the conditions of Reynold's equation. His solution was limited to a rigid shaft without mass, in the Sommerfeld full bearing of infinite length. The theory was further developed by Robertson (1933) and Swift (1937) for the Sommerfeld case and by Burwell (1947) for the short bearing. A qualitative explanation offered by Newkirk & Taylor led to the mathematical treatment of half-frequency whirl by Newkirk, Hagg, and others (1930, 1946) based solely on continuity of flow. From this point of view the whirling of a rigid shaft is caused by the component of fluid pressure at right angles to the line of centers. Its frequency must equal the mean velocity of the film, normally half the speed of the journal.

Stodola (1925) and Hummel (III: 1926) showed how to predict oil-film critical speeds for the rigid shaft. Stodola's theory makes use of the journal-center locus and the load-eccentricity curve. It applies to any bearing for which these curves are known. Stodola solved the problem for a half-bearing of infinite length by calculating the spring stiffness of the film as a function of the eccentricity ratio. The spring constant is defined as the increment of radial load divided by the increment of eccentricity; meaning by radial load, the component along line of centers. Thus the spring constant is equal to the slope of the curve for radial load against eccentricity. Two frequencies of vibration are found at each eccentricity. The two are equal for an eccentricity ratio near 0.7, and imaginary below that limit. At eccentricity ratios approaching unity, Stodola's lower frequency is given by

$$n = 0.159(g/c)^{1/2}, \tag{3}$$

where c is the radial clearance. It will be 41 per cent higher at $\epsilon = 0.7$. A good account of Stodola's theory is given by Biezeno & Grammel in Vol. III of 1954, pages 200–205. The calculation of spring and damping constants is elaborated by Pestel (1954).

It was shown by Kapitza (1939) how the effect of damping, whether by windage or bearing friction, may carry the rotor safely through its critical speed. This had been accomplished in the Parsons turbine by means of a floating sleeve (Hodgkinson, VII: 1929).

Hagg to Poritsky. Hagg gave a theory of the rigid shaft with oil-film damping (1946, 1948). His applications include the bearing of 120-deg arc with $L/D = 0.4$, the axially grooved bearing, and the tilting pad. He calculates the whirl amplitude of an unbalanced flexible shaft, mentioning subharmonic resonance. In his treatment of partial bearings he sets up a criterion for stability in terms of the rotor mass, damping constant, and film stiffness constants. He shows that the fluid pressure resultant comes closer to the line of centers in a fixed-pad (axially grooved) bearing than in a conventional full bearing, and coincides with the line of centers in a tilting-pad bearing, hence the greater stability of these types. Hagg used an energy method in preference to differential equations in his theory of the tilting-pad bearing.

Reference should be made here to papers by Khanowitch (1947) and Manabe (1949). Khanowitch studied the influence of bearing geometry on vibrational stability. Manabe found appropriate solutions for the Reynolds equation applicable to radial shock, unbalance, and rotation starting from rest. The corresponding trajectories and orbits are shown by diagrams for the rigid shaft in a Sommerfeld bearing. The effect of frictional resistance on shaft vibration was studied by Fujii (1951). He showed how such resistance could be either a damping or exciting factor; that the direction of whirl can be opposite to that of shaft rotation, and that its frequency need not be equal to the natural frequency of the rotor.

The theory of the rigid journal supported by an oil spring was further developed by Korovchinskii (1933), starting from Lagrange's equations and leading to a criterion of stability for the Sommerfeld full bearing.

Flexible shaft theory was studied by Hagg & Warner, and by

Poritsky, in their papers of 1953. The first-named investigators developed a criterion for stability with the aid of an electrical analog. It was seen that rotor flexibility reduces the region of stable operation. Their analog circuit was arranged so that its constants represent those in the differential equations of the mechanical system, namely the rotor mass m, shaft stiffness k, damping constants and film stiffness constants. Observations on the electrical system led to a stability criterion such that when k is set equal to ∞, it reduces to Hagg's criterion for the rigid shaft (1946). Needs' data for the 120-deg bearing serve to reduce the equation to a more specific form, from which a chart for the stable and unstable regions can be plotted. Charts are constructed for two L/D ratios. In each chart the parameter cN^2/g is plotted against the Sommerfeld number for constant values of cN_0^2/g, where N is the running speed and N_0 the rotor critical speed, or natural frequency on simple supports. Values read from these charts are found to be in reasonable accord with steam turbine tests.

Poritsky saw that the imperfection in earlier theories of full bearing whirl, due to neglect of cavitation, could be compensated by assuming a radial force on the journal proportional to its eccentricity. The equations of motion then lead to a stability condition

$$m\omega^2\left(\frac{1}{k} + \frac{1}{K_1}\right) < 4\,, \tag{4}$$

where ω is the running speed in radians per unit time, k the shaft stiffness, and K_1 the component of oil film stiffness measured along the line of centers. Now the factor in parentheses represents the net displacement of the rotor per unit force and is therefore the reciprocal of $m\omega_c^2$, where ω_c is the true critical speed of the rotor. Because of the yielding of the oil film, ω_c must be somewhat less than the natural frequency ω_0 on simple supports. Substituting and solving for ω leads to the criterion $\omega < 2\omega_c$. In this way he showed that the rotor is stable for speeds below approximately twice critical, and unstable at higher speeds, as observed by Newkirk and Taylor.

7. LATER CALCULATIONS

Cameron to Korovchinskii. A new light was thrown upon the rigid shaft problem by Cameron (1955) when he predicted a steady whirl in the half-bearing at eccentricity ratios less than 0.7. Earlier investigations had revealed no possibility of a closed orbit in that region. Experimental confirmation followed (1957). Cameron gives a chart

TABLE 1. Frequency Numbers from Cameron's Chart
(Steady whirl of rigid shaft in half-bearings)

L/D	n''', Cameron region			n', Stodola region		
	$\epsilon = 0.1$	0.3	0.5	$\epsilon = 0.7$	0.9	1.0
1/4	2.43	1.72	1.65	1.44	1.09	1.01
1	2.43	1.64	1.54	1.46	1.08	1.00
∞	2.42	1.55	1.42	1.50	1.07	1.00

for dimensionless whirl frequencies against ϵ. The dimensionless number chosen is $\omega\sqrt{c/g}$, where ω denotes the whirl frequency in radians per unit time. Let n', n'', and n''' denote the respective frequency numbers in the Stodola low-frequency region, the Stodola high-frequency region, and the Cameron region, that is, below $\epsilon = 0.7$. The chart shows a family of curves for each of these numbers at constant L/D, as well as for the amplification factor of the nonsteady whirl in the Cameron region. Table 1 shows values of n' read from the chart at $\epsilon = 0.7$ and above, and those of n''' below that limit. The values of n'', not shown, are higher than n' and therefore represent a safe condition. The values of n''' for ϵ less than 0.1 seem to rise without limit as ϵ approaches zero. The effect of L/D is slight within the range of Table 1. To understand the use of the table, suppose that a bearing with $L/D = 1$ has a radial clearance of 1.0 mil and is loaded to an eccentricity ratio of 0.5. Then $n''' = 1.54$, and the Cameron whirl frequency ω will be 960 rad/sec, or 9200 rpm, regardless of diameter.

Vector analysis was applied to good effect by Boeker & Sternlicht in their study of the translatory whirl of a rigid shaft in the vertical, unloaded position (1956). They showed that the shaft begins to whirl when its speed, in our notation, reaches the value

$$N_1 = (K_1/m)^{1/2}/\pi \,. \tag{5}$$

As in Eq. (4), K_1 is the radial spring stiffness of the film or component along line of centers (not the total spring stiffness); and m the mass of the rotor. Equation (5) was confirmed experimentally both for the full journal bearing, where $K_1 = 0$ at $\epsilon = 0$, and for an anti-whirl bearing with Rayleigh steps on the journal. The derivation requires a fair degree of symmetry with respect to the axis. It would therefore not apply to a partial bearing, and is limited to nearly concentric operation. The derivation leads to a stability criterion identical with Poritsky's for the rigid shaft, but without postulating his radial force. Equation (5) is a special case of (4) when $k = \infty$.

In a study of overhung turbines, Maday (1955) showed that bearing failure can occur below the ordinary critical speed. He applied the hydrodynamic theory of 180-deg offset bearings to obtain more accurate equations; and found that a particular turbine, chosen for investigation, could be operated at higher speed by reducing the clearance.

A comprehensive analysis of the full bearing with flexible shaft was published by Hori (1956). The cavitation problem was treated by assuming no film pressure in the divergent half. The bearing is taken to be of infinite length. Under these conditions the journal-center locus is nearly a semicircle. The equations of motion for small vibrations lead to a criterion for stability whereby, in our notation, $W/mc\omega^2$ must be greater than a linear function of $W/mc\omega_0^2$. Here W is the load on the journal, m the rotor mass, and c the radial clearance; ω is the angular speed of the shaft and ω_0 the critical speed on simple supports. The two constants in the expression are known functions of ϵ. The rotor will always be stable if ϵ is greater than 0.8. A stability chart is constructed showing ϵ against $W/mc\omega^2$ for constant values of $W/mc\omega_0^2$. The right and left sides of each curve are regions of stable and unstable running, respectively. Consider a horizontal full bearing of 1-mil radial clearance, loaded by rotor weight to an eccentricity of $\frac{1}{2}$, with a rotor natural frequency such that $W/mc\omega_0^2 = g/c\omega_0^2 = 5$. Then ω_0 is 278 rad/sec. From the chart, maximum safe running speed is given by $W/mc\omega^2 = 5.31$ or $\omega = 270$ rad/sec, from which $N = 2570$ rpm. The chart shows that in every case, the larger eccentricities are more favorable for stable running. For whirl amplitudes larger than the eccentricity, the vibration will diverge or converge according as ω is greater or less than $2\omega_0$. Hori then gives an equation for the relative whirl frequency in terms of two variables K'/ω_0 and ω/ω_0, where K' is a known function of the rotor mass and geometrical factors, film viscosity, and eccentricity ratio. He shows from this equation that the whirl frequency ranges from ω_0 to $\omega/2$ as K'/ω_0 varies from zero to infinity. From these considerations Hori is able to reproduce the main characteristics of the Newkirk & Taylor observations, and to explain apparent inconsistencies reported in the literature. For example, Newkirk & Lewis (1954) found that the range of stable operation could be increased by warming the oil. Pinkus had reported that he constantly observed the opposite (1953). As seen from the stability chart, low viscosity is effective in preventing the occurrence of oil whip—a high amplitude whirl. After such whip has developed, higher viscosity is effective in reducing its amplitude. These facts help to reconcile the two investigations, and were confirmed by new experiments.

Korovchinskii (1956) offers an analytical study of journal bearing stability based on methods used by Stodola, Hagg, Cheravsky, and Burgowitz. His criteria are derived from the approximate solution of nonlinear equations. Curves are drawn for the half-bearing with rigid shaft at constant L/D's such that the upper side of each curve is a region of stability, the lower side one of instability. The dimensionless number chosen as ordinate may be St, defined as a constant times g/cN in our notation; or a new stability number, twice the product of St and the Sommerfeld number. In either case the curves are plotted against ϵ. The charts appear to show that the stability of journal motion is improved by increasing the L/D ratio. Commenting on this indication, Professor Newkirk notes that machines in the field have been stabilized by *shortening* the bearing, or by cutting a central groove completely around, dividing it into two very short bearings.

Calculations can be facilitated when static characteristics are known as functions of ϵ. Korovchinski arrives at the same conclusion reached by Hori—that the journal is stable when ϵ is greater than about 0.8, except where resonance is induced by some continuous disturbance, like unbalance. Newkirk & Lewis had found whirl-impending speeds for eccentricity ratios as high as 0.95, but often found stable operation at ϵ less than 0.7.

Tipei to Sternlicht. Earlier investigations of whirl under a steady load are reviewed by Tipei in his paper of 1956, as well as in Chapter VII of his book (VI: 1957). According to W. A. Gross (1960), the stability analysis by Tipei is the most complete thus far available on laminar films. Allowing for independent motions of the journal and the bearing, he obtained an eighth degree equation. Stability is guaranteed if all real roots are negative.

Studies by Frederiksen and by Tondl were reported in 1957. Frederiksen treats oil whip of flexible rotors in full bearings as a two-dimensional problem, neglecting end leakage. He is careful to employ a semicircular journal-center locus, but falls back on classical theory for damping, as if the clearance were completely filled with oil. Deriving the fluid forces from these assumptions, he set up the equations of motion, and obtained solutions for both vertical and horizontal rotors. Graphically, the results are shown to conform with experimental findings by Newkirk and by Hagg.

Tondl calculates whirl frequency and amplitude for the rigid shaft in half-bearings and full bearings, of infinite length, in the unstable region where ϵ is less than 0.7. Separate solutions are obtained according as the whirl orbit encloses the bearing center, or not. Performance ratios are determined as functions of ϵ. Since the whirl frequencies are approximately half running speed, the principal oscillations occur at N/N_0 around 2, where N is the running speed and N_0 the natural frequency of the rotor. Time must be allowed for thermal equilibrium before the calculated results can be accurately confirmed. In slowing down, the oscillations persist to lower speeds, then suddenly disappear. Conclusions were confirmed by observations on turbo-compressors. Tondl gives references to his earlier papers, 1954–1956, which were limited to small oscillations. The year closed with Professor Newkirk's review of journal bearing instability (1957).

Constantinescu, Downham, Halton, and Orbeck offered papers in 1958. Whirling with turbulent motion was investigated by Constantinescu. He applied Tipei's criterion for stability to the full journal bearing, and thus confirmed the finding by Smith & Fuller (VII: 1956) that the range of instability increases with Reynolds' number. Downham finds the effects on true critical speed of bearing length, clearance, viscosity, and unbalance. Amplitudes are greater for dry bearings. Halton treats the rigid shaft in an end-lubricated full bearing without cavitation. Both infinite-length and short-bearing approximations are used. Halton finds half-frequency whirl under a constant load, the orbit in general elliptical. Orbeck develops a theory for vertical rotors with rigid shaft in full bearings of finite length. Solutions are derived with the aid of a load number $1/C_n$ defined by

$$C_n = S_n \left(\frac{L}{D}\right)^{\gamma}. \tag{6}$$

Here S_n is the Sommerfeld number expressed in terms of $N - 2n$ in place of N, n the whirl frequency, and γ an empirical exponent. Note that C_n reduces to Kreisle's number (XIII: 1956) when $n = 0$ and to Ocvirk's when $n = 0$ and $\gamma = 2$. Orbeck works with C_n as a function of the eccentricity ratio. Following Kreisle, he takes $\gamma = 1.73$ as a compromise between the values for infinite and zero length. From the equations of motion of the journal, neglecting cavitation, solutions are obtained for the whirl frequency and amplitude at all speeds. Frequency ratios accord with experience, and the formula shows what to expect when the bearing dimensions are varied. Amplitude distribution agrees with the Newkirk & Taylor diagram at higher speeds only.

Reports of some interest by Rippel, Schnittger, and Sternlicht appeared in 1959. Rippel gives charts and formulas for predicting the speed, N_c in revolutions per minute, at which whirl begins, consistently with Boeker & Sternlicht (1956) and with Poritsky's criterion, provided the shaft is rigid or the rotor running below its first critical. Thus

$$N_c = 10^3 k_c \left(\frac{DNZ}{m^3 W_m} \right)^{1/2}, \tag{7}$$

where D is the journal diameter in inches, N the running speed in revolutions per minute, Z the viscosity in centipoieses, and m the clearance ratio, mils per inch. The factor W_m denotes the dead weight of the rotor in pounds per bearing. The whirl coefficient k_c is to be read from a chart against known values of the bearing characteristic A, defined as the quotient of m^2W divided by D^2ZN, where W is the load on each bearing in pounds. In a horizontal machine with only a gravity load, $W = W_m$; whereas in vertical guide bearings, $W = 0$. It will be seen that A is proportional to L/D times the Sommerfeld reciprocal. In the application of Eq. (7), after looking up k_c so that all factors are known on the right-hand side, we are to note whether N_c comes out greater or less than the contemplated running speed N. If greater, we are safe; if less, a change is required in the operation or design.

The k_c chart, Rippel's Fig. 20, gives a family of curves for constant L/D. A few values are shown in Table 2. As an example take a 2-in. diameter horizontal bearing with $L/D = 1$ and $m = 1.5$ mils/in., load $W = W_m = 200$ lb per bearing. Suppose the mean viscosity is $Z =$ 1 n or 6.9 cp. From these data, by Eq. (5), N_c is $143k_c$ times $\sqrt{N}$ and A is $16.3/N$. *Case 1:* If we want to run at 4000 rpm, we find 10^3 $A = 4.08$ and the chart or table gives $k_c = 1.77$. Substituting, $N_c = 16{,}000$ rpm or $N/N_c = 1/4$ and the bearing is safe. *Case 2:* Contemplating a speed of 24,000 rpm, it is seen that 10^3 $A = 0.68$,

TABLE 2. Whirl Constant k_c from Rippel's Chart

10^3 A	$L/D = 3/5$	$L/D = 1$	$L/D = 2$
1	0.9	1.0	–
10	4.0	2.8	3.0
100	–	16	11

$k_c = 0.90$, and $N_c = 20{,}000$ rpm; from which $N/N_c = 1.2$, and whirling is to be expected.

Schnittger eliminated the principal vibrations of a 10,000-kw double-rotor gas turbine by substituting two bearings with horizontally displaced halves. After accomplishing this, he explained the result by generalizing the Boeker & Sternlicht (1956) stability equations. He determined the eccentricity locus of the new bearing experimentally. The respective parts played by the oil film stiffness and the shaft stiffness are clearly indicated. Many different aspects of whirl theory and turbine experience are brought into focus by Schnittger's investigation.

Sternlicht calculated the film stiffness and damping properties in full bearings of finite length. Negative pressures were assigned a value of zero. The film forces needed in the equations of motion of the journal are found by an iterative numerical solution of Reynolds' equation, assuming uniform viscosity. Damping factors for radial motion, as well as the radial and tangential stiffness, are plotted against eccentricity ratios up to 0.8 for L/D's from 1/2 to 3/2.

Gross to Warner. Reports by Gross, Holmes, Jennings, Landsberg, and others appeared in 1960. Gross offers a complete analysis of unsteady films and bearing systems, including instability and whirl. He treats first the whirling of a rigid shaft, indicating how the simpler equations of motion and stability criteria can be established, then goes on to the case of shaft whipping. He treats successively small and large eccentricity, first with small, then with large oscillations. Gross describes the effect of cavitation, and the persistence of whip after reducing the shaft speed. Although low L/D and low viscosity provide stability at low speeds, the reverse is true at speeds above the first critical; and turbulence should improve stability. Holmes confined his analysis at this time to a linear stability theory.

An electric analog computer was used in a study by Jennings & Ocvirk. Both transient and steady whirl paths were determined for a rigid shaft. Film forces were derived by the short-bearing approximation with a pressure film of 180 degrees. Rotor mass is shown to have an important influence on whirl amplitude. Kestens compared stability calculations with Sommerfeld and actual pressure distributions. Korovchinskii reviewed his investigations of dynamic loading and whirl in the light of other literature, advancing both subjects mathematically.

Landzberg became interested in a steam force at right angles to the shaft displacement—a force that can feed energy into a whirling rotor, producing greater instability than oil whip alone. The solution grew out of the methods used by Poritsky, Boeker & Sternlicht, and Prohl. The analysis is limited to the beginning of instability, so that small amplitudes can be assumed. Otherwise it is extremely general in form, applicable to a turbine-generator system with multiple rotors and bearings. Landzberg derives a "characteristic equation" with complex coefficients that can be simplified in each application. For simple rotors with concentrated mass, supported in symmetrical bearings, the equation agrees with Poritsky's solution. Landzberg's paper may be considered a supplement to Schnittger's investigation. Warner discussed the practical aspects of turbine bearing design (1960).

Later Studies. Journal bearing instability is reviewed in the books by Pinkus & Sternlicht (II: 1961) and Gross (IX: 1962). Poznyak (1961) studied partial bearings, offering improvements over Hagg & Sankey (1956, 1958). Lund & Sternlicht investigated attenuation (1962). See also Parzewski & Cameron, Poletskii, and Reddi & Trumpler (all 1962).

High-speed bearings were studied by Cheng & Trumpler (1963). Holmes (1963) improved his stability theory by determining the modifications of whirl frequency and amplitude brought about by the nonlinear elements. Vertical flexible rotors were studied by Martin; see paper by Sternlicht, and one on partial bearings by Warner, all in 1963. Turbine journal bearings were further investigated by D. M. Smith (1964).

Huggins retained the nonlinear terms leading to complex equations, and obtained analog computer solutions (1964) for the Ocvirk short-bearing approximation. It was found that self-sustained half-frequency oscillations can exist, and that subharmonics are likely to result from an exciting force. See also Lund (1964) on coefficients for the tilting-pad journal bearing. The stability of a shaft free from unbalance was studied mathematically by Someya (1964), who considered especially the influence of L/D. The solutions seemed to be in good accord with previous experimental results. Hydrodynamic grooved bearings were investigated by Hirs and the calculations checked experimentally (1965).

A comprehensive review of rotor behavior is given by Sternlicht in the Houston volume (1965). He shows how fluid film forces determine which of three conditions will prevail:

1. A "stable condition" in which the journal axis approaches a fixed equilibrium position.
2. A "stable whirl," the axis describing a fixed orbit, as in synchronous whirl.
3. An "unstable whirl," for example half-frequency whirl.

The distinction is emphasized between "translational whirl," in which the journal axis remains parallel to the bearing axis, and "conical whirl," in which the journal axis wobbles about a fixed point. Critical speeds, synchronous and half-frequency whirl, and resonant whip are discussed at length.

Rotor stability is further discussed in the design text by Professor Trumpler, pages 212–32, 250–4 (1966).

8. SUMMARY

The chapter starts with a brief account of nonfluid vibrations. The literature on oil film whirl experiments is then reviewed chronologically, and followed by a similar review of analytical studies.

The "Newkirk effect," vibrations associated with rotor shaft flexibility, disappears when the oil is drained out of the bearings. The "Stodola effect," associated with a rigid shaft, depends on spring stiffness of the oil film. These two effects, both discovered in 1925, are seen to be of practical importance as well as historical interest.

Procedures for minimizing whirl by favorable operating conditions and improved bearing design are described. There is need for future research in this field, but a more acute need for better coordination of experimental results, possibly with the aid of dimensional analysis.

REFERENCES

Biezeno, C. B. & Grammel, R. (1953), *Technische Dynamik,* Springer, Berlin; 2nd ed. (a) Transl., Blackie, London, see vol. 3; 180–260 (1954).

Boeker, G. F. & Sternlicht, B. (1956), "Investigation of Translatory Fluid Whirl in Vertical Machines." *T* **78,** 13–9.

Bowman, R. M., Collingwood, L. C., & Midgley, J. W. (1963), "Factors Affecting Whirl Instability of a Journal Bearing," *LWC,* Paper 2, 18–26, 147–9, 367–8. (a) 171–9, 305 (1964).

Cameron, A. (1955), "Oil Whirl in Bearings. Theoretical Deduction of a Further Criterion." *Eng.* **179,** 237–9.

Cameron, A. & Solomon, P. J. B. (1957), "Vibrations in Journal Bearings: Preliminary Observations." Paper 103, *CLW,* 191–7; 183, 756–7, 833–4.

Cheng, H. S. & Trumpler, P. R. (1963), "Stability of the High-Speed Journal Bearing under Steady Load." *JEI, T* **85,** 274–80.

Church, A. H. & Plunkett, R. (1961), "Balancing Flexible Rotors." *JEI, T* **83,** 383–9.

Cole, J. A. (1957), "Film Extent and Whirl in Complete Journal Bearings." Paper 59, *CLW,* 186–90; 183, 756, 834.

Constantinescu, V. N. (1958), Journal Bearing Stability in the Turbulent Regime (Rumanian). *Studii si Cercetari de Mecanica Aplicata* **9,** 007–1012. (a) *Rev. Mécanique Appl., Acad. RPR* **4,** 73–96 (1959).

Cooper, S. (1963), "Preliminary Investigation of Oil Films for the Control of Vibration." Paper 28, *LWC;* 305–15, 355–6, 378–9.

Dayton, R. W. & Simons, E. M. (1951). *Discrepancies Between Theoretical and Observed Behavior of Cyclically Loaded Bearings.* NACA TN 2545; 27 pp.

Demarest, K. E. & McKannan, E. C. (1963), "Vibration of Journal Bearings in Vacuum." *LE* **19,** 59–67.

Den Hartog, J. P. (1940), *Mechanical Vibrations.* McGraw-Hill, New York, 2nd ed., 324–43. (a) 4th ed., 295–9, 321 (1956).

Den Hartog, J. P. (1958), "Vibration: A Survey of Industrial Applications." *Proc. IME* **172,** 8–27. (a) *JASNE* **70,** 683–704.

Downham, E. (1957), "Theory of Shaft Whirling." *The Engineer* **204,** 518–22, 552–5, 588–91, 624–8, 660–5.

Dunkerley, S. (1894), "On the Whirling and Vibration of Shafts." *Phil. T. Roy. Soc. (London) A* **185,** 279–360.

Ehrich, F. F. (1964), "Shaft Whirl Induced by Rotor Internal Damping." *JAM* **31,**·*T* **86,** 279–82.

Frederiksen, Eyvind (1957), "Notes on Oil Whirl of Flexible Rotors Supported by Cylindrical Journal Bearings." *Ingeniøren, Internat. Ed.* **1,** 29–36.

Fujii, S. (1951), "The Roles of Resistances in the Vibration of a Shaft." *Proc. Japan Nat. Cong. AM* **1,** 599–603.

Grobel, P. (1953), "Balancing Turbine-Generator Rotors." *G. E. Rev.* **56,** 22–5.

Gross, W. A. (1960), *Film Lubrication VIII. Unsteady Bearing Films and Bearing Systems.* IBM Report RJ 117–8; 89 pp.

Hagg, A. C. (1946), "The Influence of Oil-Film Journal Bearings on the Stability of Rotating Machines." *JAM* **13,** *T* **68,** A-211-20; **14, 69,** A-77–8 (1947).

Hagg, A. C. (1948), "Some Vibration Aspects of Lubrication." *LE* **4,** No. 4, 166–9.

Hagg, A. C. & Sankey, G. O. (1956), "Some Dynamic Properties of Oil Film Journal Bearings with Reference to the Unbalance Vibration of Rotors." *JAM* **23,** *T* **78,** 302–6; **25,** *T* **80,** 141–43 (1958).

Hagg, A. C. & Warner, P. C. (1953), "Oil-Whirl of Flexible Rotors." *T* **75,** 1339–44.

Halton, J. H. (1958), "Elliptical Whirl of Flooded Journal Bearings." *Proc. Camb. Phil. Soc.* **54,** 119–27.

Hersey, M. D. (1917), "Note on the Vibration Frequencies of Elastic Systems." *JWAS* **7,** 437–45.

Hirs, G. G. (1965), "The Load Capacity and Stability Characteristics of Hydrodynamic Grooved Journal Bearings." *TASLE* **8,** 296–305.

Holmes, R. (1960), "The Vibration of a Rigid Shaft in Short Sleeve Bearings." *JME Sci.* **2,** 337–41.

Holmes, R. (1963), "Oil Whirl Characteristics of a Rigid Rotor in 360° Journal Bearings." *Proc. IME* **177,** 291–302.

Hori, Yukio (1956), "A Theory of Oil Whip," *Fifth Japan Nat. Cong. AM,* 1955; *Proc.,* 395–8. (a) *JAM* **26,** *T* **81,** 189–98 (1959).

Horsnell, R. & McCallion, H. (1963), "Prediction of Some Journal Bearing Characteristics under Static and Dynamic Loading." *LWC;* 126–38.

Howarth, H. A. S. (1937), *Radial Bearing.* U.S. Pat. 2,093,521.

Huggins, N. J. (1964), "Non-Linear Modes of Vibration of a Rigid Rotor in Short Journal Bearings." Paper 18, *LWC,* 238–45, 256–7, 306.

Hull, E. A. (1958), "Oil Whip Resonance." *T* **80,** 1490–6.

Hull, E. A. (1960), "Control of Oil Whip Frequency Ratio." *JBE, T* **82,** 894–8.

Hull, E. A. (1961), "Shaft Whirling as Influenced by Stiffness Asymmetry." *JEI, T* **83,** 219–26.

Jennings, U. D. & Ocvirk, F. C. (1962), "The Simulation of Bearing Whirl on an Electronic-Analog Computer." *JBE, T* **82,** 503–10.

Kapitza, P. (1939), Stability and Transition through Critical Revolutions of High Speed Rotors in the Presence of Friction (Russian). *Zhur. Tekh. Fiziki* **9,** No. 2. (a) Transl., *J. Physics* **1,** 7–50.

Kestens, J. (1960), "Stabilité de la position de l'arbre dans un palier à graissage hydrodynamique." *Wear* **3,** 329–57.

Khanowitch, M. G. (1947), Vibrational Stability of Bearings (Russian). *Proc. Second All-Union Conf. on Friction and Wear in Machinery,* AN, SSSR 1.

Kimball, A. L. (1924), "Internal Friction Theory of Shaft Whipping." *G.E. Rev.* **27,** 244–51. (a) **28,** 554–8 (1925).

Korovchinskii, M. V. (1953), Journal Bearing Stability (Russian). *TIM* **7,** 223–37.

Korovchinskii, M. V. (1956), "Stability of Position of Equilibrium of a Journal on an Oil Film" (Russian). *TIM* **11,** 264–323. (a) Transl., *FWM* **11,** 248–305.

Korovchinskii, M. V. (1960), Non-Steady Motion of the Journal in a Bearing (Russian). *TIM* **14,** 267–83. (a) Transl., *FWM* **14,** 243–58 (1962).

Landzberg, A. H. (1960), "Stability of a Turbine-Generator Rotor Including the Effects of Certain Types of Steam and Bearing Ecxitations." *JAM* **27,** *T* **82,** 410–6.

Lewis, J. F. & Fulton, G. B. (1956), *A Further Investigation of Disturbances Due to Oil Films in Journal Bearings.* WADC TR 56–259; 22 pp. (a) *Continued Studies of Oil Film Whirl,* 57–737 (1957); 19 pp.

Linn, F. C. & Prohl, M. A. (1951), "The Effect of Flexibility of Support upon the Critical Speeds of High Speed Rotors." *TSNAME* **59,** 536–53.

Lund, J. W. (1964), "Spring and Damping Coefficients for a Tilting Pad Journal Bearing." *TASLE* **7,** 342–52.

Lund, J. W. & Sternlicht, B. (1962), "Rotor Bearing Dynamics with Emphasis on Attenuation." *JBE, T* **84, 491**–502.

Maday, C. J. (1955), "The Effect of Bearings on the Dynamics of Overhung Turbines." *ASME* 55-SA-16; 17 pp. (a) *ME* **77,** 707.

Manabe, Daikaku (1949), Transient Behavior of Bearings (Japanese). *TJSME* **15,** No. 50, 5–11.

Martin, F. A. (1963), "Steady State Whirl in Journal Bearings for a Vertical Flexible Rotor System." *ASME* 63-LUBS-8; 8 pp. (a) *ME* **85,** 73.

Morrison, D. (1962), "Influence of Plain Journal Bearings on the Whirling Action of an Elastic Rotor." *Proc. IME* **176,** 542–53.

Newkirk, B. L. (1924), "Shaft Whipping." *G. E. Rev.* **27,** 169–78.

Newkirk, B. L. (1931), "Whirling Balanced Shafts." *Third ICAM,* Stockholm (1930); *Proc.* **3,** 105–10.

Newkirk, B. L. (1937), "Instability of Oil Films and More Stable Bearings." *GDLL* **1,** 223–6.

Newkirk, B. L. (1956), "Varieties of Shaft Disturbances due to Oil Films in Journal Bearings." *T* **78,** 985–8.

Newkirk, B. L. (1957), "Review Paper." *CLW,* 179–85.

Newkirk, B. L. & Grobel, L. P. (1934), "Oil Film Whirl—A Non-Whirling Bearing." *T* **56,** 607–15. (a) Discussion by M. Stone, 614–5. (b) *GDLL* **1,** 223–6 (1937).

Newkirk, B. L. & Lewis, J. F. (1956), "An Investigation of Disturbances Due to Oil Films in Journal Bearings." *T* **78,** 21–7.

Newkirk, B. L. & Taylor, H. D. (1925), "Shaft Whipping Due to Oil Action in Journal Bearings." *G. E. Rev.* **28,** 559–68.

Ørbeck, Finn (1958), "Theory of Oil Whip for Vertical Rotors Supported by Plain Journal Bearings." *T* **80,** 1497–502.

Parzewski, Z. & Cameron, A. (1962), "Oil Whirl of Flexible Rotors." *Proc. IME* **176,** 523–97.

Pestel, E. (1954), "Beitrag zur Ermittlung der hydrodynamische Dämpfungs- und Federeigenschaften von Gleitlagern." *Ing. Archiv* **22,** 147–55.

Pinkus, O. (1953), "Note on Oil Whip." *JAM* **20,** *T* **75,** 450–1.

Pinkus, O. (1956), "Experimental Investigation of Resonant Whip." *T* **78,** 975–83.

Poletskii, A. T. (1962), Stability of Motion of the Journal on a Lubricating Film (Russian). *TIM* **17,** 165–79. (a) Transl., *FWM* **17,** 156–69 (1965).

Poritsky, H. (1953), "Contribution to the Theory of Oil Whip." *T* **75,** 1153–61. Erratum: Linear term in Eq. 4.3 should be $2\lambda k/M$.

Poznyak, E. L. (1961), The Dynamic Properties of the Oil Film in Lubricated Bearings (Russian). *Izvestiia AN SSSR OTN Mekhanika i Mashinostroenie* No. 6, Moscow, 52–67.

Rankine, W. M. J. (1869), "On the Centrifugal Force of Rotating Shafts." *The Engineer* **27,** 249.

Reddi, M. M. & Trumpler, P. R. (1962), "Stability of the High-Speed Journal Bearing under Steady Load, 1. The Incompressible Film." *JEI, T* **84,** 351–8.

Rippel, H. C. (1959), "Lightly Loaded, High-Speed Bearings." Pages 26–8 in

Cast Bronze Bearing Design Manual, Cast Bronze Bearing Institute, Inc., Evanston, Illinois.

Robertson, David (1933), "Whirling of a Journal in a Sleeve Bearing." *Phil. Mag.* (7) **15,** 113–30.

Routh, E. J. (1877), *A Treatise on the Stability of a Given State of Motion; Particularly Steady Motion.* Macmillan, London; 108 pp.

Schnittger, J. R. (1959), "Development of a Smooth Running Double-Spool Gas-Turbine Rotor Systems." *JEP, T* **81,** 151–60.

Shawki, G. S. A. (1955), "Whirling of a Sleeve Bearing. Experiments under No-Load Conditions." *Eng.* **179,** 243–6.

Sines, George (1954), "The Dynamics and Lubrication of a Miniature Turbine Rotor on Journal Bearings." *T* **76,** 319–26.

Smith, D. M. (1964), "Dynamic Characteristics of Turbine Journal Bearings." Paper 8, *LWC,* 1963; 72–86, 150, 153, 363–6, 369–70.

Someya, Tsuneo (1964), "Stability of a Balanced Shaft Running in Cylindrical Journal Bearings." Paper 21, *LWC,* 196–214, 254–7, 306.

Sternlicht, B. (1955), "Experimental Verification of Theoretical Investigations into Half-Frequency Whirl." *ASME* 55-LUB-20; 15 pp. (a) *ME* **77,** 1099.

Sternlicht, B. (1959), "Elastic and Damping Properties of Cylindrical Journal Bearings." *JBE, T* **81,** 101–8; **82,** 249–50 (1960).

Sternlicht, B. (1963), "Stability and Dynamics of Rotors Supported on Fluid Film Bearings." *JEP, T* **85,** 331–42.

Sternlicht, B. (1965), "Influence of Bearings on Rotor Behavior." Pages 529–699 of *HS.*

Stodola, Aurel (1925), "Kritische Wellenstörung infolge der Nachgiebigkeit des Ölpolsters im Lager." *Schweizerische Bauzeitung* **85,** 265–6. (a) *Second ICAM,* Zurich; *Proc.* 201–6 (1926).

Stone, M. (1935), "Film Lubrication of Sleeve Bearings." *JAM* **2,** *T* **57,** A 59–64; **3, 58,** A 31–4 (1936).

Tipei, N. (1956), "La stabilité du mouvement dans les paliers à charge dynamique." *Rev. Mécanique Appl., Acad RPR* **1,** 115–122.

Tondl, Alês (1957), "The Motion of a Journal in a Bearing in the Unstable Region of Equilibrium Position of the Centre of the Journal." *Ninth ICAM,* Brussels, 1956; *Actes* **5,** 95–107.

Tondl, Alês (1960), Experimental Results on Bearing Stability (Rumanian). *Studii si Cercetari de Mecanica Aplicata, Acad. RPR* **11,** 603–26. (a) *Acta Technica* (Prague) **5,** No. 3, 213–45 (German).

Trumpler, P. R. (1966), *Design of Film Bearings.* Macmillan, New York; 258 pp.

Warner, P. C. (1960), "Turbine-Bearing Design." *ME* **82,** 82–5.

Warner, P. C. (1963), "Static and Dynamic Properties of Partial Journal Bearings." *JBE, T* **85,** 247–57.

Warner, P. C. & Thoman, R. J. (1964), "The Effect of the 150-Degree Partial Bearing on Rotor-Unbalance Vibration." *JBE, T* **86,** 337–47.

Wilcock, D. F. (1961), "Orthogonally Displaced Bearings—I." *TASLE* **4,** 117–23.

Chapter XIII Thin-film Research

1. Thick-film and thin-film lubrication. 2. Thin-film hydrodynamic lubrication. 3. Boundary lubrication. 4. Mixed friction. 5. Oiliness. 6. Static friction and stick-slip. 7. Lubrication in metal working. 8. Accuracy and surface finish. 9. Summary.

Rolling and sliding motions under concentrated loads at low values of ZN/P, or with insufficient oil, are accompanied by thinner films than are met in the normal operation of thrust and journal bearings. The need for special research to determine the characteristics of thin films was recognized by Albert Kingsbury, Sir William Hardy, and other pioneer investigators.

1. THICK-FILM AND THIN-FILM LUBRICATION

A distinction is commonly made between thick-film and thin-film lubrication. This terminology was used by the writer in the *American Machinist* (1929). It was stated that when conditions permit the formation of a sufficiently thick film, "we have the simplest and most desirable type of lubrication"; but "when the film thickness falls below certain limits, the action of the lubricant becomes more complex and must be dealt with under the head of thin-film lubrication." It was noted that thick-film lubrication depends on no property of the lubricant except its bulk viscosity; whereas thin-film lubrication depends on surface roughness and, in general, on some unknown factor, known as the property of *oiliness*.

In keeping with later usage, thin-film lubrication may be pictured without mentioning "oiliness" yet retaining the idea that the action of the lubricant becomes more complex the thinner the film. A new kind of complexity due to surface roughness has been introduced, with or without departure from bulk properties of the lubricant. Two kinds of thin-film lubrication are defined below: *thin-film hydrodynamic*, and *boundary* lubrication.

Although no numerical limits can be fixed, it is unlikely that thin-film effects of any kind would be noticeable at mean thicknesses much greater than $\frac{1}{10}$ mil (100 micro-inches). It is an open question how thick *boundary* films can be; opinions ranging from 25 micro-inches down to molecular magnitudes—see Dowson (X: 1965).

The term *fluid film lubrication* is often used to indicate both thick-film and thin-film hydrodynamic lubrication, whereas *imperfect lubrication* implies surface damage.

The Friction Minimum Point. When the coefficient of friction of a thrust or journal bearing is plotted against speed, load, viscosity, or temperature, it passes through a minimum point. This fact was known to Thurston nearly a hundred years ago, and more thoroughly demonstrated by Stribeck (V: 1902). It was further discussed by Gümbel (V: 1914), Leloup (1947), and Vogelpohl (1954). The high-speed branch of the curve represents hydrodynamic or thick-film lubrication. The low-speed branch represents varying degrees of thin-film lubrication. A transition occurs in the vicinity of the minimum point. When the coefficient f is plotted against ZN/P, the curves coalesce at the right of the minimum but diverge at the left (Fig. 1). If the experiment is made by varying the speed, while holding the load and viscosity constant, different curves would result from different values of the load or viscosity at low ZN/P, as noted in Chapter V. The condition represented by the family of curves at the left of the minimum is known as mixed friction or "halb-flüssige Reibung." As ZN/P approaches zero, the condition approximates one of boundary lubrication of "halb-trockene Reibung."

The friction minimum point diagram conveys a vivid picture of thin-film lubrication. Investigators have tried to represent the curves mathematically. Gümbel's method is outlined under "Mixed Friction" in this chapter. McKee and co-author (1932) found that their coefficient of journal friction could be plotted as a function of ZN/P^n, where $n = \frac{1}{2}$. It will be recalled from Chapter V that $n = 1$ for rigid surfaces and uniform viscosity.

Investigating different types of friction element, Bridgeman, McKee, & Bitner (1932) found $n = 0$, Bodart (1933) $n = \frac{1}{2}$. Leloup (III: 1954) concluded that for journal bearings $n = \frac{1}{4}$. Vogelpohl found ZN/P at the minimum point almost inversely proportional to the

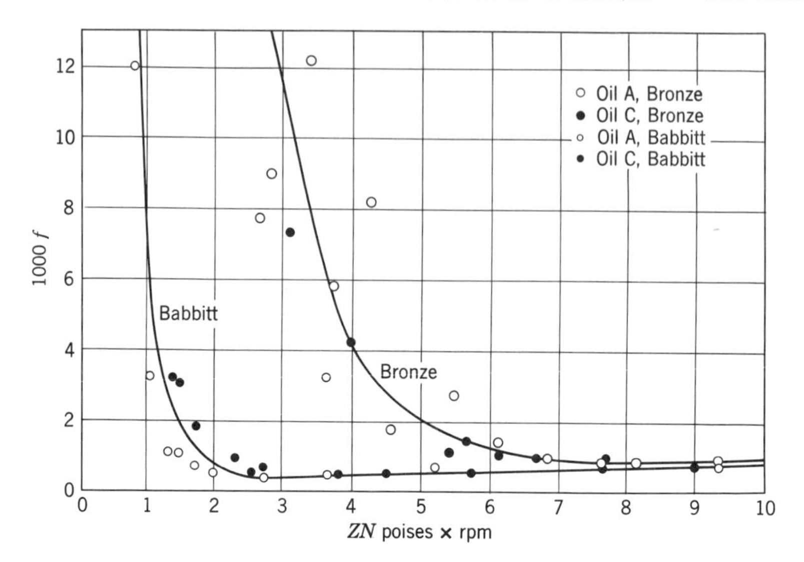

Fig. 1. Thin-film friction of mineral oils on different metals in a journal bearing under constant load (viscosity of Oil A, 44 cp and of B, 246 cp; P = 192 psi). From McKee & McKee, *J. SAE* 1932.

bearing volume, that is, to D^2L (see pages 113–5 in his book, VI: 1958). Givens and Talley fitted empirical equations to the left-hand branch (1962, 1962). Blok explained $n = 0$ by letting P enter two parameters inversely, thus canceling out (pages 194–8 of X: 1962). A similar explanation could apply to other cases where n differs from unity.

Further Experiments. Low speed tests on vertical thrust bearings by Howarth (VI: 1919) as well as tests by Tenot (VI: 1937) provide examples of the friction minimum point for flat surfaces commercially finished. The effect of roughness on the minimum point in journal bearings was shown by Burwell and co-authors (1941). The critical ZN/P shifts toward lower values with increasing smoothness, in accord with McKee's experiments on running-in. Roughness data were included in reports of thrust bearing tests by Lafoon and co-authors (VI: 1947) and by Salama (1950). The latter found that for a fixed ZN/P, the friction was least when the roughness wavelength is 900 times its amplitude.

The effects of roughness and clearance have been investigated by Kreisle (1956, 1957) and by Ocvirk & DuBois (1959). See also the papers by Iwaki & Mori (1958); by Kragel'skii & Demken, Kudinov, and Peeken (all 1960); by Denny (1961), Davies (1963), and Martin (1964). Closely related experiments showing the effect of roughness on rolling contact are reviewed in Chapter X; see, for example, Tallian and co-authors (X: 1964) as well as O'Donoghue & Cameron (X: 1965).

2. THIN-FILM HYDRODYNAMIC LUBRICATION

A concept of thin-film or microhydrodynamic lubrication influenced by roughness, elastic deformation, and nonuniform viscosity was introduced by the writer and expressed in terms of dimensionless variables (V: 1933). It was postulated that for any given surface profile, the friction coefficient and mean film thickness would depend on the pressure and temperature coefficients, as well as on the viscosity at a chosen reference point; also on the thermal conductivity of the lubricant and its heat capacity per unit volume; elastic constants of the solids; as well as on the load, the speed, and the absolute size of the bearing, or other element. Accordingly, as shown in Chapter V, the appropriate dimensionless parameters could be set up, thereby reducing the number of independent variables.

Similar ideas were soon proposed by Vogelpohl, Heidebroek, and Reissner (1936, 1937, 1941). Vogelpohl and the writer both found a significant new factor in the ratio of the temperature coefficient of viscosity, a, to the heat capacity per unit volume, q. The influence of this factor was confirmed by Voitländer's experiments (1929) in which fatty oils gave lower friction than mineral oils, and sugar-and-water solutions the lowest friction of all, when compared at the same value of ZN/P. Moreover, the dimensionless form aP/q taken as a parameter, or alternatively aE/q, helps to explain the branching out of the f, ZN/P curve at the left of the minimum. The new point of view was clarified and usefully applied by Blok in several publications (1951, 1957, X: 1962).

Rugulose Lubrication. Heidebroek's thought was carried out mathematically by A. G. M. Michell, who used the term "rugulose lubrication" to describe thin-film hydrodynamic lubrication (pages 104–8, 116–7, and 281–91 of V: 1950). Michell showed that when a tapered-land is corrugated, or grooved, in the direction of motion, there will be a minimum point on the f, ZN/P curve. This will be true even if the surfaces are rigid and without solid contact, and the film remains of a uniform viscosity. The minimum friction occurs when the least film thickness is approximately equal to the height or amplitude of the corrugations above their mean level. Although Michell's geometry was highly simplified, the amazing fact became evident that the mere roughening of surfaces will explain the occurrence of minimum friction as a purely hydrodynamic effect.

Reissner's problem dealt with corrugations transverse to the direction of motion, but was not carried far enough to reveal a minimum coefficient of friction. Such a minimum can be seen from calculations based on Ocvirk's short-bearing theory. Let the corrugations be of height ρ in a rigid bearing of mean radial clearance c. Grooving may be axial, circumferential, criss-cross, or otherwise if the top lands are of uniform height and occupy say half the bearing area. The journal surface is left smooth. Calculate eccentricity ratio ϵ as a function of the Sommerfeld number S disregarding roughness. Friction per unit area is ZU/h, where U is the surface speed and h the local film thickness in terms of circumferential position θ. Over one half of the bearing area, $h = H - \rho$; over the other half, $H + \rho$ where H denotes the mean film thickness.

Contact occurs with theoretically infinite friction at a critical $S = S_c$ such that $\epsilon = 1 - \rho/c$. The minimum coefficient f will be at an S value somewhat greater than S_c. For example, with $L/D = \frac{1}{2}$ and $\rho/c = \frac{1}{10}$, the friction factor fr/c has a minimum of about 1.5, which occurs near $S = 0.025$. The solution is independent of C/D and the size of the bearing. A more accurate solution would require eccentricity calculations based on the complete profile rather than on the mean bearing surface.

Further Mathematical Study. Professor Rightmire reported on the probable behavior of contacts (1957) and had other studies under way. See the publications by Burton, Davies, and Tipei (all 1963), as well as by Hamilton, Walowit, & Allen (1966). A note on hydrodynamic sliding friction is included by O'Donoghue & Cameron (X: 1965).

Burton studied the effects of pressure-viscosity on load capacity of surfaces with sinusoidal, deformable roughness. Davies treated the sawtooth profile. He cites confirmatory friction observations by Summers-Smith of face seals. Hamilton and co-authors calculated the load capacity of a rigid surface with flat asperities like those visualized in our discussion of the friction minimum point. Conclusions were in good accord with experiments on seals.

3. BOUNDARY LUBRICATION

Sir William Hardy, pioneer investigator of boundary lubrication, took his cue from Osborne Reynolds, who had spoken of "unknown boundary conditions" at the solid-liquid interface. Petroff and contemporaries assumed that the oil film could actually slip on the metal, and therefore put coefficients of external friction into the equations, along with viscosity. Slip is now ruled out by modern evidence, with but few apparent exceptions—see, for example, Goldsmith & Mason (1962). Perfect adhesion is taken for granted in the action of liquid lubricants.

Definition. Hardy & Doubleday (1922), in a report to the DSIR Lubrication Committee, after mentioning the kind of lubrication in which the solid surfaces are completely floated apart, state that

> There is, however, another kind of lubrication in which the solid faces are near enough together to influence directly the physical properties of the lubricant—what Osborne Reynolds called "boundary conditions" then operate, and the friction depends not only on the lubricant, but also on the chemical nature of the solid boundaries. Boundary lubrication differs so greatly from complete lubrication as to suggest that there is a discontinuity between the two states. In the former, the surfaces have the property of static friction . . . .

Boundary lubrication may then be defined as a state of lubrication in which the film is so thin that its properties are no longer the same as in bulk.

In practice it is difficult to achieve boundary lubrication, thus defined, without other interactions such as substrate contact or hydrodynamic effects. And it is commonly believed that boundary films are too thin, and too firmly adsorbed to have ordinary rheological properties such as are visualized in the homogeneous, isotropic materials of continuum mechanics. Ragnar Holm says (pages 198–9 of 1946), "Experience indicates that the last layer adheres strongly to the metal and prevents metallic contact . . . behaves as a rigid continuation of the solid body." According to Michell (page 115 of V: 1950) such films serve only to modify or "mitigate" the frictional properties of the metal.

Experimental Facts. Prompted by scientific curiosity, Albert Kingsbury, P. W. Bridgman, and others wished to learn the facts. Kingsbury's first trial indicated no systematic change in viscosity down to film thicknesses of $\frac{1}{40}$ of a mil (25 micro-inches). This was as far as he could go with a tapered-plug viscometer (1919). Professor Bridgman had suggested to the writer in 1915 that an investigation might well be made to "determine the limiting thickness below which the properties of the film would begin to differ from its bulk properties." That was essentially Kingsbury's problem. Bridgman's suggestion was passed along to Ronald Bulkley of the National Bureau of Standards. Bulkley took it up for his doctoral thesis at The Johns Hopkins University (1931). Poiseuille's law was verified both in glass and platinum capillary tubes of extraordinarily fine bore. The results indicated that there could not have been any plastic boundary film more than 1 micro-inch thick adhering to the walls of the capillary. Confirmation was soon obtained from Bastow & Bowden's experiments (1935) on the flow of liquids through fine clearances between flat plates. These results, skillful as they were, did not satisfy everyone because they were obtained only for stationary surfaces, instead of surfaces in relative motion.

Accordingly, Kingsbury initiated a new series of experiments, finally published by S. J. Needs (II: 1940). Part I, using an ultramicrometer of new design, was confined to the measurement of film thickness between flat disks that were stationary except for the motion of approach. After following Stefan's law for a considerable time, the rate of approach usually slowed down. Part II added a torque measurement accompanied by slow rotation. Both mineral and vegetable oils were compared, little difference being found apart from the influence of viscosity. In general the film was reduced to some limiting thickness under various loads up to 800 psi. The departure from Stefan's law was gradual, not indicating contact.

Thus a parallel-surface thrust bearing had been constructed, without grooves, which would support such loads indefinitely at an extremely low speed. There was no sign of scratching except by the lowest viscosity mineral oil; attributed to hard particles in the oil. Electrical resistance decreased and seemed to vanish before the minimum thickness was reached.

Film thicknesses were initially estimated by calculation from the measured torque, knowing the bulk viscosity. They were later checked by an optical interference method, utilizing a pin that projected up from a hole at the center of the lower disk. Measured thicknesses were higher than the calculated values. The two disks were cut from tool steel $\frac{7}{8}$ in. in diameter. They were finished to an unprecedented degree of flatness and smoothness by a guard-ring technique described in the reference.

Oil samples were drawn by pipette from the center of the oil drum. They were quickly transferred to the clean disk surface to avoid contamination. The oils were not filtered because (1) it was feared that such advanced filtering methods as used by Bulkley might remove the ingredients responsible for the boundary film observed; and (2) in practice new oils must be used as obtained from commercial drums. It was desired that the results of the investigation be applicable to engineering practice. Ultracentrifuging made no difference in the results.

Continuing his experiments, Needs found that the boundary film exhibited the familiar properties of plastic and thixotropic lubricants in rotational shear while refusing to flow radially outward under load as if endowed with a property that he called "directional rigidity." The same characteristic has been seen in thick-film grease-lubricated bearings. Thixotropic recovery after periods of rest is shown by Needs' Fig. 8. Variation of torque with speed is seen from the graphs in Fig. 14, indicating effective viscosities in the boundary film nearly five times the bulk value.

Confirmation and extension of Needs' experiments offer a challenge for future research. Somewhat similar investigations have been conducted by Beeck and co-authors (1940), Larsen & Perry (1945), Marcelin (1947), von Schiessl (1950), Bikerman (1956), Fuchs (1957, 1958), and Derjaguin (1957, 1963) with varying results. See also Henniker's review (1949).

A different approach was taken by A. I. Bailey in her experiments with surfaces of molecular smoothness formed by cleavage of mica, as described by Bowden & Tabor (I: 1964, Part II). In one such experiment the two surfaces were coated by Langmuir's method with monomolecular layers of stearic and other fatty acids. Upon separation, the respective sheets curled into a cylindrical form such that friction could be measured by the crossed-cylinder method. The behavior, under mechanical stress, of adsorbed layers on tungsten wires has been investigated by Professor Rightmire. The results, intended for early publication, should throw further light on boundary lubrication.

Theory. Hardy's theory of boundary lubrication, amplified by Woog, assumed that the adsorbed molecules would orient themselves with their polar groups adhering to the metal. The contact between two boundary lubricated surfaces would take place between the nonpolar groups whereby slip would be permitted, friction reduced, and the metal protected. The theory received very interesting experimental support but is now considered only a limiting case. It has been superseded in some respects by an adhesion theory described by Bowden & Tabor (I: 1950), based on dry friction studies by Holm (1938) and Merchant (1940). According to this view boundary friction involves three terms: (1) the ploughing effect of asperities; (2) the force needed for shearing off the metallic junctions formed by welding of high spots where the lubricant has broken down; and (3) the force necessary to overcome the "shear strength" of the lubricant where it has not yet broken down. A good lubricant would reduce the total area of the junctions, or the force needed to shear the lubricant, or both.

Close examination reveals that the theory applies better to static than to kinetic friction; hence the reference to "shear strength." This factor is discussed by Bowden & Tabor on pages 395–6 of Part II (1964). Adhesion theory makes use of the fact that the true contact area A is only a small fraction of the nominal or apparent area. See also D'yachenko and co-authors (1964). It derives its name from adhesion of the high spots, or "asperities," where the lubricant has been squeezed out. It is assumed that the asperities are flattened by plastic flow until the yield flow pressure p can balance the load per unit area W/A. Then $A = W/p$. Let s denote the mean shear strength of the junctions. Then when lifting and ploughing are neglected, the force required to overcome friction will be $F = sA = W(s/p)$. Since s/p is assumed constant, F will be proportional to W in agreement with the Amontons-Coulomb law. The coefficient of friction F/W will equal s/p and is independent of area. The same reasoning can be extended to the more general case where only a fraction α of the contact area is occupied by welded junctions, the remaining fraction $1 - \alpha$ representing the lubricated area. If the shear strength of the lubricant, or better its tangential resistance per unit area be denoted by s_0, we find for the coefficient of friction

$$f = \alpha s/p + (1 - \alpha)s_0/p \, . \tag{1}$$

Model friction junctions have been studied by Greenwood & Tabor (1957). Numerous modifications have been introduced to render the theory more flexible and realistic. For example, Tabor (1959) noted that the theory of plasticity requires an increase of contact area when tangential forces are applied, owing to the action of combined stresses. A mathematical correction was derived. Following up this lead, he found why the slightest contamination of the junctions would reduce the coefficient of friction from enormously high values for clean metals to values of the order of unity or less such as are frequently observed.

Though Eq. (1) is often cited as the basic formula for boundary lubrication, in our view that would be true only when $\alpha = 0$. Surely a welded junction (no lubricant or contamination at the interface) is the antithesis of lubrication.

Alternative considerations leaning more toward the Hardy-Woog concept were proposed by Cameron (1959). See also Derjaguin (1958) and Akhmatov (1963). A modified Eq. (1) was proposed by Tamai &

Rightmire, with experimental support (1965), to allow for lubricant friction in the peripheral zone of each asperity.

Dry Surfaces. It is customary to treat the dry friction of everyday experience as a case of boundary lubrication resulting from contamination by the atmosphere or by handling (Campbell, 1939, 1948; Reichenbach, 1964; Spurr, 1965). Reviews of dry friction and boundary lubrication will be found in Chapter 11 of Shaw & Macks (1949), Chapters 10 and 11 of Professor Fuller's book (I: 1956), Chapter 8 of Kragel'skii (I: 1962), and in Bisson's Chapter 2 (1964). Solid films are discussed by Braithwaite (IV: 1964) and by Rabinowicz (IV: 1965).

4. MIXED FRICTION

The minimum point and left-hand branches of the friction coefficient diagram, f against ZN/P, result from a combination of boundary and thin-film hydrodynamic lubrication. The intercept on the f-axis at $ZN/P = 0$ represents static boundary friction.

Gümbel's Explanation. If thick-film hydrodynamic lubrication prevailed exclusively, the diagram for a given bearing would be a single curve, convex upward for a partial bearing except perhaps near the origin. Gümbel showed that if a part of the load is supported by solid contact (boundary or "semidry" lubrication) and the remainder hydrodynamically, the graph should start out from the static friction intercept, sloping downward in the form of a straight line (V: 1914, VI: 1925). The reason for the downward slope is that with increasing film thickness, a progressively smaller fraction of the surface incurs the high friction of boundary contact. Eventually the down-sloping line intersects the up-coming curve. Minimum friction occurs at this point, which Gümbel and his followers called the "unlatching point." Ordinates may be added to obtain a more precise resultant curve. Gümbel applied the calculation to the 180-deg or "half" bearing with reasonable results. Owing to the simultaneous action of boundary and hydrodynamic films, the condition at the left of the minimum is commonly called "mixed friction" or "mixed film lubrication."

Thus we have two explanations for the occurrence of a minimum point: Michell's theory of "rugulose" or thin-film hydrodynamic lubrication, and Gümbel's combination of boundary with hydrodynamic lubrication. Since both theories are physically correct, a more realistic calculation might be worth undertaking in which both are applied.

Confirmations. The existence of fluid pressures under mixed film conditions has been confirmed experimentally by Altrogge (1950) and by Barwell & Milne (1952, 1956). Mixed film lubrication has been discussed by Professor Fuller (1954), with special reference to Altrogge's demonstration; and by Bondi (1954). See also Bikerman (1948); Furey (1961); as well as Archard, Dobry, Rush, and coauthors (all in 1964).

Terminology. Boundary lubrication, as noted above, is an idealized concept. It can be closely approached, but seldom realized in practice. Some like to restrict the definition to lubrication by monomolecular films on each surface, that is, by a bimolecular film separating the solids. This, again, is idealistic owing to the roughness of most surfaces. Others tell us that hydrodynamic lubrication is understood and that everything else is "boundary." We have even seen an engineer go to the blackboard and draw a minimum point curve, saying that "all lubrication at the right is hydrodynamic; all to the left, boundary." Thus a habit seems to be creeping in whereby the term "boundary lubrication" is taken as a synonym for mixed film lubrication. The term is loosely applied to the whole bearing surface even though only a small part of it is undergoing boundary lubrication. If this habit prevails, it may be necessary to identify the historic concept as "pure" boundary lubrication.

Literature. A voluminous literature has sprung up on boundary lubrication and related subjects. It is a sequel to the earlier work of Hardy (Collected Scientific Works, 1936), Woog (IV: 1926), Bowden & Tabor (I: 1950), and their associates, but is greatly in need of clarification.* Chemists of every opinion are to be congratulated for their brave attack on the moot questions of boundary lubrication, including chemisorption and chemical reactions at the bearing surfaces. Yet, for any given condition of the surfaces, lubrication is a mechanical action. The present is therefore no time for the mechanical engineers and rheologists to withdraw into a cloak of silence.

Wear and Seizure. Thus far we have spoken of thin-film lubrication as comprising thin-film hydrodynamic, boundary, and mixed film lubrication without mentioning wear. The term "imperfect lubrication" has been applied to boundary lubrication accompanied by wear or other surface damage. The onset of seizure can be delayed by the use of EP (extreme pressure) lubricants. As described in earlier chapters, such lubricants enter into chemical reaction with the frictionally

* Plans have been made by the ASME Research Committee on Lubrication for a critical appraisal of world literature on boundary lubrication by Professor W. E. Campbell of the Rensselaer Polytechnic Institute, and others, aided by an advisory board. This work is reported well underway, and should provide a starting point for new experiments.

heated rubbing surfaces. The reaction products, such as iron chlorides or sulfides, then serve to prevent welding and protect the metal surfaces. See also Phillips et al. (1962); Klaus & Bieber (1964).

Hydrodynamic Contamination. It is difficult, in practice, to isolate boundary lubrication from some slight degree of surface damage, or some slight hydrodynamic action. The last mentioned effect might be termed "hydrodynamic contamination." Chemists are trained to beware of contamination by the atmosphere, or by traces of foreign matter, but are sometimes indifferent to the subtle effects of hydrodynamic contamination.

5. OILINESS

Suppose two oils A and B give different static frictions when compared under proper test conditions. Evidently some friction-reducing property (or group of properties) other than viscosity is at work, since viscosity can hardly affect static friction. To be doubly certain, however, two oils may be compared that are of the same viscosity. If A is a fatty and B a mineral oil, A will give the lower friction. The same result is obtained from a comparison of kinetic friction if ZN/P is low enough to preclude significant hydrodynamic action. The term "oiliness" came into use to describe this mysterious property. It appeared in Archbutt & Deeley's book as early as 1900, and was shortly followed by the French and German terms "onctuosité" and "Schmierfähigkeit." See, for example, Woog (1921); Wilson & Barnard (1922), and V. Vieweg (1932).

Definitions. Herschel was the first to offer a strict definition. He said, in effect, that when two oils of equal viscosity at the test temperature and at atmospheric pressure are found to give different friction, the oil giving the lower friction will be considered to have the greater oiliness (1922). Under that definition, the change in viscosity with pressure becomes a possible factor of oiliness. This, in fact, was demonstrated experimentally by Needs at a later date (VII: 1937). An alternative form was proposed requiring that the oils under comparison have the same viscosity at the temperature *and pressure* of the film. The scientific merit of the alternative definition was clear, yet after sufficient study the RCL accepted Herschel's definition. Professor A. W. Duff had been appointed advisor to the Committee. He was a physicist of high standing, whose published comparison of viscosity-temperature formulas commanded respect (IV: 1897). Professor Duff convinced the RCL members that the alternative definition would be found unworkable in practice. It demanded too much knowledge of the pressure coefficients of the test oils and, especially, of the pressure distribution in the film.

The Heat Transfer Analogy. No definition of a purely abstract kind could be wholly satisfying. An analogous situation can be visualized in the case of heat transfer. Imagine thermal conductivity unknown, so that heat transfer by forced convection would be expected to depend only on heat capacity q per unit volume, and the mean velocity of a turbulent stream, for a given temperature drop. Curves for the heat-transfer coefficient against speed with q constant might yet differ from one liquid to another. An unsuspected property must be at work! Call it k; and define it by saying that when two liquids are compared having the same q, yet give different heat-transfer coefficients, the one giving the lower coefficient shall be regarded as having the greater k value. This, too, may be better than no definition at all, but falls short of our present understanding of thermal conductivity, and our ability to measure it. Thus in the case of oiliness, it was seen that more experimental work was needed.

Oiliness Experiments. Deeley experimented with static friction. Kingsbury's study with the "new oil-testing machine" (1903) seems to be the first kinetic test on oiliness. Kingsbury's machine made use of two opposed bearings loaded against a vertical journal by means of a helical spring. For demonstrating hydrodynamic lubrication, centrally supported brass bearings of 120-deg arc had been fitted to a journal of $1\frac{3}{8}$ in. diameter. But for showing up differences in "body," or oiliness, a much smaller journal, $\frac{1}{4}$ in. in diameter, was used so as to obtain higher bearing pressures. Each bearing block was subdivided by a wide central groove so that contact could be made only on two narrow lands to eliminate hydrodynamic action (Fig. 2). Oiliness tests were conducted at low enough speeds to bring the friction coefficients well over to the left of the minimum. It was found that sperm, olive, lard, and castor oils possessed increasing oiliness in the order named, which is the relative order of their viscosities; and that all fatty oils were of higher oiliness than mineral oils of comparable viscosity.

Kingsbury's experiments were repeated by C. W. Staples and the writer (1958) with certain differences: (1) heavier loads were applied in order to explore a wider range at the left of the minimum point, (2) pressure coefficients of viscosity were determined, (3) temperature was measured by thermocouple junctions in the bearing blocks, and (4) bearing wear was measured at the conclusion of the tests. Eight lubricants were compared, including lard and mineral oils of nearly equal viscosity. Thus Kingsbury's findings were confirmed by more complete tests.

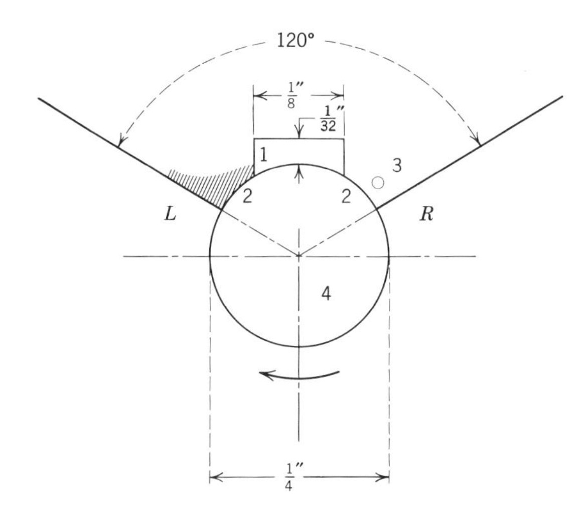

Fig. 2. Nominal contact on test journal in Kingsbury's oil-testing machine (from Hersey & Staples, Trans. ASME, 1958).

Three methods of plotting friction data were compared. The best correlation was seen by plotting f against Z_1N/P at constant values of the pressure-viscosity parameter bP, as in Fig. 3A. Here Z_1 denotes the viscosity at atmospheric pressure. The lard oil and mineral oil data had been plotted by the conventional method in Fig. 3C, indicating, apparently, that the mineral oil, Exp. 7, gave less friction than the fatty oil, Exp. 1, above $Z_1N/P = 2$ cp rpm/psi. The paradox was resolved by introducing the new parameter. The difficulty of isolating boundary lubrication either from wear or from hydrodynamic action became evident.

Countless experiments have been published in the attempt to measure oiliness, or to determine the practical laws governing boundary friction. Little or no effort has been made to coordinate the results. Many of the test methods used or proposed have, however, been listed, and guidelines indicated, in our study entitled "Logic of Oiliness" (1933).

Interpreting Test Data. Possible methods for deriving a numerical measure of oiliness were suggested by Fig. 1 of the preceding reference. Thus if the coefficients f and f' observed for the test lubricant and a reference lubricant are plotted against ZN/P at constant load, the ratio f/f' at a chosen value of ZN/P might serve. Such methods

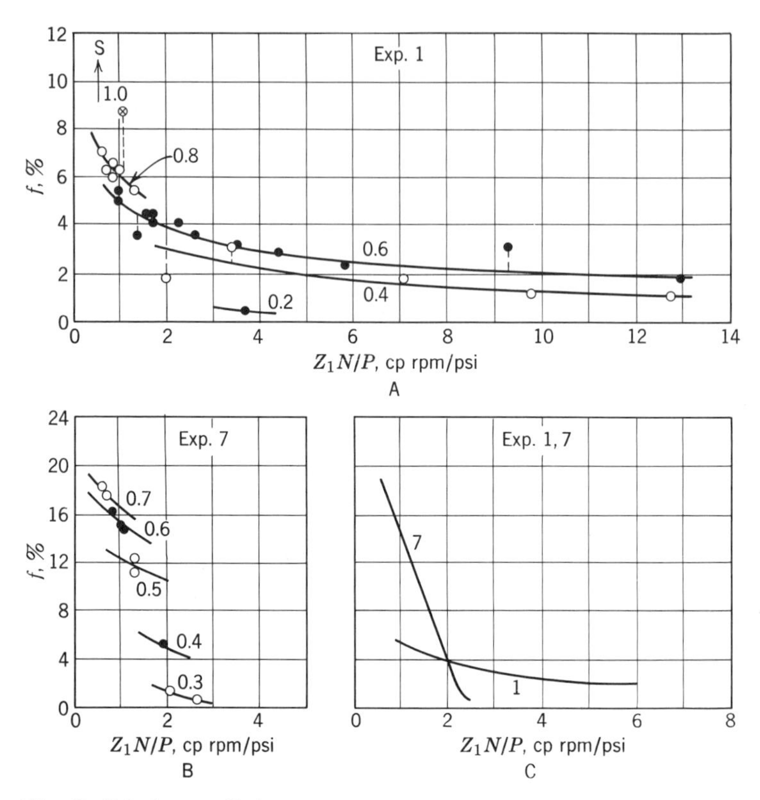

Fig. 3. Friction coefficient of fatty and mineral oils lubricating brass on steel (A, lard oil; B, light mineral oil, both at constant values of the pressure-viscosity parameter bP; C, conventional curves representing same points without regard to bP). From Hersey & Staples *Trans. ASME,* 1958).

are not free from objection, but may be of use in connection with the new techniques described by Givens and Talley (1962, 1963).

Vogelpohl was successful in explaining oiliness indications in terms of the temperature coefficient of viscosity and related properties (1936). Needs (VII: 1937) and Blok (1951) explained further results in terms of the pressure coefficient of viscosity. Thus with advancing knowledge of thin-film lubrication, less frequent use need be made of any stop-gap term like "oiliness." There is no excuse for broadening the concept to cover *all* desirable properties of a lubricant. As noted earlier (1933), "confusion will be avoided by restricting the term oiliness to its historically accepted meaning in connection with frictional resistance."

Wetting and Spreading. It was generally supposed at one time that fatty oils owed their superior oiliness to greater adhesion, indicated by spreading faster than mineral oils. Bulkley and Snyder remembered that porpoise-jaw oil was preferred for clocks and watches because it stayed where it was put. Mineral oils spread all over, leaving the pivots that required lubrication nearly starved for lack of oil. Their careful experiments (1933) proved that the spreading ability of fatty oils was a myth. These investigators went far toward determining the laws of spreading on metal surfaces, the laws of preferential adsorption, and the surface-action properties of fatty oils and fatty acids. Their report contains a review of the earlier work on angles of contact, wetting, interfacial tensions, and surface energy relations. See also the publications by Paul Woog (1929) and by W. E. Campbell (1937). Later experiments on spreading and wetting are described by Bielak, Kil', Schnell, and co-authors (1954, 1955, 1959). Dr. W. E. Campbell and associates found that the friction of a journal bearing can be reduced by nonwettable coatings on the bearing surfaces (II: 1957).

The properties of monomolecular films on solid surfaces were investigated by W. A. Zisman (1959). He gave an historical review with especial reference to surface chemistry and synthetic lubricants (1962). See also the collective volume by Zisman et al. (1964).

6. STATIC FRICTION AND STICK-SLIP

It is well known that static friction is usually greater than low-speed kinetic friction, whether the surfaces are dry or lubricated; and that static friction increases with the duration of the load. Coulomb expressed static friction as an exponential function of time. Some such effect might be expected from Stefan's law applied to lubricated surfaces. See also Burwell & Rabinowicz (1953); Spurr (1961); and pages 78–9 of Bowden & Tabor, Part II (1964). It is not so commonly known that the static coefficient varies with the rate of application of the tangential force, as found by Parker and co-authors (1950).

Stick-slip is familiar to everyone who has heard a squeaky door hinge calling for oil, or evoked a musical tone by rubbing the moistened edge of a glass goblet. It creates a problem in the operation of machine tools like planers with sliding ways, where it can often be remedied by hydrostatic lubrication, or by a fortunate selection of lubricants (Wolf, 1965).

Scientific observations were reported by Bowden & Leben (1939) Blok (1940), Khaikin and co-authors (1940). Usually in friction apparatus one member is rotated at a constant speed, while the "rider" is held by a spring of some kind, whose deflection or strain measures frictional resistance. Stick-slip can be minimized by damping, and more effectively by combining a stiff spring with a light enough rider to give a high natural frequency. Simple mathematical treatments are found in Bowden & Tabor (1950) and Barwell (1956). The rider, or slider, is dragged forward by the moving surface until the backward force exerted by the spring can overcome static friction. The rider then slides back and the process repeats. A fairly definite value can thus be obtained for the coefficient of static friction, but the kinetic value is left in doubt. The effect of elasticity in the measuring system was apparently overlooked by the earlier investigators.

Research was continued by Morgan (1941), Bristow (1942), Sinclair (1954), Cook (1958), Niemann (1964), and their associates. More advanced theories were developed by Dudley & Swift (1949), as well as by Derjaguin and co-authors (1957); Cook (above); Kosterin & Kragel'skii (1958); Rabinowicz (1959); Singh (1960); and Seirig & Weiter (1963), taking into account the variation of static friction with time elapsed under load, and with the rate of application of the pulling force. Stick-slip in machine tool operation was investigated by Voorhees (1963) with the aid of simulated slideways. A critical velocity of moving surface at which the vibrations die out was predicted by Brockley and co-authors (1965) and confirmed experimentally.

7. LUBRICATION IN METAL WORKING

Palm oil, when available, has long been a favorite in the cold rolling of steel. New lubricants had to be developed to meet the required supply. Molten glass is used in some hot-metal operations (Chapter IV).

Wire drawing through dies has been the subject of many investigations—see, for example, Wistreich (1957) and Steele (1964). The possibility of hydrodynamic lubrication was noted by Christopherson and Naylor (1955). Owing to the high pressures required, the lubricating films must be relatively thin in metal-forming processes; but not so thin as in metal cutting, which is possibly an extreme example of boundary or thin film lubrication.

Cutting fluids lubricate the passage of the chip over the top of the tool and help carry away the heat developed. Lubricating action is relatively more important at the lower speeds; cooling action at the higher speeds. Efficient cutting fluids serve to lengthen tool life, reduce power consumption, and improve the final surface finish on the work. Different fluids may be required for different materials to be cut. Various performance tests have been devised. An ASME subcommittee was set up to outline future problems (1929) and provide a bibliography of earlier work. The state of knowledge has been further reviewed by H. W. Swift (1938).

The question is often asked, how does a cutting oil penetrate to the point of the tool? Capillary action between tool and chip is apparently the answer. The "built-up-edge" shows that such action is not perfect; but the chip friction to be overcome is not limited to the edge or point of the tool. What does the lubricant do when it gets there? It may effect boundary lubrication by physical adsorption on the rubbing surfaces, but it is generally believed to react chemically with the respective metals. Solid compounds are thus formed that can protect the metal surface from seizure and minimize wear. Present-day understanding of cutting fluid action is largely due to investigations by M. E. Merchant and M. C. Shaw (1957, 1949). See also Black (1961); Norris, Eyres, & Brown (1964).

8. ACCURACY AND SURFACE FINISH

The importance of accurate workmanship is illustrated by Thurston's historic experience in comparing the friction coefficients of bearing metals submitted for test. The pendulum type of friction tester was in use, with two opposed partial bearings fitted to the journal and loaded by a spring. Albert Kingsbury, assisting Professor Thurston in 1888, discovered that differences in friction attributed to different bearing metals disappeared when sufficient care was taken in fitting the surfaces (Kingsbury, p. 957 of VI: 1950). His accurate workmanship created hydrodynamic films where none existed before.

Optically Flat Steel. Another example of accurate workmanship is found in the optically flat surfaces of tool steel, $\frac{7}{8}$ in. in diameter, skillfully produced by Sydney Needs (II: 1940). These disks would support loads of 800 psi, or more, indefinitely, with no indication of contact, while separated by a thin film of oil of uniform thickness, at low speeds of rotation. Final surfaces contained microscopic pits, but apparently no asperities.

Journal Bearing Performance. The influence of accuracy and finish has been studied more recently by Blount, Avery, DeHart, and co-authors (1957, 1963, 1964). The first two investigations deal with engine bearings and traction motor bearings, showing the beneficial effects of good conformity of surfaces in the loaded area, and of smooth surfaces generally. The third describes typical imperfections, then reports an experimental study of "out-of-roundness." Results are applied to the V-8 engine. Surface finish was described by Hemmingway as the key to bearing life (1944).

Bearing Balls. The importance of sphericity is well understood, but the effects of surface finish on ball-bearing balls have received less attention. The influence of microgeometry has been scientifically investigated by Tallian and co-authors (1964). A four-rolling-ball tester of the Barwell type was used with electrical measurement of oil film thickness and radioactive wear measurements. The results are analyzed statistically and shown to be in fair accord with elastohydrodynamic theory.

Influence of Roughness. The effects of surface roughness on friction were reviewed by J. J. Bikerman (1944); David Clayton (1945); Grunberg & Scott (1958). Cameron (1949) investigated the relation of roughness to oil film thickness at breakdown. Grunberg (1957), Spurr (1957), and their associates studied the effects of roughness on boundary lubrication and dry surfaces, respectively. Furey (1963) described roughness effects on metallic contact and friction. The influence of surface topography was brought out by F. G. Rounds in his experiments on the friction of steel against steel, using additives in the lubricant (1963). Thus it is seen how dry friction and boundary lubrication, as well as thin-film hydrodynamic lubrication, are influenced by surface finish.

Measurement and control of surface finish came to be studied as soon as its importance to industry was appreciated. Tracer-point instruments have been commercially available since the early thirties. Various techniques, such as taper sectioning, were discussed at the M.I.T. Conferences on Friction and Surface Finish (1940). Optical methods are promising—see, for example, Guild (1940) and Fried (1941). Tolansky developed multiple-beam interferometry (1946). An interference microscope is described in *Scientific Lubrication* (1957). The Institution of Production Engineers published two books on surface finish (1942). A review of the literature by J. W. Sawyer (1953) reflects the interest of Naval engineers. Recently J. B. Bryan described a microinterferometric technique for fast inspection of an entire surface (1964).

The relation of instrument readings to the actual surface profile has been studied by Tarasov, Bryan, and Selwood (1945, 1962, 1962). Definitions and standards are discussed by Becker (1950) and by Rubert (1959). American standards have been adopted for "surface roughness, waviness, and lay" (1955). Yet it is agreed that the arts of measuring surface accuracy and finish remain open for further research and improvement.

9. SUMMARY

Many investigations have been undertaken to gain a better understanding of thin-film lubrication phenomena.

Thin-film hydrodynamic or "microhydrodynamic" lubrication is believed to account for some of the observations earlier attributed to "boundary" lubrication. Thus when the bearing surfaces are separated by a lubricating film that is relatively thick compared to the roughness height, the precise quality of surface finish makes little or no difference. But when the surfaces approach almost closely enough to make contact, the irregularities act like randomized miniature bearing pads, each accompanied by wedge-shaped fluid films. Tops of high spots add to the fluid friction.

Where there is neither hydrodynamic action nor metallic contact, boundary lubrication is said to exist. At these spots the metal surfaces are protected by some kind of "adsorbed" film, perhaps not more than one or two molecules thick. Boundary lubrication has been defined as a state of lubrication in which the film is so thin that its properties are no longer the same as in bulk. Dry friction is attributed to the force required to shear the welded junctions where high spots come into contact. Boundary lubrication is explained by a similar picture except that the high spots are separated by the "low-shear strength" adsorbed film. Further study is required to clarify the literature on boundary lubrication, only a glimpse of which could be given here.

The familiar left-hand branch of the curve for coefficient of friction plotted against speed, or against ZN/P, is taken to represent "mixed friction," a combination of boundary and microhydrodynamic friction. Under severe operating conditions "imperfect lubrication" may occur with incipient wear and seizure.

Brief descriptions are given of oiliness, static friction, stick-slip, and lubrication in metal working. The chapter concludes with research on accuracy and surface finish: the influence of these factors on lubrication performance, and methods for defining and measuring surface finish.

REFERENCES

Akmatov, A. S. (1963), *Molecular Physics of Boundary Lubrication* (Russian). Fiz. Mat. Lit, Moscow; 472 pp.

Altrogge, Wilhelm (1950), *Zur Gesetsmässigkeit der Misch-Reibung*. Dr. Ing. Dissertation, T. H. München; 47 pp.

American Standards Association (1955), *Surface Roughness, Waviness and Lay*. ASA B 46.1—1955; 23 pp.

Archard, J. F., Hatcher, B. G., & Kirk, M. T. (1964), "Some Experiments upon the Behaviour of Hypoid Oils in Heavily Loaded Contact." Paper No. 4, *LWG* Convention, Eastbourne, pp. 258–63.

Archbutt, L. & Deeley, R. M. (1900), *Lubricants and Lubrication*. Griffin, London; 451 pp.

ASME (1929), "Cooling and Lubrication of Cutting Tools." Prog. Rep. No. 1 of Subcom. on Cutting Fluids. *T* **51,** MSP-51-8; 47–58.

Avery, R. W. (1964), "Effect of Surface Finish on Load Capacity and Reliability of Traction Motor Support Bearings." *LE* **20,** 422–8.

Barwell, F. T. (1956), "Mixed or Thin-Film Lubrication." Pages 43–5 in *Lubrication of Bearings,* Butterworths, London.

Barwell, F. T. & Milne, A. A. (1952), "Criteria Governing Scuffing Failure." *JIP* **38,** 624–32 (Fig. 1).

Bastow, S. H. & Bowden, F. P. (1935), "The Physical Properties of Surfaces, II. Viscous Flow of Liquid Films. The Range of Action of Surface Forces." *PRS* **151,** 220–33.

Becker, H. (1950), "Method of Measurement and Definition of Surfaces." *Microtechnic* **4,** 180–4.

Beeck, O., Givens, J. W., & Smith, A. E. (1940), "On the Mechanisms of Boundary Lubrication, I. The Action of Long Chain Polar Compounds." *PRS* **177,** 90–102. II. "Wear Prevention by Addition Agents," 103–18.

Bielak, E. B. & Mardles, E. W. J. (1954), "The Rate of Spread of Liquid Pools over Horizontal Solid Surfaces and between Approaching Parallel Flat Plates." *J. Colloid Sci.* **9,** 233–42.

Bikerman, J. J. (1944), "Surface Roughness and Sliding Friction." *Revs. Modern Phys.* **16,** 53–68.

Bikerman, J. J. (1948), "The Mechanism of Friction and Lubrication." *LE* **4,** 208–16, 235, 238; **5,** 27.

Bikerman, J. J. (1956), "Drainage of Liquids from Surfaces of Different Rugosities." *J. Colloid Sci.* **11,** 299–307.

Bisson, E. E. (1964), "Boundary Lubrication." Chap. 2, pp. 15–16 in *ABT*.

Black, P. H. (1961), *Theory of Metal Cutting*. McGraw-Hill, New York; 204 pp.

Blok, Harmen (1951), "Fundamental Mechanical Aspects of Thin-Film Lubrication." *Ann. N. Y. Acad. Sci.* **53,** Art. 4, 779–804.

Blok, Harmen (1957), "Marginal and Partial Hydrodynamic Lubrication with Particular Reference to the Conference Papers." *CLW,* 198–204.

Blount, E. A. & de Guerin, D. (1957), "The Importance of Surface Finish, Loaded Area Conformity, and Operating Temperature in Small End Plain Bearings for High Duty Two-Stroke Engines." Paper 80, *CLW*, 224–9; 202; Pl. 14.

Bodart, E. (1933), "Contribution à l'étude du frottement dans la phase onctueuse et semi-fluide." *RUM* (8) **9,** Nos. 11, 12.

Bondi, A. (1954), "Remarks on Mixed Film Lubrication." Pages 31–40 in *Fundamentals of Friction and Lubrication in Engineering*. ASLE, Chicago.

Bowden, F. P. & Leben, L. (1939), "The Nature of Sliding and the Analysis of Friction." *PRS* **169,** 371–91. (a) *Phil. Trans. Roy. Soc. London* (A) **239,** 1–27.

Bowden, F. P. & Tabor, D. (1939), "The Area of Contact between Stationary and between Moving Surfaces." *PRS* **169,** 391–413. (a) Holm, R., *Wiss. Veröff.* Siemens-*Werke* **7,** 217–71 (1929); **17,** 43 (1938); **20,** 68–84 (1941).

Bridgeman, O. C., McKee, S. A., & Bitner, F. G. (1932), "The Effect of Viscosity on Friction in the Region of Thin Film Lubrication." *Proc. API* **13** (III), 154–8.

Bristow, J. R. (1947), "Kinetic Bearing Friction." *PRS* **189,** 88–102. (a) *Proc. Phys. Soc. London B,* **63,** 964–5 (1950).

Brockley, C. A., Cameron, R., & Potter, A. F. (1966), "Friction Induced Vibration." *JBE, T* 88 (in press). (a) *ASME* 65-Lub-5; 7 pp. (1965).

Bryan, J. B. (1964), "Microinterferometric Scanning for Full Surface Metrology." *ASME* 64-WA/Prod-15.

Bryan, J. B., Boyadjieff, G. I., & McClure, E. R. (1962), "Measuring Surface Finish." *ME* **85,** 42–6.

Bulkley, Ronald (1931), "Viscous Flow and Surface Films." RP 264, *NBS JR* **6,** 89–112.

Bulkley, R. & Snyder, G. H. S. (1933), "Spreading of Liquids on Solid Surfaces. The Anomalous Behavior of Fatty Oils and Fatty Acids with Experiments Leading to a Tentative Explanation." *J. Am. Chem. Soc.* **55,** 194–208. (a) *Nature* **138,** 407.

Burton, R. A. (1963), "Effects of Two-Dimensional, Sinusoidal Roughness on Load-Support Characteristics of a Lubricant Film." *JBE, T* **85,** 258–64.

Burwell, J. T., Kaye, J., van Nymegen, D. W., & Morgan, D. A. (1941), "Effects of Surface Finish." *JAM* **8,** *T* **63,** A—49–58.

Burwell, J. T. & Rabinowicz, E. (1953), "The Nature of the Coefficients of Friction." *JAP* **24,** 136–9.

Cameron, A. (1949), "The Surface Roughness of Bearing Surfaces and its Relation to Oil Film Thickness at Breakdown." *Proc. IME* **161,** 73–9.

Cameron, A. (1959), "A Theory of Boundary Lubrication." *TASLE* **2,** 195–8.

Campbell, W. E. (1937), "Non-Spreading Oils." *Bell Tel. Record* **15,** 149–53.

Campbell, W. E. (1939), "Studies in Boundary Lubrication, I. Variables Influencing the Coefficient of Friction between Clean and Lubricated Metal Surfaces." *T* **61,** 633–41.

Campbell, W. E. & Thurber, E. A. (1948), "Studies in Boundary Lubrication, II. The Influence of Adsorbed Moisture Films on the Coefficient of Static Friction between Lubricated Surfaces." *T* **70,** 401–6. (a) III, 491–8.

Christopherson, D. G. & Naylor, H. (1955), "The Promotion of Fluid Lubrication in Wire Drawing." *Proc. IME* **169,** 643–54, 666–77.

Clayton, David (1945), "A Short Review of Surface Finish in Relation to Friction and Lubrication." *Proc. IME* **153,** 332–4.

Davies, M. G. (1963), "The Generation of Pressure between Rough Fluid-Lubricated, Moving Deformable Surfaces." *LE* **19,** 246–52.

DeHart, A. O. & Smiley, J. O. (1964), "Imperfect Journal Geometry—Its Effect on Sleeve Bearing Performance." *SAE Paper* 873 C; 36 pp. (a) *JSAE* **72,** 147 (Aug.)

Derjaguin, B. V. (1963), editor, *Research in Surface Forces.* Consultants Bureau, New York; 190 pp.

Derjaguin, B. V., Karassev, V. V., Zakhavaeva, N. N., & Lazarev, V. P. (1958), "The Mechanism of Boundary Lubrication and the Properties of the Lubricating Film . . . ." *Wear* **1,** No. 4, 277–90.

Derjaguin, B. V., Push, V. E., & Tolstoi, D. M. (1957), "A Theory of Stick-Slip Sliding of Solids." Paper 13, *CLW,* 257–68; 241, 281.

Dobry, A. (1964), "The Transitions between Boundary, Mixed, and Hydrodynamic Lubrication." *Wear* **7,** 290–7.

Dudley, B. R. & Swift, H. W. (1949), "Friction Relaxation Oscillations." *Phil. Mag.* **40,** 849–61.

D'yachenko, P. E., Tolkacheva, N. N., Andreev, G. A., & Karpova, T. M. (1963), *True Contact between Touching Surfaces* (Russian), AN, SSSR; 94 pp. (a) Transl. by Consultants Bureau, New York: 75 pp. (1964).

Feng, I-Ming (1955), "Critical Thickness of Surface Film in Boundary Lubrication." *ASME* 55-A-84; 3 pp.

Fried, Bernard (1941), "Reflectometer for Measuring the Smoothness of Metal Surfaces." *Ohio State University, EES News* **13,** 10–14 (Oct.). (a) Schwartz, W. M., **14,** 3–4 (Apr. 1942).

Fuchs, G. I. (1957), Flow of Liquids in Narrow Clearances between Adherent Plane Surfaces of Solids (Russian). *Doklady AN, SSSR* **113,** 635–8.

Fuller, D. D. (1954), "Mixed Film Conditions in Lubrication." *LE* **10,** 256–61.

Furey, M. J. (1961), "Metallic Contact and Friction between Sliding Surfaces." *TASLE* **4,** 1–11. (a) With J. K. Appeldoorn, *TASLE* **5,** 149–59 (1962).

Furey, M. J. (1963), "Surface Roughness Effects on Metallic Contact and Friction." *TASLE* **6,** 49–59.

Givens, J. W. & Talley, S. K. (1962), "Friction Measurement with the Tapered Spindle Top." *LE* **18,** 443–9.

Goldsmith, H. L. & Mason, S. G. (1962), "The Flow of Suspensions through Tubes, I." *J. Colloid Sci.* **17,** 448–76.

Greenwood, J. A. & Tabor, D. (1957), "The Properties of Model Friction Junctions." Paper 92, *CLW,* 314–7; 240, 588, 758, 760, 812, 815, 817.

Grunberg, L. & Campbell, R. B. (1957), "Metal Transfer in Boundary Lubrication and the Effect of Sliding Velocity and Surface Roughness." Paper 65, *CLW,* 291–301; 242, 319, 590, 760, 762, 806; Plates 10–11.

Guild, J. (1940), "Optical Smoothness Meter for Evaluating the Surface Finish of Metals." *J. Sci. Instr.* **17,** 178–5.

Hamilton, D. B., Walowit, J. A., & Allen, C. M. (1966), "A Theory of Lubrication by Microirregularities." *JBE, T* **88,** 177–85.

Hardy, Sir William B. (1936), *Collected Scientific Works.* Cambridge University Press, 922 pp.

Hardy, W. B. & Doubleday, Ida (1922), "Boundary Lubrication.—The Paraffin Series." *PRS* **100,** 550–74.

Heidebroek, E. (1937), "The Nature of Limit Friction." *GDLL* **1,** 133–7.

Hemmingway, E. L. (1944), "Surface Finish . . . Key to Bearing Life." *Mach. Design* **16,** 123–8 (Nov.); **17,** 168–70 (Jan. 1945).

Henniker, J. C. (1949), "The Depth of the Surface Zone of a Liquid." *Revs. Modern Phys.* **21,** 322–41.

Herschel, W. H. (1922), "Viscosity and Friction." *TSAE* **17,** pt. 1, 282–320.

Hersey, M. D. (1929), "Fundamental Action of Lubricants." *Am. Machinist* **70,** 919–21, 975–7 (June 13 and 20).

Hersey, M. D. (1933), "Logic of Oiliness." *ME* **55,** 561–6.

Hersey, M. D. (1933), "Thin Film Lubrication of Journal Bearings." *JWAS* **23,** 297–305.

Hersey, M. D. & Staples, C. W. (1958), "Experiments on Imperfect Lubrication." *T* **80,** 1104–7.

Holm, Ragnar (1938), "Über die auf die wirkliche Berührungsfläche bezogene Reibungskraft." *Wiss. Veröff. Siemens-Werk* **17,** No. 4, 38–42, 199–212, 224–54. (a) *Electric Contacts.* H. Gebers, Stockholm, 1946; 398 pp.

Institution of Production Engineers (1942), *Surface Finish.* Am. edition, ASME, New York, 240 pp. (a) Schlesinger, Georg, IPE, London; 231 pp.

Iwaki, A. & Mori, M. (1958), "On the Distribution of Surface Roughness When Two Surfaces are Pressed Together." *Bull. JSME* **1,** No. 4, 329–37.

Khaikin, S., Lissovsky, L., & Solomonovitch, A. (1940), Dry Frictional Forces (Russian). *Zhurn. Fizik,* **2,** 253.

Kingsbury, Albert (1896), "Experiments on the Friction of Screws." *T* **17,** 96–116.

Kingsbury, Albert (1903), "A New Oil Testing Machine and Some of Its Results." *T* **24,** 143–60.

Kingsbury, Albert (Chairman) (1919), "Report of Subcommittee on Lubrication." *ME* **41,** 537.

Klaus, E. E. & Bieber, H. E. (1964), "Effect of Some Physical and Chemical Properties of Lubricants on Boundary Lubrication." *TASLE* **7,** 1–10.

Kol', J. A. & Hughes, U. (1955), Study of Spreading of the Oil Film in Sliding Bearings (Russian). *Mashinostroenie,* No. 10; 34–49.

Kosterin, I. I. & Kragel'skii, I. V. (1958), Relaxation Oscillations in Elastic Friction Systems (Russian). *TIM* **12,** 119–43. (a) Transl., *FWM* **12,** 111–34 (1960).

Kragel'skii, I. V. & Demken, L. B. (1960), Determination of the Actual Area of Contact Surfaces (Russian). *TIM* **14,** 37–62. (a) Transl., *FWM* **14,** 30–53.

Kreisle, L. F. (1956), "Very Short Journal Bearing Hydrodynamic Performance . . . Approaching Marginal Lubrication." *T* **78,** 955–63.

Krisle, L. F. (1957), "Predominant-Peak Surface Roughness, a Criterion for Minimum Hydrodynamic Oil-Film Thickness of Short Journal Bearings." *T* **79,** 1235–46.

Kudinov, V. A. (1960), Hydrodynamic Theory of Semifluid Friction (Russian). *Proc. Third All-Union Conf. on Friction and Wear* **2,** 161–70, 198–207; AN SSSR, Moscow.

Larsen, R. G. & Perry, G. L. (1945), "Investigation of Friction and Wear Under Quasi-Hydrodynamic Conditions." *T* **67,** 45–50.

Leloup, L. (1949), "Le frottement oncteux des paliers lisses." *Rev. générale mécanique,* Jan.–Feb. (a) *RUM* **3,** 373–419 (Oct. 1947).

LeMay, I. & Vigneron, F. R. (1965), "Initial Studies on the Use of Rapeseed Oil as a Cold Rolling Lubricant." *LE* **21,** 276–81.

Lenning, R. L. (1960), "The Transition from Boundary to Mixed Friction." *LE* **16,** 575–82.

Marcelin, André (1947), "Limite à l'épaisseur des films lubrifiantes. Relation entre la viscosité et l'épaisseur des films liquides." *CR* **225,** 225–6; **228,** 650 (1949).

Martin, F. A. (1964), "Minimum Allowable Oil Film Thickness in Steadily Loaded Journal Bearings." Paper 16, *LWC,* 161–7, 169–70, 299, 304.

Massachusetts Institute of Technology (1940), *Proc. Conferences on Surface Finish;* 244 pp.

McKee, S. A. & McKee, T. R. (1932), "Journal Bearing Friction in the Region of Thin Film Lubrication." *JSAE* **31,** T 371–7.

Merchant, M. E. (1940), "The Mechanism of Static Friction." *JAP* **11,** 230.

Merchant, M. E. (1957), "Cutting Fluid Action and the Wear of Cutting Tools." Paper 99, *CLW,* 566–74; 495, 598, 789–92, 800; Pl. 17–8.

Morgan, F., Muskat, M., & Reed, D. W. (1941), "Friction Phenomena and The Stick-Slip Process." *JAP* **12,** 743–52. (a) With Sampson, **14,** 684–700 (1943).

Needs, S. J. (1940), "Boundary Film Investigations." *T* **62,** 331–45.

Niemann, G. & Ehrlenspiel, K. (1964), "Relative Influence of Various Factors on Stick-Slip of Metals." *LE* **20,** 84–6.

Norris, R. H., Eyres, A. R., & Brown, A. C. (1964), "Some Aspects of Improved Straight-Cutting Oil Developments." Paper 32, *LWC* (1963), 347–54; 355–7, 380–1.

Ocvirk, F. W. & DuBois, G. B. (1959), "Surface Finish and Clearance Effects on Journal-Bearing Load Capacity and Friction." *JBE, T* **81,** 245–53.

Parker, R. C., Farnsworth, W. & Milne, R. (1950), "The Variation of the Coefficient of Static Friction with the Rate of Application of the Tangential Force." *Proc. IME* **163,** 176–84.

Peeken, H. (1960), "Über den Einfluss der Unterteilung von Schmierflächen auf die Tragfähigkeit von Schmierfilmen." *Ing. Arch.* **29,** 199–218.

Phillips, K. F., et al. (1962), *Non-Conventional Lubricants and Bearing Materials Such as Used in Nuclear Engineering. LWG,* Manchester (April); 95 pp.

Rabinowicz, Ernest (1959), "A Study of the Stick-Slip Process." Pages 149–64 of *Symposium on Friction and Wear,* ed. by R. Davies; Elsevier, Amsterdam.

Reichenbach, G. S. (1964), "The Importance of Humidity in Friction Measurement." *LE* **20,** 409–13.

Reissner, H. (1941), "On Lubricant Flow with Periodic Distribution between Prescribed Boundaries." *Th. von Karmán Anniv. Volume,* California Institute of Tech., Pasadena; 310–16.

Rightmire, B. G. (1957), "Probable Behavior of Contacts in the Sliding Process." Paper 51, *CLW,* 281–5; 241, 589, 763.

Rounds, F. G. (1964), "Effects of Additives on the Friction of Steel on Steel. I. Surface Topography and Film Composition Studies." *TASLE* **7,** No. 1, 11–23.

Rubert, M. P. (1959), "Confusion in Measuring Surface Roughness." *Eng.* **188,** 393–5.

Rush, W. F. & Krueger, R. H. (1964), "The Frictional Behavior of Materials and Synthetic Lubricants in Sliding Systems." *TASLE* **7,** 107–210.

Salama, M. E. (1950), "Effects of Macro-Roughness on Performance of Parallel Thrust Bearings." *Proc. IME* **163,** 149–61.

Sargent, L. B., Jr. (1965), "A General View of Metal Working Lubrication." *LE* **21,** 282–6.

Sawyer, J. W. (1953), "Review of Surface Finish Literature." *JASNE* **65,** 400–24 (May).

Schiessl, Sibylle von (1950), "Versuche zur Klärung der Gültigkeitsgrenze der Hydrodynamik in dünnen Schmierölschichten." *Chemische Technik* **2,** No. 3, 91–7.

Schnell, Erhard (1959), "Slippage of Nonwettable Surfaces." *JAP* **27,** 1149–52.

Scientific Lubrication (1957), "Surface Finish Interference Microscope." **9,** 26–7 (Sept.).

Seirig, A. & Weiter, E. (1963), "Frictional Interface Behavior under Dynamic Excitation." *Wear* **6,** No. 1, 66–77.

Selwood, A. (1962), "The Topography of Rough Surfaces." *Wear* **5,** No. 2, 148–57.

Shaw, M. C. (1942), "The Chemico-Physical Role of the Cutting Fluid." *Metal Prog.* **42,** 85–8.

Shaw, M. C. & Macks, E. F. (1949), "Boundary Considerations." Pages 438–531 in *ALB.*

Singh, B. R. (1960), "Study of Critical Velocity of Stick-Slip Sliding." *JEI, T* **82,** 393–8.

Spurr, R. T. (1961), "The Ringing of Wine Glasses." *Wear* **4,** No. 2, 150–3.

Spurr, R. T. (1965), "The Friction of Contaminated Metal Surfaces." *Wear* **8**, 264–9.

Spurr, R. T. & Newcomb, P. T. (1957), "The Friction and Wear of Various Materials Sliding against Unlubricated Surfaces of Different Types and Degrees of Roughness." Paper 28, *CLW*, 269–75; 242, 587–8, 594, 802–3, 806, 812, 822.

Steele, M. C. (1964), "An Evaluation of Boundary Behaviour in Lubrication with Special Reference to Drawing Lubricants." *Sci. Lubn.* **17**; 14, 16, 18 (May).

Swift, H. W. (1938), "Functions of Cutting Fluids." *JIP* **24**, 662–71, 673–91.

Tabor, D. (1959), "Junction Growth in Metallic Friction: The Role of Combined Stresses and Surface Contamination." *PRS* **251**, 378–93.

Talley, S. K. & Givens, J. W. (1962), "Oiliness Measurement in the Metahydrodynamic Region." *LE* **18**, 396–403. (a) *LE* **19**, 194–200 (1963).

Tamai, Y. & Rightmire, B. G. (1965), "Mechanism of Boundary Lubrication and the Edge Effect." *JBE, T* **87**, 735–40.

Tarasov, L. P. (1945), "Relation of Surface-Roughness Readings to Actual Surface Profile." *T* **67**, 189–96.

Tipei, N. (1963), "Effects of Microgeometry of Surfaces on Lubrication." *Rev. Mécanique Appl.* (Bucharest) **8**, No. 6, 981–96.

Tolansky, S. (1946), "Further Interferometric Studies with Mica. New Multiple Beam Fringes and Their Application." *PRS* **186**, 261–71.

Tomlinson, G. A. (1929), "A Molecular Theory of Friction." *Phil. Mag.* (7) **7**, 905–39.

Vieweg, V. (1932), "Die Messung der Schmierfähigkeit von Ölen." *Tech. Mech. u. Thermo., VDI* **1**, 101–5.

Vogelpohl, G. (1936), "Hydrodynamische Theorie und halb-flüssige Reibung." *ZAMM* **16**, 371–2. (a) *Oel u. Kohle* **12**, 943–6. (b) *GDLL* **2**, 442–51 (1937).

Vogelpohl, G. (1954), "Die Stribeck-Kurve als Kennzeichen des allgemeinem Reibungsverhaltens geschmierter Gleichflächen." *ZVDI* **96**, 261–8.

Voitländer, R. (1929), "Untersuchungen an einem neuen Apparat zur Beurteilung der Schmierfähigkeit von Ölen." *Mitt. Hydraulisches Inst.,* TH München, No. 3, 145–59. See also Nos. 4, 5.

Voorhees, W. G. (1963), "Investigation of Stick-Slip in Simulated Slideways." *LE* **19**, 457–62.

Wilson, R. E. & Barnard, D. P., 4th (1922), "Methods of Measuring the Property of Oiliness." *IEC* **14**, 683–95.

Wistreich, J. G. (1957), "Lubrication in Wire Drawing." Paper 4, *CLW*, 504–11; 323, 496, 790, 798–9, 837.

Wolf, G. J. (1965), "Stick-Slip and Machine Tools." *LE* **21**, 273–5.

Woog, Paul (1921), "Sur l'onctuosité des corps gras." *CR* **173**, 303–6.

Woog, Paul (1929), "De l'extension des lubrifiantes sur les surfaces solides. Influences moléculaires. Rôle de la photolyse." *CR* **189**, 977–9. (a) **181**, 772 (1925); **186**, 71 (1928).

Zisman, W. A. (1959), "Friction, Durability and Wettability Properties of

Monomolecular Films on Solids." Pages 110–48 of Symposium, *Friction and Wear;* R. Davies, editor, Elsevier, Amsterdam.
Zisman, W. A. (1962), "Historical Review, Lubricants and Lubrication." Chap. 2, pp. 6–60 in *Synthetic Lubricants,* ed. by R. C. Gunderson & A. W. Hart; Reinhold, New York, 1962.
Zisman, W. A. et al. (1964), *Contact Angle, Wettability, and Adhesion* (F. M. Fowkes, Symposium Chairman). Am. Chem. Soc., Washington, D. C.; 389 pp.
Zlatin, N. & Snider, R. E. (1965), "Critical Aspects in the Selection of Cutting Fluids for Difficult-to-Machine Materials." *LE* **21,** 287–92.

Chapter XIV

epilog: Looking Back and Looking Ahead

The oil-shed fallacy. Lubrication engineering. Friction in space equipment. Properties of lubricants. Thin film lubrication. Need for coordinating the test data. Standards of Performance. Field research. Lubrication education.

Although a remarkable change in the understanding of lubrication problems resulted from the work of Tower and Reynolds, some fifty years passed before this new knowledge became widely appreciated.

The Oil Shed Fallacy. As recently as 1936 the situation was reflected in an article by the present writer prepared at the invitation of the *Technology Review*. The following is quoted from that article:

> Go to the executives of almost any manufacturing corporation—with a proper introduction—and ask how they are handling their vibration problems, new lightweight alloys, or calculation of electrical constants. In five minutes a number of Ph.D.'s, or electrical and mechanical engineers, will be sitting around the conference table with you.
>
> Inquire about lubrication, however, and the vice-president calls for the chief metallurgist, who doesn't know. "Let's see, now," they ask each other, "who has the key to the oil shed?" After Mike has been sent for and is getting his shirt on, somebody remembers the purchasing agent. But when all has been said and done, it is unanimously agreed that Mike has the only *real dope* in the organization, and that he learned it from the labels on the oil barrels.
>
> Isn't there a fallacy here—a fallacy in the organization chart? Why call

upon doctors of science to study the shearing stresses in steel shafting, leaving the shearing stresses in the oil film to the doctors of janitorial service? Why the mahogany conference table for structural analysis and the tin oil shed for lubrication?

Lubrication Engineering. Upgrading the art of selecting lubricants from a private game of chance between oil salesmen and the purchasing agent, to a rational process, has been accomplished by lubrication engineers. For this purpose a branch of technology called "lubrication engineering" has been evolved out of mechanical engineering, chemistry, and metallurgy; spiked with a bit of mathematics, and garnished with horse sense. It is an outgrowth of earlier work by the ASME RCL. This activity has been fostered by the American Society of Lubrication Engineers, with able support coming up from the ASME Lubrication Division, and the IME Lubrication and Wear Group. Periodicals like *Lubrication Engineering, Schmiertechnik, Scientific Lubrication,* and *Wear* are to be congratulated on widening the base of the population segment that is beginning to understand the science of lubrication, which, when combined with the art of lubricating, constitutes lubrication engineering.

Friction in Space Equipment. Looking ahead, we naturally think of space problems as the most challenging. Owing to the evaporation of liquid lubricants in the space vacuum, attention is being centered on solid lubricants, and on specially treated bearing materials that may not require lubrication in the usual sense. Another approach is the use of hermetically sealed, or partially sealed, enclosures within which gyroscopes and other equipment can operate for the required time at essentially atmospheric pressure.

Properties of Lubricants. A far more complete understanding of the physical and chemical properties of lubricants is needed not only for space problems but for nuclear reactor applications, and for everyday industrial machinery. For example, we need to know how the viscosity of lubricating oils is affected by high shear stresses combined with high pressures. How are such properties to be measured under transient conditions, as in gear lubrication? Can still better lubricants be developed for use at extremely high or low temperatures?

Thin Film Lubrication. There are fascinating new problems in "thin-film hydrodynamic" lubrication, described briefly in Chapter XIII. Here the films are too thin and irregular for very accurate measurement. Could something be learned experimentally from highly magnified, dynamically and thermally similar models? Above all, it remains to confirm and greatly extend the experiments by Needs and others on the mechanical characteristics of true boundary films.

Need for Coordinating the Test Data. It will be helpful in future work if the test data already accumulated on bearing and gear lubrication are coordinated, in the sense of reducing the results to the same units and definitions, for more effective comparison and application. Graphical representation in terms of dimensionless variables can be recommended.

Standards of Performance. It would be useful if we could learn by experiment precisely what load capacity, friction, oil flow, and temperature rise to expect from standardized machine elements, operating under specified conditions with a lubricant of known properties. Close tolerances will be required on geometrical and operating factors to facilitate comparison with theory. The research could be extended to include the approximate effect of deviations from the assumed condition. Performance investigations of this kind have been conducted by the National Bureau of Standards in various lines of applied technology.

Field Research. Not everything can be learned in the laboratory. Manufacturers of large machinery often make full-scale tests on their product after its installation in the customer's plant. Such tests may be only for compliance with the contract, or they may be for research information. Might not a survey of what has already been learned about lubrication in this way lead to a more definite philosophy of field research? New light could be thrown upon obscure questions by combining field research with standards of performance.

Once upon a time the writer served as division head in the research laboratory of a leading oil company. The division was asked to develop an oil that would not damage the product of a certain textile mill by flying out of the bearings. Arrangements were initiated for investigating properties like thixotropy; but before going too far, the writer wished to contact the customer and find out what really happened. Since it was company policy to isolate the researcher from the customer, an intermediary determined the facts. The trouble was no fault of the oil, or even of the bearings. The operatives had often neglected to hold their oil cans right side up when crossing over the goods from one bearing to another! This episode may be considered a minor example of "field research."

Lubrication Education. Looking ahead, we can see that the advancement of research and application in the field of lubrication rests upon a broad base of scientific and technical education. It is a kind of pyramid, the attainable height depending on the extent of the base. The pyramid should not be truncated, but should be carried up to a peak of specialized knowledge within the engineering school.

Similar views were expressed by the editorial in *Mechanical Engineering* (I: 1934) entitled "Lubrication in Engineering Schools." The question was raised whether sufficient heed is being paid to the importance of lubrication as a factor in machine design and maintenance. A useful activity of the ASLE, under its Educational Committee, has been the program of short courses in fundamental and advanced lubrication offered to registered applicants at its local and national meetings.

A contrast can therefore be seen with the time when lubrication was not recognized as a teachable subject. An increasing number of the leading universities in the United States and abroad now offer instruction in the science of lubrication, with facilities for graduate study and research. The look ahead reveals a promising and uncrowded field as far as we can see into the future.

Appendix

A. Abbreviations. B. Bibliography details. C. Commercial viscosities. D. Practical conditions. E. Lubrication in space.

A. ABBREVIATIONS

The following abbreviations are frequently used in the reference lists:

ABT	*Advanced Bearing Technology.* Bisson & Anderson (1964).
ALB	*Analysis and Lubrication of Bearings.* Shaw & Macks (1949).
AMR	*Applied Mechanics Reviews.*
AN	Akademiia Nauk.
API	American Petroleum Institute.
ASLE	American Society of Lubrication Engineers.
ASME	American Society of Mechanical Engineers.
ASTM	American Society for Testing and Materials.
CLW	*Conference on Lubrication and Wear, IME, Proceedings of.*
CR	*Comptes Rendus* (Paris).
DSIR	Department of Scientific and Industrial Research (London).
Ed.	Edition; edited by.
EES	Engineering Experiment Station.
Eng.	*Engineering.*
FI	The Franklin Institute, Philadelphia.
FWM	*Friction and Wear in Machinery.*
GDLL	*General Discussion on Lubrication and Lubricants, IME, Proceedings of.*

H&H	Hersey & Hopkins (1954), *Viscosity of Lubricants under Pressure.* ASME, New York.
HS	Houston Symposium of 1963 (1965).
IAE	Institution of Automobile Engineers.
IBM	International Business Machines, Inc.
ICAM	International Congress of Applied Mechanics.
ICE	Institution of Civil Engineers.
ICG	*International Conference on Gearing, IME, Proceedings of.*
IEC	*Industrial and Engineering Chemistry.*
IME	Institution of Mechanical Engineers.
IP	Institute of Petroleum.
ISGB	*First International Symposium on Gas-Lubricated Bearings,* Washington, D. C., 1959 (1961).
J	Japan; Journal.
JAM	*Journal of Applied Mechanics.*
JAP	*Journal of Applied Physics.*
JBE	*Journal of Basic Engineering.*
JEI	*Journal of Engineering for Industry.*
JEP	*Journal of Engineering for Power.*
JR	*Journal of Research.*
JSME	Japan Society of Mechanical Engineers.
LE	*Lubrication Engineering.*
LWC	*Proceedings, Lubrication and Wear Convention, IME.*
LWG	Lubrication and Wear Group, IME.
ME	*Mechanical Engineering.*
NACA	National Advisory Committee for Aeronautics.
NASA	National Aeronautics and Space Administration.
NBS	National Bureau of Standards, Washington, D. C.
OTN	Otdelenie Tekhnik, Nauk.
Phil. Mag.	*London, Edinburgh and Dublin Philosophical Magazine.*
PMM	*Prikladnaia Matematika i Mekhanika.*
PRS	*Proceedings of the Royal Society (London) A.*
RCL	Research Committee on Lubrication, ASME.
RCP	*Rolling Contact Phenomena,* Elsevier, ed. by J. B. Bidwell.
Res.	Research.
RUM	*Revue Universelle des Mines.*
RV	*The Role of Viscosity in Lubrication* (ASME 1960).
SAE	Society of Automotive Engineers.

SSSR	USSR.
T	*Transactions* (ASME unless otherwise shown).
TIM	*Trenie i Iznos v Mashinak* (Friction and Wear in Machines), Institut Mashinovedentiia, AN, SSSR.
TM	*Technical Memorandum.*
TN	*Technical Note.*
TTM	*Teknisk Tidskrift, Mekanik.*
VDI	Verein deutscher Ingenieure.
Z	*Zeitschrift.*
ZAMM	*Zeitschrift für angewandte Mathematik und Mechanik.*
ZAMP	*Zeitschrift für angewandte Mathematik und Physik.*
Zh	*Zhurnal.*

Names of units are abbreviated as follows:

Å	Ångstrom or "tenth-meter" (10^{-10} m, 10^{-7} mm).
Btu	British thermal unit.
C	temperature Celsius (formerly Centigrade).
F	temperature Fahrenheit.
K	absolute temperature, Kelvin.
N	newton (unit of force in the MKS system).
R	absolute temperature, Rankine.
atm	atmosphere (metric or standard).
cm	centimeter.
cp	centipoise.
dC	temperature difference, Celsius.
dF	temperature difference, Fahrenheit.
fpm	feet per minute.
hr	hour.
in.	inch.
kg	kilogram force.
kgm	kilogram mass.
lb	pound force.
lbm	pound mass.
m	meter.
mm	millimeter (10^{-3} m).
mp	millipoise (10^{-1} cp).
mph	miles per hour.
n	norton (10^{-6} reyn).
psi	pound per square inch.
psia	same, absolute.

psig	same, gage.
rad	radian.
rms	root mean square.
rpm	revolutions per minute.
rps	revolutions per second.
sec	second.
μ	micron (10^{-3} mm).
μ-in.	micro-inch (10^{-6} inch, 10^{-3} mil).
$\mu\mu$	micromillimeter (10^{-6} mm).

For further information see first reference at end of Appendix.

B. BIBLIOGRAPHY DETAILS

There are different ways of arranging a bibliography. It can be presented in bits at the end of each chapter, as we have done, or in one list at the end of the book. Our references, as a rule, will be found in the most appropriate chapter for a given subject. Roman numerals in the text show the chapter where a reference can be found if not in the same chapter.

References can be listed alphabetically, as is done here, or chronologically, which is more informative. The alphabetical arrangement, however, helps the reader to locate a reference that he remembers having seen before; and avoids the need for a name index at the end of the book. Closely related references are indicated by (a) and (b) under the principal reference.

Titles are translated only when appearing in a language not familiar to most American readers. Translated titles are not put in quotation marks. Names of journals are abbreviated as described in the preceding section.

Dates of publication are those of a book copyright, or a paper in its final library form, if such a date is actually printed. Thus the *Proceedings* of the International Symposium on Lubrication and Wear, held at Houston in 1963, are referenced under the copyright date, 1965. Yet those of the Conference on that subject held by the Institution of Mechanical Engineers in 1957 are listed under that year, since there is no printed record of the date of publication. United States patents are listed under the year the patent was granted. British patents, although customarily identified only by the year of application, are here referenced under the year of issue.

Bibliographies should be read straight through as diligently as the text of each chapter, not just held incommunicado for later use, since they tell an essential part of the story.

C. COMMERCIAL VISCOSITIES

Punch a hole in the bottom of a tin can. Fill can with oil. Record the time t required for a known volume to flow out. The kinematic viscosity can then be approximated by Herschel's formula (1919),

$$K = At - (B/t)\,, \tag{1}$$

after calibration to determine the constants. This is the principle of commercial efflux viscometers. A glass funnel will do for high-viscosity oils, but a graduated pipette is better. Pipettes served as commercial viscometers until superseded by the Saybolt, Redwood, and Engler.

In the Saybolt Universal, $A = 0.22$ and $B = 180$, or nearly so, when t is given in seconds and K in centistokes. If the oil had no viscosity at all, it would require a time equal to the square root of B/A, or nearly 29 sec, to run out. Thus the commercial viscometer is not at all sensitive to viscosity differences at low viscosity, but gives efflux times almost proportional to viscosity at the higher values.

Conversion of Viscometer Readings. A simple rule for converting Saybolt readings into true viscosity for an oil of specific gravity 0.90 is that the number of centipoises will be approximately one-fifth the number of Saybolt seconds (Hersey II: 1935, pages 690–1). This applies only to Saybolt times greater than about 150 sec. For any specific gravity σ, the viscosity in centipoises will then be $1.11\ \sigma t/5$ (see Fig. 1). Similar rules could be set up for the Redwood and Engler.

Principal limitations of the commercial instruments are (1) that they apply only to liquids of fairly high viscosity, and (2) that they require a knowledge of the density or specific gravity of the oil. The use of a commercial viscometer is like watching a baseball game through a knothole while holding a grandstand ticket in your pocket. The late A. G. M. Michell writes on page 43 of his book (I: 1950):

> Of these commercial viscometers it has been well said by Hersey (*Theory of Lubrication*, 1938, pp. 29–30):
>
> "The commercial viscometers so widely used in the petroleum industry are fitted with outlet tubes that are too short to be described as capillaries . . . . The standardization of such instruments has brought about a reasonable degree of uniformity, but at the risk of perpetuating the complications . . . ."

Finding his words quoted by the eminent Australian authority, the writer felt more than ever convinced of their truth.

After years of standardization, followed by interchange of samples, the savants discovered an odd fact. They discovered that oil *contracts*

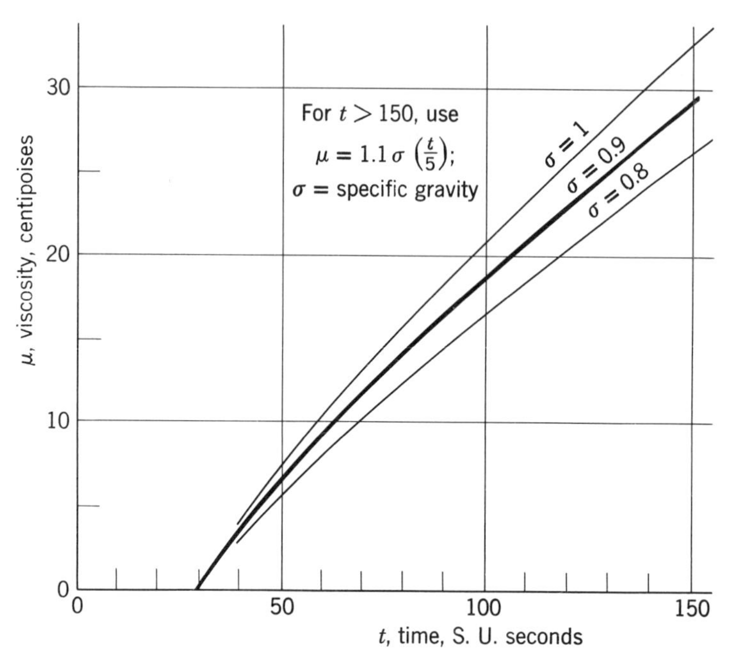

Fig. 1. Approximate conversion of Saybolt seconds to viscosity in centipoises (specific gravity 0.8 to 1.0). From Hersey, 1935, 1936.

when passing out of the viscometer at 100 or 210 F into the 60-ml receiving flask at room temperature! Consequently, the ASTM conversion tables and Saybolt viscosity-temperature charts all had to be corrected and reprinted. Separate tables are now published for converting Saybolt seconds at 100 and 210 F, respectively.

And at about that time a redetermination was made on the absolute viscosity of water (Chapter IV). All the oil viscosities and calibration constants had to be reduced about 0.3 of 1 per cent. No wonder the laboratory use of commercial viscometers has been widely discontinued, and replaced by the use of more scientific instruments. Yet the petroleum industry and its customers still transact business in Saybolt seconds, going to all the trouble of converting good kinematic readings back into fictitious SUS.

The Viscosity Index. The commonly used Dean & Davis viscosity index (IV: 1929, 1940) is defined by the ratio of $L - U$ to $L - H$, multiplied by 100. Here L is the viscosity of a reference oil of low VI, H a reference oil of high VI, and U the unknown; all at 210 F. Originally L, H, and U were measured in Saybolt seconds. An improvement was introduced by changing to kinematic values. In order to find the "VI" of an oil having a kinematic viscosity U at 210 F, we look up the corresponding H and L in the ASTM table (see page 53 in Fuller's I: 1956) and apply the above formula. The H and L values come from two series of petroleum oils having, respectively, good and poor temperature characteristics, defined as 100 VI and 0 VI: each series extending over a wide range of viscosities at 210 F.

The VI system is further discussed by Zuidema (1960) with especial reference to its deficiencies when applied to VI's much above 100, the traditional value for Pennsylvania oils.

Classification of Viscosity Grades. Several attempts have been made to agree upon a limited number of viscosity grades for commercial use. The SAE numbers are satisfactory for automotive lubricants except that they are based on Saybolt seconds instead of true viscosity (see recent editions of the Society of Automotive Engineers' *Handbook*). The definitions allow a range of values under each "SAE No." *Multigrade* oils have low enough slopes on the ASTM chart to qualify simultaneously as summer and winter grades.

A similar system designed for industrial lubricants has been recommended by cooperating bodies and described by John Boyd (1964).

D. PRACTICAL CONDITIONS

While hydrodynamic principles can be very simply stated one at a time, their application requires experience—see Hodgkinson (1929), Gordon (1933), pages 213–22 of Professor Fuller's book, and his lecture at Houston (pages 855–77; 1965). A dozen or more factors may be listed that require special consideration in practice. For references see Chapter VII except where otherwise indicated.

Geometrical Imperfections. The journal may be out-of-round, tapered or bent, or too rough. The bearing may be misshapen or improperly grooved. The clearance may not be within design tolerances. Some of these conditions have been treated analytically by Tipei (VI: 1957), as well as by Ocvirk & DuBois (1959) to determine their possible influence.

Misalignment. When a shaft is supported in two or more bearings, the bearings must be correctly lined up. A manufacturer was asked for design data on a grinding machine. Blueprints were cheerfully furnished. "Now, what are your loads?" he was asked. "That, we don't know," came the surprising reply. "But surely you must have some idea of the rotor weight and pressures between wheel and work. We are making up problems for students and need to know the Sommerfeld number," the questioner persisted. "Yes indeed," came the voice of experience, "but those are not the important loads. The real loading is due to misalignment." It was brought out that misalignment is not necessarily objectionable, since it offers a handy way of preventing "oil film whirl." Misalignment conditions have been investigated by McKee (II: 1932), Schiebel (VI: 1933), Sassenfeld & Walther (1954), DuBois (1957), Khrisanova (1959), and others.

Uncertain Load Direction. Trouble can result from the use of partial or split bearings when the line of action of the load is not correctly judged. Professor Norton enjoyed telling his classes about the failure of an important bearing in a New England rubber mill. Finally, after conferences and visits, it was found that the resultant load came at the split between the two halves. Here, the metal had been relieved to provide oil channels—see Norton's discussion of Bradford's paper, pages 880–882 (1940). An M.I.T. professor found the same explanation for the chronic failures in a powerful diesel engine. Now the professors know where to look, and the designers are rapidly learning.

Starting Friction. Starting up under load requires extra driving torque and may damage the bearing surfaces if they have been standing long. Rolling contact bearings avoid this difficulty. Another remedy is found in the admission of oil under pressure from an external source —the familiar "hydrostatic lift," described by Gordon (1933) and by Fuller (II: 1947, pp. 117–22).

Running-In. The classical investigation by S. A. McKee (1927) confirmed the benefits to be derived from running-in. With continued operation the friction coefficient minimum point moves toward lower values of ZN/P, apparently approaching a limit. This raises the factor of safety, but does not insure that the bearing will remain run-in for new values of the separate variables.

Deformation and Expansion. The effects of elastic deformation were early recognized by Falz (1926), Dettmar (1927), Buske (1951), and others. The theory has been developed by Raven & Wehe (1959) and by Sugimoto (1959). When a journal bearing is first warmed up, it may expand inward, and thus decrease the clearance. Seizure has been known to occur. Later, it can expand outward. These conditions must be watched in aluminum and other bearings with high expansion rates. Deformations caused by unequal heating sometimes lead to nonuniform wear (Karelitz, 1926). Sizer (1939) described the effect of practical conditions on grinding machines. Fuller (1965) discussed the effects of thermal deformation and residual stress.

Overheating. Hotboxes have been caused by thermal expansion, excessive speed, faulty assembly, insufficient cooling, and failure of the oil supply, especially in railroad freight car service (Keller, 1955). A general discussion of overheating is given by Thoma (III: 1938). See also Chapter I.

Insufficient Oil. In the writer's experiments on a small bearing (1923), electrical resistance observations offered a sensitive indication of each drop of oil added after the supply had been interrupted. The flow through an ordinary oil hole is appreciably restricted, as shown by Muskat & Morgan (III: 1939). The minimum feed rate needed to maintain fluid film lubrication was investigated by Fuller & Sternlicht (II: 1956), and starved bearings were discussed by Wilcock (II: 1957). Limitations of oil supply by capillary action are reviewed by Karelitz (1937) and by Michell (I: 1950, pages 189–91).

Lubricating Systems. Lack of oil can be avoided by proper lubricating arrangements. Many such are available, from the hand oil can to forced feed (Hitchcock, 1960). Novel devices are constantly brought to our notice by advertisements and exhibitions. The more elaborate lubricating systems include filter and cooler as well as means for delivering the lubricant. At one time the art of lubricating diverted attention from the problems of lubrication. Personnel skilled in the art were called "lubricating engineers." They have been superseded by the "lubrication engineer," who is expected to understand *both* lubricating and lubrication. Oil-ring performance is reviewed in Chapter V.

Harmful Environment. It is not enough that a bearing shall be correctly designed, constructed, assembled, and run-in. Successful operation depends also on the environment. A turbine bearing may suffer from heat conduction along the shaft. When an engine is started in cold surroundings, the oil may refuse to circulate. Abrasives and other contaminants must be kept out. In a steel mill, the bearing surfaces may be exposed to water and scale at times. A remedy was found by Dahlstrom, who had the idea (1933) of a removable bearing-and-journal sleeve assembly. This unit can be slid off and put back on a tapered roll neck without exposing the mirror-finish journal sleeve. In a hemp or textile mill the atmosphere may be charged with floating lint. Again, stray electric currents can damage bearing surfaces (Wilcock & Booser, I: 1957; Kauffman & Boyd, 1959). In marine applications, salt spray can go where it is not needed (Leggett et al., 1949). In nuclear energy installations the lubricant can deteriorate under radiation. Lubrication requirements in space are reviewed by Bisson (VII: 1964) and Lewis (VII and *HS,* both 1965).

Failure of Bearing Metal. In steady load bearings the metal can be damaged by corrosion, abrasive wear, wiping, and seizure. "Trouble shooting" usually starts with an inspection of the bearing and journal surfaces (Wilcock & Booser, I: 1957). Outright bearing failures are more familiar in engine bearings, where dynamic loading leads to fatigue cracks, pitting, bond failure, loss of metal, and hints of "cavitation erosion" (Bidwell, VIII: 1954; Snapp & Hersey, VIII: 1957).

Misbehavior of Lubricating Oils. Although petroleum oils are more stable than the fatty oils of earlier times, they change with continued use, especially at elevated temperatures, hence the recent interest in synthetic lubricants (Chapter IV). Lubricating oils tend to oxidize, to lose volatile fractions, carbonize, form sludge, corrode metals, and misbehave in countless ways. These conditions lie outside the scope of the present study, but are fully discussed in publications of the American Society for Testing and Materials; Institute of Petroleum; and American Petroleum Institute.

E. LUBRICATION IN SPACE

Bearing and lubrication problems met in outer space exploration have been mentioned in Chapters IV, IX, X and XIV; by Paul Lewis (1965); and in Chapters 8–14 of Bisson & Anderson (I: 1964). In general the conditions to be dealt with include the high vacuum, high- and low-temperature extremes, and the detrimental effects of radiation.

The principal lines of approach thus far tried are (1) the use of conventional lubricants on equipment hermetically sealed, (2) gas bearings with controlled leakage, (3) use of solid lubricants in powdered form, and (4) special bearing materials and surface treatments not requiring lubrication, in the usual sense, for a limited time. See for example Freundlich & Hannan (1961); McKannan & Demorest (1964); de Laat et al. (1966).

Many well-known agencies are at work on space instrumentation and accessories. Government contract reports constantly indicate new findings. Although it is difficult to design bearings and select lubricants that are equally effective under all flight conditions, it has been found by trial that more than one combination can be installed, so that the load can be transferred at will. A report is at hand by the

Barden Corporation (1963), another by Lewis, Murray, & Peterson (1964) typifying research and development along all these lines.

A future problem is to devise lubrication systems that will permit our astronauts to make more extended visits to the moon, or a planet, without having to send for the AAA to bring them home.

REFERENCES

Appendices A, C, and E: Abbreviations, Commercial Viscosities, and Lubrication in Space

American Society of Mechanical Engineers (1941), *Abbreviations for Scientific and Engineering Terms.* ASA Z **10.1**, 7 pp. (a) *NBS Technical News Bull.* **44,** 199 (1960); **47,** 29 (1963).

Barden Corporation, The (1963), *Investigation of Unconventional Bearings.* Air Force Systems Command, Rep. No. ASD-TDR-63-700; 119 pp. + bibliography.

Boyd, John (1964), "At Last! A Viscosity System for Industrial Lubricants." *LE* **20,** 142–4. (a) *ASTM Standards* **17,** pt. 7; App. XXVIII of D-2 Report, 1963. (b) Tentative Standard, ASTM 1965.

de Laat, F. G. A., Shelton, R. V. & Kinsey, J. H. (1966), "Status of Lubricants for Manned Spacecraft." *ASLE Paper* No. 66 AM-7A2; 22 pp.

Freundlich, M. M. & Hannan, C. H. (1961), "Problems of Lubrication in Space." *LE* **17,** 72–7.

Herschel, W. H. (1919), *Standardization of the Saybolt Universal Viscosimeter.* NBS Technologic Paper 112, 2nd ed.; 25 pp.

Lewis, Paul (1965), "Lubrication in the Environment of Space." *HS,* 905–39.

Lewis, P., Murray, S. F., & Peterson, M. B. (1964), *Investigation of Complex Bearing and/or Lubrication Systems for High Speed, High Temperature Operation.* Air Force Systems Command, Rep. No. FDL-TDR-64-12; 131 pp.

McKannan, E. C. & Demorest, K. E. (1964), "Dry Film Lubrication of Highly Loaded Bearings" *LE* **20,** 134–41.

Zuidema, H. H. (1960), "Effect of Temperature on Viscosity." *RV,* 6–10.

Mathematical Symbols

The following are the symbols most frequently used in this book:

A Area; angle; a constant.
B Breadth of bearing or element in direction of motion; a constant.
C Diametral clearance; a constant.
D Diameter; any chosen linear dimension.
E Young's modulus.
F Friction force; F_0, per unit area; a function; force (in a system of dimensions).
G Compressibility thrust bearing number (Chapter IX).
H Power loss; H_0, per unit area; H_1, power input; H', rate of heat transfer.
J Mechanical equivalent of heat; an integral.
K Kinematic viscosity; compressibility; Kingsbury number, ZU/PB; a constant.
L Length at right angles to motion; length (in dimensions).
M Moment of friction; mass (in dimensions).
N Number of revolutions per unit time; N_1, whirl impending speed.
O Position of journal axis; O', of bearing axis.
P Load per unit of projected area; pressure drop in pipe or capillary of length L; P_0, limit as B/L approaches zero; P', ratio of P to p_a.
Q Flow rate, volume/time.
R Rate of shear; radius; resistance.
Re Reynolds' number.
S Sommerfeld number (Chapter V).
T Temperature elevation; absolute temperature; time (in dimensions).
U Velocity of body in x-direction; sum velocity, $U_1 + U_2$.

V Velocity in y-direction.
W Total load; W_1, per unit length, sometimes without subscript.
X Axis in direction of motion; in Chapter VI, a thrust variable.
Y Axis perpendicular to film; relative film thickness h_0/h_t.
Z Axis at right angles to X and Y; viscosity; Z_1, at atmospheric pressure and any temperature; Z_0, value of Z_1 at a standard temperature.

a Film ratio h_1/h_0; temperature coefficient of viscosity; whirl amplitude; a constant.
b Pressure coefficient of viscosity; a constant.
c Radial clearance; taper angle or tilt; pressure-temperature coefficient of viscosity; a constant; c_p, specific heat at constant pressure, c_v at constant volume.
d Ball diameter.
e Eccentricity; base of natural logarithms.
f Coefficient of friction; f_0, for concentric bearing; a function.
g Gravitational acceleration.
h Film thickness; h_0, at point of nearest approach; h_1, at leading edge; h_t, taper-height.
i Number of independent dimensionless products.
k Number of fundamental units required; thermal conductivity; a constant.
m Clearance/diameter ratio, $C/D \equiv c/r$, rotor mass; a constant.
n Number of physical quantities; whirl frequency; n_0, natural frequency; an exponent.
p Fluid pressure; p_1, inlet pressure; p_a, ambient pressure.
q Heat capacity per unit volume.
r Radius; a constant.
s A constant; also, shear strength.
t Time; temperature; t_0, solidifying temperature; t_1, inlet temperature.
u Velocity of fluid in x-direction, at a point.
v Same in y-direction.
w Same in z-direction; weight per unit volume.
x Coordinate of a point, usually in the fluid; value of a parameter.
y Coordinate at right angles to bearing surface.
z Coordinate perpendicular to x and y.

Capital Greek Letters

Δ An increment or small difference.
Θ The temperature unit (in dimensions).

Λ Compressibility bearing number (Chapter IX).
Π A dimensionless product (Chapter V).
Φ The heat unit (in dimensions).

Lower Case Greek Letters

α Volumetric thermal expansivity.
β Coefficient of pressure rise; a factor or constant.
γ Ratio of specific heats, c_p/c_v.
δ Relative film thickness $h_0/c = (1 - \epsilon)$.
ϵ Eccentricity ratio, e/c.
η Efficiency; load ratio, P/P_0.
θ Temperature; an angle.
μ Viscosity; μ_1 and μ_0, same as Z_1 and Z_0.
ν Kinematic viscosity, μ/ρ; Poisson's ratio.
π The constant $3.141\overline{6}$ (approximately 22/7).
ρ Density (mass/volume); ratio of radii; roughness height.
σ Specific gravity; surface tension.
τ Shear stress; τ_0, yield value.
ϕ An unknown function; pressure angle in gearing.
ω Angular velocity.

These symbols were chosen for convenient use in the present work but are not offered as a standard system of notation.

Letter symbols have been published by the American Society of Mechanical Engineers as follows:

Letter Symbols for Hydraulics, ASA Z 10.2-1942; 9 pp.

Letter Symbols for Heat and Thermodynamics . . . , ASA Z 10.4-1943; 19 pp.

A "Memorandum on Definitions, Symbols and Units" was published by the Institution of Mechanical Engineers, pages 1–5 in *Proceedings of the Conference on Lubrication and Wear* (1957).

Standard symbols were suggested for use in gas-bearing research on pages 529–32, Appendix A of the *First International Symposium on Gas-Lubricated Bearings,* Washington, D. C., 1959 (U. S. Government Printing Office, 1961).

The American Society of Lubrication Engineers published "American Standard Letter Symbols for Lubrication Problems," ASA Z 11.116-1962 on pages 445–6 of *Lubrication Engineering* **19** (Nov. 1963); note the "Foreword," p. 445.

Index

Because of the condensed nature of our discussions, a conventional index would be too voluminous. Except for naming the chief investigators cited in the text, this is primarily a subject index. If a name is not found in the index, look in the alphabetized reference list at the end of an appropriate chapter.